Multimedia Communications

Springer

London
Berlin
Heidelberg
New York
Barcelona
Hong Kong
Milan
Paris
Santa Clara
Singapore
Tokyo

Francesco De Natale and Silvano Pupolin (eds)

Multimedia Communications

With 253 Figures

 Springer

Francesco De Natale
Università di Cagliari, Piazza d'Armi, Cagliari I-09123, Italy

Silvano Pupolin
Dipartimento di Elettronica ed Informatica, Università di Padova,
via Gradenigo 6a, 35131 Padova, Italy

ISBN-13:978-1-85233-135-1

British Library Cataloguing in Publication Data
Multimedia communications
  1. Multimedia systems   2. Wireless communication systems
  I. Natale, Francesco De   II. Pupolin, Silvano, 1947-
  621.3'981
ISBN-13:978-1-85233-135-1

Library of Congress Cataloging-in-Publication Data
Multimedia communications / Francesco De Natale and Silvano
  Pupolin (eds.).
    p.   cm.
  Includes bibliographical references and index.
  ISBN-13:978-1-85233-135-1   e-ISBN-13:978-1-4471-0859-7
  DOI: 10.1007/978-1-4471-0859-7

  1. Multimedia systems   I. De Natale, Francesco.   II. Pupolin,
  Silvano, 1947- .
  QA76.575.M8524   1999   98-53818
  621.382--dc21   CIP

Typesetting: Camera ready by contributors

69/3830-543210 Printed on acid-free paper

Preface

M ULTIMEDIA COMMUNICATIONS *is a keyword today widely used in technical and non-technical language with several different meanings. In a strict sense, it indicates the transmission within the same communication session of several bit streams generated by as many information sources, relevant to different media. At the receiver, each source is processed and conveyed to the suitable output device, in order to be properly rendered. Current multimedia applications involve the transmission of two or more of the following information sources: video, audio, graphics, text, images, sometimes complemented by software programs. Common examples of multimedia communication applications are videoconferencing (even with the possibility of cooperative work), home entertainment (e.g., pay-per-view movies, with all the special effects that one experiences in modern cinemas), and the well-known world-wide-web distributed information system.*

Obviously, different multimedia applications require different technical solutions, and imply the integration of several experiences in various technological domains. Therefore, the purpose of this workshop was to collect in the same discussion forum a large variety of research and industrial skills that are jointly needed to make Multimedia Communications a successful system.

When we started the organization of the workshop, several researchers working in the field of Multimedia Communications Systems were interviewed, in order to determine the main issues involved in the subject. After a short time, we realized that the word 'multimedia' can have many different meanings, depending on the viewpoint. For instance, computer-scientists are mainly concerned with data base management and archiving, which is extremely important to offer multimedia services with advanced retrieval capabilities (e.g., video-on-demand), while electronic engineers are more interested by the problem of hardware architectures and input/output devices, essential for high-quality acquisition and reproduction of multimedia data.

Although all of these aspects are important in the global picture of multimedia technologies, we tried to focus on a restricted scenario, and concentrated on the most telecommunication-oriented topics. The result of this 'focus-of-attention' was the breakdown of the workshop into five Sessions, namely: Signal Processing,

Multimedia Traffic Characterization, Network Access, Terminals and Applications. A short introduction to each session is outlined here.

The first Session on Signal Processing was split into two parts, one devoted to Audio and Image processing, and the other to signal processing within the receiver of a multimedia transmission.

The focus of Session 2 was on the analysis of multimedia traffic sources. The information derived will be extremely useful in the design and performance evaluation of transport networks able to guarantee a satisfactory quality of service as perceived by the end user.

The user access to the network is the subject of Session 3. Different media exhibits completely different demands in terms of available bandwidth, bit error rate (BER) and cost. Each user should operate some choice on the minimum and average requirements of his applications, by considering the trade-off between the cost and the performance available. An up to date analysis of several access techniques related to the BER and the available bandwidth is presented.

The design of advanced multimedia terminals is another field in rapid evolution. The present standardization activities and the use of common interfaces for fixed and mobile multimedia terminals are the main issues of Session 4. Particular attention has been paid to both the hardware platforms and the choice of the operating systems which enable the downloading of application software from the telecommunication network.

Finally, Session 5 presents some interesting multimedia services and applications. In this framework, some aspects relevant to hardware architectures and computer science were also considered, which were not covered by the previous sessions. As an example, some state-of-the-art database management and information retrieval techniques are proposed that guarantee a certified quality of service to the end user.

We would like to take this opportunity to express our sincere appreciation to all the authors and session organizers who have contributed to the 10-th Tyrrhenian International Workshop on Digital Communications.

Francesco De Natale   Silvano Pupolin

September 1998

Acknowledgements

The editors are much indebted and wish to express their sincere thanks to the components of the Technical Committee of the 1998 edition of the International Tyrrhenian Workshop on Digital Communications, namely *Giovanni Cancellieri* from the University of Ancona, Italy, *Giuseppe Coppola* from Philips Research Monza, Italy, *Nikil Jayant* from Georgia Institute of Technology, USA, *Sanjit K. Mitra* from University of California Santa Barbara, USA, *Giulio Modena* from CSELT, Italy, *Chrysostomos L. (Max) Nikias* from the University of Southern California, USA, *Giovanni Pacifici* from IBM T.J. Watson Research Center, USA, *Franco Russo* from the University of Pisa, Italy, *Gianni Vernazza* from the University of Cagliari, Italy, whose precious cooperation was essential to the organization of the Workshop and to the publication of this book.

The Workshop has been technically co-sponsored by the
IEEE Communications Society

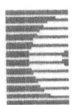

IEEE

The Workshop would not have come into being without the support of the Italian National Consortium for Telecommunications CNIT, and without the sponsorship of the following companies, which are gratefully acknowledged.

Italian Ministry of University and Scientific Research

 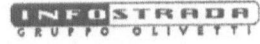

Table of Contents

Part 1

Signal Processing for Multimedia Communications

Image Compression Using Adaptive Wavelets and Trellis-Coded Quantization

Ioannis Katsavounidis   and   C.-C. Jay Kuo   and   Z. Zhang

The authors are with the Department of Electrical Engineering-Systems, University of Southern California, Los Angeles, CA 90089 E-mail: katsavou@sipi.usc.edu, cckuo@sipi.usc.edu and zzhang@milly.usc.edu

Abstract. In this work we present some new results on still image compression. The first step involves adaptive wavelet transform (AWT) as the fundamental means of de-correlation of the image pixel values. The wavelet decomposition scheme is an adaptive one, where at each point a locally optimal decision is taken regarding whether a wavelet decomposition step is to be applied and also the direction of the filter (along image lines or columns) and the choice of wavelet filter, using the optimal bit allocation formula. The same formula also determines the scanning order and the bit rate allocated for each resulting wavelet subband. The next step is the application of multistage Trellis-Coded Quantization (MTCQ) on the various subband. To do so, we use a family of MTCQ quantizers, parametrized by the target bit rate. Progressive transmission is achieved by quantization of the residual error.

1   Introduction

Wavelet Transform (WT) [7],[1],[10], [12],[9],[14] and Trellis-Coded Quantization (TCQ) [3],[13],[11] are two very important tools to achieve high-quality still image compression. Recent efforts to establish a new image compression standard (JPEG-2000) [5] for the next millenium make extensive use of both of them.

Our method looks into both WT and TCQ in an attempt to improve them. In particular, the aspect of *adaptivity* is studied for the wavelet transform, together with a multistage implementation of the optimal rate-distortion TCQ that offers the additional advantage of *progressive transmission*.

Progressive transmission is a very important and much desirable feature for lossy image compression methods. It means that the encoder is producing just a *single* coded bit-stream for a particular image, independent of the target bit rate or desired image quality. By receiving more and more of this bit-stream, the decoder can reconstruct the original image with increasing fidelity. The advantages of progressive transmission are obvious; in particular for applications over the internet, where the browser can display image data while receiving them and also for image databases that can provide variable quality images, yet maintaining a single coded version. Imposing the additional requirement of progressive transmission to an image compression method results in performance degradation. But as long as the price one has to pay for progressive transmission is not too high, it maybe this feature to decide the commercial success of a new image compression method.

Adaptivity is a long-wanted feature for image compression. It is common knowledge that there is no one single compression method that outperforms all the others; JPEG [8], for example, works well with images of human faces and smooth images in general, while it performs poorly when applied to textures and other images with lots of details. The simplest form of an adaptive scheme is to have available to both the encoder and the decoder a certain number of operation modes. In this way, the coder can choose the method that best compresses a given image; the additional information regarding the chosen method can be easily encoded on the header with just a few bits. An alternative is to use a few bits of information that describe the internal structure of the compression method used. Going back to the last example of the encoder/decoder capable of many different compression methods, instead of using the same method for the whole input signal, one may decide to partition the input signal into regions that are then compressed and transmitted with possibly different compression methods. In this way, for the smooth regions of an image one can use JPEG compression, while for texture-like regions subband coding may work better.

Independent of the adaptive scheme used, the ultimate criterion is always rate-distortion optimality. What this fundamental coding rule says, is that one should spend a given bit-budget at the parts of the signal that reduce the distortion the most; usually the parts with the highest variance.

2 Wavelet Transform

Fig. 1(a) and (b) depicts one of the JPEG-2000 test images and its pyramid wavelet transform, correspondingly. One can immediately notice that, unlike the Fourier transform, the transformed image resembles the original image, except that one can see multiple copies of the original image in various resolutions. The major advantages of wavelet transform over other transform schemes is the multiresolution property and the space-frequency structure. Traditional transforms, like the Fourier transform and its close relative, the Discrete Cosine Transform (DCT), have strong frequency characteristics, i.e. the transform coefficients have little or no similarity to particular signal values - they instead represent the entire signal *in some weighted-average fashion*. This is a direct consequence of the fact that the Fourier basis functions have an infinite region of support. What one prefers is a transform that has space-localization properties. In this way, quantization of the transform coefficients has a direct impact on the input image, or rather a part of the input image.

(a)   (b)

Fig. 1. (a) The "woman" (2048×2560) JPEG-2000 testing image; needs 5Mbytes of disk space and almost 11min. to down-load from the network with a 64Kbps modem. (b) Symmetric Bi-orthogonal 7-9 tap Pyramid Wavelet Transform decomposition applied to the "woman" image.

For this reason, the JPEG standard uses *block* DCT, which means applying DCT to 4 × 4 or 8 × 8 blocks of the input image.

Wavelets offer exactly that; having a *finite* region of support, the wavelet coefficients have local (in the spatial domain) properties, i.e. they can be used to reconstruct part of the input image. Thus, by successively applying the wavelet decomposition scheme, one can also obtain frequency-like coefficients, or rather coefficients that represent bigger and bigger parts of the input signal.

On the advantages of the wavelet transform, it should be emphasized that their complexity is $O(N)$, where N is the signal size - the FFT family is $O(NlogN)$ - and also the fact that there is a whole family of wavelet transforms that offer various energy-compaction properties for different classes of input signals. These different wavelet transforms, each characterized by two decomposition/reconstruction filters, have also different regions of support. So, for example, the Haar transform has a region of support of just 2 samples, while the 16-tap orthogonal compact wavelet (also called "Daubechies-16" [2]) has a region of support of 16 samples.

Among the various wavelet transforms, perhaps the most celebrated one is the so-called "7-9 tap bi-orthogonal wavelet" [1]. Part of its celebrity is due to the fact that many state-of-the-art image compression methods, like EZW [10], LZC [12], MTWC [5],[14] and SPIHT [11] use this wavelet transform as their first step. The shape of the 2 decomposition and 2 reconstruction filters is given on Fig. 2. This particular transform is not orthogonal, i.e.

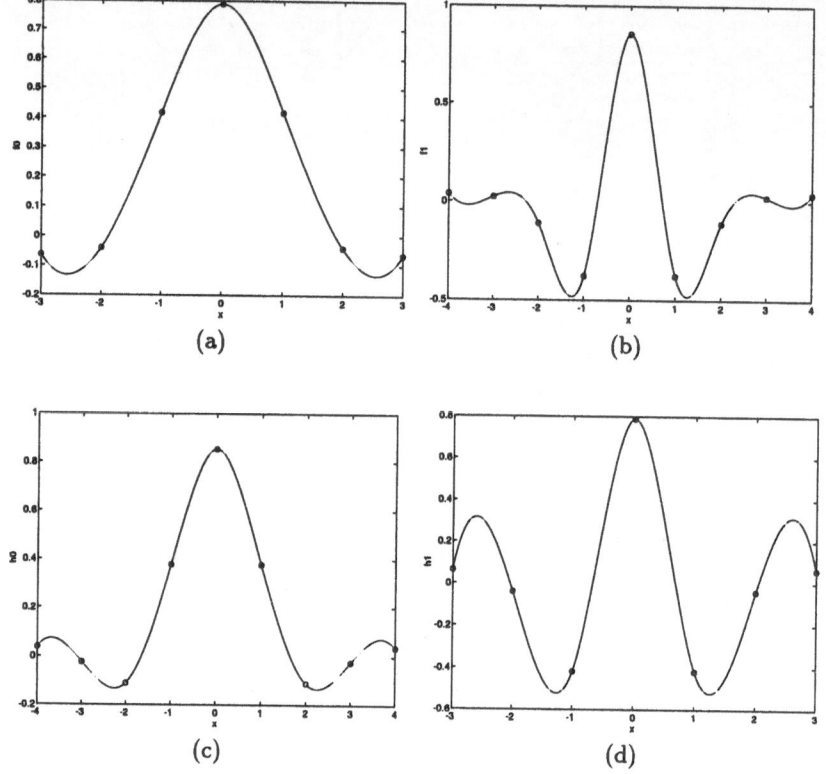

Fig. 2. 7-9 tap bi-orthogonal wavelet filters; (a) Low-pass decomposition, (b) High-pass decomposition, (c) Low-pass reconstruction, (d) High-pass reconstruction

the decomposition filters, used during the forward transform phase, and the reconstruction filters, used for the inverse transform are not the same. Still, they have a nice relationship among them, that is, if g_i, $i = -4, \ldots, 4$ and h_i, $i = -3, \ldots, 3$ are the 2 decomposition filters, the corresponding reconstruction filters are given by $g_i' = (-1)^i h_i$ and $h_i' = (-1)^i g_i$. This relationship is called "bi-orthogonality". Just like for the orthogonal wavelet transforms, there exist many bi-orthogonal wavelet transforms, each characterized by a pair of decomposition/reconstruction filters.

A fundamental question for all transforms is how to extend them from the one-dimensional case to 2 or more dimensions. The usual approach is to apply the one-dimensional filter on each dimension separably, i.e. for the case of images, we apply the one-dimensional filter along lines and the results are used as input to the one-dimensional filtering along columns. In this way, after the application of one level of separable wavelet decomposition, we obtain 4 subbands, referred to as "LL", "LH", "HL" and "HH", according to the type of filter used for each direction.

Additionally, for the multi-dimensional wavelet transforms rises the question on how to iterate the decomposition scheme; that is which of the 4 subbands mentioned above should be used as the input for the next level of wavelet decomposition. The traditional approach is to choose the "LL" subband, i.e. the low-pass filtered quarter of the original image, that in most cases resembles the original image in a sub-sampled version. This decomposition scheme is called "pyramid decomposition" since it results in subbands geometrically decreasing in size, all embedded on the original image.

What makes the pyramidal scheme work better than all other fixed-structure wavelet decomposition schemes is the fact that most real images have their energy content in the low-frequency band, since most images are smooth. Yet, there is a big class of images that has a lot of energy content in the high- and mid-frequency bands. Also, some images have most of their energy content on low frequencies along the direction of lines, but on the high frequencies along the direction of columns and vice-versa.

2.1 Adaptive Wavelet Transform

Due to the above reasons, one can consider the following decomposition scheme which when applied recursively to every subband, starting with the input image itself, forms the basis of our *Adaptive Wavelet Transform* (AWT).

- Each subband maybe split into two other subbands or left intact, depending on the estimated coding gain from each decision. If the coding gain, i.e. the reduction in number of bits to quantize the two resulting subbands compared to that needed to quantize the original subband, is more that the cost to send the splitting decision, the subband is split.
- The splitting decision is taken considering the two different directions for applying the decomposition filters (along lines and along columns) and the various wavelet filters in consideration. So, for example, with 2 different wavelet filters and 2 direction, there are 4 alternatives that need to be evaluated.

In evaluating the coding gain, a rate-distortion model is needed for the wavelet subbands. We use the generalization of the rate-distortion formula for uncorrelated Gaussian source to model the wavelet subbands as follows:

$$D(b) = \sigma^2 2^{-\alpha b} \tag{1}$$

where α is a shape parameter; $\alpha = 2$ for uncorrelated Gaussian source. Let (N, σ, b, D) be the coding point for one wavelet subband of size N coefficients, with variance σ^2, quantized using an average of b bits per coefficient and resulting in an average distortion of D. Suppose this subband is decomposed to

2 subbands, $(N_i, b_i, \sigma_i D_i)$, $i = 1, 2$. Assuming $\sigma_1 > \sigma_2$, optimal bit allocation results into

$$b_1 = \frac{1}{\alpha} \log_2 \frac{\sigma_1^2}{\sigma_2^2} + b_2 \tag{2}$$

Suppose

$$b_2 = 0 \implies b_1 = \frac{1}{\alpha} \log_2 \frac{\sigma_1^2}{\sigma_2^2} \tag{3}$$

That results in a total bit budget of:

$$N_2 b_2 + N_1 b_1 = N_1 \frac{1}{\alpha} \log_2 \frac{\sigma_1^2}{\sigma_2^2} \tag{4}$$

with a total distortion:

$$N_2 \sigma_2^2 + N_1 \sigma_1^2 2^{-\alpha \frac{1}{\alpha} \log_2 \frac{\sigma_1^2}{\sigma_2^2}} = N_2 \sigma_2^2 + N_1 \sigma_1^2 \frac{\sigma_2^2}{\sigma_1^2} = N \sigma_2^2. \tag{5}$$

To achieve the same distortion without decomposition,

$$N \sigma_2^2 = N \sigma^2 2^{-\alpha b} \implies 2^{\alpha b} = \frac{\sigma^2}{\sigma_2^2} \implies b = \frac{1}{\alpha} \log_2 \frac{\sigma^2}{\sigma_2^2} \tag{6}$$

Thus, coding gain after decomposition:

$$\begin{aligned}
g = Nb - N_1 b_1 &= N \frac{1}{\alpha} \log_2 \frac{\sigma^2}{\sigma_2^2} - N_1 \frac{1}{\alpha} \log_2 \frac{\sigma_1^2}{\sigma_2^2} \\
&= N_1 \frac{1}{\alpha} \log_2 \frac{\sigma^2}{\sigma_2^2} + N_2 \frac{1}{\alpha} \log_2 \frac{\sigma^2}{\sigma_2^2} - N_1 \frac{1}{\alpha} \log_2 \frac{\sigma_1^2}{\sigma_2^2} \\
&= N_1 \frac{1}{\alpha} \log_2 \frac{\sigma^2}{\sigma_1^2} + N_2 \frac{1}{\alpha} \log_2 \frac{\sigma^2}{\sigma_2^2}.
\end{aligned} \tag{7}$$

Experimental results indicate the value $\alpha = 2.5$ as a reasonable fit of model (1) to the wavelet subband rate-distortion performance; this value was used by our AWT scheme.

The decision to split a given subband carries some cost due to the additional information that the coder needs to transmit to the decoder regarding the decomposition scheme. For example, our TCQ coder relies on the knowledge of the variance of each subband, since the codebooks are normalized to have unit variance; the same is true for scalar quantizers, too. In this way, even though we only need to transmit one such variance if we choose *not* to split, we need to transmit two variances (for the two "children" subbands) it we *do* split. The cost is also logarithmically proportional to the number of choices for the wavelet filters, W. Here is a breakdown of the coding decision:

- 1 bit to send the spitting decision (split/don't split)
- 1 bit to send the direction of the filter (columns/lines)
- $log_2 W$ for the choice of wavelet filter.

– 2 bits to send the (no-split) decision for the 2 subbands resulting from such a plit.

In this way, we need to send $4 + \log_2 W$ bits to code a splitting decision, against only 1 bit for a no-splitting decision. Thus, the overhead is $3 + log_2 W$ bits. For our quantization scheme, we need to add to this overhead the cost of sending 2 variances instead of 1. These variances are - in general - floating point numbers, that can be efficiently coded using 16-32 bits. Thus, assuming F bits to efficiently quantize a variance and a choice among W wavelet filters, the coding cost in bits is given by

$$c = F + log_2 W + 3 \qquad (8)$$

For the suggested AWT scheme, we used $W = 2$ wavelet filters (Haar/7-9 bi-orthogonal) and quantized the variances using $F = 16$ bits. In this way, the coding cost of a splitting decision is 20 bits and the splitting strategy becomes:

split a particular wavelet subband, if the coding gain of the best decomposition scheme is higher than 20 bits.

This splitting decision is shown on Fig. 3. The final result for such an

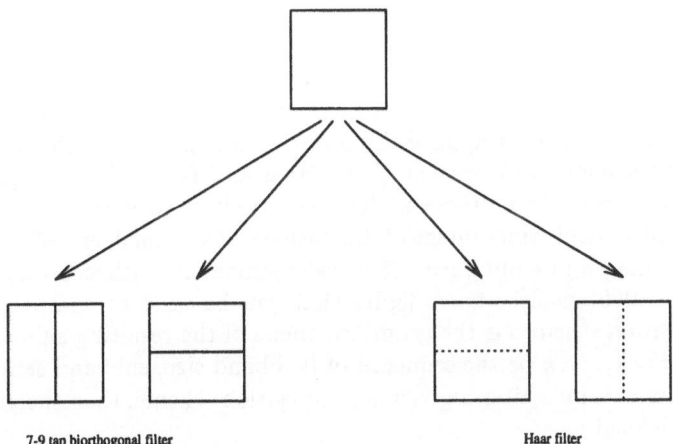

Fig. 3. 1-level Adaptive Wavelet Transform decomposition; the coder can choose between 2 different filters (Haar/7-9tap bi-orthogonal) and 2 different directions (columns/lines).

adaptive wavelet transform can be highly asymmetric among the two image dimensions (i.e. subbands that are much more "thiner" than they are "long" or "short") and with a much larger number of subbands with respect to a

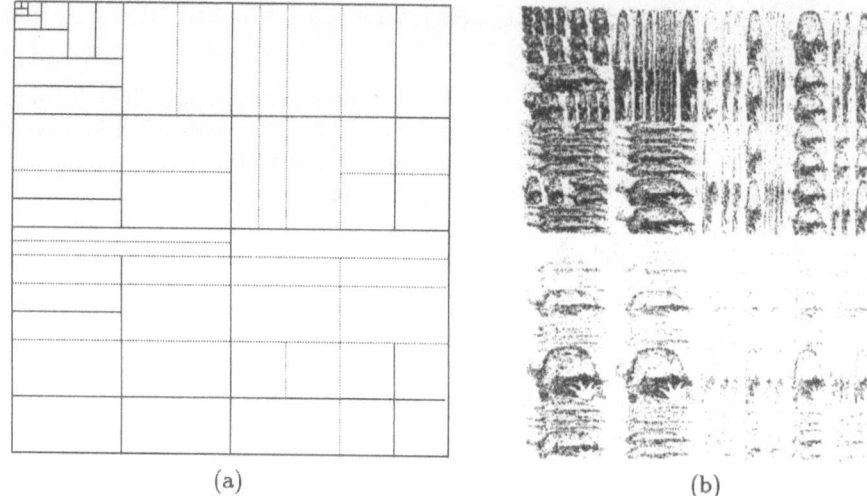

(a)   (b)

Fig. 4. Adaptive Wavelet Transform decomposition applied to the "woman" image; (a) Skeleton of the decomposition - by choosing a high coding cost for the subband variances (512 bits), only 53 subbands were created. The solid lines represent decomposition using the 7-9 bi-orthogonal filters and the dotted lines decomposition with the Haar filter. (b) Another decomposition - by choosing high coding cost for the subband variances (128 bits), 177 subbands were created.

pyramidal decomposition. This can be clearly seen on the example of Fig. 4. For our method, we have applied the adaptive decomposition scheme outlined above for the choice of 2 wavelet filters, Haar and 7-9 tap bi-orthogonal. Applying the suggested adaptive wavelet decomposition scheme to the "woman" image (a black-and-white image of dimensions 2048×2560, one of the JPEG-2000 test images,) we obtained 438 wavelet subbands, with size varying from 400 to 100,000 coefficients. A figure that can be used to compare various decomposition schemes is the geometric mean of the resulting subbands. Let (N_i, σ_i), $i = 1, \ldots, k$ be the sequence of (subband size, subband sample variance) pairs obtained after a given decomposition scheme, then the geometric mean is defined as

$$\rho = \prod_{i=1}^{k} \sigma_i^{\frac{N_i}{N}}, \qquad \text{where} \qquad N = \sum_{i=1}^{k} N_i$$

is the total number of coefficients. The geometric mean of the adaptive wavelet decomposition scheme resulted about 10% lower than the corresponding pyramid decomposition.

3   Trellis Coded Quantization (TCQ)

A *Trellis* is a state transition diagram; it is usually depicted as in Fig. 5(a), which we will also use as an example.

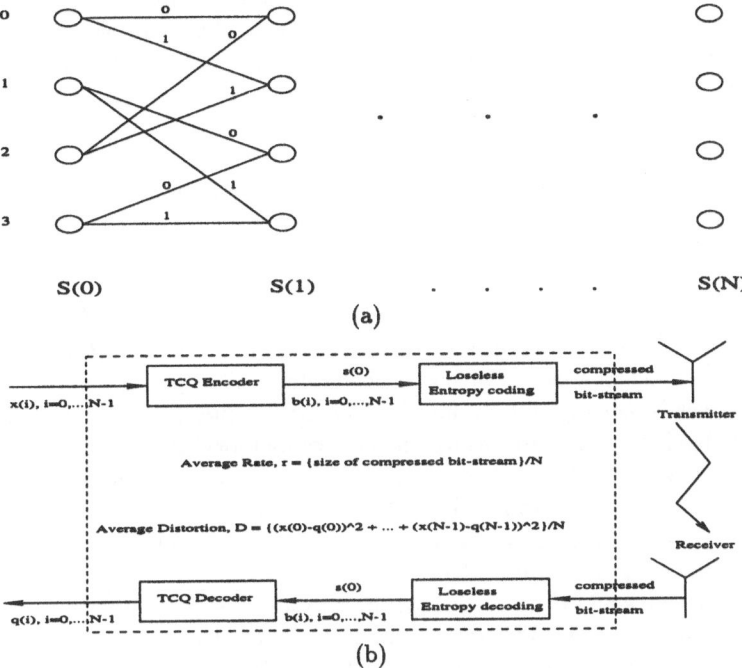

Fig. 5. (a) 4-state, binary alphabet Trellis diagram; (b) Block diagram of a TCQ encode/decoder

In this figure, the horizontal axis, indexed from $0, 1, \ldots, N$, can be interpreted as the time axis. The vertical axis, in this case indexed from $0, \ldots, 3$, represents the *states* of the Trellis, shown as empty circles. The *branches* connecting states to each other, in this case labeled with just 0's and 1's, are the *symbols* and they uniquely describe – as seen in this figure – a *state transition* and also a quantization value for each input value. Trellises are not only used in the context of quantization; they are also used in digital communication (Trellis Coded Modulation) and error correction for convolutional codes [6]. It is in particular from the area of error correcting codes that a very important algorithm, called the Viterbi algorithm, originates.

A system's block diagram of TCQ is given on Fig. 5(b). From this figure, we immediately see that the input to a TCQ encoder is a sequence of N input values, x_i. The output of the TCQ encoder is an initial state, s_0, followed by a sequence of N symbols, s_i. These symbols and initial state, after possible entropy coding and decoding, are available as the input to the TCQ decoder. The output of TCQ decoder is a sequence of N quantized values, q_i. The average *rate* and *distortion* induced by the entire TCQ encoder/decoder block are also given on Fig. 5(b).

The encoder quantizes one input value at a time, producing one *symbol* for each input value, for a total of N symbols. In this example, symbols are binary (0/1) but in general they come from a symbol set, called the *alphabet*. The alphabet size can by 3, 4 or more symbols; this controls the maximum data rate of TCQ. A binary alphabet imposes an upper limit to the average rate at 1 bit/sample, or - in general - an alphabet of B symbols allows for an average bits rate of up to $log_2 B$ bits/sample. Obviously, the bigger the alphabet size is, the higher the quality of quantization is, with a cost of higher bit rate though.

In this example, TCQ uses 4 states; for each one of these states, there is a scalar quantizer associated to it. So, when the encoder finds itself in state #0, it uses quantizer #0 etc. For the case of binary alphabet, each one of these 4 quantizers consists of just 2 real values. The quantization rule that the encoder is using to determine which of the 2 values to use for a particular input value is not as simple as for the case of memoryless scalar quantization (nearest neighbor.) This is because the choice of quantization value does not only affect the error introduced at this point of the quantization process but also the choice of quantizer – and thus the amount of quantization error – for the *following* input samples. It becomes a matter of optimizing the choice of state transitions and thus output symbols for the *entire* input sequence. This fact alone, qualifies the TCQ as a form of "sequential vector quantizer", since the input sequence is quantized as a whole. The analogy between VQ and TCQ is similar to that between block-codes and convolutional error correcting codes.

3.1 Terminology

Let us denote a trellis as (T, S, B), where $S = \{0, \ldots, |S| - 1\}$ is the set of states, $B = \{0, \ldots, |B| - 1\}$ is the alphabet and T is the state transition function. Let $s_n \in S$ denote the current state of the trellis and $b_n \in B$ the current symbol. Clearly, the trellis is a mapping $T : (SxB) \to S$, since the next state is uniquely defined by the current state and symbol, i.e.

$$s_{n+1} = T(s_n, b_n).$$

This recursion produces all states, s_n, given the initial state s_0 and the sequence of symbols, b_n.

Associated with the state transition function $T(S, B)$ is the quantizer function denoted by $Q(S, B)$. Just like any other quantizer, this is a mapping $Q : R \to C$, where C is the set of codewords, i.e. it maps any real value to one of the values of C. This is in fact done indirectly through states and symbols: $x_n \to (s_n, b_n) \to q_n$.

Here is how TCQ works:

- given the state transition and quantizer functions T and Q,
- given an input sequence $x_i, i \in \{0, \ldots, N-1\}$,
- the encoder calculates the initial state s_0 and a sequence of symbols $b_i \in B, i = 0, \ldots, N-1$,
- the decoder applies the recursion for $i = 0, \ldots, N-1$, $q_n = Q(s_n, b_n)$, $s_{n+1} = T(s_n, b_n)$.

In order to entropy-code the sequence of symbols, we use the k-th order arithmetic coding length function, calculated from the conditional probability of symbol b_n, when preceded by symbols $b_{n-1} \ldots b_{n-k}$ as follows

$$L(b_n|b_{n-1} \ldots b_{n-k}) = -log_2 p(b_n|b_{n-1} \ldots b_{n-k}).$$

Given that we use the "shift register" state transition function, i.e. $s_{n+1} = (s_n|B| + b_n)\%|S|$, where % denotes the modulo operator, we can express the arithmetic coding length function as

$$L(b_n|b_{n-1} \ldots b_{n-k}) = L(b_n|s_n) = -log_2 p(b_n|s_n)$$

where $k = log_{|B|}|S|$ is the depth of the arithmetic coding length function. For example, if $|B| = 2$ and $|S| = 4$ then $k = 2$. Thus, we are using the frequencies of each symbol, conditioned by each state.

This function can also be expressed as a 2-D array, $L(s, b)$, where $s \in S$ and $b \in B$. A similar array can be used to store the quantizer structure, $Q(s, b)$. Finally, yet another such array is used to store the Trellis structure, i.e. the state transition function, $T(s, b)$.

As explained before, TCQ calculates the sequence of symbols $b_i, i \in 0, \ldots, N-1$, together with the initial state s_0 that minimizes some error function. In order to provide optimal quantization (in the rate-distortion sense,) we use the error function

$$E = \sum_{i=0}^{N-1} L(b_i|s_i) + \lambda \sum_{i=0}^{N-1} (x_i - q_i)^2$$

where $\lambda \in [0, +\infty)$ is a Langrange multiplier that allows for different tradeoffs between bit rate and effective distortion. Clearly, for $\lambda = 0$ one looks for the sequence of symbols that result in the minimum bit rate, regardless of the distortion induced, while for $\lambda \to \infty$ we obtain the sequence that produces the minimum average distortion, regardless of the coding cost (in bits). Thus, we can completely characterize a TCQ with a quadruple, (T, Q, L, λ).

The algorithm that determines this optimal sequence is the well known Viterbi algorithm, used in a number of combinatorial optimization problems, and in particular in convolutional error-correcting codes. The Viterbi algorithm, a key element for this new method, can be explained as an optimization algorithm, providing optimal solution to a wide class of combinatorial problems. For more details, we refer to [6]. The computational complexity of the Viterbi algorithm is $|S||B|$ since at any given state we need to calculate the error function for all possible state transitions. The memory requirements depend on an additional (implementation) parameter called *depth* or *constrained length* that determines the maximum size of the buffer used to store possible paths through the Trellis. For our experiments, a constrained length of about 1000 offered quasi-optimal performance for the Viterbi algorithm.

3.2   Progressive TCQ

Rate distortion theory provides the theoretical background for optimal quantization of a known source at a given rate or distortion level. It provides a supremum to the performance that all possible quantizers can achieve for that particular source. As explained at the introduction, imposing progressivity as a feature for an image coding scheme results in some loss in absolute performance. But the advantages, especially in practical applications, are important; enough to justify 1-2 – or even more – dB performance degradation.

In order to apply TCQ in a progressive manner, there is need to further quantize the residual error(s) from previous quantization steps. That is, if we denote with V_0 the original image to be quantized, then

$$E_0 = V_0 - Q(V_0)$$

is the error introduced by the first quantization step and, in general, we define the sequence of errors as

$$E_j = V_j - Q(V_j), \qquad V_j = E_{j-1}, \qquad j = 1, \ldots, \infty$$

Other progressive image compression methods, like EZW, LZC and MTWC separate the encoding procedure in *passes*, where the coefficients are quantized over and over again, each time reducing the residual error energy. In each one of these passes, the wavelet coefficients – or the errors from some previous quantization of the wavelet coefficients – are subject to quantization with different quantizers: the compression algorithm maintains some sort of memory of the quantization and applies the appropriate quantizer to each sample.

For the case of TCQ, since we are always applying it on selected wavelet sub-bands, we need to maintain such info for each sub-band separately. We use a number, which we call the "order" of quantization. This is defined as the number of quantization passes that a particular sub-band has been undergone. At the beginning, all sub-bands have their order equal to zero. The first one, i.e. the one with the highest variance, is then quantized using a properly selected TCQ scheme. As a result, the variance of the residual error is decreased according to the rate-distortion curve of this source. At the same time, this sub-band has its order increased by one since it no longer contains wavelet coefficients, but rather residual errors from the quantization of the wavelet coefficients. When this sub-band needs to be further quantized, a different TCQ has to be chosen; one that has been trained with "order-1" residual errors, since the source statistics of quantization errors is – in general – quite different from those of the original source. It will then become an "order-2" sub-band since it will contain the residual quantization errors of the residual quantization errors of the original wavelet coefficients and so on and so forth.

Residual quantization errors tend to become less and less correlated as the quantization procedure advances until the – asymptotic – point where they can be considered uncorrelated, uniformly distributed. For this purpose, we designed a series of TCQs for each quantization pass. We fixed the quantizer parameters (# of states, alphabet size, λ) and we then trained the TCQ scheme presented above using normalized to unit variance wavelet subbands of test images. Fig. 6(a) shows the convergence behavior of the Viterbi training algorithm for the case of 3 states, ternary alphabet and $\lambda = 1.0$. Putting

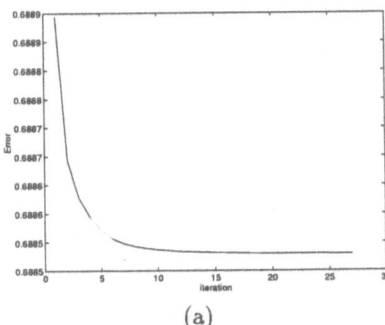
(a)

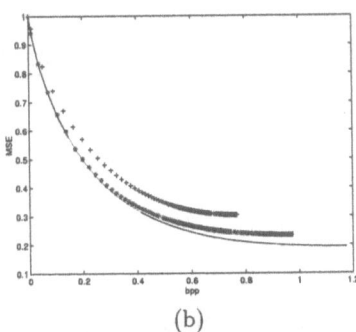
(b)

Fig. 6. (a) Convergence behavior of the Viterbi training algorithm for # of states = 9, alphabet size = 3, $\lambda = 1.0$. (b) Rate-distortion curve achieved by the Viterbi training algorithm for # of states = 1 (+'s),3 (*'s) and 9 (solid line), alphabet size = 3.

together all the bit-rate and average distortion points achieved for different values of λ, one can get the empirical rate-distortion curve. Three different rate-distortion curves obtained for ternary alphabet by varying the number of states = 1, 3 and 9 can be seen in Fig. 6(b).

Note how the performance of the TCQ scheme is improving as the number of states is growing; the price one has to pay for this improvement is higher computational complexity and memory requirements.

The procedure outlined above yielded the "order-0" codebook, obtained from the normalized wavelet subbands as training sequence. After quantization, the residual error is obtained. This is normalized again to unit variance and used as the new training sequence in order to obtain the "order-1" codebooks. Applying this procedure iteratively we were able to produce codebooks of 8 different "orders". Needless to say, the codebooks appeared to be converging to each other and to the uniform quantizer after the first 3-4 "orders". Table 1 presents the results of this procedure. We immediately see that as

Table 1. TCQ performance. Note that dB refers to the difference from the Shannon bound for uncorrelated Gaussian source

Order	Alphabet size	# of states	λ	MSE	bpp	dB	CPU (secs)	Memory depth
0	3	9	3.0	0.263072	0.500811	-2.784075	127.6	135
1	3	9	3.0	0.286963	0.744301	-0.940604	176.6	203
2	3	9	3.0	0.287686	0.942033	0.260792	225.0	258
3	3	9	3.0	0.276821	1.026650	0.603041	296.1	365
4	3	9	3.0	0.273822	1.027591	0.561392	294.7	337
5	3	9	3.0	0.272550	1.032684	0.571840	280.4	360
6	3	9	3.0	0.274408	1.025333	0.557087	301.3	362
7	3	9	3.0	0.276688	1.019747	0.559388	302.6	328

quantization proceeds, the residual error becomes less and less uncorrelated, and thus more difficult to compress.

3.3 New method

Given all the above, the new method can be outlined as follows.

1. Apply Adaptive Wavelet Transform (AWT) to the input image. Remember that the result of AWT is an array of size equal to the original containing the wavelet coefficients together with a tree structure describing the decomposition levels. The leaves of this tree also contain the variances of the various sub-bands.

2. Sort the resulting sub-bands in a descending order according to variance. Also assign order = 0 to all of them (i.e. all sub-bands contain wavelet coefficients).

3. Iterate on this ordered list of sub-bands by always applying TCQ of the appropriate order to the top one (the one with the highest variance.) After quantization, the contents of this sub-band is the residual error and thus the order of this sub-band is increased by one. We then re-order the list of sub-bands, since the residual error has a significantly smaller variance w.r.t. the original one.

4 Conclusions and Extensions

In this work, we have presented a new method for image compression, based on Adaptive Wavelet Transform and Multistage Trellis-Coded Quantization. Our goal is a method that surpasses existing state-of-the-art image compression methods, offering the additional advantage of progressive transmission and universality. This work, still in progress, has so far produced results at the same level with other compression methods, like MTWC and Marcellin's TCQ. Among the parameters that need to be optimized is the number of states for the Trellis and the choice of λ parameter for the different stages of the method.

References

1. M. Antonini, M. Barlaud, P. Mathieu, and I. Daubechies, "Image coding using wavelet transform," *IEEE Trans. on Image Processing*, Vol. 1, pp. 205–220, Apr. 1992.
2. I. Daubechies, *Ten Lectures on Wavelets*, Philadelphia, PA: SIAM, 1992.
3. T. Fisher and M. Marcellin, "Trellis-Coded Quantization of memoryless and Gauss-Markov Sources," *IEEE Trans. on Communications*, Vol. 38(1), pp. 82–93, 1990.
4. A. Gersho and R. M. Gray, *Vector Quantization and Signal Compression*, Kluwer Academic Publishers, 1992.
5. Y. B. Houng-Jyh Wang and C.-C. J. Kuo, "Multi-Threshold Wavelet Coded (MTWC)," in *JPEG2000 Document, WG1N665*, November 1997.
6. S. Lin and D. J. Costello, *Error Control Coding: Fundamentals and Applications*, Prentice Hall, 1983.
7. S. Mallat, "Multifrequency channel decomposition: the wavelet representation," *IEEE Trans. on Acoustic, Speech, and Signal Processing*, Vol. 37, pp. 2091–2110, 1989.
8. W. B. Pennebaker and J. L. Mitchell, *JPEG - Still Image Data Compression Standard*, New York: Van Nostrand Reinhold, 1993.
9. A. Said and W. Pearlman, "A new, fast, and efficient image codec based on set partitioning in hierarchical trees," *IEEE Trans. on Circuits and Systems for Video Technology*, Vol. 6, pp. 243–250, June 1996.
10. J. M. Shapiro, "Embedded image coding using zerotrees of Wavelet coefficients," *IEEE Trans. on Signal Processing*, Vol. 41, pp. 3445–3462, Dec. 1993.
11. P. Sriram and M. Marcellin, "Image-coding using wavelet transforms and entropy-constrained Trellis-Coded Quantization," *IEEE Trans. on Image Processing*, Vol. 4, pp. 725–733, June 1995.
12. D. Taubman and A. Zakhor, "Multirate 3-D subband coding of video," *IEEE Trans. on Image Processing*, Vol. 3, pp. 572–588, Sept. 1994.
13. M. M. T.R Fisher and M. Wang, "Trellis-Coded Vector Quantization," *IEEE Trans. on Information Theory*, Vol. 37(6), pp. 1551–1556, 1991.
14. H.-J. Wang and C.-C. J. Kuo, "A Multi-Threshold Wavelet Coder (MTWC) for High Fidelity Image," in *IEEE Signal Processing Society, 1997 International Conference on Image Processing (ICIP 97)*, July 1997.

Synthesis Filter Bank Optimization in 1D and 2D Separable Subband Coding

G. Calvagno, M. Cantagallo, G.A. Mian, R. Rinaldo
Dipartimento di Elettronica e Informatica
Via Gradenigo 6/a, 35131 Padova, Italy

Abstract

Subband coding is a popular and well established technique used in multimedia communications, such as audio, image and video transmission. In the absence of quantization and transmission errors, the analysis and synthesis filters in a subband coding scheme can be designed to obtain perfect reconstruction of the input signal, but this is no longer the optimal solution in the presence of quantization of the subband coefficients. In this paper, we analyze the problem of optimal design of the synthesis filters in a subband coding system. To simplify the filter design, we model the input signal as a first-order Markov process, whose correlation parameters can be easily estimated from the input data. The use of a statistical model for the input signal makes the design technique efficient and the problem of an M-subband filter bank design tractable. The approach is extended to image coding using a sub-optimal separable solution, in which the input image is modeled as a 2D separable Markov process plus additive white noise. The extension to the 2D case is not trivial, since the processing in one direction, e.g., the rows, alters the statistics of the signals for the design of the filters in the other direction, e.g., the columns. We show that the filters designed using the model can give a significant gain with respect to the perfect reconstruction solution, especially when the dither technique is used for quantization. Design examples for 1D signals and images are shown.

1 Introduction and notation

In subband coding [1] the input signal is decomposed into a set of decimated subband signals, each relative to different regions of the input spectrum. In this paper we address the problem of designing, for a given analysis filter bank and assuming uniform quantization of the subband coefficients, the set of synthesis filters that minimize the reconstruction error at the output of the subband system [2, 3, 4].

In this section, we briefly review subband coding principles and present the matrix notation that will be used in the derivation of the main results of the paper.

A two-channel subband decomposition system for 1D signals is shown in Fig. 1. Subband y_i, $i = 0, 1$, is obtained by filtering the input signal with

$h_i(n)$ and subsampling by a factor two. In the scheme of Fig. 1, $h_0(n)$ is a low-pass filter, while $h_1(n)$ has high-pass characteristics: thus, each subband y_i has one half of the input samples and is relative to details that pertain to different frequency regions of the input signal spectrum.

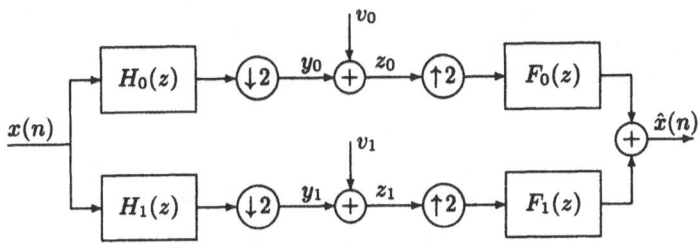

Fig. 1. Two-channel subband system.

The reconstruction is performed at the receiver by upsampling y_i, and filtering with $f_i(n)$. The outputs of the two upsample and filter sections are summed together to give the reconstructed signal $\hat{x}$. It is well known that in the absence of coding errors, i.e., if $z_i = y_i$, perfect reconstruction of the input x can be obtained by appropriate design of the analysis filters $h_i(n)$ and of the synthesis filters $f_i(n)$ [5].

The basic two-channel scheme of Fig. 1 can be used to generate finer decompositions of the input spectrum. In a uniform subband decomposition with $M = 4$ subbands, the scheme of Fig. 1 is applied to each of the subband signals as in Fig. 2. By applying the *noble identities* to interchange the role of filters and decimators [5], we obtain the scheme of Fig. 3, where the M equivalent synthesis filters can be computed from the two-channel filter bank $h_0(n)$ and $h_1(n)$ through proper interpolation and convolution [5]. In the following, for an M subband system, we will denote with h_i, $i = 0, ..., M-1$, the equivalent analysis filters, with $y_i(n)$, $i = 0, ..., M-1$, the subband signals decimated with a factor M, and with f_i, $i = 0, ..., M-1$ the synthesis filters.

The analysis and synthesis procedures of an M-subband system can be conveniently formulated in matrix notation. We assume in the following that

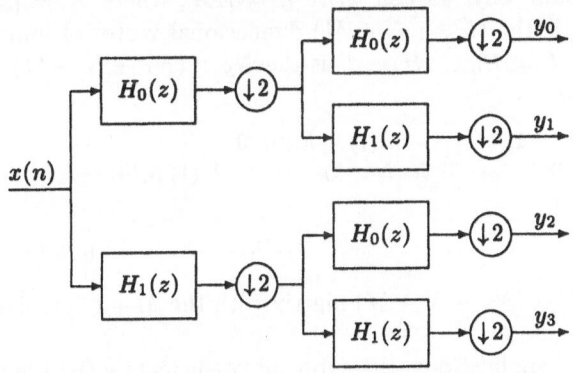

Fig. 2. A subband scheme with $M = 4$ subbands.

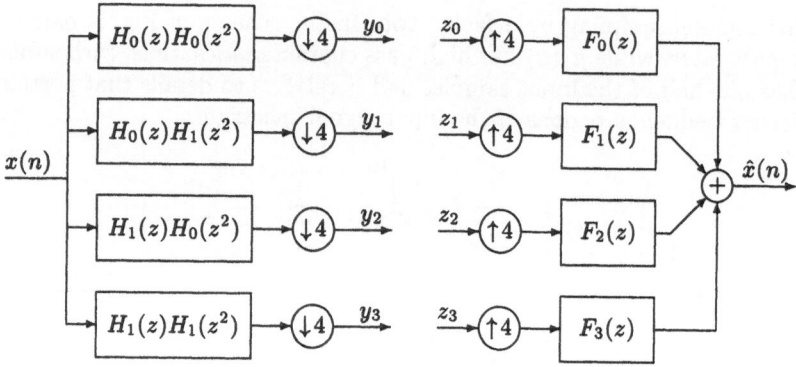

Fig. 3. Equivalent subband scheme with $M = 4$ subbands.

the synthesis filters have length $N_F = (k+1)M$, where k is an integer greater than or equal to zero, and that the length of the analysis filters is N. It is immediate to verify that the computation of M consecutive values of the reconstructed signal $\hat{x}(n)$ can be written in matrix notation as

$$\hat{x} = Fy. \tag{1}$$

In equation (1),

$$\hat{x} = [\hat{x}(0), ..., \hat{x}(M-1)]^t \tag{2}$$

is the M-dimensional column vector with the reconstructed signal values, $F = [F_0, F_1, ..., F_{M-1}]$ is the $M \times N_F$ synthesis matrix, whose rows are defined from the M polyphase components of the synthesis filters, namely

$$F_i = \begin{bmatrix} f_i(Mk) & \cdots & f_i(M) & f_i(0) \\ f_i(Mk+1) & \cdots & f_i(M+1) & f_i(1) \\ \vdots & & & \vdots \\ f_i(M(k+1)-1) & \cdots & f_i(2M-1) & f_i(M-1) \end{bmatrix},$$

and $y = [y_0, ..., y_{M-1}]^t$, with $y_i = [y_i(-k), ..., y_i(0)]$, is the N_F-dimensional column vector with the subband coefficients.

In a similar way, we can write $y = HX$, where $X = [x(-Mk - N + 1), ..., x(0)]^t$ is the $(N_F + N - M)$-dimensional vector of input signal coefficients, and $H = [H_0, ..., H_{M-1}]^t$ is the $N_F \times (N_F + N - M)$ analysis filters matrix where

$$H_i^t = \begin{bmatrix} h_i(N-1) \cdots \cdots & h_i(1) & h_i(0) & 0 & \cdots & \cdots & \cdots & \cdots & \cdots & 0 \\ 0 & \cdots & 0 & h_i(N-1) & \cdots & \cdots & h_i(1) & h_i(0) \cdots & \cdots & \cdots & 0 \\ \vdots & & & & & & & & & & \vdots \\ 0 & \cdots & \cdots & \cdots & \cdots & & & 0 & h_i(N-1) & \cdots & h_i(0) \end{bmatrix}$$

is the $(k+1) \times (N_F + N - M)$ matrix with the M-lag translated versions of the analysis filters.

In coding applications, the subband coefficients $y_i(n)$ are actually quantized before the synthesis stage. The quantization operation can be modeled by adding the quantization error to the subband coefficients, as shown in Fig. 1 for the case of a two-subband system. Thus, equation (1) has to be modified as

$$\hat{x} = Fz, \tag{3}$$

where $z = y + v$ and $v = [v_0, ..., v_{M-1}]^t$, $v_i = [v_i(-k), ..., v_i(0)]$, with $v_i(n)$ the noise component in subband i.

Interestingly enough, when quantization noise is added to the subband coefficients, the analysis filters that minimize the MSE between the original and reconstructed signals are no longer given by the perfect reconstruction filter pair. In this paper, we are interested in designing the optimal reconstruction filters, given the analysis filters and a statistical model for the input and the quantization noise.

2 Determining the optimal filters

In the following, we will consider the case of paraunitary analysis filters, but the discussion can be easily adapted to other classes of filters. In this case, both the analysis and synthesis perfect reconstruction filters have the same length N. Assuming a stationary input signal $x(n)$, the error component in the reconstructed signal $\hat{x}(n)$ is cyclostationary with period M, due to the presence of the interpolators in the synthesis stage. Therefore, the objective is to minimize the average power

$$\text{MSE} = \frac{1}{M}\text{trace}(\text{E}[ee^t]), \tag{4}$$

where $e = \hat{x} - x$ is the error between the reconstructed signal vector (2) and an appropriately defined original signal vector. In (4), $\text{E}[\cdot]$ denotes statistical expectation.

In the definition of x, we have to take into account the delay produced by the causal analysis and synthesis filters. In particular, if the synthesis filter length is chosen to be equal to the analysis filter length N, we define

$$x = [x(-N + 1), ..., x(-N + M)]^t,$$

to take the delay into account. In the presence of quantization noise, we may want to increase the synthesis filter length to allow more freedom in the MSE minimization procedure. Choosing a synthesis filter of even length $N_F > N$, we define

$$x = [x(-N/2 - N_F/2 + 1), ..., x(-N/2 - N_F/2 + M)]^t.$$

As a matter of fact, the optimal solution in the absence of coding error will be in this case an $(N_F - N)/2$-tap delayed version of the perfect reconstruction filters, namely

$$[f_{i,\text{opt}}(0), ..., f_{i,\text{opt}}(N_F - 1)] = [0, ..., 0, f_i(0), ..., f_i(N - 1), 0, ..., 0],$$

with $(N_F - N)/2$ leading and trailing zeros. Thus, we take into account the additional $(N_F - N)/2$-tap delay in the definition of x.

Using equation (3), it is easily verified that the minimization of (4) is equivalent to the minimization of $\text{trace}(FR_{zz}F^t - R_{zx}^tF^t - FR_{zx})$, where $R_{zz} = \text{E}[zz^t]$ and $R_{zx} = \text{E}[zx^t]$. The solution is obtained by standard optimization methods and yields the optimal solution [8]

$$F^t = R_{zz}^{-1}R_{zx}. \tag{5}$$

Using relation $z = HX + v$, matrices R_{zz} and R_{zx} can be computed from the statistics of the input signal and of the noise. In the case of a quantization error v uncorrelated with the input, we have in particular

$$R_{zz} = HR_{XX}H^t + R_{vv}, \qquad R_{zx} = HR_{Xx}, \tag{6}$$

where $R_{XX} = \text{E}[XX^t]$ and $R_{Xx} = \text{E}[Xx^t]$.

For filter design, we will assume in the following that the input signal $x(n)$ is a first-order gaussian Markov process with autocorrelation function $r_x(k) = \sigma_x^2\rho^{|k|}$. Furthermore, we will assume that the quantization error in each subband is white and uncorrelated with the input. These hypotheses make the optimal reconstruction filter design problem particularly simple using (5) and (6). However, using a uniform quantizer with step Δ in each subband, it is well known that the quantization error has variance $\Delta^2/12$ and is practically white and uncorrelated with the input only at high bit rates, i.e., for small values of the quantization step. This hypothesis be made exact at all bit rates by using the subtractive dither technique in the quantization procedure [9]. Figure 4 shows the subtractive dither quantization scheme, where a white random process d, independent of the input and uniformly distributed in $[-\Delta/2, \Delta/2]$, is added before quantization and subtracted after it. It can be shown that the quantization error $v = Q[y+d] - (y+d)$ results to be uniformly distributed in $[-\Delta/2, \Delta/2]$, that it is white, and uncorrelated with y. Thus, if we use the dither technique in the subband quantizers, the hypotheses made for the design of the filters are exactly valid. Furthermore, it is well known that, for audio and image coding, the dither technique can give superior subjective quality results, because it decorrelates the reconstructed signal error with the signal itself.

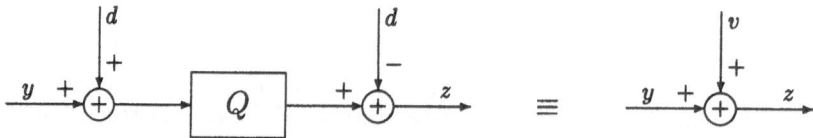

Fig. 4. The dither technique and the equivalent additive quantization scheme.

3 Extension to 2D separable systems

The results for 1D systems outlined above can be extended to 2D systems, where the analysis filter bank is arranged in a separable scheme as shown in Fig. 5 for a 2×2 subband system. The optimal 2D synthesis filters can be designed using the same optimization technique of the previous section [4], but the resulting filters are not separable in general. This approach is inconvenient in many aspects, because of the computational complexity of the filter design and implementation.

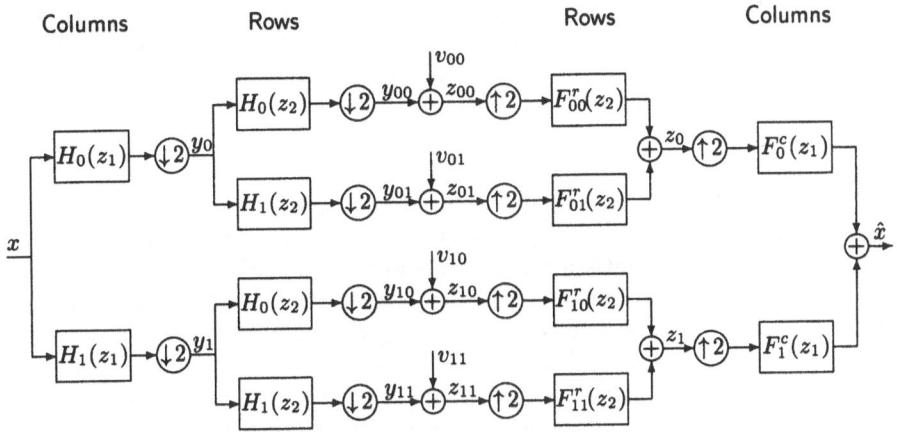

Fig. 5. 2D separable 2×2 subband system.

In this paper we refer to a separable synthesis filter bank and separately optimize the row and column synthesis filters. In particular, we suppose that the input image $x(m, n)$ is a 2D random separable Markov process plus additive white noise. Its autocorrelation function has the form

$$r_{xx}(h, k) = (\sigma_x^2 - \sigma_w^2)\rho_r^{|h|}\rho_c^{|k|} + \sigma_w^2\delta(h)\delta(k), \tag{7}$$

where ρ_r and ρ_c are the row and column correlation coefficients, respectively, σ_x^2 is the image variance and σ_w^2 is the noise variance. The correlation parameters ρ_r and ρ_c can be easily estimated by sample averages.

Although used by many authors, we found that a simpler 2D separable Markov model for images, without any additive noise, can not describe accurately the signal characteristics. Indeed, the separable model gives an acceptable approximation of the 2D input spectrum at low frequencies and along the vertical and horizontal frequency axes, but severely underestimates the signal power in the other frequency regions. In our experiments we found that the inclusion of the additive noise in the model is essential for optimal image filter design.

The noise variance σ_w^2 was estimated on the basis of the 2D periodogram of several test images. We observed that the periodogram amplitude at high vertical and horizontal frequencies is about 30 to 40 dB lower than the amplitude at zero frequencies. In our model the spectrum value at zero frequencies is given by

$$R_{xx}(\omega_x, \omega_y)|_{\omega_x=0, \omega_y=0} = (\sigma_x^2 - \sigma_w^2)\frac{1+\rho_r}{1-\rho_r}\frac{1+\rho_c}{1-\rho_c} + \sigma_w^2.$$

This value has to be compared with the amplitude of the spectrum at high frequencies, that we approximate with the constant σ_w^2.

Values of σ_w^2 in the range $0.1\sigma_x^2$ to $0.5\sigma_x^2$ give a good match to the attenuations observed for real images. Moreover, we found that the filter design procedure is not very sensitive to the exact value of σ_w^2, and we chose $\sigma_w^2 = 0.3\sigma_x^2$ for the experiments described below, corresponding to an attenuation of about 34 dB.

The determination of the optimal separable 2D filters is inherently a nonlinear problem. Following the scheme of Fig. 5, our suboptimal approach to the design of the separable synthesis filter bank splits the procedure into two steps. First, we determine the optimal reconstruction row filters for each subband, based on the statistics of the subband quantization noise and of the inputs y_i. Once these row optimal filters are determined, the column reconstruction filters can be designed by mean square error minimization, leading to a linear system of equations. This second step requires a careful analysis of the second order statistics of the signals in the system, also taking into account the nontrivial effect of the previously designed row reconstruction filters.

For the 2D filter design, we model the quantization noises $v_{ij}(m, n)$ in Fig. 5 as white uncorrelated random processes with variance $\Delta^2/12$. We also assume the noise uncorrelated with the input process. These hypotheses are exactly met when the subtractive dither technique is used in the quantization procedure.

The optimal row synthesis matrices $F_{r,i}$, $i = 0, \ldots, M-1$, can be designed using the procedure described in the previous section for input processes y_i. The correlation along each row of y_i can be easily computed from the 2D correlation $r_{xx}(h, k)$ in (7) and the column analysis filter coefficients h_i, namely

$$r_{y_i y_i}(0, k) = E[y_i(m, n)y_i(m, n+k)] = \alpha_i(\sigma_x^2 - \sigma_w^2)\rho_c^{|k|} + \sigma_w^2\delta(k), \qquad (8)$$

where

$$\alpha_i = \sum_{k_1}\sum_{k_2} h_i(k_1)h_i(k_2)\rho_r^{|k_1-k_2|}.$$

In the derivation of (8), we use the fact that the energy of the filters h_i is equal to one. Note that along each row, the processes y_i are the sum of a Markov process and white noise, although they have in general different variances.

Similarly to what happens in the 1D case, the output of the 2D separable system of Fig. 5 is cyclostationary in the row and column direction with period M. Let us define the output matrix

$$\hat{X} = \begin{bmatrix} \hat{x}(0,0) & \hat{x}(0,1) & \cdots & \hat{x}(0,M-1) \\ \hat{x}(1,0) & \hat{x}(1,1) & \cdots & \hat{x}(1,M-1) \\ \vdots & \vdots & \ddots & \vdots \\ \hat{x}(M-1,0) & \hat{x}(M-1,1) & \cdots & \hat{x}(M-1,M-1) \end{bmatrix}.$$

The column causal filter design consists in minimizing the quadratic norm $\|\hat{X} - X_D\|_2$, where X_D is the matrix of appropriately delayed input coefficients, namely

$$X_D = \begin{bmatrix} x(D,D) & x(D,D+1) & \cdots & x(D,D+M-1) \\ x(D+1,D) & x(D+1,D+1) & \cdots & x(D+1,D+M-1) \\ \vdots & \vdots & \ddots & \vdots \\ x(D+M-1,D) & x(D+M-1,D+1) & \cdots & x(D+M-1,D+M-1) \end{bmatrix},$$

where $D = -N/2 - N_F/2 + 1$. It is easy to show that the matrix $\hat{X}$ can be obtained from the quantized subband coefficient matrices

$$Z_{i,j} = \begin{bmatrix} z_{ij}(-k,-k) & \cdots & z_{ij}(-k,0) \\ \vdots & \ddots & \vdots \\ z_{ij}(0,-k) & \cdots & z_{ij}(0,0) \end{bmatrix}$$

by relation

$$\hat{X} = F_c \begin{bmatrix} Z_{0,0} \cdots Z_{0,M-1} & 0 & \cdots & 0 \\ 0 & Z_{1,0} \cdots Z_{1,M-1} & \cdots & 0 \\ \vdots & \vdots & \ddots & \vdots \\ 0 & 0 & \cdots Z_{M-1,0} \cdots Z_{M-1,M-1} \end{bmatrix} F_r^t \quad (9)$$

where $F_r = [F_{r,0}, F_{r,1}, \ldots, F_{r,M-1}]$ is the $M \times MN_F$ synthesis matrix, with $F_{r,i}$ already computed for each subband in the first step of the design procedure. Referring to Fig. 5, the process $z_{ij}(m,n)$ is the sum of the signal component $y_{ij}(m,n)$ and of the quantization component $v_{ij}(m,n)$. In turn, the signal component $[Y_{i,0} \cdots Y_{i,M-1}]$ in $[Z_{i,0} \cdots Z_{i,M-1}]$ of (9) can be obtained from the input matrix X, via relation

$$[Y_{i,0} \cdots Y_{i,M-1}] = H_i^t X H^t, \quad (10)$$

where we used the same notation of the previous section and

$$X = \begin{bmatrix} x(-Mk-N+1,-Mk-N+1) & \cdots & x(-Mk-N+1,0) \\ \vdots & \ddots & \vdots \\ x(0,-Mk-N+1) & \cdots & x(0,0) \end{bmatrix}$$

is the matrix of input coefficients.

In summary, by using (9) and (10), we have a linear expression relating the input signal and quantization noises to the output of the subband system $\hat{X}$. This linear relation can be equivalently written in the form

$$\hat{X} = F_c([B_0 \text{vec} X, B_1 \text{vec} X, \ldots, B_{M-1} \text{vec} X] +$$
$$[C_0 \text{vec} V, C_1 \text{vec} V, \ldots, C_{M-1} \text{vec} V])$$
$$\triangleq F_c Z \tag{11}$$

where $\text{vec} X$ is the column vector of length $(Mk+N)^2$ obtained by stacking the columns of X [10], and $\text{vec} V$ is the column vector of length $M^2(k+1)^2$ with the quantization noise components in (9) relative to the M^2 subbands. We can interpret $[B_0 \text{vec} X, B_1 \text{vec} X, \ldots, B_{M-1} \text{vec} X]$ as the signal component at the output of the row reconstruction stage in the subband system of Fig. 5, just before column synthesis by F_c. Similarly, $[C_0 \text{vec} V, C_1 \text{vec} V, \ldots, C_{M-1} \text{vec} V]$ is the noise component before column synthesis.

The optimal reconstruction filter matrix F_c is determined by solving the problem

$$\min_{F_c} \|\hat{X} - X_D\|_2 = \min_{F_c} \text{trace}[(F_c Z - X_D)(F_c Z - X_D)^t]$$

that is formally identical to the minimization performed for the 1D design procedure of the previous section. In the hypothesis of uncorrelated signal and quantization noise, the solution of this problem is therefore

$$F_c^t = R_{ZZ}^{-1} R_{ZX_D}$$

where

$$R_{ZZ} = \sum_{i=0}^{M-1} B_i R_{XX} B_i^t + \sum_{i=0}^{M-1} C_i R_{VV} C_i^t$$

$$R_{ZX_D} = E[\,[B_0 \text{vec} X, B_1 \text{vec} X, \ldots, B_{M-1} \text{vec} X] X_D^t\,]$$

with $R_{XX} = E[\text{vec} X \text{vec} X^t]$ and $R_{VV} = E[\text{vec} V \text{vec} V^t]$.

4 Results

In this section we present some experimental results relative to the design of 1D and 2D separable reconstruction filters. In all the following design examples, we assume that the analysis filter bank is obtained starting from the orthogonal paraunitary 8 taps FIR filter of Daubechies [11]. In the coding procedure, all the subband coefficients are quantized using the same uniform quantizer with quantization step Δ, and coding results at different rates are obtained by varying the values of Δ. We assume that the quantization noise is zero mean with a white spectrum, and that the noise components in all the subbands are uncorrelated with each other. Therefore, the autocorrelation matrix R_{vv} of the noise components results to be diagonal, with the elements of the main diagonal given by the variances of the noise components in the subbands. In the filter design, we suppose that the input signal is gaussian and is Markov.

We considered three different approaches to the filter design, by selectively applying subtractive dither to the quantization of the subbands.

Case 1: Dither. We apply the subtractive dither quantization technique to all the subbands. The variances of the noise components result to be all equal to $\Delta^2/12$.

Case 2: Adaptive dither. As it is well known, the typical input signals to a subband coding systems have a low-pass characteristic, and the high frequency subbands result to have small power. For these subband, when the quantization step Δ is high, i.e., at low rates, the use of subtractive dither would increase the quantization error power. In this approach, we apply the dither *only* to those subbands with power greater than $\Delta^2/12$. The quantization noise components of the high energy (low frequency) subbands result therefore uncorrelated with the input signal, while for the low energy (high frequency) subbands this is not true in general. For the filter design, we *assume* all the noise components uncorrelated with the subband coefficients. The variances of the noise components are equal to $\Delta^2/12$ for the high energy subbands, while for the low energy subbands they can be computed from the autocorrelation of the input signal and the analysis filters.

Case 3: No dither. In this approach the coefficients of all the subbands are quantized without the use of dither. Therefore, the quantization noise is in general correlated with the subband signals. Nevertheless, to simplify the computation of the reconstruction filters, we *assume* that the noise is white and uncorrelated with the subband coefficients.

4.1 1D system design example

In these example we considered a 1D system with 2 subbands, with synthesis filters of length 10. The input is a 10000 sample realization of a first order Markov process with autocorrelation coefficient $\rho = 0.95$. Fig. 6 compares the SNR relative to the perfect reconstruction filter bank with the designed optimal filter bank. The dither technique was applied for quantization and the bit rate is calculated as the entropy of the quantized coefficients. As seen, the gain given by the optimal filter is appreciable, especially at low bit rates.

4.2 2D system design examples

The first example considers the basic 2×2 separable subband system. The input is a realization of a 500×500 separable Markov process with unit variance and correlation parameters $\rho_r = \rho_c = 0.95$. The filters were designed using

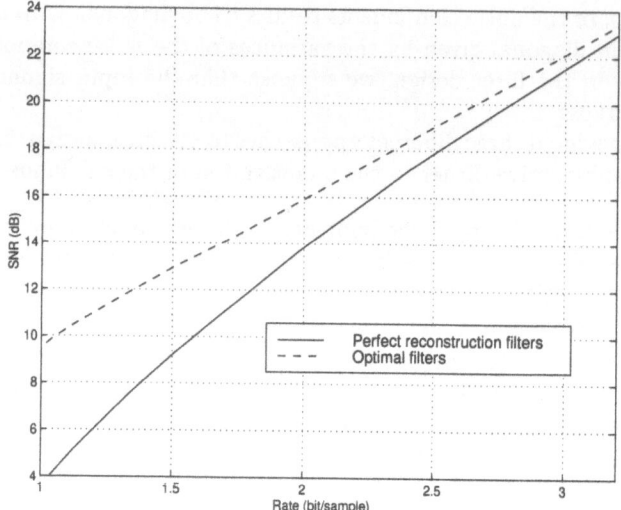

Fig. 6. SNR versus rate for a 2 subband system.

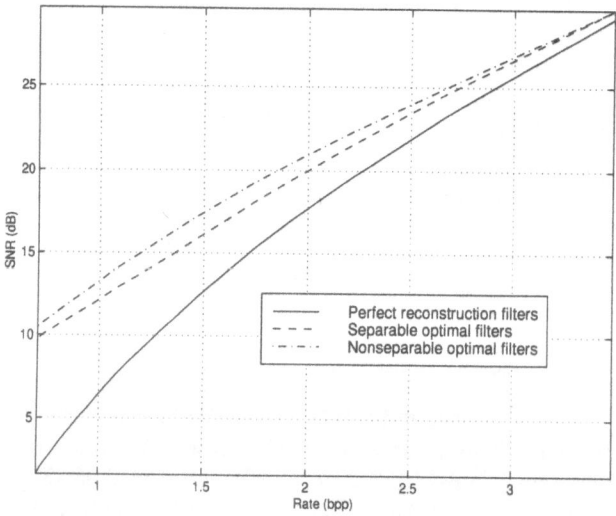

Fig. 7. SNR versus rate for a 2×2 subband system.

the approximate technique described in section 3 and have length 10 (in this case the model noise variance σ_w^2 was set to zero). The dither technique was applied for quantization. Fig. 7 shows the SNR for the perfect reconstruction and the designed optimal filters. The figure also shows the SNR corresponding to the 10×10 nonseparable filter. As seen, the performance of the separable optimal filters is close to that of the nonseparable ones, beside the reduced complexity of the separable solution.

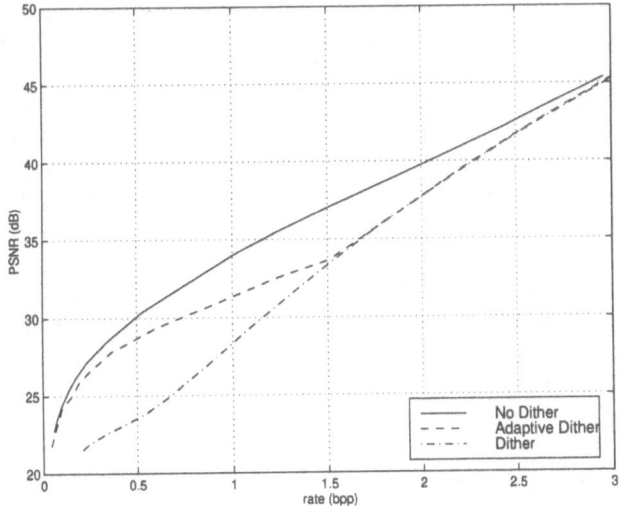

Fig. 8. PSNR versus rate for an 8 × 8 subband system.

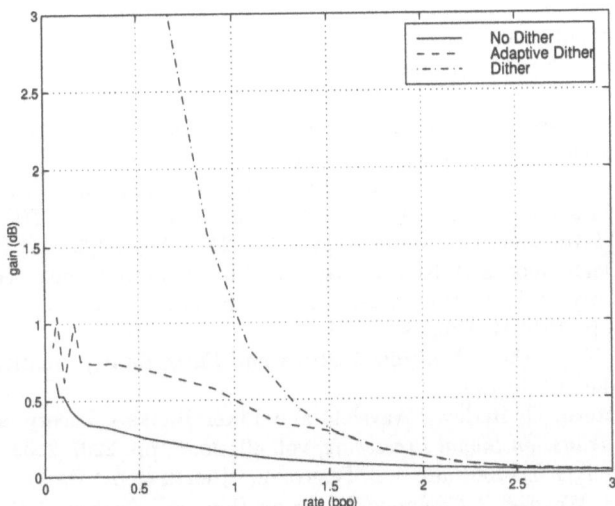

Fig. 9. Noise gain versus rate for an 8 × 8 subband system.

The second example is relative to the application of the design procedure to an 8×8 separable subband system using as input the 256×256 test image "Lenna". For the design of the filters, the input image is modeled as a random Markov process plus additive white noise with estimated normalized correlation parameters $\rho_r = 0.91$ and $\rho_c = 0.96$. The length of the synthesis filters is 56. Note that the perfect reconstruction equivalent filters would have length 54. Fig. 8 shows the PSNR of the reconstructed image for the 3 cases of dither, adaptive dither and no-dither. Fig. 9 shows the gain with respect to the perfect reconstruction filter bank in the 3 cases.

To compare the effects of the three different design approaches on the visual quality of the reconstructed images, Fig. 10 shows the three corresponding coded images and reconstruction error images obtained using the same quantization step $\Delta = 25$. As it can be seen, the error image obtained in the case of quantization with dither is practically uncorrelated with the input image, while in the case of no dither the error presents some features that resemble the image of "Lenna". The case with adaptive dither gives an error image very similar to that obtained by using dither in all the subbands.

We note that the combined use of adaptive dither and of the proposed optimal filter design procedure gives good results both in terms of PSNR and visual quality. From Fig. 8, it can be seen that adaptive dither has comparable performance with the case of no dither in terms of PSNR, while allowing a superior visual image quality. Also, notice that in this case the use of the proposed filter optimization procedure gives considerable PSNR gain with respect to the usual perfect reconstruction solution.

References

1. J.W. Woods and S.D. O'Neil, "Subband coding of images," *IEEE Trans. on Acoustics Speech and Signal Processing*, vol. 34, no. 5, pp. 1278–1288, Oct. 1986.
2. Jelena Kovačević, "Subband Coding Systems Incorporating Quantizer Models," *IEEE Trans. on Image Processing*, vol. 4, no. 5, pp. 543–553, May 1995.
3. N. Uzun and R.A. Haddad, "Cyclostationary Modeling, Analysis, and Optimal Compensation of Quantization Errors in Subband Coders," *IEEE Trans. on Signal Processing*, vol. 43, no. 9, pp. 2109–2119, Sept. 1995.
4. A.N. Delopoulos, S.D. Kollias, "Optimal filter banks for signal reconstruction from noisy subband components", *IEEE Trans. on Signal Processing*, vol. 44, no. 2, pp. 212-224, Feb. 1996.
5. P.P. Vaidyanathan, *Multirate Systems and Filter Banks*, Prentice Hall, A.V. Oppenheim Ed., 1993.
6. M. Vetterli, C. Herley, "Wavelets and Filter Banks - Theory and Design," *IEEE Trans. on Signal Processing*, vol. 40, no. 9, pp. 2207–2232, Sept. 1992.
7. A. Cohen, I. Daubechies, J.C. Feauveau, "Biorthogonal Bases of Compactly Support Wavelets," *Communications on Pure and Applied Mathematics*, vol. 45 no. 5, pp. 485-560, June 1992.
8. D.G. Luenberger, *Optimization by vector space methods*, New York: John Wiley and Sons, 1969.
9. N.S. Jayant and P. Noll, *Digital Coding of waveforms*, Englewood Cliffs, N.J.: Prentice-Hall, 1984.
10. R.A. Horn, C.R. Johnson, *Topics in Matrix Analysis*, Cambridge University Press, 1991.
11. I. Daubechies, "The wavelet transform, time-frequency localization and signal analysis", *IEEE Trans. on Information Theory*, vol. 36, no. 5, pp. 961-1005, Sept. 1990.

Fig. 10. Reconstructed images and corresponding error images for a fixed value of the quantization step, $\Delta = 25$. Top: no dither. Center: adaptive dither. Bottom: dither.

Multiple Descriptions
Source-Channel Coding Methods for Communications

Jelena Kovačević and Vivek K Goyal

Bell Labs, Innovations for Lucent Technologies
600 Mountain Avenue
Murray Hill, NJ 07974, USA

Abstract. We describe multiple description transform coding (MDTC), a joint source-channel coding method for robust data communication. MDTC achieves robustness and graceful degradation in the presence of erasures by representing a source using several descriptions. Then, if one of the descriptions is lost, we can reconstruct from those received. After the theoretical foundation is laid out, applications to robust audio and image transmission are presented.

1   Introduction

In recent years the following problem has received considerable attention: Suppose we are transmitting data over network links of different capacities. Such networks are called heterogeneous networks. A typical scenario might involve a high-capacity network link switching over to a wireless link where packets have to be dropped to accommodate the lower capacity of the latter link. Packets could also be lost in the network due to transmission errors or congestion. If packet retransmission is not an option (for example, due to real-time constraints), one has to devise a way of getting meaningful information to the recipient despite the loss.

This problem finds its natural solution in the so-called multiple-description framework. The idea is to send multiple descriptions of a single source to the receiver. Each of the descriptions can be lost with a certain probability. If all the descriptions are received, we want a high-fidelity estimate of the original data. If only some of the descriptions are received, we want to be able to reconstruct the transmitted data as well as possible. These requirements imply that each of the descriptions should individually be good and thus close to the original data. If this is true, however, the descriptions are very similar so receiving more descriptions will add little extra information. For the descriptions to be collectively good and not add much to the data rate, they must be relatively independent, but such descriptions cannot be individually good. These conflicting requirements will lead to the trade-offs typical in the design of joint source-channel coding systems.

The original multiple description problem with two descriptions and three receivers was posed by Gersho, Witsenhausen, Wolf, Wyner, Ziv and Ozarow at the 1979 IEEE Information Theory Workshop. It can be formulated as the search for achievable quintuples $(R_1, R_2, D_0, D_1, D_2)$ where $R_i, D_i, i = 1, 2$ are the rates and distortions for the ith channel, respectively, and D_0 is the distortion when both descriptions are received. Interestingly, to this day, the problem has been solved only for the memoryless Gaussian source with squared error distortion (Ozarow [1]). El Gamal and Cover [2] construct an achievable rate region for a memoryless source. To show how difficult the problem is, consider, for example, the special case of an i.i.d. binary source with $R_1 = R_2 = 1/2$, $D_0 = 0$ and $D_1 = D_2 = D$. How small can D be? A naive approach of sending alternate source symbols over the two channels results in $D = 0.25$; evaluating the distortion-rate function at rate $1/2$ implies that $D \geq 0.11$. In between these extremes lies the minimum D. In 1980, Wolf, Wyner and Ziv [3] showed that $D \geq 1/6 \approx 0.16$ [3], and the following year Witsenhausen and Wyner [4] improved the lower bound to $D \geq 1/5 = 0.2$. In 1982, El Gamal and Cover [2] gave the upper bound $D \leq (\sqrt{2} - 1)/2 \approx 0.207$. Finally, in 1983, Berger and Zhang [5] closed the gap and showed that $D = (\sqrt{2} - 1)/2$.

Vaishampayan [6] has worked extensively on designing multiple description scalar quantizers as well as on applying the multiple description framework to speech, image and video coding.

We consider multiple description joint source/channel transform coding for erasure channels. This is motivated by the fact that the limitations of separate source and channel coding (spurred by Shannon's famous "separation principle") have lead many researchers to the problem of designing joint source-channel (JSC) codes. An examination of Shannon's result leads to the primary motivating factor for constructing joint source-channel codes: The separation theorem is an asymptotic result which requires infinite block lengths (and hence infinite complexity and delay) at both source coder and channel coder; for a particular finite complexity or delay, one can often do better with a JSC code. JSC codes have also drawn interest for being robust to channel variation.

We propose two methods: In the first, a block of n independent, zero-mean Gaussian variables with different variances are transformed to a block of n transform coefficients in order to create a known statistical correlation between transform coefficients. The transform coefficients from one block are distributed to different packets so in the case of a packet loss, the lost coefficients can be estimated from the received coefficients. The redundancy comes from the relative inefficiency of scalar entropy coding on correlated variables. This method is a generalization of the technique proposed in [7, 8] for two channels. The second method uses a *deterministic* redundancy between descriptions, introduced by representing the source data via a frame. That is, we use a frame expansion from C^n to C^m $(m > n)$ as a computationally simple approach to generalized MD coding [9].

2 Multiple Descriptions Using Statistical Redundancy

In this first method (see [10, 11]), a block of k independent, zero-mean variables with different variances are transformed to a block of k transform coefficients in order to create a known statistical correlation between transform coefficients. The coding of a source vector x proceeds as follows:

1. x is quantized with a uniform scalar quantizer with step size Δ: $x_{q_i} = [x_i]_\Delta$, where $[\cdot]_\Delta$ denotes rounding to the nearest multiple of Δ.
2. The vector $x_q = [x_{q_1}, x_{q_2}, \ldots x_{q_k}]^T$ is transformed with an invertible, discrete transform $\hat{T} : \Delta Z^k \to \Delta Z^k$, $y = \hat{T}(x_q)$. The design and implementation of $\hat{T}$ are described below.
3. The components of y are independently entropy coded.
4. If $n < k$, the components of y are grouped to be sent over the n channels.

The discrete transform $\hat{T}$ is related to a continuous transform T through "lifting." Starting with a linear transform T with determinant one, the first step in deriving a discrete version $\hat{T}$ is to factor T into a product of upper and lower triangular matrices with unit diagonals $T = T_1 T_2 \cdots T_m$. The discrete version of the transform is then given by

$$\hat{T}(x_q) = \left[T_1 \left[T_2 \ldots [T_m x_q]_\Delta \right]_\Delta \right]_\Delta . \tag{1}$$

The lifting structure ensures that the inverse of $\hat{T}$ can be implemented by reversing the calculations in (1):

$$\hat{T}^{-1}(y) = \left[T_m^{-1} \ldots [T_2^{-1} [T_1^{-1} y]_\Delta]_\Delta \right]_\Delta .$$

When all the components of y are received, the reconstruction process is to (exactly) invert the transform $\hat{T}$ to get $\hat{x} = x_q$. The distortion is precisely the quantization error from Step 1. If some components of y are lost, they are estimated from the received components using the statistical correlation introduced by the transform $\hat{T}$. The estimate $\hat{x}$ is then generated by inverting T. The reader is referred to [10, 11] for the algebraic details.

The optimal design of the transform $\hat{T}$ for Gaussian sources, where arbitrary (unequal, dependent) packet loss probabilities are allowed, is discussed in [10, 11]. Here we consider the simpler case where packet losses are i.i.d. and the transform is implemented as parallel and/or cascade combinations of 2-by-2 transforms. It is shown in [10, 11] that for coding a two-tuple source over two channels, where each is equally like to fail, it is sufficient to consider transforms of the form

$$T_a \stackrel{\text{def}}{=} \begin{bmatrix} a & 1/(2a) \\ -a & 1/(2a) \end{bmatrix} . \tag{2}$$

We use this as a building block to form larger transforms as in the cascade structures [11]. The cascade structure simplifies the encoding, decoding, and design when compared to using a general $k \times k$ transform. Empirical evidence suggests that for $k = 4$ and considering up to one component erasure, there is no performance penalty in restricting consideration to cascade structures.

3   Multiple Descriptions Using Frames

Robustness to lost packets comes from redundancy in the source representation. In the previous technique, the redundancy is *statistical*: the distribution of one part of the representation is reduced in variance by conditioning on another part. The second method that we consider (see [9]) uses a *deterministic* redundancy between descriptions.

Consider a discrete block code which represents k input symbols through a set of n output symbols such that any k of the n can be used to recover the original k. (For concreteness, this may be a systematic (n, k) Reed-Solomon code over $GF(2^m)$ with $n = 2^m - 1$ [12].) If the k input symbols are quantized transform coefficients, this may be a good way to communicate a k-dimensional source over an erasure channel that erases symbols with probability less than $(n - k)/n$. A problem with this approach is that except in the case that exactly k of n transmitted symbols are received, the channel has not been used efficiently. When more than k symbols are received, those in excess of k provide no information about the source vector; and when less than k symbols are received, it is computationally difficult to use more than just the systematic part of the code.

An alternative to (discrete) block coding was proposed in [9]. A linear transform from R^k to R^n, followed by scalar quantization, is used to generate n descriptions of a k-dimensional source. These n descriptions are such that a good reconstruction can be computed from any k descriptions, but also descriptions beyond the kth are useful and reconstructions from less than k descriptions are easy to compute.

Assume that we have a tight frame $\Phi = \{\varphi_m\}_{k=1}^n \subset R^k$ with $\|\varphi_m\| = 1$ for all m and that $y = Fx$, where F is the frame operator associated with Φ (see [13] and references therein for details on frames). This vector passes through the scalar quantizer Q: $\hat{y} = Q(y)$. The entropy-coded components of $\hat{y}$ can each be considered a description of x.

For simplicity, let us assume that Q is a uniform quantizer with step size Δ and that $n < 2k$. If $m \geq k$ of the components of $\hat{y}$ are known to the decoder, then x can be specified to within a cell with diameter approximately equal to Δ and thus is well approximated. Since the constraints on x provided by each description are independent, on average, the diameter is a nonincreasing function of m.

When $m < k$ components of $\hat{y}$ are received, R^k can be partitioned into an m-dimensional subspace and a $(k - m)$-dimensional orthogonal subspace such that the component of x in the first subspace is well specified. With a mild zero-mean condition on the component in the latter space, a reasonable estimate of x is easily computed. For any m, estimating x can be posed as a simple least-squares problem.[1]

Let us analyze the distortion when e of the descriptions are erased. Let E denote the index set of the erasures, that is, suppose $\{\langle x, \varphi_m \rangle\}_{m \in E}$ are erased. If $\Phi' = \Phi \setminus \{\varphi_m\}_{m \in E}$ is a frame, the minimum MSE estimate $\hat{x}$ is obtained with the dual frame of Φ'; otherwise, x can only be estimated to within a subspace and distributional knowledge is needed to get a good estimate.[2] We model the quantizer as an additive white noise source,[3] so $\eta = \hat{y} - Fx$ has independent components and is independent of x with $E|\eta_m|^2 = \sigma^2$.

When there are no erasures, the error between the reconstructed signal $\hat{x}$ and source x is due only to the quantization noise. The MSE in this case is $\text{MSE}_0 = k^2 \sigma^2 / n$ [13]. Now suppose (renumbering, if necessary) $E = \{1, 2, \ldots, e\}$ and let F_e denote the frame operator associated with Φ'. The MSE can then be written as

$$\text{MSE}_e = \sigma^2 \sum_{m=e+1}^{n} \|(F_e^* F_e)^{-1} \varphi_m\|^2$$

$$= \left[1 - \frac{e}{n} + \frac{k}{n^2} \sum_{i,j=1}^{e} B_{ij}^{(e)} C_{ij}^{(e)} \right] \text{MSE}_0,$$

where $B^{(e)}$ and $C^{(e)}$ depend on the inner products between erased vectors. A simple special case is when the erased components are pairwise orthogonal. In this case, (3) reduces to $\text{MSE}_e = (1 + e/(n-k))\text{MSE}_0$.

The coding method proposed here could be viewed as a signal-domain alternative to a (discrete) rate-(k/n) block channel code. Though discussion here was limited to the general framework and the distortion with different numbers of erasures, preliminary calculations and simulations indicate potential for the proposed method. Compared to the use of a discrete channel code, the proposed method seems to give better performance at high rates and to be less sensitive to knowledge of the erasure probability. Minimum norm reconstructions can be computed in polynomial time for any e.

4 Applications

In this section, we give examples of the MDTC applied to robust audio as well as robust image transmission systems. They are meant to illustrate the potential of the proposed method; for more details, the reader is referred to [15, 16].

---

[1] For $m \geq k$, a better estimate can be found by exploiting the boundedness of the quantization error [13, 14].

[2] Extensions to where Φ' is not a frame are suggested by [10].

[3] Better reconstructions are possible when the boundedness of the quantization error is exploited [13, 14].

4.1 Multiple Description Source-Channel Coding of Audio

Audio compression uses a digital representation of audio signals to provide maximum signal quality with a given number of bits, delay and cost. Human perception plays a key role in compression of audio material. As a result, recent audio standards work has concentrated on a class of audio coders known as *perceptual coders*. Rather than trying to understand the source, they model the listener and attempt to remove *irrelevant* information contained in the input signal.

We apply the MDTC to a state-of-the-art audio coder developed at Bell Labs, the *Perceptual Audio Coder (PAC)* [17]. Instead of developing a new audio coder from scratch, we implement this new technique in a well-known and fully operational audio coder.

PAC Coder Like most perceptual coders, the PAC combines both source coding techniques to remove signal *redundancy* and perceptual coding techniques to remove signal *irrelevancy*. The PAC divides the input signal into 1024-sample-blocks of data – *frames* – that will be used throughout the encoding process. It consists of five basic parts: The *analysis filter bank* converts the time-domain data to frequency domain. The *perceptual model* computes the frequency-domain threshold of masking both from the time-domain signal and from the output of the analysis filter bank. Depending on the transform that was used previously, each 1024-block is split into a predefined number of groups of bands – *gain factor bands*. Within each factor band, a perceptual threshold value is computed. In the *quantization* process, within each factor band the quantization step sizes are adjusted according to the computed perceptual threshold values in order to meet the noise level requirements. Once a quantized representation that fits with the constraint on the coded signal bit rate has been obtained, *noiseless coding* such as Huffman coding is used to provide an efficient representation of the quantized coefficients. The *frame formatter* forms the bit stream, adding to the coded quantized coefficients the side information needed at the decoder to reconstruct the 1024-sample-block. This block is defined as the *frame*.

At the decoder, all the blocks are inverted and an error mitigation block is added between the inverse quantization and the synthesis filter bank. In this block, lost frames are interpolated based on the previous and following frames.

MD PAC Coder The only difference when compared to the PAC coder is the addition of the MDTC block together with the off-line design.

An *MD transform* block is inserted between the quantizer and the noiseless coder. Within each 1024-sample-block or eight 128-sample-blocks contained in the 1024-sample-unit-block, MDTC is applied to the quantized coefficients (integers) coming out of the quantizer. The transform is applied to pairs of quantized coefficients and produces pairs of *MD-domain quantized coefficients*, using the off-line designed side information. Within each pair, MD-domain quantized coefficients are then assigned to Channel 1 (quantized coefficient with higher variance) or Channel 2 (quantized coefficient with smaller variance)[4]. Side information contains both the way quantized coefficients have to be paired and the parameter of the transform for each pair. Then, the MD-domain quantized coefficients are passed to the noiseless coder.

We insert the inverse MDTC block between the noiseless decoder and the inverse quantizer. Within each 1024-sample-block or eight 128-sample-blocks contained in the 1024-sample-unit, the inverse MDTC function is applied to the MD-domain quantized coefficients (integers) coming out of the noiseless decoder. Then, if both channels are received, we invert the MD transform exactly, recovering perfectly the quantized coefficients. If only one of the channels is lost, we estimate its lost coefficients from their counterparts in the other channel, and invert the MDTC. If both channels are lost, we use the built-in loss mitigation feature of the PAC.

As in the encoder, side information provides the way quantized coefficients have to be paired, the parameter a of the inverse transform for each pair, and the variances to be used in the estimation of lost MD-domain quantized coefficients. Once the MDTC has been inverted according to one of these four strategies, the output quantized coefficients are simply passed to the inverse quantizer.

Since we have blocks of 1024 coefficients and we want to group them into two channels, we need to design the pairing. As described in [11], when there are $2N$ variables and two channels, the optimal pairing consists of pairing the variable with the highest variance with the one with the lowest variance, the second highest variance variable with the second lowest variance one, etc.

Therefore, we first tried pairing in the optimal fashion, that is, across all bands. According to this scheme, we can have either 1024 or 128 variables that we have to pair, leading to either 512 or 64 pairs. Of course, since factor bands may have different quantization steps, this approach implies a rescaling of the domain spanned by the variables prior to the application of MDTC, by multiplying variables by their respective quantization steps.

As will be explained later, the optimal pairing across all bands did not work well. Thus we now take the factor bands into account, and pair variables belonging to the same factor band.

The next step was to design the correlating transform T_a defined by (2). For each pair we obtain the transform parameter a and find both the optimal

---

[4] If the source is Gaussian, uncorrelated, zero mean and if the quantizer preserves the zero mean, then the outputs of the MDTC block should have the same variance. The reason why this is not the case here is that either the quantizer does not preserve the zero mean or that the source is sufficiently far away from Gaussian. We leave this investigation for future work.

redundancy allocation between pairs and the optimal a for a given set of variances and their pairing. After the application of MDTC, we pass the two channels (the MD-domain quantized coefficients) from the 1024 or 128-sample-block to the noiseless coder of the PAC. We do not entropy code each set separately. We do this for convenience and do not optimize the codebooks for our MD-domain quantized coefficients. Since we lose by doing it, we feel that the comparison between the SD PAC and MD PAC is fair. It is part of future work for us to actually separate the channels and optimize codebooks. From one set of variables, the MDTC scheme produces two distinct channels that have to be sent separately through a network. Since these two channels have to be sent separately, side information of the original frame has to be doubled and sent with each channel.

Experimental Results Here we discuss the experiments we did to compare the MD PAC to the original single description version – SD PAC. Since we were interested in Internet audio applications, we selected a bit rate of 20 kbps. Our experiments with the MD PAC were all done with a small amount of redundancy, $\rho = 0.1$ bits per variable.

EXP 1: The first experiment is to compare SD PAC and MD PAC at the same bit rate when no frames are lost. Since we are introducing redundancy, the MD version should sound slightly worse.

EXP 2: Then, still without losing anything, we increase the bit rate in the MD PAC until we reach the same quality as in SD PAC. Here we can see the price we pay in bits for robustness; these bits are wasted when no data is lost.

EXP 3: Finally, we compare the MD PAC and SD PAC at various loss rates. The SD PAC uses frame interpolation to recover from lost frames. If frame information from only one channel is lost, the MD PAC uses statistical redundancy. If frame information from both channels is lost, that is, we lost the whole frame, then we use the SD PAC error recovery scheme.

In what follows, P_i is the loss rate of Channel i, $i = 1, 2$. We also define the overall loss rate as $P_{tot} = (P_1 + P_2)/2$. For example, if $P_1 = 20\%$, then 20% of half-frames corresponding to Channel 1 are lost.

Pairing Across All Bands When we performed Experiment 1, we were quite disappointed. The quality degradation in the MD PAC was extreme. Here is an explanation why:

After applying MDTC to the quantized coefficients, the MD-domain outputs were simply passed to the original PAC noiseless coder. Since the correlating transforms have been designed to produce two equal-rate outputs, we are introducing nonzero values at the positions where the noiseless coder is expecting zeros. Thus, modifying the input to the noiseless coder in such a way led to ineffective coding, resulting in quality degradation. A solution to this problem would be to design and optimize new entropy coders to be used for the MD-domain quantized coefficients. We leave this for future work.

Pairing Within Factor Bands We now restrict ourselves to pairing variables belonging to the same factor band. Throughout the rest of the paper, we will be pointing the reader to the results of our experiments. The audio files are provided in three formats (*aiff*, *wave* and *next*) and can be accessed on the Web at: `http://cm.bell-labs.com/who/jelena/` under `Interests/MD/AudioDemo/DemoList.html`.

We performed **EXP 1** at 20 kbps. The quality degradation due to the redundancy was very low, though noticeable for expert listeners (listen to Files 1 and 2 under "No losses").

In **EXP 2**, the difference was hardly noticeable (listen to Files 1 and 3 under "No losses"). The price we pay is an extra 2 kbps for the MD PAC.

EXP 3 was performed at various loss rates. We started with $P_1 = 100\%$ and $P_2 = 0\%$, and then $P_1 = 0\%$ and $P_2 = 100\%$. A very annoying high-frequency artifact appeared in the decoded files. The higher the loss rates, the more present this artifact was. It turns out that the high-frequency artifact came from overvaluation of variables within a particular factor band, the one where the variances drop to very low levels (the limit of the spectrum that will be coded at a given bit rate). This overvaluation seems to stem from the huge difference of variances of the variables within this factor band, leading to a very small transform parameter a.

To improve the estimation within this factor band, we first set the variables belonging to this factor band to zero. The artifact disappeared. However, since the resulting decoded files lost their highest frequencies, they sounded quite unnatural (listen to File 3 under "Loss rate = 50%").

Note that above, we simulated only the extreme case when one entire channel is lost. We now discuss the results when losses are spread over both channels. The SD PAC error recovery scheme is particularly effective below $P_{tot} = 10\%$. However, for higher loss rates, the interpolation process becomes less and less sufficient, and gaps appear in the music (listen to File 4 under "Total loss rate = 50%").

On the other hand, the MD PAC estimation process seems to have a better behaviour at high loss rates, typically over 80% of one of the channels (listen to Files 1, 2, 3 and 4 under "Total loss rate = 50%"). Even if the estimation tends to be noisy as we have seen previously, the result is satisfactory. However, for loss rates around 50%, the noisy effects are annoying. This might be due to the fact that we are jumping from the spectrum of a perfectly received frame to one estimated in the case of a loss. Furthermore, it seems that the estimated spectrum is not only noisy but also biased from the original one. Moreover, audio is inherently nonstationary, and we are using only one set of estimated variances for an entire audio file. This leads to the differences between original and estimated spectra. As future work, we will implement an adaptive scheme where the variances are estimated on shorter pieces of music. At low loss rates, the behaviour of the MD PAC coder is effective, since the previously described effect is hardly present (listen to Files 1,2 and

3 under "Total loss rate = 5%"). At the loss rate of $(P_1, P_2) = (20\%, 20\%)$, which means that the percentage of loss of both half-frames is $P_{12} = 4\%$, the quality is satisfactory, even though we are mixing spectra from three sources: received, estimated and interpolated (listen to Files 1,2 and 3 under "Total loss rate = 20%"). However, at the loss rate of $(P_1, P_2) = (40\%, 40\%)$, $P_{12} = 16\%$, the decoded file is loaded with annoying artifacts coming from this mix as well as clipping due to the high P_{12}.

4.2 MDTC of Images

MDTC of Images Using Statistical Redundancy To demonstrate the efficacy of the correlating transform method for image coding, we consider the case of coding for four channels. This method is designed to operate on source vectors with uncorrelated components. We (approximately) obtain such a condition by forming vectors from DCT coefficients separated both in frequency and in space. A straightforward application proceeds in the following steps:

1. An 8-by-8 block DCT of the image is computed.
2. The DCT coefficients are uniformly quantized.
3. Vectors of length 4 are formed from DCT coefficients separated in frequency and space. The spatial separation is maximized, that is, for 512×512 images, the samples that are grouped together are spaced by 256 pixels horizontally and/or vertically.
4. Correlating transforms are applied to each 4-tuple.
5. Entropy coding akin to that of JPEG is applied.

The system design is completed by determining which frequencies are to be grouped together and designing a transform for each group. This can be done based on training data. Even with, say, a Gaussian model for the source data, the transform parameters must be numerically optimized.[5]

We have simulated an abstraction of this system. If we were to use precisely the strategy outlined above, the importance of the DC coefficient would dictate allocating most of the redundancy to the group containing the DC coefficient. Thus for simplicity we assume that the quantized DC coefficient is communicated reliably through some other means. We separate the remaining coefficients into those that are placed in groups of four and those that are sent by one of the four channels only. The optimal allocation of redundancy between groups is difficult, so we allocate approximately the same redundancy to each group. For comparison we consider a baseline system that also communicates the DC coefficient reliably. The AC coefficients for each block are sent over one of the four channels. The rate is estimated by sample scalar entropies.

---

[5] In the case of pairing transforms as in [8], the optimal pairing and allocation of redundancy between the pairs can be found analytically [11].

Simulation results for the standard 512×512 'Lena' image are given in Fig. 1. As desired, the MD system gives a higher quality image when one of four packets is lost at the expense of worse rate-distortion performance when there are no packet losses.

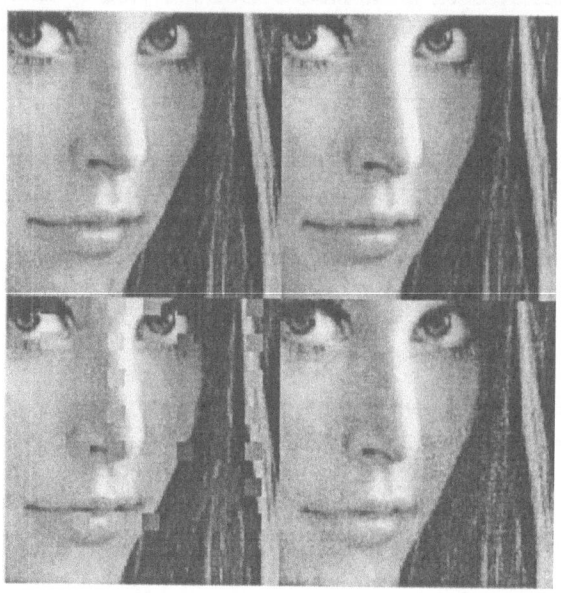

Fig. 1. Results for correlating transform method at 1 bpp. Top row: no packet losses; bottom row: one packet lost. Left column: baseline system; right column: MD system.

The results presented here are only preliminary because we have applied the techniques of [10] without much regard for the the structure and properties of images. The transform design is based on high-rate entropy estimates for uniformly quantized Gaussian random variables. Effects of coarse quantization, dead zone, divergence from Gaussianity, run length coding, and Huffman coding are neglected. Incorporating these will require a refinement of the theory and/or an expansive numerical optimization. Aside from transform optimization, this coder could be improved by using a perceptually tuned quantization matrix as suggested by the JPEG standard. Here we have used a constant quantization matrix for simplicity. With this type of tuning it should be possible to design a system which, say, performs precisely as well as the system in [7] when two or four of four packets arrive, but which performs better when one or three packets arrive.

A full image communication system requires packetization. We have not explicitly considered this, so we do not produce four streams with precisely the same number of bits. The expected number of bits for each stream is equal because of the form of (2). In contrast, with the transforms used in [8] one must multiplex the streams to produce packets of approximately the same size.

MDTC of Images Using Frames As an example, we consider a frame alternative to a $(10, 8)$ block code. For the 10×8 frame operator F we use a matrix corresponding to a length-10 real Discrete Fourier Transform of a length-8 sequence [13]. This can be constructed as $F = [F^{(1)} \; F^{(2)}]$, where

$$F_{ij}^{(1)} = \frac{1}{2} \cos \frac{\pi(i-1)(2j-1)}{10} \text{ and}$$

$$F_{ij}^{(2)} = \frac{1}{2} \sin \frac{\pi(i-1)(2j-1)}{10}, \; 1 \le i \le 10, \; 1 \le j \le 4.$$

In order to profit from psychovisual tuning, we apply this technique to DCT coefficients and use quantization step sizes as in a typical JPEG coder. The coding proceeds as follows:

1. An 8-by-8 block DCT of the image is computed.
2. Vectors of length 8 are formed from DCT coefficients of like frequency, separated in space.
3. Each length 8 vector is expanded by left-multiplication with F.
4. Each length 10 vector is uniformly quantized with a step size depending on the frequency.

The baseline system against which we compare uses the same quantization step sizes, but quantizes the DCT coefficients directly and then applies a systematic $(10, 8)$ block code which can correct any two erasures. We assume that if there are more than two erasures, only the systematic part of the received data is used. (Maximum likelihood decoding would perform somewhat better, but is complex. In practice, one often discards the entire codeword if there are too many erasures.)

We have simulated the two systems with quantization step sizes conforming to a *quality* setting of 75 in the Independent JPEG Group's software. For the 'Lena' image, this corresponds to a rate of about 0.98 bpp plus 25% channel coding. In order to avoid issues related to the propagation of errors in variable length codes, we consider an abstraction in which sets of coefficients are lost. The alternative would require explicitly forming ten entropy coded packets. The reconstruction for the frame method follows a least-squares strategy. For the baseline system, when eight or more of the ten descriptions arrive, the block code insures that the image is received at full fidelity. The effect of having less than eight packets received is simulated using the following combinatorial result: With $e > n - k$ erasures distributed uniformly in a systematic (n, k) code, the probability that m data symbols are erased is

$$\binom{n}{e}^{-1} \binom{k}{m} \binom{n-k}{e-m} \text{ for } e - (n-k) \le m \le \min(e, k).$$

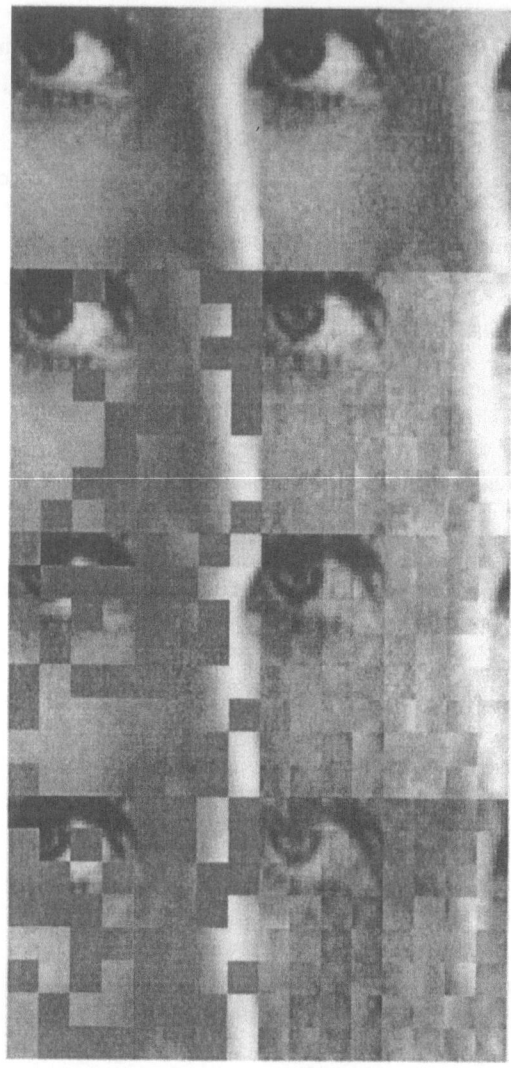

Fig. 2. Results for frame method at 1 bpp. Left column: baseline system; right column: MD system. From top to bottom, number of packets received is 8, 7, 6, and 5.

Results are shown in Fig. 2 for five through eight received packets. As expected, the frame system has better performance when less than eight packets are received. The performance of the MD system degrades gracefully as the number of lost packets increases.

References

1. L. Ozarow, "On a source-coding problem with two channels and three receivers," *Bell Sys. Tech. J.*, vol. 59, pp. 1909–1921, December 1980.
2. A.A. El Gamal and T. M. Cover, "Achievable rates for multiple descriptions," *IEEE Trans. Inform. Th.*, vol. 28, pp. 851–857, November 1982.
3. J. Wolf, A. Wyner, and J. Ziv, "Souce coding for multiple descriptions," *Bell Sys. Tech. J.*, vol. 59, pp. 1417–1426, October 1980.
4. H. Witsenhausen and A. Wyner, "Souce coding for multiple descriptions II: A binary source," *Bell Sys. Tech. J.*, vol. 60, pp. 2281–2292, December 1981.
5. T. Berger and Z. Zhang, "Minimum breakdown degradation in binary source encoding," *IEEE Trans. Inform. Th.*, vol. 29, pp. 807–814, November 1983.
6. V. A. Vaishampayan, "Design of multiple description scalar quantizers," *IEEE Trans. Inform. Th.*, vol. 39, pp. 821–834, May 1993.
7. Y. Wang, M. T. Orchard, and A. R. Reibman, "Multiple description image coding for noisy channels by pairing transform coefficients," in *Proc. First IEEE SP Workshop on Multimedia Signal Processing*, (Princeton, NJ), pp. 419–424, June 1997.
8. M. Orchard, Y. Wang, V. Vaishampayan, and A. Reibman, "Redundancy rate-distortion analysis of multiple description coding using pairwise correlating transforms," in *Proc. IEEE Conf. on Image Proc.*, (Santa Barbara, CA), October 1997.
9. V. K Goyal, J. Kovačević, and M. Vetterli, "Multiple description transform coding: Robustness to erasures using tight frame expansions," in *Proc. IEEE Int. Symp. on Inform. Th.*, (Boston, MA), August 1998.
10. V. K Goyal and J. Kovačević, "Optimal multiple description transform coding of Gaussian vectors," in *Proc. Data Compr. Conf.*, (Snowbird, UT), pp. 388–397, March 1998.
11. V. K Goyal and J. Kovačević, "Multiple description source-matched channel coding of a Gaussian source," *IEEE Trans. Inform. Th.*, 1998. To be submitted.
12. S. Lin and D. Costello, *Error Control Coding: Fundamentals and Applications*. Englewood Cliffs, NJ: Prentice-Hall, 1983.
13. V. K Goyal, M. Vetterli, and N. Thao, "Quantized overcomplete expansions in $\mathcal{R}^n$: Analysis, synthesis and algorithms," *IEEE Trans. Inform. Th.*, vol. 44, pp. 16–31, January 1998.
14. S. Rangan and V. K Goyal, "Recursive Consistent Estimation with Bounded Noise," *IEEE Trans. Inform. Th.*, July 1998. Submitted.
15. R. Arean, J. Kovačević, and V. K Goyal, "Multiple description source channel coding of audio," *IEEE Trans. on Speech and Audio Proc.*, July 1998. Submitted.
16. V. K Goyal, J. Kovačević, R. Arean, and M. Vetterli, "Multiple description transform coding of images," in *Proc. IEEE Conf. on Image Proc.*, (Chicago, IL), October 1998.
17. D. Sinha. Technical Report, Bell Labs 1997.

A Robust Spatial Filtering Method for DOA Estimation of Coherent Sources in Impulsive Interference

Marilli Rupi, Panagiotis Tsakalides, Enrico Del Re and Chrysostomos L. Nikias

Università di Firenze, Dipt. Ingegneria Elettronica, Laboratorio Elaborazione Numerica dei Segnali e Telematica; University of Southern California, Dept. Electrical Engineering, Signal & Image Processing Institute

Abstract. Multipath propagation is one of the dominant environmental influences on the performance of wireless communication systems. Meanwhile, heavy-tailed interference from neighboring cells may further degrade communication quality in mobile systems. This paper employs notions of spatial diversity using an array of sensors to address the inter-signal coherence problem. At the same time, the paper proposes the use of fractional lower-order moments to mitigate the effects of the heavy-tailed background noise environment. The improved performance achieved by the combination of these two processing modules is demonstrated via Monte Carlo simulations.

1 Introduction

The rapidly increasing demands for personal mobile communications have created an avid interest in new array processing algorithms and antenna architectures. Our choice for a particular antenna array model depends on its capability to discriminate different incoming signals, which can be generated from various digitally modulated sources. Specifically, in the wireless communications, the transmitted signal arrives to the receiver via multiple paths, due to some physical phenomena such as diffraction and reflection, caused by obstacles present in the line of view between the transmitter and the receiver. Multipath can affect the incoming signal and destroy completely the information sequence. Moreover, relative motion between the transmitter and the receiver further degrade the desired signal, due to an undesired Doppler shift. Finally, coherent interference can arise when "smart"jammers deliberately redirect scaled and delayed replicas of the same signal to the receiver. The array antenna is able to process the incoming signal and improve its quality by applying techniques based on interference cancellation and spatial diversity.

In recent years, considerable effort has been spent in developing high resolution techniques for estimating the direction of arrival of multiple signals using antenna arrays. The eigen-based class of methods has been proven to be an effective means for this goal, even when the signal sources are partially correlated [1]. However, when some of the signals are perfectly correlated, eigen-based or subspace techniques can not be further used since fall the algorithm hypotheses. Several alternatives have been proposed to overcome this problem, using ideas such as sub-aperture sampling of spatial smoothing which essentially decorrelate the coherent signals [2, 3]. In any case, all these done works dealt with Additive White Gaussian noise.

Although the Gaussian distribution plays a significant role in a mobile communication environment, mainly due to its capability to lead to simple solutions, in certain practical applications it is not an accurate model for the noise. For these reasons we need to use more realistic non-Gaussian models, such as the complex Symmetric Alpha-Stable ($S\alpha S$) distribution, which is a class of heavy-tailed random variables holding, as particular cases, the Gaussian and the Cauchy distributions.

A brief review of the $S\alpha S$ family is undertaken in Section 2. In Section 3, we apply the Spatial Smoothing Scheme to the ROC-MUSIC, introduced in [4] to address the problem of high relosution DOA estimation of narrow-band coherent sources in the presence of impulsive noise. In Section 5, we demonstrate the improved performance of the proposed method via simulation examples.Finally, in Section 6, we summarize the results and present avenues of future research.

2 Symmetric Alpha Stable mathematical analysis

The $S\alpha S$ family of distributions shares many properties with the popular Gaussian random variables, including the stability property and the central limit theorem. The main difference between the two distributions concerns the tail of the stable density, which is heavier than the tail of the Gaussian density. This characteristic makes the $S\alpha S$ variables suitable for modeling signals and noise of an impulsive nature. The impulsive component of the noise has been found to be significant in many communications problems. The atmospheric noise, e.g., may be considered as the result of a large number of independent sources (mainly thunderstorms) which shows such statistical behavior.

The $S\alpha S$ processes are defined by four parameters: the characteristic component $0 < \alpha \leq 2$ (the smaller the α, the heavier the tails of the density), the skewness parameter $-1 \leq \beta \leq 1$ (the distribution is symmetric when $\beta = 0$), the dispersion $\gamma > 0$(which plays a similar role of variance for the Gaussian case), and the location parameter $-\infty \leq a \leq \infty$ (which is the mean when $1 < \alpha \leq 2$ and the median when $0 < \alpha < 1$). All these parameters appear in the definition of the *characteristic function*:

$$\varphi(t) = e^{\jmath at - \gamma |t|^\alpha [1 + \jmath \beta sign(t) \omega(t, \alpha)]} \tag{1}$$

where $\omega(t, \alpha)$ is $tan\frac{\alpha\pi}{2}$ if $\alpha \neq 1$ and $\frac{2}{\pi}log|t|$ if $\alpha = 1$, and $sign(t)$ is $|t|$ if $t \neq 0$ and 0 if $t = 0$ (Gaussian and Cauchy distributions have $\alpha = 2$ and $\alpha = 1$ respectively). The main characteristic associated with $S\alpha S$ processes is the capability to possess finite pth order moments only for $p < \alpha$, so it is clear that for all non Gaussian $S\alpha S$ variables finite second or higher-order statistics do not exist. Hence, it is necessary to define a dual tool, namely the fractional lower-order statistics (FLOS) [5].

3 Problem Statement

We consider a uniformly-spaced linear array antenna consisting of N elements, in which impinge $Q < N$ signals (with λ being the carrier wavelength), that are located at different angles with respect to the orthogonal axis $\{\vartheta_k; \ k = 1, \ldots, Q\}$ and have constant velocities with respect to the receiver $\{\nu_k; \ k = 1, \ldots, Q\}$ corresponding to Doppler frequencies $\{f_k = \nu_k/\lambda; \ k = 1, \ldots, Q\}$. Since we assume the signal bandwidth to be narrow as compared to the inverse of the travel time across the array, it follows that, by using a complex envelop representation, the array output can be expressed as [6]:

$$\mathbf{x}(t) = \mathbf{V}(\vartheta)\mathbf{s}(t) + \mathbf{n}(t), \tag{2}$$

where

- $\mathbf{x}(t) = [x_1(t), \ldots, x_N(t)]^T$ is the array output vector;
- $\mathbf{s}(t) = [s_1(t), \ldots, s_Q(t)]^T = \mathbf{Af}$ is the signal vector received by the array;
- $\mathbf{A} = diag(a_1, a_2, \ldots, a_N)$ is the $N \times N$ matrix of the attenuation factors;
- $\mathbf{f} = [e^{j2\pi f_1}, e^{j2\pi f_2}, \ldots, e^{j2\pi f_N}]^T$ is the Doppler shift vector, and it is a function of the speed velocity and the wavelength λ: $f_k = \nu_k/\lambda$;
- $\mathbf{V}(\vartheta)$ is the *space-time steering matrix*, whose $r - th$ column vector $\mathbf{v}(\vartheta_r)$ is $[1, e^{-j2\pi(d/\lambda)sin\vartheta_r}, \ldots, e^{-j(N-1)2\pi(d/\lambda)sin\vartheta_r}]^T$
- $\mathbf{n}(t) = [n_1(t), \ldots, n_N(t)]^T$ is the noise vector, of impulsive nature, assumed to be uncorrelated with the signals.

In this paper, we assume that the Q signal waveforms are fully coherent, phase-delayed amplitude-weighted replicas of one of them. Also, the noise vector $\mathbf{n}(t)$ is a complex isotropic $S\alpha S$ random process with $1 < \alpha \leq 2$ and zero location parameter. The noise is assumed to be independent of the signals and it has covariation matrix $\Gamma_N = \gamma_n\mathbf{I}$.

4 Subspace Methods based on FLOS

We start our analysis by first considering the case of independent or uncorrelated incoming signals. In impulsive noise environments, the concept of the covariation matrix has been introduced in [4] to characterize the correlation properties of the signal noise field. In addition, application of eigendecomposition methods to the covariation matrix resulted to robust DOA estimates of independent signal sources in $S\alpha S$ noise.

The covariation matrix, Γ_X, of the observation vector process $\mathbf{x}(t)$ is defined as the matrix whose elements are the covariations $[x_i(t), x_j(t)]_\alpha$ of the components of $\mathbf{x}(t)$ [4]. We obtain the following expression for the covariations of the sensor measurements:

$$[x_i(t), x_j(t)]_\alpha = \sum_{k=1}^{Q} v_i(\vartheta_k)v_j(\vartheta_k)^{<\alpha-1>}\gamma_{s_k} + \gamma_n\delta_{i,j} \quad i,j = 1,\ldots,N. \quad (3)$$

where we define the signed power operation as $a^\alpha = |a|^\alpha sign(a)$ if a is real or $a^\alpha = |a|^{\alpha-1}a^*$ if a is complex. In matrix form, (3) gives the following expression for the covariation matrix of the observation vector

$$\Gamma_X \triangleq [\mathbf{x}(t), \mathbf{x}(t)]_\alpha = \mathbf{V}(\vartheta)\Gamma_S\mathbf{V}^{<\alpha-1>}(\vartheta) + \gamma_n\mathbf{I}, \quad (4)$$

where Γ_S is the covariation matrix of the incident signals and the $(i,j)th$ element of matrix $\mathbf{V}^{<\alpha-1>}(\vartheta)$ is $|[\mathbf{V}(\vartheta)]_{j,i}|^{\alpha-2}[\mathbf{V}(\vartheta)]_{j,i}^* = [\mathbf{V}(\vartheta)]_{j,i}^*$, and thus, the covariation matrix can be written as

$$\Gamma_X \triangleq \mathbf{V}(\vartheta)\Gamma_S\mathbf{V}(\vartheta)^H + \gamma_n\mathbf{I} \quad (5)$$

Clearly, when $\alpha = 2$, i.e., for Gaussian distributed noise, the expression for the covariation matrix is identical to the well-known expression for the covariation matrix: $\mathbf{R}_X = \mathbf{V}(\vartheta)\Sigma\mathbf{V}^H(\vartheta) + \sigma^2\mathbf{I}$, where Σ is the signal covariation matrix and σ^2 is the variance of the Gaussian noise variable. Observing (5), we conclude that standard subspace techniques can be applied to the covariation of the observation vector to extract the bearing information. In practice, we have to estimate the covariation matrix from a finite number of array sensor measurements. Besides, in a real communication model, the statistics of the transmitted signal $s(t)$ are not alpha-stable. Hence, in the following, we will assume $s(t)$ to be a white, zero-mean, random process. In this contest, we define the fractional lower-order cross correlation of two random processes $x_i(t)$ and $x_j(t)$ as:

$$< x_i(t), x_j(t) >_p = E\left[x_i(t)\, x_j(t)^{<p-1>}\right] \quad (6)$$

According to (6), we can see the presence of the operator expectation, which is a linear operator, and the introduction of the fractional pth power, which is, conversely, not linear. Even in presence of independent sources, it is not possible to find a closed-form expression suitable for this problem, that could be used for write the covariation of random variables, not necessarily only $S\alpha S$.

Under the hypothesis of uncorrelated noise and signals, independent with each other and zero-mean, we obtain that

$$< x_i(t), x_j(t) >_p = \sum_{k=1}^{Q} v_i(\vartheta_k)v_j(\vartheta_k)^{<p-1>} E\left[s_k(t)\ s_k(t)^{<p-1>}\right] +$$

$$+E\left[f(\sum_{k=1}^{Q} v_{r_i}(\vartheta_k)s_k, \sum_{k=1}^{Q} v_{r_j}(\vartheta_k)^a s_k^b, n_{\alpha_j}^c)\right] + \gamma_{n_\alpha}\delta_{i,j}(7)$$

where $f(x)$ is a function of x, which is given by all the contributions containing the fractional products between the signal $s_k(t)$ and the noise source n_{α_j}. The elements a, b, c are three fractional values strictly less than $p-1$, with the constraint $b \neq 0, c \neq 1$. Observing (7), we see that the use of FLOS for the non-$S\alpha S$ noise case, has caused the introduction of cross-terms involving correlation functions between the signal and noise components. The extent(?) to which the existence of these cross-terms affects the performance of eigen-based methods is studied in the simulation section.

4.1   Subspace Methods based on FLOS & Spatial Smoothing

Second-order or fractional lower-order methods for bearing estimation require a non coherent environment in which all Q sources are statistically independent. In a realistic mobile communication scenario, this hypothesis does not hold, due to the reflection of the transmitted signal which arrives at the receiver via correlated multi-paths. Furthermore, the relatively motion between receiver and transmitter makes the Doppler shift another important factor to consider during a wireless communication. Hence, Fading and Doppler, adversely, affect the performance of standard bearing estimation methods and create the need to introduce methods that can address the signal coherence problem.

For dealing with coherence phenomena in an impulsive noise background, we introduce a *Spatial Smoothing* version of the FLOS-based subspace algorithm. The basic principle states to divide the array of dimension $N > Q$ into uniformly overlapping sub-arrays of dimension $P > Q$, in such a way that each one shares with an adjacent sub-array all but one of its sensors. Calling as the rth sub-array the one consisting of the elements $(r, r+1, ..., r+P-1)$, we can write the covariation of the ith with the jth elements of this sub-array as

$$<x_i(t), x_j(t)>_p = E\left[\left(\sum_{k=1}^{Q} v_{r_i}(\vartheta_k)s_k + n_{\alpha_i}\right)\left(\sum_{h=1}^{Q} v_{r_j}(\vartheta_h)s_h + n_{\alpha_j}\right)^{<p-1>}\right] \quad (8)$$

for each $i, j = 1, ..., P$ and for all sub-arrays $r = 1, ..., N - P + 1$. This expectation, this time, has inside the cross terms between $s_k s_h = c_k c_h s(t)^2$, and all the values which contain fractional powers of the product between the noise and the signal. Due to the linearity property with respect to the first term, and the non-linearity of the pth fractional power, we can write

$$
< x_i(t), x_j(t) >_p = \sum_{h,k=1}^{Q} v_{r_i}(\vartheta_k) v_{r_j}(\vartheta_h)^{<p-1>} E\left[s_k(t)\, s_h(t)^{<p-1>}\right] +
$$

$$
+\gamma_{n_\alpha} \delta_{i,j} + \Phi_{i,j} \tag{9}
$$

$$
\Phi_{i,j} = E\left[g\left(\sum_{k=1}^{Q} v_{r_i}(\vartheta_k) s_k, \sum_{h=1}^{Q} v_{r_j}(\vartheta_h)^a s_h^b, n_{\alpha_j}^c \right) \right] \tag{10}
$$

with $g(x)$ a function of the variable x. The term $\Phi_{i,j}$ is what we call *corruption factor*, and is a finite quantity, since all the fractional powers (a, b, c, d) are strictly less then $p - 1$ and $(n_{\alpha_i}, n_{\alpha_j})$ are $S\alpha S$ noise components that have finite moments of order $p < \alpha$. Then, it is possible to find an equivalent expression for the FLOS matrix of the rth sub-array

$$
\Gamma_{X_P}^{(r)} \triangleq \mathbf{V}_P(\vartheta) \mathbf{D}^{r-1} \Gamma_S \mathbf{D}^{r-1<p-1>} \mathbf{V}_P^{<p-1>}(\vartheta) + \gamma_{n_\alpha} \mathbf{I}_P + \Phi_P^{(r)} \tag{11}
$$

where $\mathbf{V}_P(\vartheta)$ is the set of steering vectors for a sub-array of length P and $\mathbf{D}$ is a diagonal matrix whose lth element is equal to $e^{-j2\pi(d/\lambda)sin(\vartheta_l)}$. The (i,j)th element of $\mathbf{V}_P^{<p-1>}(\vartheta)$ is the (j,i)th element of $\mathbf{V}_P(\vartheta)$ to the signed power of $<p-1>$. Since the elements of $\mathbf{V}_P(\vartheta)$ and $\mathbf{D}^{r-1}$ have unit magnitude, it follows that

$$
\Gamma_{X_P}^{(r)} = \mathbf{V}_P(\vartheta) \mathbf{D}^{r-1} \Gamma_S \mathbf{D}^{r-1^H} \mathbf{V}_P(\vartheta)^H + \gamma_{n_\alpha} \mathbf{I}_P + \Phi_P^{(r)} \tag{12}
$$

The effect of this corruption matrix is that of a correlated noise field which will adversely affect the performance of the method. However, as we demonstrate in the simulation section, the advantage to use the FLOS formulation in the presence of heavy-tailed environment noise outweighs the negative effect of the induced corruption factors.

The FLOS matrix is evaluated as the average of the sub-array FLOS matrices.

$$
\bar{\Gamma}_X = \frac{1}{N - P + 1} \sum_{r=1}^{N-P+1} \Gamma_{X_P}^{(r)} \tag{13}
$$

Using (12) and (13) we can write the FLOS matrix as

$$
\bar{\Gamma}_X = \mathbf{V}_P(\vartheta) \bar{\Gamma}_S \mathbf{V}_P(\vartheta)^H + \bar{\Gamma}_{n_\alpha} + \frac{1}{N - P + 1} \sum_{r=1}^{N-P+1} \Phi_P^{(r)} \tag{14}
$$

where

$$\bar{\Gamma}_S \triangleq \frac{1}{N-P+1} \sum_{r=1}^{N-P+1} \mathbf{D}^{r-1} \Gamma_S \mathbf{D}^{r-1^H} \tag{15}$$

$$\bar{\Gamma}_{n_\alpha} \triangleq \frac{1}{N-P+1} \sum_{r=1}^{N-P+1} \gamma_{n_\alpha} \mathbf{I}_P \tag{16}$$

As a result of the spatial smoothing operation the matrix in (15) is non singular [2]. Hence, we can apply an eigen-decomposition of $\bar{\Gamma}_X$ to achieve robust DOA estimation in a fading Doppler environment with heavy-tailed noise. However, we have to consider the effect of the corruption matrix Φ. This matrix corrupts the eigen-based methods in such a way that the resulting eigenvectors of $\bar{\Gamma}_X$ can not perfectly span the two original orthogonal signal and signal-plus-noise sub-spaces. In other words, we can say that the corruption matrix is in effect a result of an induced colored noise component which, as we demonstrate from the experimental results, is not disruptive to the algorithms in terms of their performance.

5   Experimental Results

In this section, we show the results on the resolution capability of ROC-MUSIC Smoothing versus ROC-MUSIC, MUSIC and MUSIC Smoothing as a function of the noise characteristic exponent α and the angle separation. The array is linear with eight sensors spaced half wavelength apart, with overlapped sub-array of dimension five. Four signals impinge on the array from directions $\vartheta = [30°, -40°, 60°, -15°]$. As constant power attenuation factors, we used $\{-3dB, 0dB, -2dB, -6dB\}$ for the four paths. The noise follows the isotropic stable distribution. The transmitted waveform is a QPSK signal, filtered with a squared root raised cosine. The received signal is filtered with the same matched squared root raised cosine filter. The Doppler effect is due to the different motions of incoming signals. Here we considered velocities of $70Km/h$, $60Km/h$, $50Km/h$ and $40Km/h$, carrier frequency of $900MHz$ with relative shift Doppler given as $\nu_i/\lambda, i = 1, \ldots, Q$.

Since the alpha stable family for $\alpha < 2$ determines processes with infinite variance, we cannot define an ensemble signal-to-noise ratio. But for finite-sample realizations, we can define the *Effective SNR (ESNR)* to be the ratio of the signal power over the noise power:

$$ESNR = 10 \log\left(\frac{\sum_{t=1}^{M} |s(t)|^2}{\sum_{t=1}^{M} |n(t)|^2}\right). \tag{17}$$

The parameters $p_1, p_2 < \alpha$ in the estimation of the covariation matrix [4] were set equal to $p_1 = 1.1$ and $p_2 = 1.2$. For all these simulations, the number of signals is assumed to be known. In every experiment, we perform 10 Monte

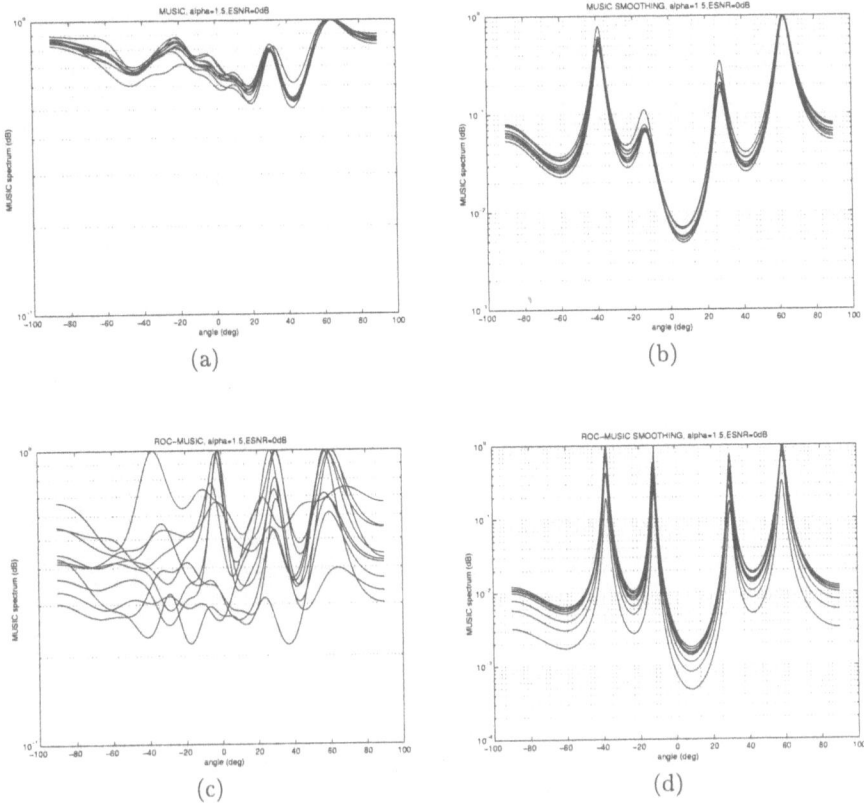

Fig. 1. (a) MUSIC, (b) MUSIC Smoothing, (c) ROC-MUSIC and (d) ROC-MUSIC Smoothing for $\alpha = 1.5$ and $ESNR = 0dB$.

Carlo runs of 10000 snapshots available to the algorithms. As we can see in Figure 1, for a fairly impulsive noise environment case ($\alpha = 1.5$, $ESNR = 0$ dB), the MUSIC Smoothing method exhibits low-resolution performance and cannot resolve all four moving signals. Besides, even when the statistic behavior of the noise is close to Gaussian ($\alpha = 1.85$), the MUSIC Smoothing method still cannot clearly resolve all four multipaths (cf. Fig. 2(d)).

On the other hand, the ROC-MUSIC Smoothing method exhibits higher stable resolution capability, showing better performance for non-Gaussian additive noise environments ($\alpha = 1.5$, cf. Fig 1(d)), and at the same time, performing well in quasi-Gaussian interference ($\alpha = 1.85$, cf. Fig. 2).

In Fig. 3 we demonstrate the algorithmic performance with respect to

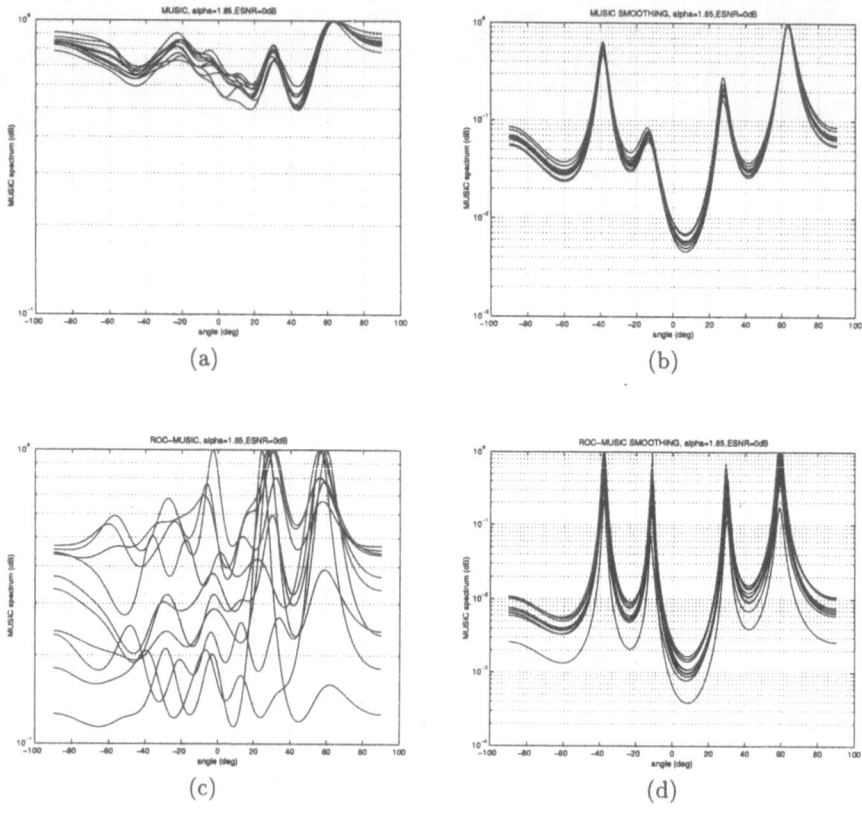

Fig. 2. (a) MUSIC, (b) MUSIC Smoothing, (c) ROC-MUSIC and (d) ROC-MUSIC Smoothing for $\alpha = 1.85$ and $ESNR = 0dB$.

the spatial angle separation of the two closely spaced incoming signals from directions $\vartheta = [15^{\circ}, -40^{\circ}, 40^{\circ}, -30^{\circ}]$, for ESNR= -8 dB and $\alpha = 1.5$. As expected, the resolution capability improves with increased angle separation between the two paths. But the ROC-MUSIC Smoothing algorithm requires a lower angle separation threshold than the MUSIC Smoothing algorithm and it is able to resolve the two paths from -40° and -30°, unlike the MUSIC Smoothing algorithm. In Fig. 3, a 20 antenna array with sub array length of 12 has been considered for better resolution.

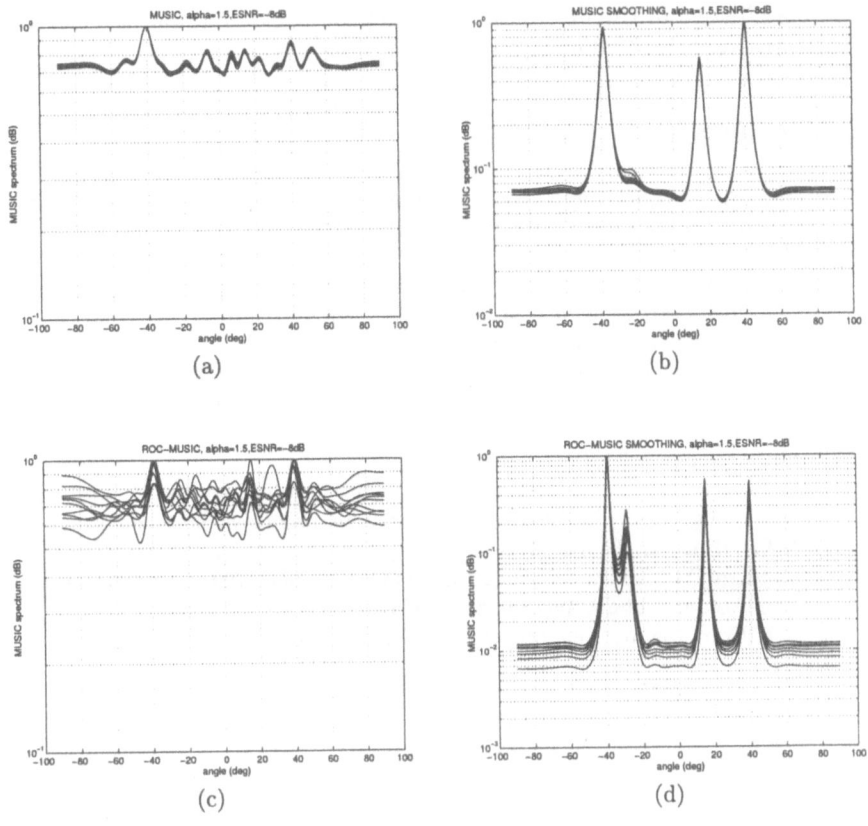

Fig. 3. (a) Music, (b) MUSIC Smoothing, (c) ROC-MUSIC and (d) ROC-MUSIC Smoothing for $\alpha = 1.5$ and $ESNR$ of $-8dB$.

6   Conclusion

Conventional high-resolution eigen-decomposition techniques perform poorly in coherent receiving environments. Spatial Smoothing methods proposed in the past address the signal coherency problem but fail to operate reliably in a non-Gaussian noise. The method proposed in this paper is able to overcome degradation in the performance due to both multipath and impulsive noise environments The new algorithm is based in spatial Smoothing of the covariation matrix of an antenna array and it is shown to exhibit better high-resolution performance in a wide range of noise environments without considerably increasing the complexity of the system. Several limitations need to be addressed in the future, including unequally spaced non-linear arrays, and correlated additive noise structures.

References

1. B.C. Ng, M.H. Er, and C. Kot. A music approach for estimation of directions of arrival of multiple narrowband and broadband sources. *Signal Processing*, November 1994.
2. T.-J. Shan, M. Wax, and T. Kailath. On spatial smoothing for direction-of-arrival estimation of coherent signals. *IEEE Transections on Acoustics, Speech, and Signal Processing*, August 1985.
3. T. Kailath V. U. Reddy, A. Paulray. Performance analysis of optimum beam-former in the presence of correlated sources and its behaviour under spatial smoothing. *IEEE Trans. on Acoustic, Speech and Signal Processing*, July 1987.
4. P. Tsakalides and C.L. Nikias. The robust covariation-based music (roc-music) algorithm for bearing estimation in impulsive noise environments. *IEEE Trans. on Signal Processing*, July 1996.
5. M. Shao C.L. Nikias. *Signal Processing with Alpha-Stable Distributions and Applications*. John Wiley & Sons, 1995.
6. R.T. Compton Jr. *Adaptive antennas*. Prentice-Hall, 1988.

Signal Processing Considerations for Immersive Audio Rendering

A. Mouchtaris, J.–S. Lim, T. Holman, and C. Kyriakakis
Immersive Audio Laboratory
Integrated Media Systems Center
University of Southern California
3740 McClintock Ave., EEB 432
Los Angeles, CA 90089

Abstract

Immersive audio systems are being envisioned for applications that include tele-conferencing and telepresence; augmented and virtual reality for manufacturing and entertainment; air traffic control, pilot warning, and guidance systems; displays for the visually-impaired; distance learning; and professional sound and picture editing for television and film. In this paper we examine signal processing issues that pertain to immersive audio rendering over loudspeakers. We propose two methods that can be used to implement the necessary filters for generating virtual sound sources based on synthetic head-related transfer functions with the same spectral characteristics as those of the real source. Furthermore, several factors are presented that pertain to high quality immersive audio reproduction at the desktop including acoustical and psychoacoustical considerations and the importance of dynamically adapting to listener's head movement.

1. Introduction

Digital audio workstations are increasingly being used to manipulate, edit, and render multichannel audio program material. This is largely due to advances both in main CPU computational power, as well as in special-purpose audio DSP's. Many sound editing operations that could previously only be performed in expensive calibrated dubbing stages are now routinely performed on digital audio workstations. In addition to accurate reproduction of the *measurable* characteristics of sound, multichannel and emerging 3-D audio program material also requires accurate *spatial reproduction* of sound in order to create a seamless aural environment and achieve good sound localization relative to visual images. For such material, a mismatch between the aurally-perceived and visually-observed positions of a particular sound causes a cognitive dissonance that can seriously limit the desired suspension of disbelief [1].

Applications for high-quality desktop audio include professional sound editing for film and television, immersive telepresence, augmented and virtual reality, distance learning, and home entertainment. Such a wide variety of applications has led to an equally wide variety of interrelated, and at times conflicting, system requirements that arise from fundamental physical limitations as well as current technological drawbacks [2]. For example, while there have been advances in sound recording and reproduction technologies, as well as in the understanding of human sound perception mechanisms, these have not yet been combined in such a way as to achieve accurate synthesis of fully 3-D auditory scenes. Furthermore, many acoustical and psychoacoustical issues that pertain to sound reproduction in large rooms have not yet been correctly translated to the desktop environment.

In this paper we discuss the acoustical and signal processing limitations that are inherent in loudspeaker-based immersive audio systems and present a set of solutions for realizing such systems.

2. Immersive Audio Rendering

Immersive audio seeks to render virtual sound sources from a particular direction using a set of loudspeakers. The challenge is to reproduce the sound pressure level that would reach the eardrum if the sound were actually coming from the direction of the virtual sound source. In order to achieve this, the key characteristics of human sound localization that are based on the spectral information introduced by the pinnae must be considered [3-7].

The implementation described here involves the design of filters that alter nondirectional (monaural) sound in the same way as the pinnae. Early attempts in this area were based on analytic calculation of the attenuation and delay caused to the soundfield by the head, assuming a simplified spherical model of the head [8, 9]. More recent methods are based on the measurement (or simulation) of the impulse response of the pinnae for each desired direction. In this paper we propose a method that is based on measurements of the impulse response for each ear using a microphone placed inside the ear of a human subject or a mannequin. The impulse response is called the Head Related Transfer Function (HRTF) both in the time and frequency domains and it contains spectral information critical for sound localization. The main advantage of this method compared to analytical models is that it accounts for the pinnae, the imperfections of the shape of the human head, and the effects of the upper body. Several practical problems that arise when trying to implement HRTF filters for immersive audio rendering are examined.

When rendering immersive audio using two loudspeakers, direction dependent spectral information is introduced to the input signal due to the fact that the sound is generated from a fixed direction (the direction of the loudspeakers). In addition, the loudspeakers generally do not have an ideal flat frequency response and therefore must be compensated to reduce frequency response distortions.

Another issue for loudspeaker-based immersive audio arises from the fact that each ear receives sound from both loudspeakers resulting in acoustic crosstalk. In order to deliver the desired signal to each ear it is necessary to implement a crosstalk cancellation algorithm. A method for delivering binaural sound over loud-speakers was initially proposed by Schroeder and Atal [10] and later refined by Cooper and Bauck [11].

3. Signal Processing Issues

3.1 Non-Minimum Phase Inverse Filters for Crosstalk Cancellation

Crosstalk cancellation can be achieved by eliminating the contralateral terms H_{RL} and H_{LR} (Fig. 1), so that each loudspeaker is perceived to reproduce sound only for the corresponding ear. Note that the ipsilateral terms (H_{LL}, H_{RR}) and the contralat-eral terms (H_{RL}, H_{LR}) are just the HRTF's associated with the position of the two loudspeakers with respect to a specified position for the listener's ears. This im-plies that if the position of the listener changes then these terms must also change so as to correspond to the new listener position. This can be achieved through

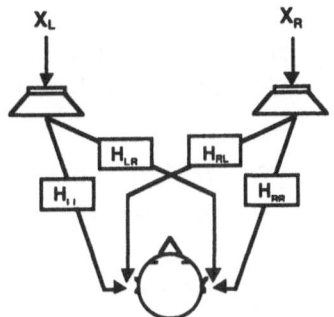

Figure 1. Loudspeaker-to-ear transfer functions in a loud-speaker-based rendering system.

tracking of the listener's head position in three-dimensional space [2, 12]. There are a number of different methods for the solution of this problem, usually at-tempting to approximate H_{LL}, H_{RR}, H_{LR}, and H_{RL} using a model. One simple method is to model the head as a sphere and then calculate the ipsilateral and con-tralateral terms [11, 13]. Another method approximates the effect of the head with a low-pass filter, a delay, and a gain (less than 1) [14].

While both of these methods have the advantage of low computational cost, the approximations involved introduce distortions particularly in the perceived timbre of virtual sound sources behind the listener. In our approach we use a different method. In order to describe the procedure mathematically we use matrix notation and represent the loudspeaker-ear system as a two input-two output system in which we need to process the two channels simultaneously. We assumed that the loudspeakers are placed symmetrically with respect to the median plane.

In a later section we describe a method for tracking the position of the listener's head. The filter coefficients in this case can be computed in real time in response to changes in the listener's position. In order to achieve real-time operation it is therefore critical for the filter implementation algorithms not to be computationally intensive.

We define H_i as the ipsilateral (same-side loudspeaker-to-ear transfer function) term in the frequency domain, H_c as the contralateral (opposite-side loudspeaker-to-ear transfer function) term, H_L as the virtual sound source HRTF for the left ear, H_R as the virtual sound source HRTF for the right ear and S as the monaural input sound. Then, in matrix notation, the signals E_L and E_R at the left and right eardrums respectively are

$$\begin{bmatrix} E_L \\ E_R \end{bmatrix} = \begin{bmatrix} H_L & 0 \\ 0 & H_R \end{bmatrix} \begin{bmatrix} S \\ S \end{bmatrix} \tag{1}$$

The introduction of the contralateral and ipsilateral terms from the physical system (taking into account the assumptions made before) will introduce an additional transfer matrix

$$\begin{bmatrix} E_L \\ E_R \end{bmatrix} = \begin{bmatrix} H_i & H_c \\ H_c & H_i \end{bmatrix} \begin{bmatrix} H_L & 0 \\ 0 & H_R \end{bmatrix} \begin{bmatrix} S \\ S \end{bmatrix} \tag{2}$$

In order to deliver the signals in (1) given that the physical system results in (2), pre-processing must be performed to the input S. In particular, the required pre-processing introduces the inverse of the matrix associated with the physical system as shown below

$$\begin{bmatrix} E_L \\ E_R \end{bmatrix} = \begin{bmatrix} H_i & H_c \\ H_c & H_i \end{bmatrix} \begin{bmatrix} H_i & H_c \\ H_c & H_i \end{bmatrix}^{-1} \begin{bmatrix} H_L & 0 \\ 0 & H_R \end{bmatrix} \begin{bmatrix} S \\ S \end{bmatrix} \tag{3}$$

We see that (1) and (3) are essentially the same. Solving (3) we find

$$\begin{bmatrix} E_L \\ E_R \end{bmatrix} = \begin{bmatrix} H_i & H_c \\ H_c & H_i \end{bmatrix} \begin{bmatrix} 1 & -\dfrac{H_c}{H_i} \\ -\dfrac{H_c}{H_i} & 1 \end{bmatrix} \begin{bmatrix} \dfrac{H_L}{H_i} & 0 \\ 0 & \dfrac{H_R}{H_i} \end{bmatrix} \begin{bmatrix} S \\ S \end{bmatrix} \tag{4}$$

assuming that $1/(1-H_c^2/H_i^2) \approx 1$. This assumption is based on the fact that the contralateral term is of substantially less power that the ipsilateral term because of the shadowing caused by the head. The terms H_L/H_i and H_R/H_i in (4) correspond to the speaker position inversion. That is, the actual position of the speakers is inverted because it adds undesired spectral information to the binaural signal. The matrix

$$\begin{bmatrix} 1 & -\dfrac{H_c}{H_i} \\[2ex] -\dfrac{H_c}{H_i} & 1 \end{bmatrix} \tag{5}$$

corresponds to the actual crosstalk cancellation. In the approach described here, the crosstalk cancellation and the inversion of the loudspeakers' position are closely connected, but it is important to state the difference between these two terms. Finally, the required filters F_L and F_R for the left and right channel are

$$F_L = \frac{H_L}{H_i} - \frac{H_c}{H_i} \frac{H_R}{H_i},$$

$$F_R = \frac{H_R}{H_i} - \frac{H_c}{H_i} \frac{H_L}{H_i}. \tag{6}$$

Alternatively, a filter can be designed for the case that the input is the binaural signal S_b instead of the monaural signal S. In this case, convolution with the pair of the HRTF's H_L and H_R is not needed since the binaural signal already contains the directional HRTF information. In this case the terms H_R and H_L are simply unity. Obviously, the filters that must be designed in this case for the left and right channel are identical (because of the symmetry assumption) and equal to

$$F_L = F_R = \frac{1}{H_i} - \frac{H_c}{H_i} \frac{1}{H_i} \tag{7}$$

The above analysis has shown that crosstalk cancellation and loudspeaker position inversion require the implementation of preprocessing filters of the type H_x/H_i (in which H_x is 1, H_L, H_R or H_c), which we will denote as H_{inv}. There are a number of methods for implementing the filter H_{inv}. The most direct method would be to simply divide the two filters in the frequency domain. However, H_i is in general a non-minimum phase filter, and thus the filter H_{inv} will be unstable. A usual solution to this problem is to use cepstrum analysis in order to design a new filter with the same magnitude as H_i but being minimum phase [15].

Here, we propose a different procedure that maintains the HRTF phase information. We first find the non-causal but stable impulse response, which also corresponds to H_x/H_i assuming a different Region of Convergence for the transfer function. Then, a delay is introduced in order to make the filter causal. The challenge is to make the delay small enough to be imperceptible to the listener. We describe below our methods for finding this non-causal solution.

62

3.2 LMS Implementation of HRTF Filters

Based on the discussion in the previous paragraph and taking into consideration the need for adding a delay in order for the preprocessing filter to be feasible (*i.e.* causal), as explained in 1.1, we conclude that the relationship between the filters H_i, H_x and the preprocessing filter H_{inv} can be represented as in the block diagram in Fig. 2.

The problem of defining the filter H_{inv} so that the mean squared error $E[e(n)]$ is minimized, can be classified as a combination of system identification (with respect to H_x) and inverse modeling (with respect to H_i) problem and its solution can be found using adaptive methods such as the LMS algorithm [16].

More specifically the taps of the filter H_{inv} can be defined based on the weight adaptation formula

$$h_{inv}(n+1) = h_{inv}(n) + \mu h_i(n)e(n) \qquad (8)$$

in which,

$$e(n) = h_x(n-d) - h_{inv}(n) * h_i(n) \qquad (9)$$

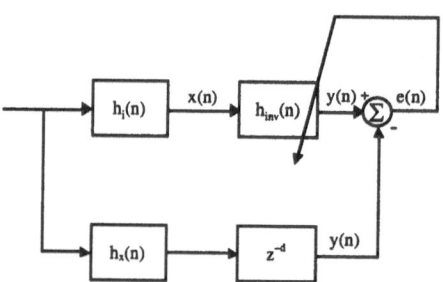

Figure 2. LMS block diagram for estimation of the inverse filters required for implementation of the crosstalk cancellation and virtual source HRTF rendering.

The filter length, as well as the delay d, can be selected based on the minimization of the mean squared error. Moreover, progressive adaptation (decrement) of the step size μ could lead to faster convergence as well as less misadjustment. This method can be used either off-line or in real time to account for movement of the virtual sound source position and movement of the listener's head.

The output filter of this method is $h_{inv}(n)$, which in the frequency domain is equal to H_x/H_i. If the desired output is $1/H_i$, $h_x(n)$ can be chosen to be the impulse sequence. The result is an FIR filter.

3.3 Least-Squares Method Implementation of HRTF Filters

Another way of approaching the problem is to notice that in the frequency domain ideally we must design a filter H_{inv} that satisfies

$$H_i H_{inv} = H_x \tag{10}$$

in which as discussed above H_x is 1, H_L, H_R or H_c. We will refer to the filter $H_{inv}H_i$ as the *cascade* filter. In the time domain (10) becomes a convolution relation

$$\sum_{m=0}^{M} h_i(n-m)h_{inv}(m) = h_x(n-d) \tag{11}$$

in which d is the delay introduced to satisfy the causality requirement. This equation is solved in the Least-Squares sense, that is we calculate h_{inv} such that

$$\min_{h_{inv}(m)} \sum_{n=0}^{N} \left| \sum_{m=0}^{M} h_i(n-m)h_{inv}(m) - h_x(n-d) \right|^2 \tag{12}$$

The above equation can be rewritten in matrix notation as

$$\min_{\mathbf{h}_{inv}} \left\| \mathbf{H}\mathbf{h}_{inv} - \mathbf{h}_x \right\|^2 \tag{13}$$

in which H is a Toeplitz matrix that can be easily derived from (11). The solution to (13) in the Least-Squares sense is

$$\mathbf{h}_{inv} = \mathbf{H}^+\mathbf{h}_x \tag{14}$$

in which we denote the pseudoinverse of H as H^+. The pseudoinverse of H can be found using Singular Value Decomposition (SVD) [16]. In order to avoid very small singular values, a tolerance value is introduced so that all singular values less than the tolerance to be considered equal to zero. The result is again an FIR filter.

4. Inverse Filter Performance

All of the filters that are encountered in (6) and (7) were designed using the Least-Squares and LMS methods. As discussed above a delay is introduced to the system. If we denote as d_1 the delay introduced by H_c/H_i in the upper part of (6) and as d_2 the delay introduced by H_R/H_i then in the z-domain we get

$$F_L = \frac{H_L}{H_i} z^{-(d_1+d_2)} - \frac{H_c}{H_i} z^{-d_1} \frac{H_R}{H_i} z^{-d_2} \tag{15}$$

Note that the delay for H_L/H_i in (15) must be equal to the sum of d_1 and d_2. The delay introduced by the filter F_R should also be equal to $d_1 + d_2$. A similar set of delays is used for the binaural case described in (7). The coefficients of these FIR filters were designed using Matlab. The delays and lengths for the filters used were optimized to achieve maximum Signal to Error power Ratio (SER) in the time domain between the cascade filter $H_{inv}H_i$ and H_x (or unity for the binaural input case). It is important to evaluate the error in the time-domain because a good approximation is required both in magnitude and phase response. Both methods worked successfully with a number of different measured HRTF's.

For the monaural input case, an inverse filter of 200 taps was designed, that introduced a delay of 70 samples. The SER for this case was 41.5 dB in the time domain for the Least-Squares method and 33 dB for the LMS method. In Fig. 3 a comparison is made between the magnitude of a particular measured HRTF (0° azimuth and 0° elevation) and the HRTF generated using our inverse filter. Of course, since the approximation of the two filters is made in time domain, their phase responses are also almost identical.

It should be noted that for frequencies above 15 kHz, the associated wavelengths are of the order of 15 mm. In this range it is impossible to place the listener's ears in the desired location for which the filters have been designed. For this reason the degradation of the normalized error above 15 kHz (Fig. 3) is acceptable.

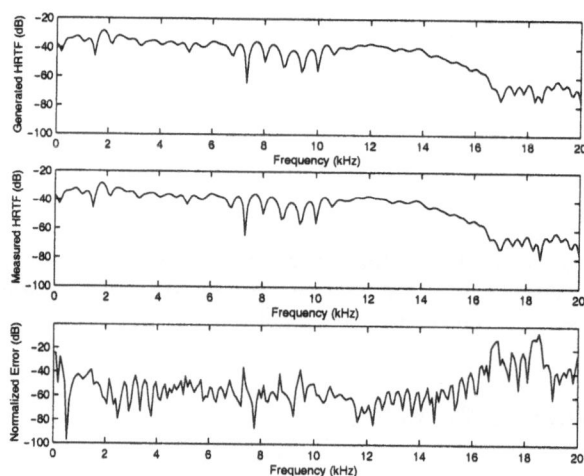

Figure 3. The HRTF generated from the inverse filter using the Least-Squares method is shown in the upper plot. The measured HRTF (0° azimuth and 0° elevation) is shown in the middle. The normalized error between the two is approximately –50 dB between 20Hz and 15 kHz.

If inversion of the type $1/H_i$ is required (binaural input), the cascade filter should be of exactly all-pass response. This case proved to be more demanding than the monaural input case. The SER in the time domain is now 28 dB, but for a higher filter length of 400 taps and 160 sample delay. Alternatively, we can use a cascade filter of the form H_a/H_i where H_a has an all-pass response up to 15 kHz. Using this approximation, the resulting filters gave significantly better performance. For the Least-Squares method the SER in time domain increased to 78 dB for the same filter length and delay (400 taps, 160 sample delay). For the LMS case, the SER increased to 52 dB. In listening tests there was no perceptible difference using this method.

5. Acoustical Considerations

A significant amount of work in the area of high quality sound production and reproduction in large rooms has originated from the film industry. A well-defined set of standards has been developed for sound monitoring conditions in dubbing stages to ensure the transparent reproduction of program material in theaters. Such standards include loudspeaker positioning for multichannel monitoring, loudspeaker frequency response and directivity requirements, precise sound pressure level calibration, control of room acoustics parameters (such as reverberation time and discrete reflections), and background noise levels. Meeting these standards ensures that material produced in one professional dubbing stage can be monitored under identical conditions in another dubbing stage or in a movie theater. The design challenge in desktop audio systems is to successfully map these standards onto the desktop environment through appropriate acoustical and psychoacoustical scaling and system design.

5.1 Desktop Systems

In a typical desktop sound monitoring environment delivery of stereophonic sound is achieved through two loudspeakers that are typically placed on either side of a video or computer monitor. This environment, combined with the acoustical problems of small rooms, causes severe problems that contribute to audible distortion of the reproduced sound [17]. While an experienced professional can identify and correct for such problems during the monitoring stage, any changes made are permanently recorded and appear as errors during playback in a different environment. Among these problems the one most often neglected is the effect of discrete early reflections. The effects of such reflections on sound quality have been studied extensively [17-20] and it has been shown that they are the dominant source of monitoring non-uniformities when all the other standards discussed above have been met. These non-uniformities appear in the form of colorations (frequency response anomalies) in rooms with an early reflection level that exceeds –15 dB spectrum level relative to the direct sound for the first 15 ms [21] (Fig. 4). Such a high level of reflected sound gives rise to comb filtering in the frequency domain that in turn causes noticeable changes in timbre. The perceived effects of such distortions were not quantified until psychoacoustic experiments

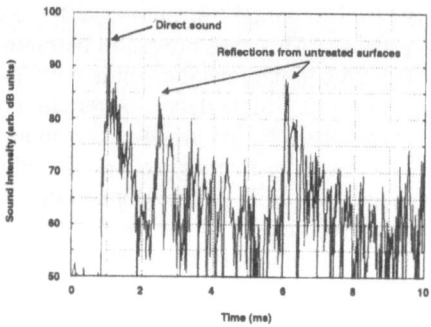

Figure 4. Desktop sound system time response that violates the requirements for low-level early reflections. The early reflection peaks at 2.5 ms and 6 ms give rise to a spectrum level that is above the –15 dB criterion.

demonstrated their importance.

A potential solution that alleviates the problems of early reflections in small rooms is near-field monitoring. In theory, the direct sound is dominant when the listener is very close to the loudspeakers thus reducing the room effects to below audibility. In practice, however, there are several issues that must be addressed in order to provide high quality sound. One such issue relates to the large reflecting surfaces that are typically present near the loudspeakers. Strong reflections from a console or a video/computer monitor act as baffle extensions for the loudspeaker resulting in a boost of mid-bass frequencies. Furthermore, even if it were possible to place the loudspeakers far away from large reflecting surfaces, this would only solve the problem for middle and high frequencies. Low frequency room modes do not depend on surfaces in the local acoustical environment, but rather on the physical size of the room. These modes produce standing waves that give rise to large variations in frequency response. Finally, another factor that has a negative effect on the quality of reproduced sound relates to the physical size of the loud-speakers. Typical two-way designs in which the tweeter is physically separated from the woofer exhibit strong radiation pattern changes in the crossover frequency range. Amplitude and phase matching in this frequency range becomes critical and as a result such speakers are extremely sensitive to placement and typically produce a flat frequency response for direct sound *in one exact position*. This limitation makes typical two-way speakers unsuitable for near-field monitoring.

5.2 Design Requirements

In order to address the problems described above, a set of solutions has been developed for single listener desktop reproduction that delivers sound quality equivalent a calibrated dubbing stage [12, 17]. These solutions include:

Direct-path dominant design. By combining elements of psychoacoustics in the system design, it is possible to place the listener in a direct sound field that is dominant over the reflected and reverberant sound. The colorations that arise due to such effects are eliminated and this results in a listening experience that is dramatically different than what is achievable through traditional near-field monitoring methods. The design considerations for this direct-path dominant design include the effect of the video/computer monitor that extends the loudspeaker baffle, as well as the large reflecting surface on which the computer keyboard rests.

Correct low-frequency response. There are severe problems in the uniformity of low-frequency response that arise from the standing waves associated with the acoustics of small rooms. Such anomalies can give rise to variations as large as ±15 dB for different listening locations in a typical room. The advantage of desktop audio systems lies in the fact that the position of the loudspeakers and, to a large extent, the listener are known *a priori*. It is, therefore, possible to use equalization to produce very smooth low-frequency response. One fundamental limitation imposed by small room acoustics is that this can only be achieved for a relatively-small volume of space centered around the listener. One possible solution to this problem can be found by tracking the listener's position and adjusting the equalization dynamically.

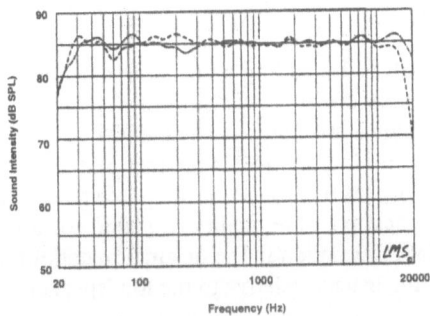

Figure 5. Frequency response of desktop loudspeaker system designed using the direct-path dominant and correct low-frequency response guidelines described in the text. The response has been corrected with minimal parametric equalization and is relatively flat (± 2 dB) from 30 Hz to 20 KHz. The solid line represents the on-axis response and the dotted line represents the 10° off-axis (vertical) response.

6. Desktop Immersive Audio System With Head Tracking

In large rooms multichannel sound systems are used to convey sound images that are primarily confined to the horizontal plane and are uniformly distributed over the audience area. Typical systems used for cinema reproduction use three front channels (left, center, right), two surround channels (left and right surround), and a separate low-frequency channel. Such 5.1 channel systems are designed to provide accurate sound localization relative to visual images in front of the listener and diffuse (ambient) sound to the sides and behind the listener. The use of a center loudspeaker helps create a solid sound image between the left and right loudspeakers and anchors the sound to the center of the stage.

For desktop applications, in which a single user is located in front of a CRT display, we no longer have the luxury of a center loudspeaker because that position is occupied by the display. In such cases sound is reproduced mainly through the use of two loudspeakers placed symmetrically on either side of the CRT, two surround loudspeakers placed to the side and above the listening position. Size limitations prevent the front loudspeakers from being capable of reproducing the entire spectrum, thus a separate subwoofer loudspeaker is used to reproduce the low frequencies. The two front loudspeakers can create a virtual (phantom) image that appears to originate from the exact center of the display provided that the listener is seated symmetrically with respect to the loudspeakers. With proper head and loudspeaker placement, it is possible to recreate a spatially-accurate soundfield with the correct frequency response in *one* exact position, the sweet spot. However, even in this static case, the sound originating from each loudspeaker arrives at each ear at different times (about 200 μs apart), thereby giving rise to acoustic crosstalk. These time differences combined with reflection and diffraction effects caused by the head lead to frequency response anomalies that are perceived as a lack of clarity [17, 22].

This problem can be solved by adding a crosstalk cancellation filter to the signal of each loudspeaker as described in the previous sections. While this solution may be satisfactory for the static case, as soon as the listener moves even slightly, the conditions for cancellation are no longer met and the phantom image moves towards the closest loudspeaker because of the precedence effect. In order, therefore, to achieve the highest possible quality of sound for a non-stationary listener and preserve the spatial information in the original material it is necessary to know the precise location of the listener relative to the loudspeakers.

We have implemented a prototype multichannel desktop audio system that incorporates a novel listener-tracking method. The algorithm is based on estimating the position of the listener's face by using a p-norm ($1<p<2$) to compare successive frames in YUV-space. The tracking algorithm has been incorporated using a standard video camera connected to the Philips Trimedia board in a Pentium NT workstation. This tracking system provides us with the coordinates of the center of the listener's head relative to the loudspeakers and is currently capable of operating at 30 frames/sec.

When the listener is located at the exact center position (the sweet spot), sound from each loudspeaker arrives at the corresponding ear at the exact same time (*i.e.*, with zero ipsilateral time delay). At any other position of the listener in this plane, there is a relative time difference of arrival between the sound signals from each loudspeaker. This time difference causes the perceived location of the sound image to shift towards the loudspeaker that is closer to the listener due to the precedence effect. In order to maintain proper stereophonic perspective, the ipsilateral time delay must be adjusted as the listener moves.

The head coordinates provided from the tracking algorithm are used to determine the necessary time delay adjustment. The required relative time delay between the two channels varies from 0 µs in the center spot to 340 µs in the extreme left or right positions. The DSP board is used to delay the sound from the loudspeaker that is closest to the listener so that sound arrives with the same time difference as if the listener were positioned in the exact center between the loudspeakers. To achieve seamless operation for continuous listener movement, a linear interpolation scheme is used to address the problem of audible clicks that result from instantaneous changes in the digital delay between the two channels.

6. Conclusions

Important practical aspects in immersive audio rendering were discussed in this paper. They include inversion of non-minimum phase filters and crosstalk cancellation that is an inherent problem in loudspeaker-based rendering. Two methods were proposed to implement a set of filters that can be used to render virtual sound sources. Our simulations have shown that both methods give satisfactory results using various HRTF's. Furthermore, the acoustical and psychoacoustical requirements for high quality reproduction were examined for the desktop environment. A set of design guidelines were proposed for the design of the loudspeakers, as well as the local acoustical environment.

Acknowledgements

This research has been funded in part by the Integrated Media Systems Center, a National Science Foundation Engineering Research Center with additional support from the Annenberg Center for Communication at USC and the California Trade and Commerce Agency.

References

1   B. Shinn-Cunningham, "Adapting to discrepant information in multimedia displays," presented at 134th Meeting of the Acoustical Society of America, San Diego, California, 1997.

2   C. Kyriakakis, "Fundamental and Technological Limitations of Immersive Audio Systems," *IEEE Proceedings*, vol. 86, pp. 941-951, 1998.

3   J. Blauert, *Spatial Hearing: The Psychophysics of Human Sound Localization, Revised Edition.* Cambridge, Massachusetts: MIT Press, 1997.

4   S. Mehrgard and V. Mellert, "Transformation Characteristics of the External Human Ear," *Journal of the Acoustical Society of America*, vol. 51, pp. 1567-1576, 1977.

5   F. L. Wightman and D. J. Kistler, "Monaural sound localization revisited," *Journal of the Acoustical Society of America*, vol. 101, pp. 1050-63, 1997.

[6] T. D. Rossing, "Spatial hearing : the psychophysics of human sound localization," *American Journal of Physics*, vol. 53, pp. 926-7, 1985.

[7] H. Moller, M. F. Sorensen, and D. Hammershoi, "Head-related transfer functions of human subjects," *Journal of the Audio Engineering Society*, vol. 43, pp. 300-21, 1995.

[8] D. H. Cooper, "Calculator Program for Head-Related Transfer Functions," *Journal of the Audio Engineering Society*, vol. 30, pp. 34-38, 1982.

[9] C. P. Brown and R. O. Duda, "A Structural Model for Binaural Sound Synthesis," *IEEE Transactions on Speech and Audio Processing*, vol. 6, pp. 476-488, 1998.

[10] M. R. Schroeder and B. S. Atal, "Computer Simulation of Sound Transmission in Rooms," *IEEE International Convention Record*, vol. 7, 1963.

[11] D. H. Cooper and J. L. Bauck, "Prospects for Transaural Recording," *Journal of the Audio Engineering Society*, vol. 37, pp. 3-19, 1989.

[12] C. Kyriakakis, T. Holman, J.-S. Lim, H. Hong, and H. Neven, "Signal Processing, Acoustics, and Psychoacoustics for High Quality Desktop Audio," *Journal of Visual Communication and Image Representation*, vol. 9, pp. 51-61, 1997.

[13] J. Bauck and D. H. Cooper, "Generalized Transaural Stereo and Applications," *Journal of the Audio Engineering Society*, vol. 44, pp. 683-705, 1996.

[14] W. G. Gardner, "Transaural 3-D Audio," MIT Media Laboratory, Technical Report 342, January/February 1995.

[15] A. V. Oppenheim and R. W. Shafer, *Discrete Time Signal Processing*: Prentice Hall, 1989.

[16] S. Haykin, *Adaptive Filter Theory, 3rd Edition*: Prentice Hall, 1996.

[17] T. Holman, "Monitoring Sound in the One-Person Environment," *SMPTE Journal*, vol. 106, pp. 673-678, 1997.

[18] F. E. Toole, "Loudspeaker measurements and their relationship to listener preferences," *Journal of the Audio Engineering Society*, vol. 34, pp. 227-235, 1986.

[19] S. Bech, "Perception of timbre of reproduced sound in small rooms: influence of room and loudspeaker position," *Journal of the Audio Engineering Society*, vol. 42, pp. 999-1007, 1994.

[20] S. E. Olive and F. E. Toole, "The detection of reflections in typical rooms," *Journal of the Audio Engineering Society*, vol. 37, pp. 539-53, 1989.

[21] R. Walker, "Early Reflections in Studio Control Rooms: The Results from the First Controlled Image Design Installations," presented at 96th Meeting of the Audio Engineering Society, Amsterdam, 1994.

[22] T. Holman and C. Kyriakakis, "Acoustics and Psychoacoustics of Desktop Sound Systems," presented at Annual Meeting of the Acoustical Society of America, San Diego, California, 1997.

Real-Time Billboard Substitution in a Video Stream

G. Medioni[1], G. Guy[2], H. Rom[3], A. François[1]

[1] Integrated Media Systems Center, USC, Los Angeles, U.S.A.

[2] Intelligent Computer Solutions, Inc.

[3] Nichimen Graphics, Los Angeles, U.S.A.

Abstract

We present a system that accepts as input a continuous stream of TV broadcast images from sports events. The system automatically detects a predetermined billboard in the scene, and replaces it with a user defined pattern, with no cooperation from the billboards or camera operators. The replacement is performed seamlessly so that a viewer should not be able to detect the substitution. This allows the targeting of advertising to the appropriate audience, which is especially useful for international events.

This requires several modules using state of the art computer graphics, image processing and computer vision technology.

The system relies on modular design, and on a pipeline architecture, in which the search and track modules propagate their results in symbolic form throughout the pipeline buffer, and the replacement is performed at the exit of the pipe only, therefore relying on accumulated information. Also, images are processed only once. This allows the system to make replacement decisions based on complete sequences, thus avoiding mid-sequence on screen billboard changes.

We present the algorithms and the overall system architecture, and discuss further applications of the technology.

1. Introduction

Editing of images or image streams is fast becoming a normal part of the production process [1]. Many recent movies such as Independence Day, Godzilla, Titanic seamlessly blend live images with Computer Generated Imagery. The mixing of multiple elements is performed primarily by screen matting, in which the background is of almost constant color, generally blue or green. This approach requires a very controlled studio environment and operator intervention for optimal results.

Instead, we are investigating the use of computer vision, computer graphics, and image processing tools and techniques to develop an array of applications for the video and film industries, all involving the "intelligent," automatic manipulation of images and image streams, *based on their contents.*

We present here a system which must function without active cooperation, and in real-time (therefore automatically, without operator intervention) and in an uncontrolled environment: it receives as input a TV broadcast signal, must identify a given billboard in the image flow, track it precisely, and replace it with another pattern (fixed or animated), broadcasting the replaced signal, in real-time, with only a short constant delay.

It is a common practice to place billboards advertising various products and services during sports events. These billboards target not only the spectators at the stadium, but also the viewers of the TV broadcast of the event. This fixed advertising is, therefore, limited, as the billboards might be advertising products out of context for the TV audience, especially for international events.

The system, which replaces the billboards in the scene in such a way that it should be transparent to the viewer, allows a local TV station to plant its own advertisement billboards regardless of the original billboard, thus increasing the overall effectiveness of the advertising. An example of replacement is shown in Figure 1, where different boards are inserted for different national broadcasts.

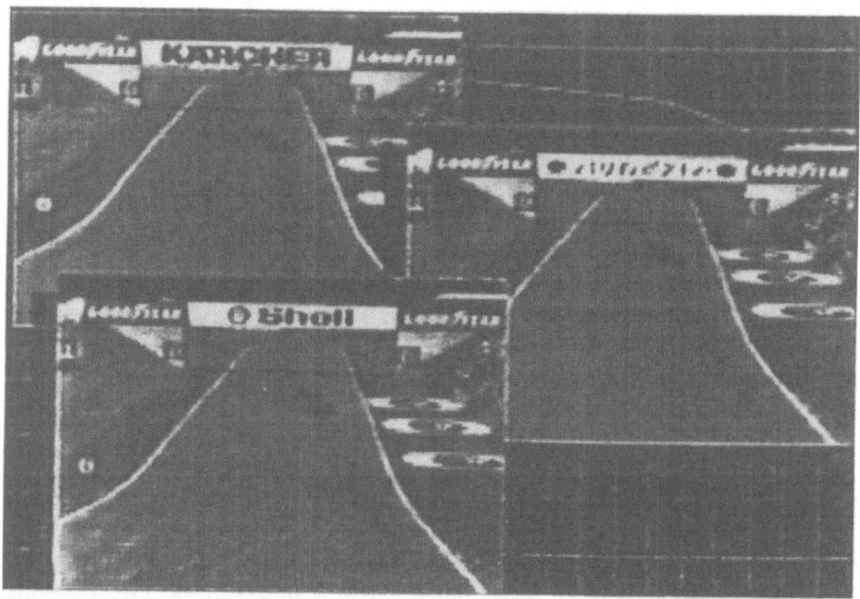

Fig. 1 The physical Shell billboard is replaced by virtual billboards, targeted for different audiences

The fundamental requirement that the system perform *on-line* in *real-time*, imposes major constraints on the design and implementation of the system. Some of the crucial constraints to keep in mind are:

• No human intervention is possible.

• No on-screen errors are permitted. The system must include quality control mechanisms to detect problems and revert to the original signal when they occur.

• Real-time implementation limits the complexity of the algorithms.

• No cooperation from the field is expected, in order to allow the system to operate independently from the imaging process (e.g. at the down link).

The contribution of such a system, for which a patent was issued [2], consists both in the design and implementation of the individual modules (Finder, Tracker, Replacer), and in the management of failure and uncertainty of each of these modules, at the system level, resulting in reliable replacement.

The structure of this paper is as follows: In the next section, we present an overview of the overall system and its components. In Section 3, we present detailed descriptions of the algorithms for the specific modules. In Section 4, we present some results demonstrating the system's capabilities, followed by a brief discussion.

2. Overall System Design

The task of the system is to find the target billboard in the scene, detect camera switches, track it throughout the sequence (between camera switches), and replace it with the new billboard. The direct naive approach would be to inspect the incoming frames, search for the billboard and replace it. Unfortunately, this approach is not sufficient, as it may be impossible to locate the billboard in the current frame, due to large focus or motion blur, or to the billboard being occluded, or to the fact that only a small part of it may be in the field of view. The billboard may therefore be found only in a later frame of the sequence, and it is not possible to start replacing at this frame, as this would be offensive to the viewer. Instead, replacement should be performed on the whole sequence to avoid billboard switches on screen.

Our system relies on a modular design, and on a pipeline architecture, in which the search and track modules propagate their symbolic, low-bandwidth results throughout the pipe, and the replacement is performed at the exit of the pipe only, therefore relying on accumulated information. This allows the system to make replacement decisions based on complete sequences, thus avoiding mid-sequence on-screen billboard changes.

The *Finder* module searches for the target billboard in the entering frames[1] and passes its results to the *Updater*, which propagates them throughout the buffer.

The *Global Motion Tracker (GMT)* module estimates the motion between the previous and current frames, *regardless of whether the billboard was found or not*. This is used as a mechanism for predicting the billboard location in the frames in which it was not found. The prediction is necessary to ensure continuity of replacement, since we do not want the billboards to switch back and forth between

---

[1] Note that since the actual system is designed to work on PAL and NTSC interlaced signals, the actual processing by all modules of the system is *field* based and not *frame* based. To avoid confusion, and since it is conceptually equivalent, we use the term *frame* throughout the paper.

the original and the new one in front of the viewer. The GMT also performs the task of camera switch detector.

The *Replacer* performs the graphic insertion of the new billboard, taking into account variations from the model due to lighting, blur and motion.

The *Updater* handles communication within the buffer and also manages the *Measure Of Belief (MOB)* associated with the information passed along, due to the MOB of each of the modules, and a decay related to the length of the propagation. The information about scene changes is also used so that the *Updater* does not propagate the predictions beyond the scene change marker.

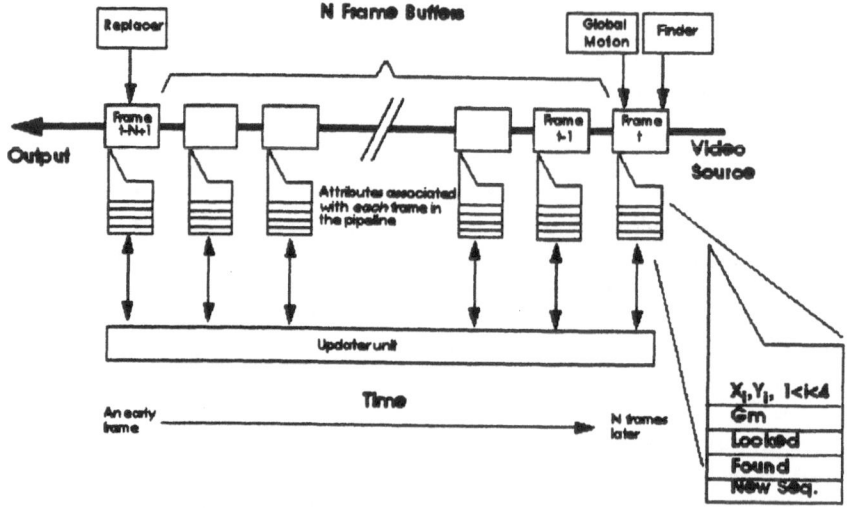

Fig. 2 A block diagram of the system

Figure 2 presents the overall system architecture. As the frame at time *t* comes in from the video source on the right, the Finder searches for the billboard. At the same time, the Global Motion Tracker (GMT) computes the camera motion between the previous and current frames, and stores it in an attribute record. If the billboard is found, its four coordinates are recorded in the attributes record, and the Updater unit predicts the location of the billboard in all the previous frames from the first frame of the sequence to frame *t-1*, based on the computed motion, and updates the attribute records accordingly. As the frame is about to be displayed, the Replacer module performs the insertion.

To better understand the system design, let us examine its operation in two different possible scenarios:

First, consider the simple case where the billboard is clearly visible in the first frame of the sequence. In this case, it would be found easily by the Finder. From then on, using the motion computation from the Global Motion Tracker, the Finder has a good prediction for the billboard location in the following frames, so it is also found in every frame. The Updater plays a minor role in this scenario.

The Replacer module inserts the new billboard in place of the target, as described in Section 3.4.

Now, consider the case where the billboard is slowly entering into view, as a result of a pan or zoom. In this case, the billboard is not found initially by the Finder. As the frames continue to come in, the Global Motion Tracker computes the camera motion between frames. As discussed later in Section 3.2, the motion can be recovered regardless of whether the billboard was found or not. The camera motion parameters found are stored in the frame attribute record to be accessed by the Updater. Now, assume the billboard is finally found in some frame, t, of the sequence. We do not want to suddenly start replacement at this point, since this would be offensive to the viewers. Instead, the Updater module uses the motion parameters computed earlier, to predict the location of the billboard in all the frames from the first frame of the current sequence up to frame $t-1$. Since this is a very simple computation (not image based), involving low bandwidth communication, it can be performed for the whole buffer in one frame time. As the images reach the end of the buffer, for each we have the location of the billboard, either directly from the Finder, if it was found in this frame initially, or via a prediction from the Updater, using the motion information.

The combined use of the Global Motion Tracker, the delay buffer and the Updater mechanism, allow the system to, in essence, go back in time without having to process the images again, and to use information from the current frame to locate the billboard in earlier frames. This enables the system to perform well under varied conditions, such as occlusion and entering billboards. The system is also very robust to failure of specific modules, as it can overcome failure in some frames by using information from the other frames of the sequence. It is important to note that each image is processed once only, and that each module works at frame rate, thus the system works in real-time, introducing only a constant delay.

This design can guarantee that no offensive substitution will take place as long as a whole sequence fits in the buffer. Otherwise, in case of a problem occurring after replacement is started, a smooth fade back to the original billboard is used. In practice, a buffer of the order of 3 seconds (180 fields in NTSC), can cover a large percentage of sequences in which the billboard is present.

3. The Components

In this section, we present detailed descriptions of the algorithms for the various modules of the system. We first discuss the Finder, which is responsible for searching and localizing the billboard in the incoming frame. We then discuss the Global Motion Tracker, responsible for computing the camera motion between frames, whether or not the billboard was found. Next, we discuss the Updater, which is responsible for maintaining the overall consistency of replacement throughout the whole sequence, including propagating the billboard location to frames where it was not found, and making the decision on whether or not to replace the sequence. Finally, we discuss the Replacer responsible for the actual insertion of the new billboard in place of the target billboard.

3.1 The Finder module

The task of the Finder is to examine the incoming frames and find the position of the target billboard if it is present in the scene. The Finder consists of the following five sub-modules, which are described further in the following subsections (see Figure 3):
• **Interest Point Operator** - Extracts "interesting" points (corners, or other "busy" formations) in the image.
• **Color-based Point Filter** - Selects the interest points which are most likely to come from the target billboard based on color information.
• **Point Matcher** - Finds a set of corresponding points between model interest points and image interest points.
• **Precise Lock-in** - Given the correspondences from the matcher, finds the precise (to a sub-pixel resolution) location of the billboard.
• **Predictor** - Predicts the location of a billboard in frame *t+1*, assuming board was found in frame *t*.

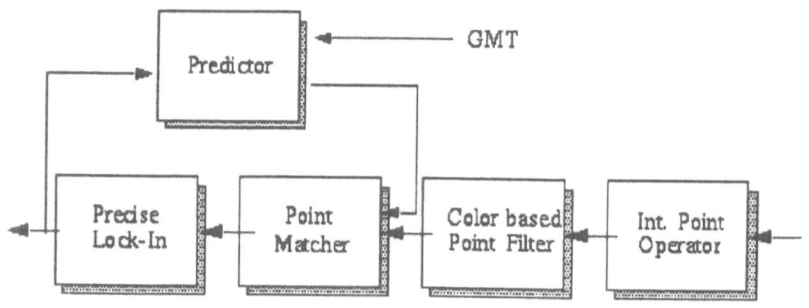

Fig. 3 The Finder module - All five sub-modules

Interest Point Operator Sub-module. The purpose of this module is to detect and locate "interesting points" on the billboard. Interest points are points which are distinct from their surrounding, typically corners and vertices, that are very often seen on letters on a billboard. The set of points found by this module on any frame will then be compared to the set of points found on the model (processed in exactly the same way) to try to find a match.
Because of the unstructured environment of the problem, we require the detected points to be very stable to changes in absolute illumination, contrast, and scale, and also to be sparse.
After experimenting with many existing approaches [3,4,5], we have chosen the one described in [6], as it only involves first derivatives of the images.

Figure 4 shows an example of a field from a tennis match with the feature points detected by the algorithm. This method has proven to be reliable in most situations. It is important to note that the different numerical values used

Fig. 4 An example of the feature points detected in a given field

(thresholds, window sizes) have been set once and for all and should satisfy most types of images. These values have been optimized for the case of a billboard in full focus and of 'reasonable' size, reasonable size being not too small because we would not find enough points for the matcher module to work, and not too big because we may find too many points, without being able to choose the most relevant ones for the matcher.

The Color-based Point Filter Sub-module. Due to the computational complexity of the Point Matcher, discussed below, it is important to limit the number of points it considers. The Color-based Point Filter module is designed to select the interest points which are most likely to be located on the target billboard.
The method is based on the observation that interest points are usually junctions of two different colors. Therefore, during the off-line preprocessing stage, the user selects the possible pairs of colors in the model (in most cases there are only one or two possible pairs). During the on-line processing, a window around each feature point is tested. For each pixel in the window, the distance from its color to the closest color of the model is recorded. These distances are summed over the window. We assign to the feature point a score inversely proportional to the computed distances. We also require the window to have both colors of the color pair. The matcher then uses the top ranked feature points. Figure 5 shows the points selected by the Color-based Point Filter from the points found in Figure 4.

Point Matcher Sub-module. The task of this module is to match the points found in the current image by the interest point operator with the pre-stored interest points of the model. This is done in order to find the transformation between the model and the image and, therefore, the position of the billboard in the image. The direct exhaustive attempt to match every point of the model with every point of the image leads to unacceptable complexity. Instead, we use the original affine-invariant matching technique proposed by Lamdan and Wolfson [7]. We provide a brief description of the algorithm here.
We assume the image coordinates to be affine-transformed coordinates of the model coordinates. The goal of the matcher is to find an affine transform, which

Fig. 5 The feature points selected by the Color based filter

maps the model points to the image points. Any three points define a *base*, relative to which all other points have some relative coordinates. A set of any three matching pairs of points (matching bases) uniquely determines the six parameter affine transform between the model and the image.

The algorithm consists of two components. An off-line pre-processing of the model points, which is time consuming, but since it is off-line time is less of an issue, and an efficient on-line process. During the pre-processing step, the coordinates of all the model points with respect to all possible bases (triplet of model points) are computed and stored in a hash table. The coordinates are used as the index into the hash table, and the base as the value.

During the on-line processing, the feature points of the current image are considered. A base (triplet of points) is selected at random, and for all feature points in the image, their coordinates are computed with respect to the selected base. The coordinates are used to index into the hash table and vote for the model bases in that entry. If any one of the model bases receives a sufficient number of votes, it is considered a possible match. The two bases (the one selected from the image and the one receiving most votes) define a transformation between the model and image. This transformation is applied to all model points, and, if a sufficient number of the transformed points indeed find a match, then the process ends, otherwise, a new base is selected.

To facilitate real-time processing and robustness, several additions were made to the algorithm: First, we only consider *stable* bases [8]. If the three base points are too close together, or nearly collinear, then the resulting transformation is very unstable (a small error in the correspondence of the base points could result in large errors for the other points). Therefore, we filter out such unstable bases, both in the pre-processing and the on-line steps. We also do not process all possible bases during the on-line processing. If a match is not found after a fixed number of random trials (set to 40), then the matching fails for the current frame.

Fig. 6 Outline of the location of the billboard as estimated by the Point Matcher

Figure 6 presents the location of the billboard as found by the Point Matcher on the example shown earlier. Note that although the billboard was found correctly its positioning is not perfect and requires refinement as discussed below.

Precise Lock-in Sub-module. The Precise Lock-In sub-module is designed to provide a very accurate, sub-pixel localization for the billboard, given a set of corresponding points between the model and the image, as generated by the Matcher. It also has a complete list of feature points in the model, an image of the target model, and the current image.

The initial results from the point matcher are used to transform the image of the model to the predicted location in the image.

For each of the feature points, pre-detected in the model, we perform a simplified form of correlation, the Sum of Squared Differences (SSD) for a *10x10* window around the point, and in a *10x10* window.

Points with a minimum SSD greater than a threshold are discarded (a fixed threshold has been used in all experiments).

The best match is extracted with sub-pixel precision by fitting a second order polynomial in x and y around the best pixel match, and a least square estimate of the transformation between model and image is recomputed with only the validated matches.

A second pass is then performed using this new updated transformation, to produce the final transformation estimate. The error is passed to the Updater, which transforms it into a normalized measure of belief, to be used in the decision to replace.

Predictor sub-module. The predictor is designed to compute a simple prediction of location from one frame to the next. Assuming the matcher was able to find the billboard at frame t, the predictor, with the knowledge of the motion parameters, predicts the location at frame t+1. It is possible to carry the prediction over many frames, as long as the matcher found the board in some earlier frame. A measure of belief (MOB) is attached to every prediction, and when this MOB drops below a certain threshold, the matcher is advised by the Updater not to use this prediction. At a sequence break (also reported by the GMT), the predictor resets itself.

The matcher uses the prediction as a Focus of Attention mechanism, when such prediction exists. This setup allows for a tracking mode, and improves the reliability of the overall system by keeping measures of belief high. It was found that if the predictor is within 5 pixels of the corners of the billboard, the PLI is able to lock-in, regardless of whether a point match was found or not. This performs as an implicit tracking mode.

3.2 The Global Motion Tracker

Global motion tracking is the process of detection of the camera motion between frames and the registration of the scene. The Global Motion Tracker computes estimates of the global camera motion parameters between consecutive frames. An excellent recent survey of optical flow computation schemes was performed by Barron et al. [9]. Motion recovery is an active research area in computer vision and many approaches have been taken. One approach is to rely on the matching of feature points between consecutive frames [10]. However, these approaches tend to be computationally expensive and unreliable in the presence motion blur. Another approach is based on the computation of the optical flow [11] for every pixel. These methods are very computationally expensive and difficult to implement in real-time. They are also very sensitive to noise. Since we are interested in the motion of the camera and not in a per pixel motion, we take a global approach, and use an iterative least squares technique on all pixels of the image [12,13]. This results in an efficient and robust method that produces accurate results.

The Global Motion Tracker module operates on two consecutive frames. The images are first smoothed to increase the spatial correlation. The spatial and temporal derivatives are then computed. Using this information, estimates to the motion parameters are computed. The parameters recovered depend on the motion model used, as discussed below. Using these estimates, Frame $t+1$ is warped towards Frame t, and the process is repeated. Since Frame $t+1$ gets closer to Frame t at every iteration, the computed motion parameters get more and more accurate. The process is stopped when the change in the computed motion parameters has converged to close to zero. The accumulated parameters are then reported to the Updater. If, however, the motion parameters do not converge after a given number of maximum iterations, then the process is stopped with a report of zero reliability.

We have implemented the algorithm at multi-resolution levels. A Gaussian pyramid [3] is created from each frame. At the beginning of a sequence, the above algorithm is applied to the lowest resolution level. The results from this level are propagated as initial estimates for the next level up, and so on until the highest level. This allows for recovery of larger motions. Within a sequence, the results from the previous pair of frames are usually good estimates for the current pair, therefore the algorithm is applied to one lower resolution level only, with the previous results as initial estimates. This allows fast recovery of the motion while accommodating for large changes within the sequence.

An improvement to the global motion algorithm allows for more accurate and stable results in the presence of moving obstacles in the scene. This is obtained by scaling the coefficients of the motion equations inversely proportional to the temporal derivatives. Moving obstacles will not register when the images are warped according to the camera motion. Therefore, the pixels corresponding to obstacles will have high temporal derivatives and consequently will have less weight in the coefficients. The improved results allow for longer propagation of estimates along the sequence.

The complete algorithm was applied to a very large set of sequences. The parameters recovered are highly accurate and allow for the accurate propagation of the billboard location for a few seconds, in the absence of confirmation by the Finder. The algorithm fails when the motion is drastically different from the modeled motion, or when the image is uniform (e.g. an image of the sky).

3.3 The Updater

The Updater's task is to collect data from all the other modules, and to correct missing or inaccurate information within a processed sequence. We can visually think of the system as a circular buffer, holding a frame and a frame attribute in each of its cells. The updater manipulates these attribute records only, which are composed of a small number of parameters. The Updater processes *all* attribute records in the buffer in one frame time.

The structure of the attribute record is as follows:

• A global motion transformation between self and next + measure of belief (MOB).

• A transformation between next and self + MOB (not the same as the above).

• A flag representing the position of the frame in the sequence (first, last etc.)

• A linked list of hypotheses, each containing 4 corners in image coordinates + a MOB, currently restricted to just one hypothesis.

The complexity of the Updater is linear in the size of the buffer and of the sequence length.

A MOB value of 1 (or close to 1) denotes high certainty of the input parameters. A low MOB value (or 0) means that the input parameters are unreliable, and should not be used unless some reinforcement is given by other modules.

For example, when the Finder fails to find the billboard in a certain frame, it assigns a zero certainty to the data it hands to the Updater. The Updater then interpolates the position of the billboard in this frame using data from the Global Motion module and/or previous frames, and/or future frames. The Updater changes the MOB accordingly.

The decision of whether to transmit a given sequence is made as late as possible. The latest is when the first frame of the sequence in question is about to exit the buffer. The Updater then evaluates the global 'goodness' of the sequence, currently taken as the minimum MOB across all frames belonging to that sequence, and if it passes a fixed threshold, a decision is made to transmit-with-replacement.

3.4. The Replacer

Given the coordinates of the billboard corners in the current image, the Replacer module replaces the image contents within these corners (the billboard) with the new desired contents (usually a new billboard), using graphics tools [14].
The important steps in this process are listed below.

• From an initial model of the new billboard, a set of models each successively smaller than the previous one is generated. This step is done off-line, prior to actual billboard replacement and this hierarchy of images is stored. This set of models is generated so that special problems occurring due to large changes in size during warping, e.g. aliasing, can be taken care of off-line.

• From the corner information, we estimate the size of the area in which the new billboard is to be placed. The closest sized model that exists in the hierarchy of the images is chosen and warped to fit the shape of the original billboard.

• Prior to replacement, we anti-alias the borders so that there are no "jagged" edges.

4. Results and discussion

It is difficult to properly illustrate the behavior of the system on sequences in the traditional printed medium. The system has been in commercial exploitation since 1995 by Symah Vision, a subsidiary of the Lagardère group. Many improvements have been made to the system described here, while the conceptual design is the same. This is illustrated in the video sequences shown during the presentation, and the frames extracted from video streams as shown in Figures 7 and 8.

The successful aggregation of computer vision, computer graphics and image processing techniques should open up a wide avenue for other applications, which are either performed manually currently, or simply abandoned as too difficult.
On a different note, it is important to note that such a system also casts some doubts as to the authenticity of video documents, as predicted in fiction such as Rising Sun. It shows that digital video documents can be altered, just like audio and photo documents.

Fig. 7 The physical Aguila billboard, targeted for Spanish audiences, is replaced by a virtual Amstel billboard, targeted for northern European audiences.

Fig. 8 The Gaz de France logo (top) is replaced by an EMB logo (bottom)

References

1. R. Fielding, 'The Technique of Special Effects Cinematography', Focal/Hastings House, London, 3rd edition, 1972, pp. 220-243.

2. G. Medioni, G. Guy, H. Rom, 'Video Processing System for Modifying a Zone in Successive Images', U.S. Patent # 5,436,672, July 1995.

3. P.J. Burt, 'Fast Filter Transforms for Image Processing', Computer Graphics and Image Processing, Vol. 16, pp. 20-51, 1981.

4. K. Rangarajan, M. Shah and D. VanBrackle, 'Optimal Corner Detector', Computer Vision, Graphics and Image Processing (48), No. 2, Nov.1989, pp. 230-245.

5. A.J. Noble, 'Finding Corners', Image and Vision Computing (6), No. 2, May 1988, pp. 121-128.

6. R.M. Haralick and L.G. Shapiro, Computer and Robot Vision, Volume II, Addison-Wesley, 1993.

7. Y. Lamdan, H.J. Wolfson, 'Geometric Hashing: A General and Efficient Model-Based Recognition Scheme', Proc. of ICCV88, pp. 238-249.

8. M.S. Costa, R.M. Haralick and L.G. Shapiro, 'Optimal Affine-Invariant Point Matching', Proc. of ICPR90, Vol-I, pp. 233-236.

9. J.L. Barron, D.J. Fleet and S.S. Beauchemin, 'Performance of Optical Flow Techniques', IJCV(12), No. 1, February 1994, pp. 43-77.

10. Q. Zheng and R. Chellappa, 'Automatic Feature Point Extraction and Tracking in Image Sequences for Unknown Camera Motion', Proc. of ICCV93, pp. 335-340.

11. B.K.P. Horn, Robot Vision, MIT Press, Cambridge, MA, 1986.

12. J.R. Bergen, P.J. Burt, R. Hingorani, and S. Peleg, 'A Three-Frame Algorithm for Estimating Two-Component Image Motion', IEEE Transactions on Pattern Analysis and Machine Intelligence, PAMI, Vol. 14, Num. 9, pp. 886-896, Sep. 1992.

13. H. Rom and S. Peleg, 'Motion Based Segmentation', International Conference on Pattern Recognition, Atlantic City, New Jersey, pp. 109-113, 1990.

14. G. Wolberg, Digital Image Warping, IEEE Computer Society Press, 1990.

Signal Processing in Satellite Systems for Multimedia Services

M. Luglio[1], F. Mazzenga[2] and F. Vatalaro[1]

[1]Dipartimento di Ingegneria Elettronica, Università di Roma Tor Vergata, Italy
[2]CoRiTel – Consorzio di Ricerca sulle Telecomunicazioni, Roma, Italy

Abstract

Among possible transmission media, in the very near future satellite systems are expected to provide cost-effective multimedia services directly to the user. To provide these services in a flexible and efficient way the concept of bandwidth on-demand on a call-by-call basis must be implemented. This feature heavily relies upon digital signal processing techniques. This paper first reviews services expected via satellite and the main multimedia systems presently under development. Then it provides an overview of some of the possible choices in the area of digital on-board processing techniques, including multiplexing and demultiplexing, modulation and demodulation, interleaving and deinterleaving, forward error correction coding and decoding, digital beam forming, and switching. Finally, it gives a look into the fast developing subject of software radio satellite technologies.

1. Introduction

The fast-growing need for multimedia services including Internet web browsing, bulk data transfer, and video services, is pushing network designers and operators to deploy versatile and efficient broadband networks. Among possible transmission media, in the very near future satellite systems are expected to provide cost-effective multimedia direct-to-user (DTU) services (see Table 1).

Service	Application	User Typology	Bandwidth	Interactivity
Messaging	PC networking	Business	Medium	High
	E-mail	Business	Small	No
	Paging	Residential, Business	Small	No
Information retrieval	Data base access	Business	Large	Medium
Telephony	Voice, data, etc.	Residential, Business	Medium	High
Video-communication	Videophone	Residential, Business	Medium	High
	Videoconference	Business	Very large	High
Video-information (VOD)	Telemedicine	Residential	Medium	High
	Tele-education	Residential, Business	Medium	Medium

Table 1 – Satellite multimedia services and typical applications.

To provide these services in a flexible and efficient way the concept of bandwidth on-demand on a call-by-call basis must be implemented. This feature heavily relies upon Digital Signal Processing (DSP) techniques, both on board and at ground. Interactive multimedia and personal services require to a satellite system:

- wide bandwidths to accommodate high capacities and data rates,
- high degree of flexibility on-board and intelligence at terminal level to allow enhancement in personal mobility and to combat transmission impairments.

The challenge of providing multimedia services to fixed users through satellite systems is to ensure the same quality provided by terrestrial networks in such a way that the user should not realize which medium is presently used. A more demanding objective, pursued by some of the most advanced projects, is to extend satellite multimedia services to also ensure terminal mobility [1].

Basic technologies to provide multimedia and personal services via satellite have been developed in the past years. They are: On-Board Processing (OBP) to demodulate, route and re-modulate signals on board; multibeam antennas, to enhance spectrum and power efficiency; use of high frequencies such as the Ka band (20/30 GHz) and higher frequencies to reduce antennas and terminal sizes and to exploit larger bandwidths. In addition to the above technological advances, efforts are being dedicated to the adaptation of the Asynchronous Transfer Mode (ATM) standard to satellite systems [2, 3]. In fact, ATM offers integration of voice, video and data services and the flexibile access to bandwidth on-demand with guaranteed quality of service.

The main limitations of satellites are the relatively long transmission delays, link noise and interference, and the impairments due to bad weather conditions. They impact on the selection of the most appropriate multiple access (MA) techniques, on the selection of the satellite constellation, on the definition of the on-board processing and switching architectures. In particular link noise and weather conditions influence the reliable data transmission due to the presence of burst of errors. The effects of link noise can be mitigated using efficient coding schemes and the burst of errors can be compensated using appropriate interleaving. Finally, the long delay caused by satellite communications has a significant impact on delay-sensitive services such as real time voice and video.

Orbit constellations for multimedia satellite applications include both the Geostationary Earth Orbit (GEO) and circular non-GEO constellations located at Low-altitude Earth Orbits (LEOs) or Medium-altitude Earth Orbits (MEOs). LEO constellations are located at altitudes around 1000 km (below the first Van Allen belt). MEO constellations are located at altitudes around 10000 km (above the first and below the second Van Allen belt). For GEO satellites the one-way propagation delay is about 125 ms. In this case the delay due to on-board processing may be considered negligible. As a consequence of the large distance between user stations and the GEO satellite and considering the practical power limitations and the antenna sizes of the user terminals, LEO and MEO constellations may be preferable. In fact, for LEO and MEO constellations one way transmissions delay is reduced to 10-15 ms and 55-65 ms, respectively.

However if Inter-Satellite Links (ISL) are included the overall transmission delay increases.

This paper provides an overview of some of the possible choices in the area of digital OBP techniques for different MA techniques. Section 2 collects the main satellite multimedia services, and identifies the main proposed systems for provision of DTU multimedia services. Section 3 describes the characteristics of the main multiple access candidates. Section 4 provides a payload reference configuration and describes the functions of main on-board subsystems. Section 5 considers possible trends in application of software radio concepts to satellites. Finally, Section 6 provides paper conclusions.

2. Satellite multimedia services and systems

2.1 Services

Using satellites, multimedia services can be provided over wide geographical areas including remote, rural, urban and inaccessible areas both for fixed and mobile users. Where terrestrial networks are lacking or insufficiently deployed the global coverage capability of the satellite is crucial to bring emerging wideband and conventional narrowband services directly to the user.

Table 2 shows some of the multimedia applications supportable by a satellite system, along with reference quality, maximum delay and data rate both in the forward link (gateway-to-terminal) and in the return link (terminal-to-gateway). Any combination of services and data rates can be supported thanks to resource allocation flexibility, based on direct exchange of data packets and on circuit switched data adopting data rate on-demand. The data rate on-demand is a basic feature which allows to flexibly adapt call-by-call data rate to the type of service and to the user profile compromising cost and quality [1].

Application	Quality (BER)	Max end-to-end delay	Data rate	
			forward link	*return link*
PC networking	10^{-6}	200 ms	64 kb/s	64 kb/s
E-mail	10^{-4}-10^{-6}	5 min	1-5 kb/s	1-5 kb/s
Paging	10^{-4}-10^{-6}	5 min	1-5 kb/s	1-5 kb/s
Data Base access	10^{-6}	500ms (file transfer)	2 Mb/s	100 kb/s
Telephony	10^{-3} - 10^{-4}	250 ms	64 kb/s	64 kb/s
Videophone		200 ms	64 kb/s - 1 Mb/s	64 kb/s - 1 Mb/s
Videoconference	10^{-6}	200 ms	64 kb/s - 2 Mb/s	64 kb/s - 2 Mb/s
Tele-medicine	10^{-6}	200 ms	64 kb/s - 2 Mb/s	64 kb/s - 2 Mb/s
Tele-education	10^{-6} (data)	200 ms (1s for data)	1 Mb/s	64 kb/s

Table 2 - Requirements for some applications supported by a satellite system.

2.2 Systems

The success of ITALSAT satellites in Italy [4] and of the American ACTS (Advanced Communications Technology Satellite) [5] has pushed toward the realisation of even more advanced systems for multimedia services. In fact, these two systems have demonstrated the feasibility of advanced on-board technologies, such as baseband processing and routing, steerable and multibeam antennas, and both high power generation at 20 GHz and low noise reception at 30 GHz. In particular, through ACTS extensive field trials are being performed with both land mobile and aeronautical terminals [6]. Through ITALSAT operational experience is being gained in support to the terrestrial network for new high data-rate services (e.g., videoconference) and for the reallocation of communication resources in a fast and flexible way [7].

Recently both GEO and LEO programs for multimedia satellite networks have been started at Ka-band. Examples of these systems are Spaceway [8], Astrolink [9], EuroSkyWay [10], and Teledesic [11]. Table 3 shows the main characteristics of these systems. Teledesic is the first Ka-band LEO broadband satellite network to be proposed with data-rates up to 2 Mbit/s per user and up to 150 Mbit/s for connections with external gateways. For an updated view on Ka-band systems under development, see [12].

System	Orbit	Min user data rate	Max user data rate
Astrolink	GEO	16 kbit/s	9.2 Mbit/s
Euroskyway	GEO	16 kbit/s	2.048 Mbit/s
Spaceway	GEO	16 kbit/s	1.544 Mbit/s
Teledesic	LEO	16 kbit/s	2.048 Mbit/s

Table 3 - Main system characteristics.

3. Multiple access techniques

Two fundamental design criteria in payload design are efficient use of both power and spectrum. The following MA techniques can be considered among the candidates for multimedia satellite communications:

- Single Channel Per Carrier (SCPC),
- Multi Frequency - Time Division Multiple Access (MF-TDMA),
- Code Division Multiple Access (CDMA) or MF-CDMA.

In the down-link, possible distribution strategies are:

- Time Division Multiplexing (TDM),
- Code Division Multiplexing (CDM).

In general, the flexible match between the above multiple access and distribution techniques asks for the adoption of OBP techniques. OBP is also required by the need to ensure flexible routing between different spots of the same satellite and/or ISLs.

SCPC is a kind of frequency division multiple access (FDMA) suitable for satellite networks involving a large number of small capacity users. Usage of SCPC in up-link and TDM in down-link allows constant output levels both at the terminal and on-board the satellite, so optimizing the use of power [13]. This is an advantage of SCPC over TDMA, which is however strongly counterbalanced by the disadvantage of a reduced flexibility in data-rate, so that the bandwidth on-demand objective is difficult to be met. The use of TDM in down-link allows the satellite transmit amplifier to be driven close to saturation and is also compatible with the data-rate flexibility requirement.

Time division multiple access has become the reference case in satellite networks for multimedia applications. We can distinguish between conventional TDMA and MF-TDMA. In conventional high-bit-rate TDMA the satellite transponder is shared by many stations using timed (i.e. not overlapped) transmission of bursts. The same nominal center frequency is used. MF-TDMA is a low-bit-rate (typically from 2 Mbit/s to 15 Mbit/s) TDMA system utilizing a number of carriers in a single transponder. MF-TDMA terminals can be implemented using variable-rate modems and processors to provide services flexibility. To support services characterized by different data rates a variable amount of time slots, possibly over different carrier frequencies, can be allocated to each user.

CDMA is a promising technique for satellite portable and mobile communications [14]. It is accomplished by associating to each signal a signature sequence (spreading code). In accordance to the degree of coordination among the transmitting users a CDMA system can be asynchronous or quasi-synchronous [15]. In the down-link synchronous CDM can be adopted. CDMA can support multimedia services by assigning to each user a number of distinct spreading codes.

4. Payload reference configuration

As previously underlined, the main constraints of a satellite system are the limitation in power generation on-board and the large transmission delays. In [16] techniques based on multi-rate computing concepts are reported to achieve meaningful savings both in power consumption and in processing delay. In particular, some devices largely utilized in multimedia communications, such as processors for the Discrete Cosine Transform (DCT), Reed-Solomon co-decs and Viterbi decoders, are analyzed in terms of delay. Power savings are obtained at the expense of an increased silicon area occupied by the devices.

A satellite-based multimedia system must adopt intelligent user terminals, including high definition screens, digital phones, multimedia computers able to handle several multimedia services. On the other hand, if aiming at a large market, the terminals must be very low cost, consume low power, and have small dimensions. Thus, the payload architecture must be as complex as possible at the baseband stages and DSP stages, implementing all the possible solutions both at technical and algorithmic level.

From the previous discussion on MA techniques, due to higher flexibility, hybrid schemes such as MF-TDMA and MF-CDMA should be preferred for multimedia satellite transmissions. In addition, when ATM is considered, due to statistical multiplexing, cells pertaining to different users can be transmitted on the same physical channel. Then, switching must be performed on-board on a a cell-by-cell basis so that the payload must be able to mo/demodulate the incoming signals and to perform co/decoding operations.

Multi-spot coverage is considered in both the up-link and the down-link. A block diagram of a digital OBP payload is shown in Fig. 1. This scheme is assumed as the reference payload for the purpose of this paper. The functions and some possible implementation schemes of the baseband subsystems indicated in Fig.1 will be discussed in the following Sections.

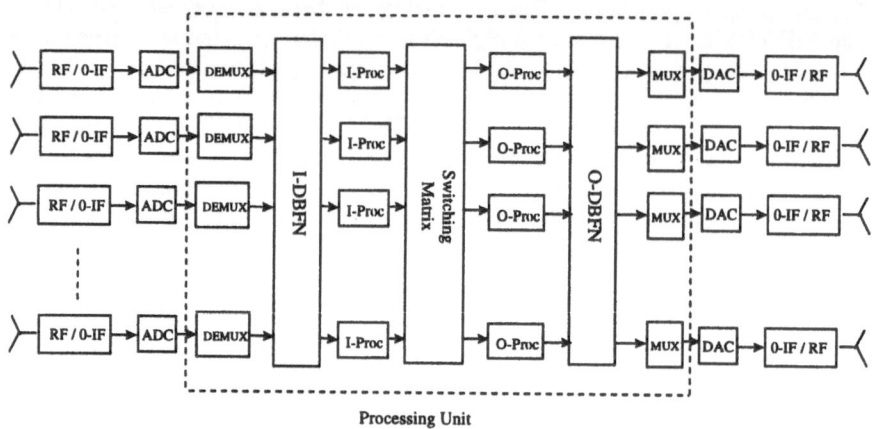

Processing Unit

Fig. 1 - Block diagram of the OBP reference payload assumed in this paper.

4.1 Demultiplexer and multiplexer

Basically, the operation of a demultiplexer (DEMUX) is to separate multiplexed channels and to supply each of them to the input of the demodulator [17-21]. Demultiplexers are required for SCPC, as well as for hybrid multiple access schemes, such as MF-TDMA and MF-CDMA, in order to recover each subcarrier signal or the individual TDMA and CDMA signals.

Digital demultiplexing algorithms can be devised both in the time domain and in the frequency domain. A more detailed classification divides them in the direct method or per-channel approach, polyphase network approach, and tree approach.

In the direct method the FDM signal filtering is performed by means of individual band-pass filters. In general, the direct method is of little practical interest due to its complexity, which increases with the number of carriers [17]. To adapt a per-channel scheme to signals with different bandwidths the coefficients of the filters must be accordingly changed. If the set of filters is obtained through frequency shifting of a single low-pass prototype filter, complexity of the direct method is greatly reduced. In fact, using the results of the polyphase network theory [17, 18] the filter bank can be decomposed into two subsystems with reduced complexity: a first filter bank implementing the polyphase decomposition of the low-pass filter prototype followed by a subsystem performing the discrete Fourier transform (DFT) operation [17, 18]. If the number of sub-channels is a power of two, the FFT algorithm can be used. The polyphase approach is the most suitable candidate in terms of computational complexity when demultiplexing involves hundreds of carriers. In general, a polyphase filter bank is easily reprogrammable to account for multiple user signal with different bandwidths integer multiple of a fixed quantity.

In [22] a programmable demultiplexer based on polyphase decomposition [23] has been proposed. The proposed architecture is composed of two cascaded stages. The first stage uses a per-channel approach and the second stage uses a polyphase approach. Programming of the demultiplexer is performed at software level. The first stage is used to separate three or more groups of carriers. Such filters can be easily adapted to the required traffic pattern. The second stage is a polyphase demultiplexer that is used to separate carriers in the same group. To properly operate it is necessary that carriers composing the up-link signal are arranged in groups of users with equal bandwidth within each group. The bandwidths can be different from one group to another, and the number of users per group can be variable. This grouping requirement is not a severe operational constraint in most practical situations.

Finally, the multistage demultiplexing technique [17, 18] is appropriate when the number of channels to be extracted is an integer power of two. In this case the input signal band is first splitted into two bands by half band filters and subsequently decimated by two. Both filter outputs are again split by two half-band filters and decimated by two leading to a division into four bands of the input signal. In general after L filtering stages and subsequent decimation, 2^L channels are obtained. Due to the inherent symmetry the amount of processing is greatly reduced. This architecture also ensures a certain degree of flexibility in bandwidth assignment, because channels with different bandwidths can be extracted from the intermediate stages. Also, a multistage demultiplexer can be adapted to signals with different bandwidths by changing the coefficients of the half-band filters.

Digital demultiplexers can be also based on FFT/IFFT frequency domain filtering. Basically, this consists in convolving the composite signal with a suitable bank of filters using overlap-and-save or overlap-and-add techniques [24]. More specifically, to obtain the signal centered around a given carrier the input signal is first transformed to the frequency domain using FFT. Then, the transformed signal is multiplied by the frequency response of the bandpass filter. The product is then transformed back via IFFT, so recovering the waveform in the time domain. To manage signals with different bandwidths IFFT operates on each individual carrier, performing transforms of different sizes corresponding to the different involved bandwidths. A digital multi-carrier demultiplexer for the SCPC access scheme operating in the frequency domain has been developed by COMSAT [25]. It permits the on-board processor to accommodate different types of multi-channel FDMA carriers by simply changing its computational rules and its organization.

Performance evaluations of the demultiplexing schemes described above are reported in [17-21], [25]. It must be remarked that usually the sampling rate at the output of the demultiplexer may be unsuitable for the following demodulation process, where an integer number of samples per data symbol is required. So, a sampling rate conversion filter may be necessary to interface the demultiplexer and the demodulator.

When the bandwidth of the signal(s) to be demultiplexed is too wide, a two stage analog-digital demultiplexer can be adopted. In the first stage analog filters with tunable center frequencies are used to perform a first coarse demultiplexing operation. Then, digital techniques are used on the coarse demultiplexed signals to perform fine demultiplexing [26].

The multiplexer performs the complementary operation to demultiplexing, i.e. each multiplexer stacks the incoming signals in frequency to the desired carrier position. In principle it consists of a bank of digital, possibly programmable, harmonic oscillators. More efficient structures can be obtained on the basis of digital interpolation and decimation operations [27].

4.2 Input processors

The input processor (I-Proc) shown in Fig. 1 basically performs baseband demodulation, deinterleaving, FEC decoding of the signals at the output of the digital beam-forming network. Deinterleaving and decoding operations may be performed using standard algorithms presented in the literature [28]. Recently, some implementations of these circuits (e.g., Reed-Solomon and Viterbi decoders) based on a polyphase approach have been proposed in order to reduce computational complexity and power consumption [29]. In the following we only describe some architectures for the implementation of the digital demodulators.

Signals obtained out of the demultiplexer can be demodulated by means of digital demodulators. In principle, at least one demodulator is required for each demultiplexed sub-channel. When possible, in order to reduce complexity, part of the demodulation process could be included in the demultiplexing stage.

When TDMA signals are considered, the same demodulator can be shared among all the users. If Quadrature Phase Shift Keying (QPSK) modulation is used, the demodulator consists of the filter matched to the symbol pulse waveform followed by a digital interpolation filter [30] which is used to adjust the sampling point. In [31] a simple VLSI digital PSK demodulator able to adapt to different bit rates has been proposed for TDMA satellite communications. If pulse matched filtering is incorporated in the demultiplexing stage (see [25]), only the digital interpolation filter need to be used.

Conventional demodulators for each demultiplexed CDMA signal can be implemented as the cascade of the Chip Matched Filter (CMF) followed by a descrambler and the despreader [32]. The CMF is shared by all the CDMA channels and the matched filtering operation can be carried out with analog components (for instance with SAW filters) directly at IF [32, 33]. This can be advantageous when a low complexity architecture of the digital section is required. For subsequent processing, after the CMF, only one sample per chip is usually required. The despreading system consists of a bank of code matched filters which is a FIR filter with coefficients ±1 equal to the despreading sequence. The output of each FIR filter is decimated to the symbol rate.

In asynchronous DS-CDMA mutual interference is unavoidable, and the performance of a conventional demodulator can be severely degraded. To obtain a bit error rate comparable to that of other access schemes extra bandwidth can be used, so degrading spectrum efficiency. To reduce the effects of mutual interference without increasing bandwidth, optimum and sub-optimum Multi-User Detection (MUD) subsystems [34, 35] have been proposed and analyzed. Optimum MUD for an asynchronous transmission channel [34] involves dynamic programming algorithms. This adds significant complexity comparable to the use of a high gain FEC coding. To avoid this disadvantage several sub-optimum MUD subsystems have been proposed. Among them an interesting MUD system based on the minimum ouput energy (MOE) criterion only requiring the knowledge of the desired user signature sequence has been proposed in [36]. An adaptive filter whose coefficients are derived solving the MOE optimization problem replaces the code-matched filter. Coefficients of the MOE filters can be adaptively adjusted without any training sequence. The coefficients are updated every symbol period.

4.3 Digital beam-forming networks

A Beam-Forming Network (BFN) implements the spot synthesis function. It can be realized with analog or digital techniques. In principle, it can be placed at RF, IF or baseband. The increase in complexity versus the number of spots for a payload based upon a DBF antenna is not as fast as with an analog beam forming antenna. Digital Beam-Forming Networks (DBFNs) have been proposed to generate a very large number of spots (in the limit one for each user) [26], [37].

The basic idea for a DBFN is the same as for an analog RF or IF beam-forming. A signal impinging on array antenna will arrive at different time instants on the different array elements in relation to its Direction Of Arrival (DOA). The function of the beamformer is to phase shift each signal on every radiating element for an exact amount related to the direction of arrival of the wanted signal and then to add all the signals. In this case only the signals arriving from the desired direction will be added in phase while the interfering signals will be added incoherently and will be subjected to less enhancement (or to cancellation) depending on their DOA.

The input DBFN in Fig.1 performs such function in the digital domain on the sampled complex envelope at the output of each demultiplexing stage. The output DBFN weights the complex envelope of the signals to be transmitted in order to build the wanted spot pattern. Among the advantages of the DBFN, note that it can be used to easily reconfigure the Earth coverage in order to adapt the capacity of the single satellite link to the actual time-variant traffic pattern. This feature is very important especially for multimedia services. In addition we can observe that in the limiting case a single channel per beam could be supported. The calibration and compensation of phase and amplitude errors originating in the analog part of the antenna, the LNA, the up-down conversion stages etc., is easier. Active interference reduction techniques can be easily implemented too, in order to reduce co-channel interference due to signals using the same frequency but arriving from different beams on different directions [38, 39]. Finally, digital beam-forming related techniques allow to find the DOA of different signals so allowing to correct the pointing of the beams especially for mobile users. Being multipath effects less important than in terrestrial systems, with satellites DOA estimation can be more precise.

4.4 Switching matrix

The switching subsystem routes bit streams coming from the uplink beams to the assigned downlink beams. It is not the purpose of this paper to discuss on the possible architectures of the switching units for multimedia systems, particularly for ATM. However it should be observed that in ATM terrestrial switching modularity drives the project in order to easily update the plant with an obvious increase in the overall complexity of the switching unit. In the satellite case the on-board switching architectures can be greatly simplified since they will be not subjected to any change during their lifetime. In general there will be a mapping between the switch input ports and the geographic distribution of the beams. Whether this mapping is one-to-one or not is related to the level of multiplexing in the satellite front-end and the allowable processing loads on the switching input ports. For further information refer to [40] and to references therein.

4.5 Output processors

Output processors (O-Proc) perform FEC encoding operations, interleaving and baseband modulation of the bit streams at the output of the switching matrix. A digital implementation of the modulator for QPSK to be used in TDM transmissions is described in [41]. The single bit stream is first serial-to-parallel converted and the resulting bit sequences are then shaped using a bandlimited pulse shape. Then, digital up-conversion to IF may be performed.

For CDMA some digital schemes for the baseband spread-spectrum modulator have been described in [32] for the cases of QPSK and double BPSK. In this last case, similarly to the digital QPSK modulator, the incoming binary bit stream is split into two parallel streams each one being spread with the respective chip sequence. The resulting sequences are then shaped by means of a Nyquist root raised cosine filter. Both baseband components can be digitally up-converted to IF and the resulting stream is input to the digital-to-analog converter. An analog anti-aliasing filter is used before the following up-conversion to RF stage.

4.6 Synchronization issues

The schemes discussed in the previous Sections must be supplemented with appropriate circuits to perform carrier, frequency and also code-time acquisition in the CDMA case.

Coherent and not coherent demodulation strategies could be adopted. Coherent demodulation can be used to reduce the signal-to-noise ratio required to achieve a specified BER. As a consequence both carrier and clock recovery circuits are required for each demultiplexed channel. Carrier phase and symbol timing estimates can be independently derived by suitable techniques or can be jointly estimated using, for example, a maximum a posteriori probability method.

In the case of TDMA carrier recovery and timing synchronization can be obtained using digital algorithms. An extensive collection of synchronization algorithms can be found in [42] for some of the modulation formats such as BPSK or QPSK. Some examples of suitable choices are now reviewed.

In principle carrier phase recovery can be obtained using a digitized versions of the Costas loop. Feedback loop based architectures are chosen for their potentially low computation rate. Another method for carrier recovery is based on the nonlinear estimation technique proposed in [43]. This approach gives a good estimate of the carrier phase, is less sensitive to finite arithmetic, and requires a very short acquisition time.

The selection of the timing recovery scheme is closely related to the bit rate of the application and to the modulation format. Clock recovery schemes are described in [42] and for example the "clock estimate method" seems to be a good choice since the estimation procedure is independent on the carrier phase.

A leading criterion to choose carrier and clock recovery schemes is the degree of integration of the demultiplexing and the demodulation functions in order to reduce the overall computational complexity. For example in [17] the adoption of the algorithm in [43] and the "clock estimate method" lead to a good degree of integration of such functions thus maintaining the overall system complexity at an acceptable level. Another example where the two functions are well integrated is given by the system developed by Comsat [25].

In the case of CDMA, as indicated in [32], automatic gain control and automatic frequency control can be effectively performed with standard analog low complexity IF loops that can be integrated in the analog receiver front end. Carrier phase correction can be implemented at IF with a numerically controlled oscillator driven by an error signal obtained from the processed baseband signal. However in a spread spectrum system the most important synchronization operations are related to the code acquisition and tracking. The Digital Delay Lock Loop (DDLL) seems to be suitable for code tracking when Nyquist raised-cosine pulse functions are used for chip shaping. Operations of the DDLL are insensitive to carrier phase error and data modulation due to the presence of square modulus operation. DDLL operates at chip rate so that ASIC implementation may be required. The function of initial code acquisition may be performed with a standard serial acquisition subsystem as indicated in [32].

5. Software radio satellite

As the technology continues its rapid transition from analog to digital, more radio system functions tend to be software implemented. The baseband schemes used to perform mu/demultiplexing, mo/demodulation and synchronization, described in the previous Section, are mainly implemented through ASIC components and their operations can be controlled by means of suitable microcontrollers. This is especially true for spread-spectrum systems where high bandwidths are involved and relatively high sampling rates need to be considered.

In a software radio system the above functions are completely implemented via software by means of general programmable digital signal processors. Software radios are becoming more practical as the costs per million instructions per second of digital signal processors continue to drop.

The two advantages of software radio are flexibility and concentration. Flexibility consists in the ability to switch channels, change the modulation schemes or receive different type of modulated signals. This feature render software radio very attractive for satellite systems. In fact, for a single satellite to be at the same time a backup unit for different existing or future terrestrial networks it must be able to treat different signals with different modulation formats and protocols (e.g. GSM, AMPS, wideband CDMA for the 3rd generation cellular system). Concentration means that multiple channels share the same front-end radio stage while having low-cost DSP functions performed for each one.

The software radio satellite architecture is similar to that in Fig. 1, but the processing unit is entirely implemented in software. The sampling rate at the output of the analog-to-digital converters does not need to be specific to an application (i.e., it does not need to be a multiple of a given symbol rate). The only requirement for the sampling rate is that it must be sufficiently high in order to manage different channels with different modulation formats and data rates. To increase flexibility digital IF to baseband conversion front-end is preferred since the trend of software radio is to incorporate as many functions as possible moving towards the RF stage.

In advanced applications software radio techniques are not limited to transmit and receive functions. They must be able to characterize the available transmission channels, to electronically steer transmit beams, to select the appropriate transmit power levels, etc. For this reason among the functions implemented in the digital signal processors contained in the OBP part of the scheme in Fig.1 we can distinguish [44]:

a) *Real time processing functions*. They include functions related to mo-demodulation operations such as co/decoding, pulse shaping, digital IF up-conversion and down-conversion, synchronization, etc. They must be able to operate with very low sampling time.

b) *Environment management functions*. They are set to continuously characterize the propagation environment in frequency, time and space domains. This characterization includes channel identification and estimation of other parameters such as channel interference levels depending on the adopted multiple access modulation formats and the transmission power levels. Normally such parameters are needed in times on the order of hundred microseconds up to hundred milliseconds. The interface between these functions and those described in point a) is very important, since it is necessary to synchronize them.

c) *On-line and off-line management functions*. They comprise functions to perform on-line and off line analysis of the quality of the link or to realize an enhanced beam-former or equalizer that may be necessary to reduce interference. Some of these enhancement functions may also be included in those described in above point b).

6. Conclusions

Satellite systems are expected to provide soon cost-effective multimedia services directly to the user. To provide these services in a flexible and efficient way the concept of bandwidth on-demand on a call-by-call basis must be implemented. This is a feature that heavily relies upon DSP techniques. After a short review of services expected via satellite and on the main multimedia systems under development, this paper provided an overview of digital satellite OBP techniques, which are crucial to the above objective. Main topics of concern for the DSP are in the areas of mux/demux, mo/dem, interleaving/deinterleaving, FEC co/decoding, digital BFNs, and switching. A final Section mentions possible trends in software radio concepts for satellites, a topic that is expected to be further developed in the near future.

References

[1] G. Losquadro, M. Luglio, F. Vatalaro, A. Paraboni, "An advanced satellite system to provide interactive multimedia mobile services", European Trans. Telecommunications (ETT), Vol. 8, N. 4, July-August '97, pp. 379-391.

[2] I. F. Akyildiz, "Satellite ATM networks: a survey", IEEE Comm. Mag., Vol. 35, N.7, July 1997, pp. 30-43.

[3] M.Wittig, "Large capacity multimedia satellite systems", IEEE Comm. Mag., Vol. 35, N. 7, July 1997, pp. 44-49.

[4] F. Marconicchio, G. Morelli, F. Valdoni, "ITALSAT, an advanced on board processing satellite communication system", Proc. 14th AIAA, Washington (DC), USA, March 22-26, 1992, pp. 832-841.

[5] W. F. Cashman, "ACTS multibeam communication package: description and performance characterization", Proc. 14th AIAA, Washington (DC), USA, March 22-26, 1992, pp. 1151-1161.

[6] F. Gargione, B. Abbe, M. J. Agan, T.C. Jedrey, P. Sohn, "Mobile experiments using ACTS", Space communications, Vol. 13, N. 3, 1995, pp. 193-223.

[7] S. De Padova, F. Marconicchio, A. Saitto, F. Valdoni, "Integration and exploitation aspects for a switching satellite system in the Italian public network", Proc. 1st ECSC, Munich, Germany, November 28-30, 1989, pp. 21-27.

[8] E. J. Fitzpatrick, "Spaceway providing affordable and versatile telecommunication solutions", Proc. Ka Band Utilization Conf., Rome, Italy, Oct. 12-15, 1995, pp. 213-230.

[9] E. Elizondo, R. V. Gobbi, A. Modelfino, F. Gargione, "Evolution of the Astrolink system", Proc. 3rd Ka Band Utilization Conf., Sorrento, Italy, Sept. 15-18, 1997, pp. 3-7.

[10]   G. Losquadro, M. Marinelli, "The EuroSkyWay system for interactive multimedia and the relevant traffic management", Proc. 3rd Ka Band Utilization Conference, Sorrento, Italy, Sept. 15-18, 1997, pp. 17-24.

[11]   M. A. Sturza, F. Ghazvinian, "Teledesic satellite system overview", Proc. Ka Band Utilization Conf., Rome, Italy, Oct. 12-15, 1995, pp. 231-238.

[12]   Proc. 4th Ka Band Utilization Conf., Venice, Italy, Nov. 2-4, 1998.

[13]   F. Carassa, "New satellite systems and higher frequencies utilization", Alta Frequenza, Vol. LVI, N. 1-2, 1987.

[14]   K. S. Gilhousen, I. M. Jacobs, R. Padovani, L. A. Weaver, "Increased capacity using CDMA for mobile satellite communications", IEEE J. Sel. Areas Comm., Vol. 8, May 1990, pp. 503-514.

[15]   R. De Gaudenzi, C. Elia, R. Viola, "Bandlimited quasi-synchronous CDMA: a novel satellite access technique for mobile and personal communication systems", IEEE J. Sel. Areas Comm., Vol. 10, N. 2, Feb. 1992, pp. 328-343.

[16]   K. J. R. Liu, An-YEU WU, A. Raghupathy, J. Chen, "Algorithm-based low-power and high-performance multimedia signal processing", Proc. IEEE, Vol. 86, N. 6, June 1998, pp. 1155-1202.

[17]   E. Del Re, R. Fantacci, "Alternatives for on board digital multicarrier demodulation", Int. J. Sat. Comm., 1988, Vol. 6, pp.267-281.

[18]   W. H. Yim, C. C. D. Kwan, F. P. Coakley, B. G. Evans, "Multi-carrier demodulators for on board processing satellites" Int. J. Sat. Comm., 1988, Vol. 6, pp. 243-251.

[19]   P. Cangiane, G. Caso, H. Courtois, R. Rey, "Multi-channel demultiplexer-demodulator", Proc. AIAA-94, pp.822-830.

[20]   S. Kato, T. Arita, K. Morita, "Onboard digital signal processing technologies for present and future TDMA and SCPC systems", IEEE J. Sel. Areas Comm., SAC-5, N. 4, May 1987.

[21]   S. Kato, K. Ohtani, T. Kohri, M. Morikura, M. Umheira, S. Kubota, "On board digital signal processing technologies for Present and Future SCPC Systems", Int. J. Sat. Comm., 1988, Vol. 6, pp. 289-300.

[22]   X. Y. Guo, G. Maral, M. Bousquet, "A programmable demultiplexer using a polyphase approach for regenerative satellites", Proc. AIAA-92, pp.1227,1233.

[23]   P. P. Vaidyanathan, "Multirate digital filters, filter banks, polyphase networks, and applications: a tutorial", Proc. IEEE, Vol. 78, N. 1, Jan. 1990, pp.56-93.

[24]   A. V. Oppenheim, R. W. Schafer, *Digital Signal Processing*, Prentice Hall 1975.

[25]   S. Sayegh, M. Kappes, J. Thomas, J. Snyder, "An overview of COMSAT work in multicarrier demultiplexers/demodulators", Proc. AIAA-92, pp.1234-1244.

[26]   M. Lisi, M. Piccinni, A. Vernucci, "Payload design alternatives for geostationary personal communications satellites", Proc. EPMS-94, Springer-Verlag, 1994.

[27] R. E. Crochiere, L. R. Rabiner, "Interpolation and decimation of digital signals: a tutorial review", Proc. IEEE, Vol. 69, N. 3, March 1981, pp.300-331.

[28] J. G. Proakis, *Digital Communications*, Mc Graw Hill, 3rd Ed. 1997.

[29] K. J. R. Liu, A. Y. Wu, "Algorithm-based low power and high performance multimedia signal processing", Proc. IEEE, Vol. 86, N. 6, June 1998, pp. 1155-1202.

[30] F. M. Gardner, "Interpolation in digital modems - Part I: fundamentals", IEEE Trans. Comm., Vol. 41, N. 3, March 1993, pp.501-507.

[31] F. Hansen, J. H. Thomsen, F. L. Jacobsen, K. Olsen, "VLSI digital PSK demodulator for space communication", European Trans. Telecommunications (ETT), Vol. 4, N. 1, Jan-Feb. 1993.

[32] R. De Gaudenzi, T. Garde, F. Giannetti, M. Luise, "An overview of CDMA techniques for mobile and personal satellite communications", Proc. EPMS-94, Springer-Verlag, 1994.

[33] F. Coakley, W.H. Yim, "A hybrid CDMA scheme and demodulator structure for Ka band operation", 10th ICDSC, 15-19 May, 1995.

[34] S. Verdu, "Minimum probability of error for asynchronous Gaussian multiple access channel", IEEE Trans. Info. Th., Vol. IT-32, pp. 85-96, Jan. 1986.

[35] S. Verdu, "Adaptive multiuser detection", Proc. IEEE Int. Symp. Spread Spectrum Th. and Applications, Oulu, Finland, July 1994.

[36] M. Honig, U. Madhow, S. Verdu, "Blind adaptive multiuser detection", IEEE Trans. Info. Th., Vol. 41, N. 4, July 1995, pp.944-960.

[37] J. Ventura-Traveset, M. Hallraiser, J. Stojkovic, F. Petz, "A digital transparent satellite payload concept for personal mobile satellite communications: a VLSI technology review", Proc. EPMS-94, Springer-Verlag, 1994.

[38] A. J. Paulraj, C. B. Papadias, "Space-time processing for wireless communications", IEEE Sign. Proc. Mag., Vol. 14, N. 6, Nov. 1997, pp.49-83.

[39] L. C. Godara, "Applications of antenna arrays to mobile communications. Part II: beam-forming and direction of arrival considerations", Proc. IEEE Aug. 1997, Vol. 85, N. 8, pp.1195-1247.

[40] T. Inukai, B. A. Pontano, "Satellite on-board baseband switching architectures", European Trans. Telecommunications (ETT), Vol. 4, N. 1, Jan-Feb. 1993.

[41] R. Steele, *Mobile Radio Communications*, J. Wiley, 1992.

[42] F. M. Gardner, "Demodulator reference recovery techniques suited for digital implementation", Final report ESTEC contract. N. 6847/86/NL/DG.

[43] A. J. Viterbi, A. M. Viterbi, "Nonlinear estimation of PSK-modulated carrier phase with application to burst digital transmission", IEEE Trans. Info. Th., Vol. IT-29, N. 4, July 1983.

[44] J. Mitola, "The software radio architecture", IEEE Comm. Mag., May 1995, pp. 26-38.

A Sketch-Based Approach for the Recovery of Transmission Errors in Coded Images

F.G.B. De Natale, D.D. Giusto and G. Vernazza
DIEE - University of Cagliari, Italy

Abstract

The transmission of coded visual information over packet networks introduces fidelity problems related to the loss of frames during transmission. In this paper, a new concealment technique is presented, which aims at restoring the lost visual information by using a synthetic reconstruction of the high-frequency content of the damaged blocks. The method is funded on the theory of sketch-based encoders: for each block to be interpolated, the sketch information of the available surrounding blocks is extracted and propagated to the missing area. Then, the low-pass content is easily interpolated from the sketch.
The proposed method uses only the spatial correlation, and has been applied with good results in the transmission of both video and still pictures.

1. Introduction

Visual communications are becoming the most important source of traffic in modern data networks. Images and video sequences are used in a wide range of applications, from home entertainment to complex industrial systems, and represent, due to the high demand in terms of bandwidth, the crucial point for the overall performance of multimedia applications and services. Recently developed broadband networks, as satellite links or ATM networks, provide suitable infrastructures for such applications. Nonetheless, they pose other challenging problems in the definition of robust protocols, able to provide a guaranteed QoS.

In this framework, a major concern is to prevent the image quality degradation due to possible data losses occurred during transmission. For instance, in high-speed packet switching the loss of a packet (or cell, in ATM terminology), is in general related to link congestion problems caused by traffic peaks [1]. In the particular but quite common case that the transmitted data are in compressed format, transmission errors can drastically compromise the decoding process. In fact, due to the extensive use of block-based approaches in image coding (e.g., JPEG and MPEG), every data corruption, independently of the cause and of the extent, may result in a missing or faulty sequence of blocks.

Specific procedures, targeted to the transmission of visual data, have been proposed in recent years, with a twofold goal: to make the transmission more robust, and to minimise effects of errors and losses on the reconstructed picture quality. The solutions adopted can be grouped into two categories: packet-loss resilient coding techniques, which make possible to decode the stream also in the presence of errors, and error concealment algorithms, which mask the effect of corrupted blocks by using temporal or spatial interpolation.

Error concealment techniques can be applied in addition or substitution to error resilient coding. For instance, the recovery problem of subband coded images has been extensively studied [2,3] due to the wide use of image subband coding for transmission in packet network. In [4] an error recovery scheme is proposed that [5] exploits a layered coding strategy to approximate the partially lost transform coefficients at the reception side. To achieve this goal, the available DCT coefficients and the pixel-wide external boundary of the damaged block are used to predict the lost coefficients, based on a maximally smooth criterion. The major advantage of the approaches based on smooth interpolation is that they allow to define simple numerical optimisation criteria that are easy to calculate and ensure a low average estimation error.

Nevertheless, smooth reconstruction presents the drawback of a unpleasant visual effect, due to the low-pass ant blocky appearance. Low pass or blurring mainly affects the visual quality of static images, while tiling or blocking produces undesired effects in video sequences. Both phenomena can be corrected by applying some high frequency information recovery procedure for lost blocks belonging to textured or edged image areas. In this sense, the use of contour information is very significant, for image edges are proven to be fundamental for the preservation of both the semantics and the perceived quality of an image.

This work addresses the problem of concealing lost image data by using a spatial interpolation technique based on the restoration of the contour geometry and the relevant grey level information. The basis of the proposed approach can be found in the theory of sketch coding, which provides a very efficient image representation methodology [6]. Owing to this scheme, high-quality grey level images can be reconstructed from the edge-related information, which is the only information transmitted. The quality of the resulting image depends on the accuracy in the detection and representation of image edges.

3. Sketch-Based Error Concealment

The compact representation of image information is one of the main goals of second-generation image coding techniques, developed during the past decade. The efficient description of visual data implies the use of the basic primitives that can be extracted from an image, such as contours, regions or textures. An in-depth analysis of such approaches can be found in [7], where the new contour/texture oriented methods are critically compared to the classical first-generation coding techniques, based on local operators.

Among other descriptors, edges are proven to be particularly significant for what concerns the preservation of the semantic content of an image. Moreover, they are very important for the human perception of the visual information, due to the high sensitivity of the human visual system to edges and contours. Based on this, contour-based approaches have been proposed as post processing techniques to reduce block artefacts exploiting edge information [8]. In particular, sketch-based image coding provides a very efficient representation of an image, by exploiting the image contour information. It extracts from the image to sources of information: the geometry of the image edges and the relevant gray level information across the edge. The ensemble of these two data constitutes the image sketch, and is a perceptually very rich representation of the original image.

Starting from the sketch, a non-unique approximation of the original image can be obtained, which accuracy is mainly related to the perceptual correctness and appearance of the reconstructed image.

The proposed concealment approach is based on the aptitude of sketch information to represent the essential content of visual data. The key concept is that, if the sketch information in the lost area can be recovered, it would be possible to reconstruct with good accuracy the whole block. The main advantage of passing through the sketch information is clearly related to higher simplicity of interpolating linear structures (like contours) instead of bi-dimensional data, in particular in the case of discontinuous regions. The recovering process should be based on spatial correlation present on the sketch information of natural images.

Fig. 1 shows a global scheme of the proposed method. Three are the main processing steps: extraction of the sketch from the area surrounding the block to be recovered, interpolation of the sketch within the lost area, and final interpolation of the block.

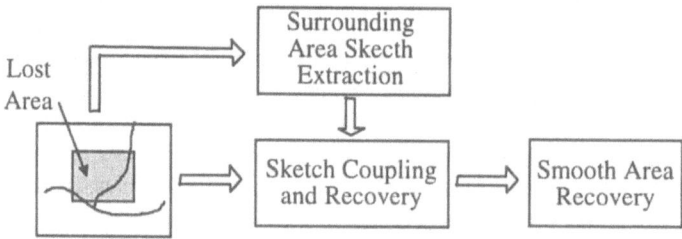

Fig. 1 Sketch-based recovery procedure

The first step starts with the detection of image edges in a region of interest, which is required to determine the contour geometry. The region of interest includes the blocks adjacent to the lost area. For the purpose of reducing the computation, only the direct neighbors are used, although in principle it was possible to consider also more distant blocks. Moreover, it must be observed that with a block dimension of 16×16 pixels (equal to the macro-block dimension in MPEG), the correlation at a distance of two blocks is in general negligible.

Once the edges are extracted, a subset of them is selected that potentially extend into the missing area, and the relevant gray level information is computed to complete the sketch data. Then, the candidate edges are conveniently coupled considering several contiguity criteria, and each couple is connected across the lost block. Finally, the gray level information of the interpolated curves is reconstructed to complete the sketch approximation, and the smoothed areas are restored by a bi-dimensional interpolation.

3.1 Extraction and Coupling of the Damaged Sketch Segments

The accuracy of the edge extraction process is very important, for it affects the performance of the whole method. Edge extraction is intended to locate the gray level discontinuities, that denote the object contours on the area surrounding the block affected by losses or errors. In our scheme, the Canny edge extractor was chosen for several reasons [9]:
- it allows to detect one pixel wide contour lines (good localization property);
- it extracts edges with good connectivity (thanks to hysteresis thresholding);
- it provides smoothed edges that approximate the natural course of object shapes.

The filter was applied to the neighboring blocks on a 16 pixels wide area around the lost block, typically sufficient to recover the missing sketch information without causing an excessive computational complexity.

Among the edges detected in the surrounding area, the ones that end on the border of the missing image area are the selected candidates for the following interpolation procedures. To this purpose, the border is analyzed and, starting from each edge point found, the corresponding edge segment is followed backwards to the beginning or till a maximum number of points is reached

The obtained set of contour segments is then integrated with the gray level information to build the damaged sketch set (DSS), which represents the starting point for the recovery of the missing sketch data, performed during the following phases. The detection of the gray levels across each contour segment is a straightforward operation. The edge chain is followed from one end to the other, and for each point the luminance values at the left and right sides on the perpendicular direction are stored.

The following processing phase aims at restoring the portion of the sketch belonging to the lost image area, by connecting and interpolating the sketch segments belonging to the DSS. To perform this task, it is necessary to analyze the visual information contained in the DSS, to predict the structure of the sketch in the missing area, localizing the possible connections among edge couples. To identify the most likely pairs (or, in rare cases, the links among three or more segments), several parameters can be taken into account, including shape, distance, gray level intensities, edge strength, and so on.

The problem of linking fragmented edges, commonly known as edge grouping, is quite common in computer vision applications, due to the importance of having well structured contours for subsequent high-level processing (e.g., object recognition and scene interpretation). Most part of edge grouping approaches are based on clustering techniques, and use a set of features computed on the available edge segments. Obviously, the selected features must show a high regularity within the group of edges belonging to the same object and be sufficiently distant for different objects. For instance, a global feature frequently used is the convexity, which is typical of a group of segments belonging to a closed object shape, thus enforcing a closure constraint on the edge group [10]. Other approaches exploit local measures, and connect couples of segments that show common properties [11].

The edge linking procedure to be applied in block recovery presents some similarities with edge grouping, but is different in several respects: first, edge interruptions are usually quite large, and no information at all is available within the lost area; second, the shape and dimension of the missing region is known a-priori, thus allowing some assumption on the missing segment; third, it is not needed to group edges, but simply to find the most likely coupling among a set of interrupted segments. Therefore, the results achieved in edge grouping can be only partially exploited in our problem, and some specific approaches need to be defined. In particular, in lost block restoration:

- global features are not useful: comparative measures are more significant;
- proximity criteria are no more significant, for the distance between two edge segments only depends on their position on the lost block and does not affect the coupling probability;
- due to the large gaps, some additional information is needed to achieve a reliable edge coupling.

Based on the above considerations, some typical properties of natural images have been identified that proved to be useful in sketch prediction:

- *smoothness of the area encompassed by edges*: the intensity of the areas on each edge side shows slow luminance variations (high spatial correlation of gray levels on the edge sides);
- *smoothness of the edges*: natural object's contours show a slow curvature variation (continuity and low values of the curve derivatives);
- *closure of the edges*: in natural images, rarely contour segments terminate abrupt (continuity and convexity of the curves).

Properties (i) and (ii) take into account the luminance and the geometrical information respectively contained in each DSS element. They were quantified with local measures and used to evaluate the similarity among sketch segments. Property (iii) is a more global parameter, and was used to formulate the coupling strategy.

In order to provide a measure of the similarity between each DSS element couple i, j, a general of cost function model $H_{i,j}$ was defined, as follows:

$$H_{i,j} = \sum_k \alpha_k f_{i,j}^k \qquad (1)$$

where each cost $f_{i,j}^{\wedge}$ (k=1, 2, ...) represents the k-th feature measured on the sketch couple i,j, and is weighted by a parameter α_k. Each cost $f_{i,j}^{\wedge}$ is in the range [0,1], and decreases with the probability of i and j belonging to the same shape. In addition, the weights must fulfill the normalization condition so that $H_{i,j}$ is in [0,1]. The specific implementation of the cost function $H_{i,j}$ for the block recovery problem, includes two terms, which are directly related to the above defined properties. The smoothness of the area encompassed by edges (property (i)) implies that the gray level intensity along each side of the contour has low-pass characteristics. Consequently, given two segments in the DSS, if the intensity of the pixels at the corresponding sides have matching values (the required gray level information is contained in the sketch points) the segments are likely to belong to the same contour. The cost term associated to each pair i, j can be expressed by the following expression:

$$f_{i,j}^1 = \sqrt{\left(m_{l,i} - m_{r,j}\right)^2 + \left(m_{r,i} - m_{l,j}\right)^2} \qquad (2)$$

where $m_{s,i}$ and $m_{r,i}$ represent the mean luminance intensity along the left and right sides respectively of the i-th sketch segment.

Property (ii) concerns the smoothness of the reconstructed edge portion, which can be forced by posing some constraints on the contour shape. The new term introduced to take into account this property is the measure of the total curvature of the recovered edge segment. A simple estimate of this parameter is the measure of the variation of the tangent angle at the two extremes of the segments to be connected. Given two elements i and j of the DSS, ending at the points A and B respectively, the total curvature can be expressed as:

$$f_{i,j}^2 = \left|\delta_i\right| + \left|\delta_j\right| \qquad (3)$$

where δ_i is the angle between the tangent of the edge segment i at the point A and the rectilinear segment AB (see Fig. 2), and analogously for δ_j.

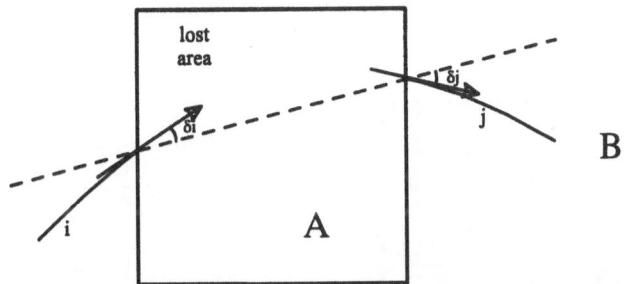

Fig. 2 Computation of the total curvature of the reconstructed edge portion

Since the two measures $f_{i,j}^{1}$ and $f_{i,j}^{2}$ have different ranges and variance, they are normalized to the relevant standard deviations computed inside the DSS.

Unfortunately, the reliability of the two measures within the coupling procedure is affected by the characteristics of the area where the loss occurred. In particular, the analysis of most critical samples revealed that blocks belonging to textured or noisy regions constantly present higher cost values. This situation alters the values in a global sense, but has in general a low impact on the success of the coupling procedure, due to the locality of the operation. Moreover, it was verified that noise typically affects at a greater extent the cost associated to edge geometry, while textures mainly alter the cost related to gray levels.

106

To appropriately take into account this phenomenon, it is necessary to introduce correct weights in the cost function for each edge couple i,j. Depending on the reliability of each cost term, a confidence parameter α_k is introduced to balance its contribution to the cost function $H_{i,j}$. The degree of confidence associated to each cost term is peculiar to the relevant similarity measure, and is defined as follows: the gray level continuity measure $J_{i,j}$ is weighted proportionally to the smoothness (low texture) of the area surrounding the lost block (e.g., the normalized standard deviation of the gray levels around the edge); the curvature measure $J_{i,j}$ is weighted by a measure of the direction variation of the edge segments to be linked (e.g., the normalized variance of the point derivatives along the curve).

It was experimentally found that the introduction of the adaptive weighting parameters allows to notably increase the percentage of correct coupling: on a test set of one thousand samples, randomly chosen on a set of natural images, the percentage of correct coupling was 85% with constant weighting parameters and over 93% with adaptive weights.

Having defined the cost function, the coupling strategy can be easily outlined based on property (iii) of natural edges. Since the abrupt termination of an edge is highly improbable, the coupling procedure consists in linking every edge couple, with the exception of odd DSSs, where one edge remains isolated. In the last case a special treatment is required for the spare edge in the recovery phase. The possible coupling configuration theoretically are in combinatorial number, but only a subset of them represents acceptable solutions (e.g., some configuration implies edge crossings).

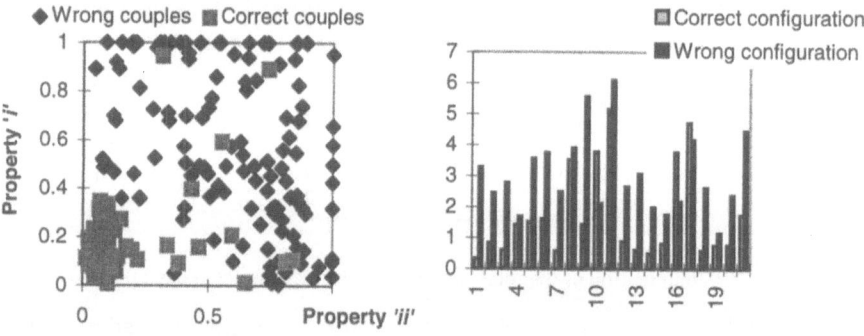

Fig. 3 Left, cost vectors of the edge couples extracted from the test image Maskerade; Right, comparison between correct configuration and least-cost wrong configuration.

Then, the M potential configurations are selected, a global configuration cost H_s^m is computed for each of them, given by the sum of the link costs Π_{ij} included in the m-th configuration. Finally, the minimum cost configuration is selected and used in the recovery phase.

To show the discrimination capability of the two cost terms $J_{i,j}$, $J_{i,j}$, and of the global cost function, Fig. 3 shows the distribution of the values assumed by in the test image "Maskerade", where the loss of 70 randomly distributed blocks of 16×16 pixels was simulated. On the left graph, the cost vectors associated to the correct edge couples are represented differently from the incorrect ones, and form a cluster near the origin of the axes (low cost values) with good compactness and separation properties. On the right chart, the final result achieved for the 70 blocks of the "Maskerade" test image is represented.

3.2 Sketch Reconstruction and Block Recovery

The reconstruction of the sketch inside the lost image area includes the interpolation of the damaged contours, and the prediction of the relevant gray level information at the contour sides.

The recovery of the geometrical contour structure is accomplished by a polynomial interpolation applied to the segment couples defined in the previous step. Furthermore, single segments remaining after the coupling procedure are extended into the missing area. The maximum number of points available for the interpolation is equal to 2×m, where m is the maximum number of edge points considered for each segment (12, in our tests). Since this set of points is overabundant for an efficient interpolation, the edge segments are previously under-sampled, choosing a subset of n points (typically, n=6). An effective solution is to use a cubic spline, which accomplishes the interpolation of successive couples of points by a third degree polynomial, so as to avoid critical oscillations. The spline has the further advantage to pass through the selected points, thus reducing the risk of geometrical distortions.

In mathematical terms, the third order function $y = f(x)$ that passes through the selected $n+1$ points is obtained with n functions defined in each interval as follow:

$$S_i(x) = a_i + b_i x + c_i x^2 + d_i x^3 , \quad x \in \left[x_{i-1}, x_i\right] \tag{4}$$

where $i = 1, 2, \ldots, n$ denotes the interval where $S_i(x)$ has to be applied. These functions are required to be continuous also in the first and second derivatives in the whole domain of $f(x)$, and in particular, in the conjunction points, thus making possible the computation of the coefficients a,b,c,d through the solution of a linear system in $4n$ unknowns.

Once the geometrical information has been recovered, the gray level values associated to the edges should be predicted to complete the reconstruction of the sketch. The two pixel-wide strips along the recovered edge is interpolated by means of a cubic spline, this time applied to the gray level as a function of the contour length. Finally, the smooth areas encompassed by the recovered sketch data mist be reconstructed. To this end, there are two main possibilities: if there are no strict time constraints, a maximally smooth filtering can be applied; otherwise, a faster procedure that gives sufficiently accurate results consists in applying a one-shot interpolation. To each non-contour pixel the weighted mean of the nearest known pixels in the four main directions (N,W,S,E) is assigned. The weight associated to each direction simply depends on the distance between the pixel to be recovered and the nearest available pixel in that direction.

4. Experimental Results and Conclusions

The proposed method has been tested with several pictures and video frames, characterised by different resolutions and complexity. The cell loss was simulated by a regular distribution of lost macro-blocks (16×16 pixel blocks) over the whole image, in order to test the behaviour of the concealment technique applied to areas with different frequency and luminance characteristics. The average loss rate was approximately the 15%. In Fig. 4 the results on the test image 'Maskerade' are shown. The proposed method is also compared to a classical maximally-smooth interpolation in order to show the improvement in the reconstruction of high frequencies (see details).

Test performed on video sequences were applied by only considering I-frame losses. Fig. 5 show the result of the smooth recovering procedure and of the proposed method, respectively, applied to an I-frame of the video sequence "Salesman" with a block loss rate of 15,2%.

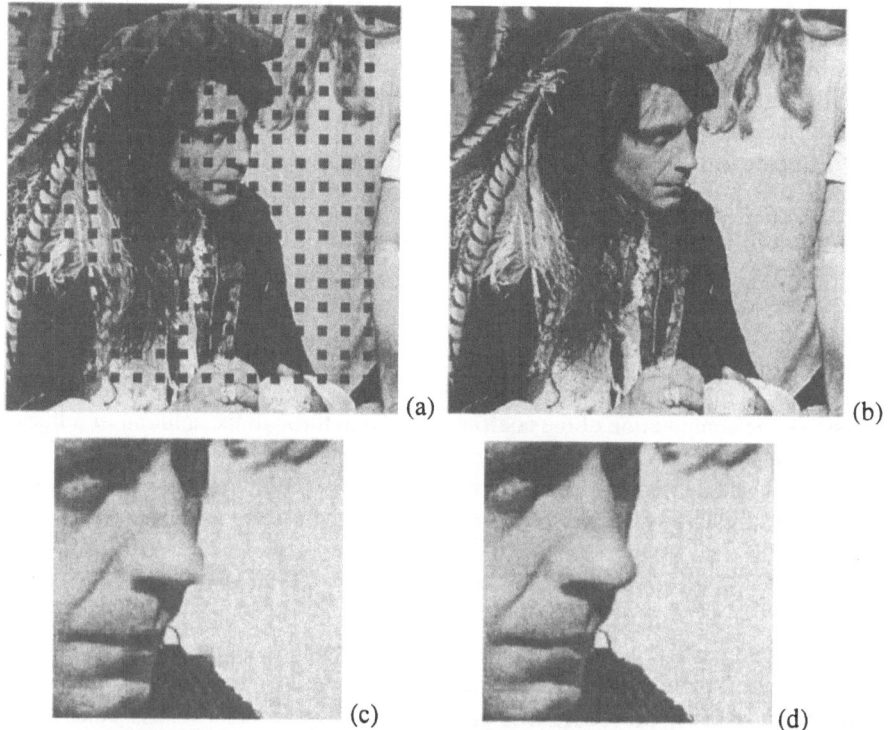

(a) (b) (c) (d)

Fig. 4 Results on "Maskerade" test image: (a) damaged image with 15% block loss, (b) sketch-based concealment, (c-d) detail comparison: smooth interpolation vs. our method

(a)   (b)

Fig. 5 Results on "Salesman" test video frame: (a) damaged with 15,2% block loss, (b) sketch-based concealment

References

1. Hong, T.D.Suda, *Congestion Control and Prevention in ATM Networks,* IEEE Network Magazine, vol. 3, no. 4, pp. 10-16, July 1991.
2. Y. Wang, V. Ramamoorthy, "Image reconstruction from partial subband images and its application in packet video transmission", Signal Processing: Image Communication, vol. 3, nos. 2-3, pp. 197-230, June 1991.
3. S.S. Hemami, R.M. Gray, "Subband-coded image reconstruction for lossy packet networks", IEEE Trans. Image Processing, vol. 6, no. 4, pp. 523-539, April 1997.
4. Y. Wang, Q.F. Zhu, L. Shaw, "Maximally Smooth Image Recovery in Transform Coding", IEEE Trans. Communications, vol. 41, no. 10, pp. 1544-1551, October 1993.
5. M. Ghanbari, "Two layer coding of video signals for VBR networks", IEEE J. Selected Areas on Communication, vol. 7, no. 5, pp. 771-781, June 1989.
6. S. Carlsson, "Sketch based coding of grey level images", Signal Processing, vol. 15, pp. 57-83, July 1988.
7. M. Kunt, A. Ikonomopoulos, M. Kocher, Second-generation image-coding techniques, IEEE Proceedings, April 1985.
8. Y. Itoh, An edge-oriented progressive image coding, IEEE Trans. CAS for Video Technologies, April 1996.
9. J. Canny, "A Computational Approach to Edge Detection", IEEE Trans. on Pattern Analysis and Machine Intelligence, Vol. 8, No. 6, Nov. 1986, pp 679-698.
10. D. W. Jacobs, "Robust and Efficient Detection of Salient Convex Group", IEEE Trans. on Pattern Analysis and Machine Intelligence, Vol. 18, No. 1, Jan. 1996, pp 23-37.
11. R. Mohan, R. Nevatia, "Perceptual organization for scene segmentation and desciption", IEEE Trans. on Pattern Analysis and Machine Intelligence, Vol. 14, No. 6, June 1992, pp 616-635.

An Algorithm for Fast Segmentation of Color Images

L. Lucchese[1,2] and S.K. Mitra[1]

[1]Department of Electrical and Computer Engineering, University of California, Santa Barbara, California
[2] Department of Electronics and Informatics, University of Padova, Italy

Abstract

A new algorithm for real-time segmentation of color images is presented. The algorithm is based on a palettized representation of the image and consists of the following steps: down-sampling the lattice upon which the image pixels are defined, blurring the image through low-pass filtering in order to average colors lying in almost uniform regions, exploiting the well-established and fast methods for color quantization in order to compute color clusters in the RGB space. Computer simulation results of the proposed method are included.

1 Introduction

Color image segmentation has been receiving an increasing attention only in recent years, as color images convey more information than gray-level images and allow one to obtain more meaningful and robust segmentation. In addition, several misclassification problems that may arise in segmenting gray-level image are easily avoided by resorting to color information. However, a color image is a three-component vector signal and increases the computational load compared to that required in the processing of a gray-level image.

Existing algorithms for color image are either suitable extensions of techniques adopted for gray-level image segmentation or *ad hoc* techniques accounting for specific knowledge inherent color [1]. They can be broadly and usefully divided into the following categories: pixel-based, area-based, edge-based and physics-based techniques [1]. Most of these techniques are highly computationally intensive and not appropriate for real-time applications.

In this work we present a supervised pixel-based algorithm for fast segmentation of color images which resorts to palettized format of images, down-sampling, low-pass filtering and color clustering techniques, reducing significantly the computational load in real-time or near real-time applications.

Section 2 introduces the RGB and the palettized representations of color images. Section 3 describes the various phases of the proposed procedure for color image segmentation and presents some simulation results. Section 4 draws the conclusions.

2   Representation of color images

Color images are usually represented in terms of their red, green and blue components (RGB format) for purposes of storage and display. An alternative and often useful way of treating color images is in terms of their palettized representation. The relationship between the RGB and the palettized representations of color images is straightforwardly derived as follows. Let a color image be denoted as a vector function

$$f(n) = \left[f_r(n)\ f_g(n)\ f_b(n) \right]^T \in \mathbb{R}^3, \quad n \in \mathcal{N}, \tag{1}$$

where $f_r(.), f_g(.), f_b(.)$ are, respectively, the red, green and blue components (monochromatic primary colors) of $f(n)$. These functions are defined upon a finite orthogonal 2-D lattice $\mathcal{N}$ of size $N_1 \times N_2$ and take their values respectively within three sets R, G and B according to

$$f_r(.) : \mathcal{N} \longrightarrow R \doteq \{r_1, r_2, \ldots, r_R\} \in [0, 1], \tag{2}$$

$$f_g(.) : \mathcal{N} \longrightarrow G \doteq \{g_1, g_2, \ldots, g_G\} \in [0, 1], \tag{3}$$

$$f_b(.) : \mathcal{N} \longrightarrow B \doteq \{b_1, b_2, \ldots, b_B\} \in [0, 1]. \tag{4}$$

The functions $f_r(n)$, $f_g(n)$ and $f_b(n)$ will be also referred to as matrices, since they will be numerically processed as arrays of the three primary colors. Let us suppose that the elements of R, G and B can combine in at most P ways; this means that the color image $f(n)$ contains P different colors. As an example, Fig. 1(a) shows a color image $f(n)$ of size 303×243 with $P = 256$ colors. The color image $f(n)$ can be also represented in palettized[1] (or indexed) format [2] as $\{p(n), C\}$

$$f(n) \xrightarrow{\mathcal{P}} \{p(n), C\}, \tag{5}$$

$$p(.) \in \{1, \ldots, P\}, \quad C = \left[c_1^T \ldots c_p^T \ldots c_P^T \right]^T = \left[r\ g\ b \right] \in \mathbb{R}^{P \times 3}. \tag{6}$$

Since by hypothesis there are at most P possible combinations of the primary colors, according to Eqs. (5) and (6), each pixel of $f(n)$ is assigned an index $p(.)$ of a suitably defined list C of these P combinations. For convenience, C can be represented as a matrix whose rows $c_p = [r_p\ g_p\ b_p], r_p \in R, g_p \in G, b_p \in B, p = 1, \ldots, P$, are the P triplets of primary colors giving the whole set of P colors belonging to the image.

The inverse operation of de-indexing of a color image, which converts its palettized representation $\{p(n), C\}$ back to the RGB format $f(n)$, can be expressed as

$$\{p(n), C\} \xrightarrow{\mathcal{P}^{-1}} f(n). \tag{7}$$

---

[1] Throughout this work, the vectors c's in the color palette matrices are referred to as row-vectors and their transposition is adopted only for compactness' sake. Their red, green and blue components r, g and b are instead regarded, within the palette matrices, as column-vectors.

The use of the palettized representation allows one to perform processing of color information, without taking care of spatial information, by working with a single matrix C of size $P \times 3$ instead of three matrices $f_r(n)$, $f_g(n)$ and $f_b(n)$ of size $N_1 \times N_2$, where usually $P << N_1 \times N_2$. In the case of the image of Fig. 1(a), this means working with $256 \times 3 = 768$ entries instead of $400 \times 600 = 7.2 \times 10^5$.

3 Segmentation Algorithm

The segmentation algorithm presented in this work consists of four steps: down-sampling, low-pass filtering, color quantization and color matching in the $L^*u^*v^*$ color space. There is a fifth optional post-processing step represented by a median filtering. They are examined in details in the following sections.

3.1 Down-sampling

A filtering of color images needs to be performed on each of the three color components and is thus computationally very intensive. For example, for filtering a color image of size $N_1 \times N_2$ with a FIR filter with squared support mask $F \times F$, without taking into account possible optimizations, requires about $3N_1N_2F^2$ multiplications (and a similar number of additions). However, if this image is down-sampled by a factor of M in both horizontal and vertical directions for each color component, the number of required multiplications decreases to $3\frac{N_1N_2}{M^2}F^2$ (and similarly for the number of needed additions). If the regions of the color image to be segmented are large enough to retain most of the color information also after down-sampling, i.e. there is a sufficient spatial adjacency among pixels within regions having "similar" colors[2], down-sampling can reduce considerably the computational burden for filtering of color images.

The down-sampling of the original color image $f(n)$ of size $N_1 \times N_2$ onto a sublattice $\mathcal{S}_{M \times M}$ of $\mathcal{N}$, by taking one pixel every M in both horizontal and vertical directions, results in

$$y(s) \doteq f(n)|_{n=s} = \left[\, f_r(s)\ f_g(s)\ f_b(s)\,\right]^T, \quad s \in \mathcal{S}_{M \times M} = sublattice(\mathcal{N}). \quad (8)$$

This reduces the number of pixels from N_1N_2 to about $\frac{N_1N_2}{M^2}$.[3] Note that the down-sampling of the color image can be readily accomplished in the palettized format but the subsequent filtering operation needs spatial information so that the RGB and the palettizzed formats are computationally equivalent.

---

[2] The delicate concept of color similarity can be mathematically formulated only in an appropriate uniform color space [3, 4].

[3] If the down-sampling rate M does not divide exactly both N_1 and N_2, the final number of sampled pixels is slightly less than $\frac{N_1N_2}{M^2}$.

3.2 Low-pass filtering

Each one of the three color components of the RGB representation of $y(\mathbf{s})$ is then low-pass filtered by means of the Gaussian FIR filter

$$
h(\mathbf{s}) = \begin{cases} \dfrac{e^{-\frac{1}{2\sigma^2}\mathbf{s}\cdot\mathbf{s}}}{\displaystyle\sum_{\mathbf{s}\in mask(S)} e^{-\frac{1}{2\sigma^2}\mathbf{s}\cdot\mathbf{s}}}, & \text{if } \mathbf{s} \in mask(S) \subset \mathcal{S}_{M\times M}, \\ 0, & \text{elsewhere,} \end{cases} \tag{9}
$$

where

$$
mask(S) \doteq [-S,\dots,0,\dots,S] \times [-S,\dots,0,\dots,S], \tag{10}
$$

which gives the blurred image

$$
g(\mathbf{s}) = \begin{bmatrix} g_r(\mathbf{s}) \\ g_g(\mathbf{s}) \\ g_b(\mathbf{s}) \end{bmatrix} = \begin{bmatrix} f_r(\mathbf{s}) * h(\mathbf{s}) \\ f_g(\mathbf{s}) * h(\mathbf{s}) \\ f_b(\mathbf{s}) * h(\mathbf{s}) \end{bmatrix}. \tag{11}
$$

The number of pixels averaged according the Gaussian mask through the FIR filtering is therefore $F^2 = (2S+1)^2$; the larger the value of S, the higher the computational complexity. The low-pass filtering aims to smooth colors within quasi-uniform regions; this means that values too large for S could undesirably mix colors of neighboring small regions. Nonetheless, this can be prevented by trimming the aperture σ of the Gaussian filter to small values in order to weigh the colors of the pixel under processing and its closest neighbors much more than the farthest ones. Moreover, the parameter S has to be chosen by taking into account also the down-sampling factor M, since the down-sampling makes less heavy the filtering computation but eliminates a considerable number of pixels belonging to the same quasi-uniform regions. Therefore there exists a trade-off among the choices of the three parameters S, σ and M, and they do hinge on the particular image to be segmented. The image of Fig. 1(a) has been down-sampled according to Eq. (8) with $M = 4$ and filtered according to Eqs. (9)-(11) with $S = 4$ and $\sigma = 0.4$ yielding the image $g(\mathbf{s})$ shown in Fig. 1(b). Similarly to Eqs. (5)-(6), the image $g(\mathbf{s})$ is expressed in palettized form as $\{t(\mathbf{n}), C'\}$ through

$$
g(\mathbf{s}) \xrightarrow{\mathcal{P}} \{t(\mathbf{n}), C'\}, \tag{12}
$$

where

$$
t(.) \in \{1,\dots,T\}, C' = \begin{bmatrix} c_1'^T & \cdots & c_t'^T & \cdots & c_T'^T \end{bmatrix}^T = \begin{bmatrix} r' & g' & b' \end{bmatrix} \in \mathbb{R}^{T\times 3}. \tag{13}
$$

The color palette C' of Eqs. (12)-(13) contains many more colors than the original color palette C'; in fact, although the down-sampling operation considerably reduces the number of pixels and therefore the number of colors of the original image, the Gaussian filtering replaces the color at any pixel with the weighted sum of the colors of a set of pixels centered around it. Even though this process, in general, enlarges the palette creating $T >> P$ new colors, it benefits the segmentation task in two ways.

First, the blurring effect due to the low-pass filtering compensates by smoothing for differences that may normally arise among pixels belonging to the same colored patch. Second, the blurring populates the RGB space with new colors which accumulate towards the clusters representing the main colored regions in the image to segment; this roughly allows for the spatial interaction of colors. Nevertheless, the blurring introduces also false and spurious colors nearby the edges between different regions; but they are statistically considerably less than the colors belonging to the approximately uniform colored regions and they can be easily absorbed in the following color quantization procedure.

3.3   Color quantization

The color image quantization problem consists of reducing the color palette of a given image to a smaller one and in assigning each pixel of the original image to the new palette. The aim is that of designing a color palette allowing almost no perceivable differences between the original image and the quantized one.

The segmentation goal is quite different since it requires to partition an image into connected regions featuring almost homogeneous colors. This may be pursued by adopting an appropriate uniform color space, by seeking in it certain clusters corresponding to the main colored regions in the image and by mapping all the pixels into the set of colors related to these clusters [5]. The mapping operation is accomplished by minimizing the distances, usually according to an Euclidean metric that can be associated to such uniform spaces, among the color of each pixel and those characterizing in some way (e.g. center of mass or mean value) the clusters.

One may think to exploit the techniques of color quantization in order to find out these clusters in the color space; indeed, if the final number of colors is considerably smaller than the initial one, color quantization can be easily thought of as a rough segmentation.

However, if color quantization algorithms are applied to images *sic et simpliciter* without any further provision, the outcoming segmentation performs rather poorly. The low-pass Gaussian filtering preceding color quantization plays a key role in overcoming this poor segmentation together with the subsequent color matching procedure performed in the CIELUV space.

The technique adopted here for color quantization is later briefly examined. Quantization of colors is generally performed in the RGB color space since it has been proven that no particular advantage is gained by using other color spaces [5]. Among the most popular quantization algorithms there are the splitting algorithms [5, 6] that divide the RGB cube into disjoint regions and select a representative color from each region as a new palette color;

they differ in the methods adopted to split the color space. The minimum variance quantization method proposed by Heckbert [5] cuts the color cube in the red, green and blue directions until a chosen number of non-empty regions is obtained; it then uses the average color in each region to create the new reduced palette.

The increased palette C' comprises a large number of colors; but, as highlighted above, they are more clustered than the colors of the original palette C around certain accumulation points in the RGB space related to the main colored patches in the image. This fact allows an easy color quantization of the RGB cube.

Therefore the $T \gg P$ colors in C' are grouped into $Q < P$ clusters according to the minimum variance quantization method [5] which yields a new color palette

$$C'' = \left[c_1''^T \ldots c_q''^T \ldots c_Q''^T \right]^T = \left[r'' \ g'' \ b'' \right] \in \mathbb{R}^{Q \times 3}. \qquad (14)$$

Of course, the number of quantized colors Q (or clusters in the RGB space) defines the number of phases in the final segmented image so that a smaller value of Q results in a coarser segmentation. The parameter Q is therefore chosen by the user according to the desired segmentation resolution.

3.4 Color matching in the $L^*u^*v^*$ space

The RGB space is hardware-oriented and is commonly used for representing color images in acquisition and display devices; moreover, as just pointed out, the RGB space is the best one for color quantization. But unfortunately it is not perceptually uniform [4]. In it, distances among colors do not correspond to analogous perceptual differences among them. Several perceptual uniform spaces have been defined where the perceived color differences are measured by Euclidean distances [3, 4]. The $L^*u^*v^*$ space (or CIELUV for easy reference) has been chosen for our segmentation task.

The two color palettes C and C'' are represented in the CIELUV color space according to the following steps. For brevity's sake, let us consider only C. The RGB coordinates of each color c_p of C must be first converted into the XYZ color system [3, 4] as

$$\begin{bmatrix} X_p \\ Y_p \\ Z_p \end{bmatrix} = \begin{bmatrix} 0.490 & 0.310 & 0.200 \\ 0.177 & 0.813 & 0.011 \\ 0.000 & 0.010 & 0.990 \end{bmatrix} c_p^T, \quad p = 1, \ldots, P. \qquad (15)$$

From X_p, Y_P and Z_p the chromaticity coordinates (u_p, v_p), $p = 1, \ldots, P$, are computed as

$$\begin{cases} u_p = \frac{4X_p}{X_p + 15Y_p + 3Z_p} \\ v_p = \frac{9Y_p}{X_p + 15Y_p + 3Z_p} \end{cases} \qquad (16)$$

Finally, the coordinates (L_p^*, u_p^*, v_p^*), $p = 1, \ldots, P$, in the CIELUV space are obtained as

$$
L_p^* = \begin{cases} 116 \left(\frac{Y_p}{Y_n}\right)^3 - 16, & \text{for } \frac{Y_p}{Y_n} > 0.008856, \\ 903.3 \left(\frac{Y_p}{Y_n}\right), & \text{for } \frac{Y_p}{Y_n} \le 0.008856, \end{cases} \qquad \begin{aligned} u_p^* &= 13 L_p^*(u_p - u_n), \\ v_p^* &= 13 L_p^*(v_p - v_n). \end{aligned} \qquad (17)
$$

Notice that Y_n, u_n and v_n are, respectively, the values of Y, u and v for a suitably chosen reference white[4]. The P row-vectors $l_p \doteq [L_p^* \ u_p^* \ v_p^*]$ are ordered into the matrix L as

$$
L = \left[l_1^T \ \ldots \ l_p^T \ \ldots \ l_P^T \right]^T = \left[L^* \ u^* \ v^* \right] \in \mathbb{R}^{P \times 3}. \qquad (18)
$$

An analogous matrix L'' is built out of the colormap C'' as

$$
L'' = \left[l_1''^T \ \ldots \ l_q''^T \ \ldots \ l_Q''^T \right]^T = \left[L^{*''} \ u^{*''} \ v^{*''} \right] \in \mathbb{R}^{Q \times 3}. \qquad (19)
$$

The P colors of C are matched with the Q colors of C'' by minimizing the Euclidean distance between their representations in the CIELUV space, i.e. between each row of L and each row L'', according to the following procedure:

$$
C''' = \left[c_1'''^T \ \ldots \ c_p'''^T \ \ldots \ c_P'''^T \right]^T := 0_{P \times 3}
$$
for $p = 1$ to $p = P$
    $e = [e_1 \ldots e_q \ldots e_Q]^T := [0 \ldots 0 \ldots 0]^T$
    for $q = 1$ to $q = Q$
        $e_q = \|l_p - l_q''\|_2 = \sqrt{(L_p^* - L_q^{*''})^2 + (u_p^* - u_q^{*''})^2 + (v_p^* - v_q^{*''})^2}$
    end
    $q_m = \arg\min_q \{e\}$
    $c_p''' = c_{q_m}''$
end

3.5 Post-processing and discussion of results

The color matching procedure of the previous section returns a colormap $C''' \in \mathbb{R}^{P \times 3}$ with the same size as C but the color set of C''; this redundancy is avoided through a simple procedure that reduces the colors in C''' to a minimal set yielding the final palette $D \in \mathbb{R}^{Q \times 3}$ as

$$
D = reduce\{C'''\}, \ D = \left[d_1^T \ \ldots \ d_q^T \ \ldots \ d_Q^T \right]^T = \left[r_s \ g_s \ b_s \right] \in \mathbb{R}^{Q \times 3}. \qquad (20)
$$

Therefore, the segmented image in indexed form is given by

$$
\{z(n), D\} = re\text{-}index\{p(n), D\}. \qquad (21)
$$

---

[4] Throughout this work the values adopted are $X_n = Y_n = Z_n = 1$, whence $u_n = \frac{4}{19}$ and $v_n = \frac{9}{19}$.

The two operations expressed by Eqs. (20) and (21) have been separated for convenience but as a matter of fact they are performed at the same time, since the merging of the rows of C''' with the same colors implies an immediate re-arranging of the references to them in the matrix $p(n)$. It is worth observing though that the segmentation is carried out through Eqs. (20) and (21) by simply re-arranging the indexing relationship (re-indexing) between the original indexed image $p(n)$ and the new color palette D.

Fig. 2(a) shows the segmented image $\{z(n), D\}$ obtained from the original image of Fig. 1(a) with $Q = 3$ colors. It can be seen that there are some ouliers within uniform regions. In order to remove these outliers, a "colorizing procedure" is applied, as suggested in [7]. This processing is accomplished by means of a median filter [8]; in [7] the median filtering, a non-linear method with heavy computational load, is applied to each one of the three components of the color space. However, the palettized representation allows one to apply the median filtering directly to the indexed representation $\{z(n), D\}$ of the segmented image as

$$m(n) = median\{z(n - k), k \in \mathcal{W}\}, \tag{22}$$

where $\mathcal{W}$ is a suitably chosen window. For our purposes $\mathcal{W}$ has been chosen as a 3×3 window, since larger masks would require more computations and would excessively distort the image. The final result of the segmentation procedure is the trichromatic vector function $s(n)$ obtained through the de-indexing of $\{m(n), D\} \xrightarrow{\mathcal{P}^{-1}} s(n)$ and expressed as

$$s(n) = \left[s_r(n)\ s_g(n)\ s_b(n) \right]^T \in \mathbb{R}^3, \quad n \in \mathcal{N}. \tag{23}$$

Fig. 2(b) shows the final result of the segmentation procedure after applying the median filter. It may be seen that the smallest sets of pixels of Fig. 2(a) have been filtered out and that smoother regions have been obtained.

Since the median filtering is the most expensive phase in the whole segmentation procedure and since, as just explained above, it smoothes small-scale features, it may be skipped if there is no particular need for smoothness and almost real-time performance is requested.

As a conclusive remark, we have to say that most of the color segmentation techniques avalaible in the literature lack of an analysis of their computational complexity so that testing their actual performances for a comparison would imply to implement them. Moreover, this is made even more impractical by the fact that almost all these techniques depend on parameters, constants, thresholds which are usually fixed on the basis of a few experiments [1].

It is also worth observing that in many applications fast (or even real-time) segmentation algorithms are not required and a great many segmentation techniques do not care about computational speed. This was instead the primary aim of the proposed method which has been proven to segment images in a very short time, depending of course on their dimensions and on a small set of algorithmic parameters.

4 Conclusions

We have presented a pixel-based technique for color image segmentation. The main purpose of our work was cutting down the computational complexity for real-time or near real-time implementation. This motivated the incorporation of the following steps in the proposed procedure: using palettized formats for representing color images, down-sampling the images, accounting for spatial color interactions through low-pass filtering and clustering colors in the *RGB* space by taking advantage of an algorithm in use for color quantization. Besides the *RGB* color space for quantization, the CIELUV uniform color space has been adopted for matching colors.

It has been highlighted how the segmentation is finally performed by simply rearranging the references to the new color palette of the original indexed image. The use of a median filtering has been proposed for smoothing the segmented image and removing outliers while preserving spatial resolution. However, this non-linear operation is the most time consuming of the whole algorithm and it can be easily avoided in strictly real-time applications. The effectiveness of the proposed algorithm has been tested on several images and some results have been included.

There are some algorithmic parameters, such as down-sampling rate, Gaussian filter support size and aperture, number of desired colors/segments in the segmented image and median filter support size, that need to be tuned on the particular image under segmentation, even though most of them suit a large number of images. Current work aims to develop automatic setting of these parameters towards unsupervised color image segmentation.

Acknowledgements

This work was supported in part by the University of Padua and the "A. Gini" Foundation, Italy, in part by a University of California MICRO grant with matching support from Xerox Corp., and in part by the Alexandria Digital Library project at the University of California, Santa Barbara under NSF grant #IRI-94-1130.

References

1. W. Skarbek and A. Koschan, "Colour Image Segmentation - A Survey," Technical Report 94-32, Technical University Berlin, October 1994.
2. R. Balasubramanian, J. Allebach, "A New Approach to Palette Selection for Color Images," *Journal of Imaging Technology*, Vol.17, No.6, pp.284-297, Dec. 1991.
3. G. Wyszecki and W.S. Stiles, *Color Science: Concepts and Methods, Quantitative Data and Formulae*, Wiley: New York, 1982.
4. R.W. Hunt, *Measuring Color*, Ellis Horwood Ltd. Publ.: Chichester, England, 1987.

5. P.S. Heckbert, "Color Image Quantization for Frame Buffer Display," *Computer Graphics*, Vol.16, No.3, pp.297-303, 1982.

6. M.T. Orchard and C. Bouman, "Color Quantization of Images," *IEEE Trans. on Signal Processing*, Vol.35, No.12, pp.2677-2690, Dec. 1991.

7. J.R. Smith and S.F. Chang, "Single Color Extraction and Image Query," *Proc. of ICIP'95*, Washington, D.C., Vol.III, pp.528-531.

8. A.K. Jain, *Fundamentals of Digital Image Processing*, Prentice-Hall: Englewood Cliffs, New Jersey, 1989.

(a)   (b)

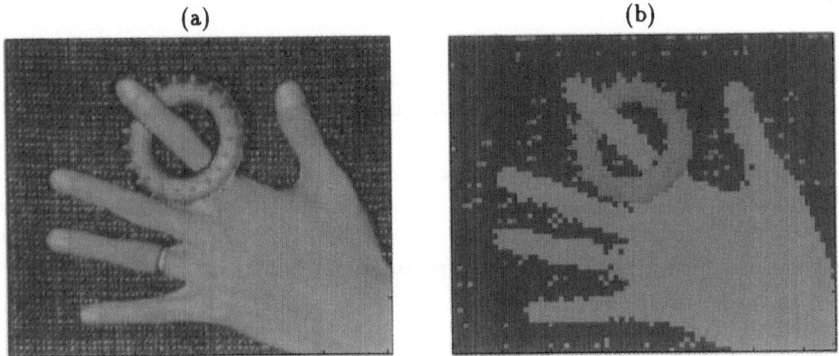

Fig. 1. (a) Original color image $f(n)$ (303×243 pixels and $P = 256$ colors); (b) down-sampled ($M = 4$) and low-pass filtered ($S = 4$, $\sigma = 0.4$) version $g(s)$ of $f(n)$.

(a)   (b)

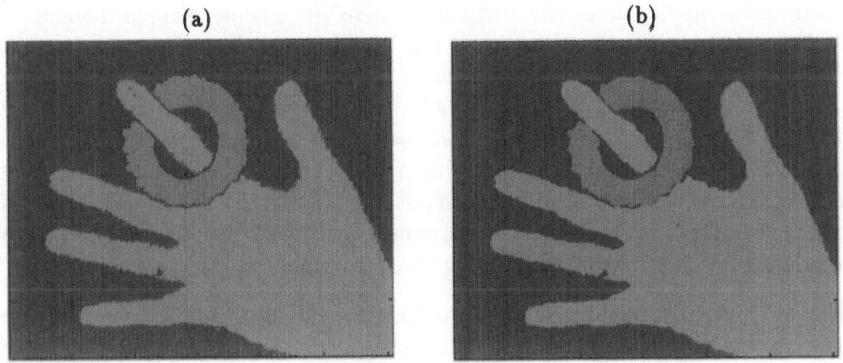

Fig. 2. (a) Segmentation of the image of Fig. 1(a) with $Q = 3$ colors; (b) segmented image after median filtering.

Multi-Camera Strategies for High-Accuracy 3D Reconstruction

Federico Pedersini, Augusto Sarti and Stefano Tubaro

Dip. di Elettronica e Informazione (DEI), Politecnico di Milano
Piazza L. Da Vinci 32, 20133 Milano, Italy
E-mail: pedersin/sarti/tubaro@elet.polimi.it

Abstract

In this paper we present our global approach to accurate 3D reconstruction of a scene using a calibrated multi-camera system. In particular, we illustrate a simple and effective adaptive technique for the self-calibration of CCD-based multi-camera acquisition systems. We also propose a general and robust approach to the problem of close-range partial 3D reconstruction of objects from stereo-correspondences. Finally, we introduce a method for performing an accurate patchworking of the partial reconstructions, based on 3D feature matching.

1. Introduction

The growing interest of the telecommunication's market in multi-media technology has determined in the past few years, an incredibly fast proliferation of methods for the 3D reconstruction of objects from the analysis of camera images. Most of these techniques are aimed at the problem of *Content Creation* for a wide range of multi-media applications. In particular, there is a considerable number of applications in which the accuracy of the 3D reconstruction plays a crucial role. For example, applications of close-range digital photogrammetry aimed at the study and the preservation and restoration of 3D works of art require effective methods for accurate, quantitative, reproducible and repeatable 3D reconstruction. In this case, in fact, suitable 3D modeling methods should be sufficiently accurate as to match the performance of the methods that are commonly adopted for the 3D relief of works of art; and to guarantee that such measurements will be reproducible and can be repeated along time for monitoring purposes.

The most popular non-invasive approaches to 3D reconstruction of mid-sized unstructured objects are based on stereo-correspondences. Such methods consist of the detection and the matching of image features (e.g. points, edges) [1] or luminance profiles [3]. When the camera parameters (position, orientation and other intrinsic physical parameters) are known (*calibrated* case), the process of determining the correspondences is helped by some rigidity constraints such as the coplanarity of corresponding visual rays (epipolar constraint), and the 3D coordinates of the features can be determined through geometric triangulation [1,2]. When, on the contrary, the camera parameters are not known (*uncalibrated* analysis), the determination of the feature correspondences becomes more difficult. In this case, in fact, the determination of the correspondences becomes a statistical matching process based on heuristic rules, while the epipolar constraints is now used for the joint estimation of camera pose and 3D coordinates of the features. As we are interested in high-accuracy applications of 3D reconstruction, we focus on calibrated methods.

The problem of matching ambiguity is present in all correspondence-based methods (feature, and area matching) that use an uncalibrated approach or a calibrated pair of cameras. In order to overcome this difficulty, we adopt a calibrated set of three or more cameras mounted on a rigid frame. This way we can exploit the multi-ocular invariance [1], according to which each one of three corresponding points in three views is bound to lie on the intersection of the epipolar lines (i.e. the views of the optical rays) associated to the other points. This is, in fact, quite a strong constraint and can be used for all calibrated feature matching methods.

All calibrated 3D reconstruction methods are critically dependent on the accuracy with which the camera parameters are known. In the past few years several approaches to the calibration problem have been proposed [9,10,11]. The camera characteristics are computed through a proper processing of the image of a test object (calibration target-frame) placed in the scene. What we developed is an advanced photogrammetric method that jointly estimates the camera parameters and the geometry of the calibration target-set in a more accurate fashion (*self-calibration*). This method is based on a *multi-camera, multi-view* calibration approach, and performs an accurate self-calibration on the multi-camera system from the analysis of several views of a simpler calibration target-frame, such as a marked planar surface (a printed sheet of paper glued on a glass surface) or some other even simpler structure. In fact, not only is this technique able to estimate the camera parameters, but it can also determine the 3D position of the targets on the calibration frame, which can be just roughly known or not known at all. Finally, we developed method for making the calibration robust against the inevitable parameter drift that takes place during the acquisition process. Such method detects and tracks some "safe" features that are naturally present in the scene, and use their image coordinates for making the calibration process adaptive.

Automatic 3D reconstruction systems based on stereo-matching can only reconstruct the *visible* portion of surface. Such systems, in fact, typically provide a description of just the *front side* of the imaged scene or, when the surface is too large to fit simultaneously in all views, of just a limited portion of it. In conclusion, in order to obtain a complete scene reconstruction through stereometry, it is necessary to observe the scene from several significant viewpoints and put together the final reconstruction like a *patchwork* of partial reconstructions. In order to be able to merge 3D data coming from different reconstructions, we need to accurately estimate the rigid motion that the acquisition system undergoes between two partial reconstructions. In order to do so, we perform detection and tracking of some image features throughout the acquisition process, and use the location of such features for estimating the camera motion. Scene features that can be quite safely detected are luminance edges [6].

122

These features are more likely to be naturally present in the scene and rather easy to detect, which makes them good candidate features for egomotion estimation. Our approach to egomotion estimation is based on the analysis of 3D contours in the imaged scene. Being the method based on a calibrated multi-ocular camera system [9,11], the estimation is performed entirely in the 3D space. In fact, all edges of each one of the multi-views are previously localized, matched and back-projected onto the object space [12]. Roughly speaking, the method searches for the rigid motion that best merges the sets of 3D edges that are extracted from each one of the multiple views.

In this paper we present the above solutions of ours to the calibrated reconstruction approach through adaptive self-calibration, local stereo-matching approach and global patchworking, with the goal of obtaining a high-accuracy reconstruction of unstructured 3D objects.

2. Calibration

Camera calibration is usually carried out through the analysis of the views of a test object (calibration *target-set*), which usually consists of a set of *fiducial marks* (*targets*), positioned within the 3D volume that is being imaged by the camera system. If the geometrical characteristics of this target-set are only partially known or not known at all, then the calibration process must include the refinement or the blind estimation of the 3D coordinates of the targets. What we developed is an advanced calibration method that jointly estimates the parameters of the multi-camera acquisition system and the geometry of the calibration target-set in an accurate fashion (*self-calibration*). The method is based on the analysis of several views of a simple calibration target-frame. The method is robust against the inevitable parameter drift that takes place during the acquisition process. Such method, in fact, detects and tracks some features that are naturally present in the scene, and use their image coordinates for making the calibration process adaptive.

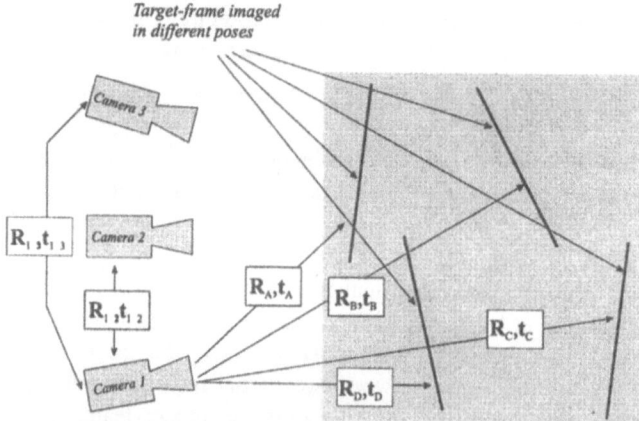

Fig. 1. General scheme for the multi-view multi-camera approach to self-calibration.

Calibration strategy - The calibration target-set that we use for calibration is planar as the pixel size is assumed known [9]. A planar target-set is much simpler to build with respect to a 3D target-frame as it can be easily constructed, for example, by gluing a laser-printed sheet of paper on a rigid planar surface. This procedure also gives us some *a-priori* information on the coordinates of the targets (and their uncertainty), relative to a frame attached to the surface. A 3D calibration target-frame, on the other hand, would require an accurate 3D measurement of the coordinates of the targets (generally through photogrammetric techniques [11]).

The main drawback of 2D target-sets is the fact that they can only occupy a rather limited volume of the 3D scene. It is well-known, in fact, that a reliable camera calibration can only be performed if the targets are not only numerous enough, but also well-distributed in the 3D space that will later be occupied by the object to be measured. In order to overcome this limitation, we proceed by virtually *enlarging* the planar target-set through the acquisition of several of its views (see Fig. 1). The poses of the target-frame are chosen in such a way that the union of all targets will fill-up the volume of interest in a rather uniform fashion. This strategy, quite clearly, modifies the calibration problem as the relative motion that the target undergoes between acquisitions is not known and needs to be *a-posteriori* determined. In order to do so, the position and the orientation of the target-set (relative to the world reference frame) will be added to the model parameters that need to be estimated for each pose of the target-frame. Quite clearly, the approach requires at least 2 views for the self-calibration problem to admit solution. An example of application of our self-calibration approach is reported in Fig. 2.

Adaptive calibration - In order to extract 3D information from the scene views the *camera parameters* must be known with good accuracy throughout the whole

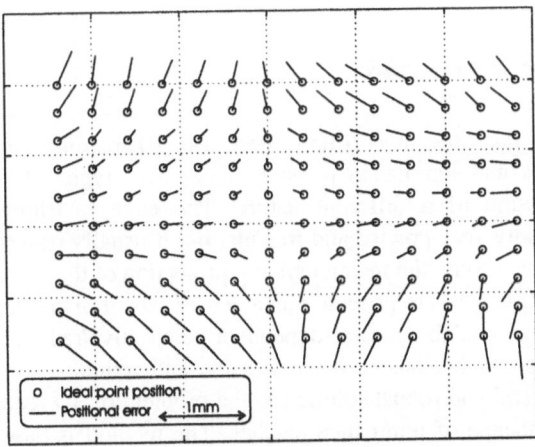

Fig. 2. *A-priori* coordinates of the fiducial points of the target-set (laser-printed circles on a sheet of A4 paper, glued to a flat surface) and corresponding *a-posteriori* corrections estimated through self-calibration. The orientation of the (magnified) correction vectors denotes the deformation of the sheet of paper due to the action of the dragging mechanism of the laser printer.

acquisition campaign. As *camera calibration* is performed before the beginning of an acquisition session, problems of parameter drift may occur. In fact, when long video sequences are acquired, the stability of the camera parameters measured at the beginning becomes a crucial problem as mechanical shocks, vibrations or thermal effects on cameras and supports, can cause small variations of the initial camera set-up. This drift of the camera parameters leads to significant 3D reconstruction errors, as the 3D back-projection is rather ill-conditioned with respect to the camera parameters. In order to overcome this problem, we detect and track any changes in the acquisition system and, whenever possible, we apply an on-the-fly correction of the camera parameters. By doing so, the calibration holds accurate throughout the acquisition campaign.

Our approach does not need any targets to be placed in the scene or any *a-priori* knowledge, but exploits luminance features that are already present in the scene (e.g. corners and spots) which can be located in the image with high precision. After the localization process, which is performed with sub-pixel accuracy, a matching operation is performed among the n sets (n being the number of cameras) of feature points, which returns a set of n-tuples of homologous points. The matched n-tuples will be then back-projected into the 3D scene space. If the camera parameters change, then the back-projection will be affected by larger errors, with respect to the predicted pre-calibration accuracy. A proper analysis of the magnitude and the temporal changes of the back-projection error allows us to reveal and characterize any incidental modifications of the camera parameters. Furthermore, if the set of matched n-tuples is informative enough, the proposed technique allows to accurately measure the occurred modification and, therefore, to re-calibrate the system.

3. Partial Reconstruction

Image features that are most often used for 3D reconstruction are points, luminance edges and luminance patches. These two types of features tend to provide information of a different nature. The edge matching/backprojection process is generally very precise and reliable, but it usually results in a sparse set of 3D points. Conversely, the matching/back-projection of the luminance profile of small image patches tends to provide much denser sets of 3D points but it is rather sensitive to the unavoidable viewer-dependent perspective/radiometric distortions, therefore this approach tends to be less stable and reliable. For this reason we developed a general and robust solution to the problem of 3D reconstruction from stereo correspondence of luminance patches. The method is largely independent on the camera geometry, and employs a calibrated set of three or more standard TV-resolution CCD cameras, which provides enough redundancy for removing possible matching ambiguities. The robustness of the approach can also be attributed to the *physicality* of the matching process, which is actually performed in the 3D space rather than on the image plane. In order to do so, besides the 3D location of the surface patches, it estimates their local orientation in 3D space as well, so that the geometric distortion of the luminance patch can be included in the model. Finally, the method takes into account the viewer-dependent radiometric distortion.

Edge-based approach - As a preliminary step we perform partial reconstruction from edge matching, in order to obtain reliable and accurate 3D data to begin with. Furthermore we can use the same type of features for egomotion estimation as well. In fact, partial reconstruction is based on 2D edge matching (stereo correspondence on the image planes), while motion estimation is based on 3D contour matching (edge correspondence in object space). It is important to emphasize the fact that, in order to be able to use edges for accurate egomotion estimation, we need them to be detected with great accuracy. We do this by first using a traditional edge detector, we then retrieve the subpixel location of the edge points through an interpolation process which takes the luminance gradient into account. Finally, a rule-based contour tracking method is employed for determining the correct connection between edge points.

The search for homologous edges on different views is performed along *epipolar lines*. Notice that using more than two cameras allows us to avoid problems of matching ambiguity. For example, with three cameras, not only can we always select the best pair of views for a specific stereo-correspondence (sharp intersection between edge and epipolar lines), but we can validate the matching through a check on the third view. In fact, the edge point must lie on the intersection of the two epipolar lines associated to the homologous edge points on the other views. Once the stereo correspondences are found, each set of corresponding contours is back-projected onto the 3D scene space by looking for the point at minimum distance from the three homologous visual rays.

Area-based approach - The luminance patches used by most area-matching techniques are normally assumed to have the same shape in all views. It is quite clear, however, that this hypothesis is acceptable only when the angles between the viewing directions of the three cameras are not too wide, which is not our case. As a consequence, we need to take into account the perspective distortion of the shape of the patch, when back-projected onto the object surface and then re-projected onto the other image plane. In order to do so, we assume the 3D surface to be locally flat, which means that it can be approximated by a plane within the back-projected surface patch.

In the other view we search, along the epipolar line, for the patch that best matches the first one. The projective distortion of the patch is accounted for by estimating, together with the position of the patch, the normal to the object surface according to which the shape and texture of the patch are most likely to warped. In practice, the minimum of a *similarity function* between a patch of the actual image and a re-projected patch after perspective warping is searched for as a function of position and local orientation of the tangent plane of the object surface. As far as the radiometric distortion is concerned, an additional pair of variables (luminance *offset* and *gain*) is included in the similarity function. If a reference patch produces reliable 3D information, then it can be used for 3D surface reconstruction. Once all reference regions have been considered, surface interpolation is carried out and the area matching process can start over with a smaller patch size. In this case the previously estimated surface can be used for initializing the search in the next step and speeding up the process.

Some experiments of 3D scene reconstruction have been carried out on several test scenes. The first test presented in this paper concerns the measurement of the accuracy of the area matching using a flat object placed at about 1.2 m of distance from the camera system. The surface reconstruction resulted to be flat with 0.1 mm of standard deviation. A second experiment concerned the 3D reconstruction of a tele-conferencing scene. The acquisition was made with a trinocular camera system at CCETT, France, within the ACTS "PANORAMA" Project. No initial reconstruction was used for area matching. Instead, progressive-scan initialization was performed. As we can see, the quality of the reconstruction greatly benefits from the fact that the geometric distortion is included in the model.

4. Egomotion through Line Matching

The egomotion estimation method that we developed and implemented for accurate patchworking purposes is organized in two mains steps. After having partitioned the available 3D contours in lines and curves, we proceed as follows:

1. **rough egomotion estimation** from straight contours based on the matching of straight contours, followed by motion estimation through minimization of the distance between homologous contours
2. **egomotion refinement** using curved contours based on the matching of curved contours followed by a motion estimation through a minimization of the distance between homologous curved contours

Notice that, as a first approximation of the egomotion is already available, the matching of curved contours is a rather simple operation compared with the matching of straight lines.

Egomotion from Straight Lines - Line matching in 3D space is performed through a hypothesize-and-test type of procedure [17]. The first step of this method consists of formulating hypotheses on the possible couplings by selecting all those that do not violate some rules of congruence based on a set of geometrical constraints. By doing so, we drastically reduce the search space over which to test for matching correctness. At this point we can proceed with an exhaustive search through the above reduced set of hypotheses and select the match that maximizes an appropriate measurement of the matching quality.

Once the matching process is complete, the egomotion estimation can be performed rather easily by searching for the rigid motion that minimizes an appropriate *merging cost* function between two sets of 3D lines that pertain two different partial reconstructions. Notice 3D contours are generally reconstructed as chains of segments whose length and fragmentation may vary quite drastically from multi-view to multi-view. We thus proceed by first determining the 3D line portions that best fit (through linear regression) the chains of fragments of edges

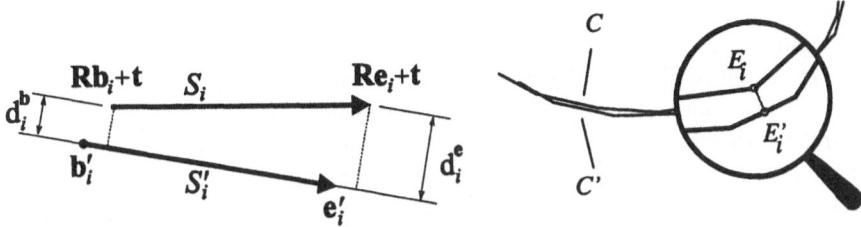

Figs. 3-4. Evaluation of the merging cost of two straight 3D contours (left). 3D curve matching: evaluation of the distance between two polylines (right).

that have been recognized as straight. Then instead of measuring the distances between extremal points of two segments, we measure the distance between the extremal points of one segment and the line that the other segment lies upon (see Fig. 3). Such distances are used for defining the merging cost as follows

$$C_s = \sum_{i=1}^{N} \left[\left(d_i^b \right)^2 + \left(d_i^e \right)^2 \right].$$

In fact, the orientation of edges is usually less sensitive to fragmentation problems than their location in the 3D space [1,17].

Egomotion Refinement from Curved Contours - As already said above, curved contours are used for improving the accuracy of the egomotion's estimate. Although a matching process is required in this case too, this step is now simplified by the knowledge of a first approximation of the camera motion, determined from straight edges. In fact, by applying the pre-determined rigid motion to the set of curved edges, we can decide whether two curved edges are matched, depending on their global distance, which can be measured, with reference to Fig. 4, as

$$d_g = \frac{1}{2}\big(d(C,C') + d(C',C)\big) \quad \text{where} \quad d(C,C') = \frac{1}{N}\sum_i \|E_i E_i'\|$$

The global cost function for motion refinement is of the form $C = C_s + kC_c$, where C_s and C_c are the merging costs associated to straight and curved contours, respectively, and k is weight for balancing the two contributes.

Examples of Application - The method has been extensively tested against convergence problems and has been applied to a series of trinocular acquisitions of real images in order to evaluate qualitatively and quantitatively the accuracy of the results and the speed of convergence. Furthermore, the performance of the proposed method has been compared with that of a previously studied method [2,8] based on point correspondences between artificially added markers. Quantitative results have been obtained by measuring the maximum thickness of the bundles of edges when superimposing different sets of them with the estimated motion parameters. The performance of the proposed method has been proven to be equal to or better than that of the point-based approach, resulting in a maximum

128

bundle size of about 100 ppm in all tests (after merging all 3D edges coming from 20 multi-views).

In Fig. 5, the results on 3D data merging are reported for an object of complex shape, in both cases of egomotion estimated through point and line correspondences. In the first case the cost function is a rigidity constraint based on the distance between reconstructed 3D points of different 3D data sets. Such points are markers that have been artificially added to the scene (white dots placed on the object's support). In the second case the egomotion is computed with the method proposed in this paper. Even though no artificially added markers have been used for the estimation, the accuracy of the estimate is comparable with that obtained through point-matching.

5. Conclusions

In this paper we presented our global approach to accurate 3D reconstruction with a calibrated multi-camera system. In particular, we presented a simple and effective technique for calibrating CCD-based multi-camera acquisition systems.

Fig. 5. One of the original views of the object, fusion of all 3D edge sets through 3D point correspondences (added marks), fusion of all 3D edge sets through 3D contour matching (natural features).

The proposed method was proven to be capable of highly-accurate results even when using very simple calibration target-sets (with little or no *a-priori* information on it) and low-cost imaging devices, such as standard TV-resolution cameras connected to commercial frame-grabbers. We also showed our approach to adaptive calibration, which proved effective for keeping track of camera parameter drift through natural feature tracking.

We also proposed and illustrated a general and robust approach to the problem of close-range partial 3D reconstruction of objects from stereo-correspondences. The method is independent on the geometry of the acquisition system which could be a set of *n* cameras with strongly converging optical axes. The robustness of the approach can be mainly attributed to the physicality of the matching process, which is virtually performed in the 3D space. In fact, both 3D location and local orientation of the surface patches are estimated, so that the geometric distortion can be accounted for. The method takes into account the viewer-dependent radiometric distortion as well. Finally, we presented a method for performing an accurate patchworking of the partial reconstructions, through 3D feature matching. The method, based on the best fusion of 3D curves, provides very accurate results even when using standard TV-resolution CCD cameras.

References

[1] N. Ayache, "Artificial vision for Mobile Robots", MIT Press, 1991.

[2] F. Pedersini, A. Sarti, S. Tubaro: "A Multi-view Trinocular System for Automatic 3D Object Modelling and Rendering". XVIII International Congress for Photogrammetry and Remote Sensing, 1996, Vienna, Austria.

[3] Y. Otha, T. Kanade, "Stereo by Intra- and Inter-Scanline Search Using Dynamic Programming", IEEE Trans. On PAMI, Vol. 7, N. 2, pp. 139-154, 1985.

[4] P. Pigazzini, F. Pedersini, A. Sarti, S. Tubaro: "3D Area Matching with Arbitrary Multiview Geometry". EURASIP Signal Processing: Image Communications. Special issue on *3D video technology*, early issues of 1998.

[5] F. Pedersini, A. Sarti, S. Tubaro: "Robust Area Matching". IEEE Intern. Conf. on Image Processing, 1997, October 26-29, 1997, Santa Barbara, CA, USA.

[6] F. Pedersini, A. Sarti, S. Tubaro: "Egomotion Estimation of a Multicamera System through Line Correspondence". IEEE Intern. Conf. on Image Processing, 1997, October 26-29, 1997, Santa Barbara, CA, USA.

[7] F. Pedersini, A. Sarti, S. Tubaro: "Automatic Surface Reconstruction of 3D Works of Art". International Conference on Electronic Imaging and the Visual Arts (EVA'97). March 19-25, 1997, Florence, Italy.

[8] F. Pedersini, A. Sarti, S. Tubaro: "3D Motion Estimation of a Trinocular System for a Full-3D Object Reconstruction". IEEE Intern. Conf. on Image Processing, September, 1996, Lausanne, Switzerland.

[9] R. Y. Tsai, "A Versatile Camera Calibration Technique for High-Accuracy 3D Machine Vision Metrology Using off-the-shelf TV Cameras and Lenses" - IEEE J. on Robotics and Automation, Vol. RA-3, No. 4, Aug. 1987, pp. 323-344.

[10] J. Weng, P. Cohen, M. Herniou, "Camera Calibration with Distortion Model and Accuracy Evaluation", *IEEE Trans. on PAMI*, Oct 1992, Vol. 14, No 10, 965-980.

[11] F. Pedersini, D. Pele, A. Sarti, S. Tubaro: "Calibration and Self-Calibration of Multi-Ocular Camera Systems". Intl. Workshop on Synthetic-Natural Hybrid Coding and Three-Dimensional (3D) Imaging (IWSNHC3DI'97). September 5-9 1997, Rhodes, Greece.

[12] F. Pedersini, S. Tubaro: , "Accurate 3D reconstruction from trinocular views through integration of improved edge-matching and area-matching techniques." *VIII European Signal Processing Conference*, September 10-13, 1996, Trieste, Italy

[13] L. Kitchen, A. Rosenfeld, "Gray-level corner detection", Pattern Recognition Letters, No. 1, 1982, pp. 95-102.

[14] G. Giraudon, R. Deriche "On corner and vertex detection", Proceedings Intl. Conf. on Computer Vision and Pattern Recognition, Maui, Hawaii, June 1991, pp. 650-655.

[15] K. Rohr, "Recognizing Corners by Fitting Parametric Models", Intl. J. of Computer Vision, Vol. 9, No. 3, 1992, pp. 213-230.

[16] J.L. Mallet: "Discrete Smooth Interpolation", ACM Transactions on Graphics, Vol. 8, No. 2, 1989, pp. 121-144.

[17] Z. Zhang, O.D. Faugeras: "3D dynamic scene analysis: a stereo based approach", Springer-Verlag, 1992.

[18] F. Pedersini, A. Sarti, S. Tubaro: "Tracking Camera Calibration in Multi-Camera Sequences through Automatic Feature Detection and Matching". IX European Signal Processing Conference, September 8 - 11, 1998, Rhodes, Greece.

Part 2

Multimedia Traffic:
Modelling, Analysis, Measurements
and Performance Evaluation

Providing Quality for Internet Video Services

G. Karlsson

Department of Teleinformatics, KTH, Sweden

Abstract

One of the most interesting improvements of the Internet today is the provisioning of services for interactive audio–visual applications. Such applications have quality requirements which place limits on both transfer delays and information losses. Our goal is to support interactive video with only small modifications to network routers and control systems. The approach we favor is to ensure the quality for a session by forward–error correction. The code strength is dynamically tuned to meet a user's quality expectation, given the experienced loss process of the transfer. To make the tuning feasible, the network state must be predictable. We accomplish this by regulation of the load through sender–based admission control. It uses end–to–end probes of the network state and a self–imposed blocking if then sender determines that a sensible transfer cannot be made. The paper outlines this procedure along with a review of source coding considerations and a description of tunable error–control coding.

1. Introduction

The expectations on the Internet are growing as its role in society is becoming more established. One of the imminent challenges is to provide some means for quality assurance for interactive applications, in addition to the best effort. The research world has dealt with the problems of giving quality guarantees for statistically multiplexed services throughout the past fifteen years of asynchronous transfer mode development. Now it appears that the broadband integrated–services digital network will be based on the internet protocol rather than ATM and we find that the same problem has to be tackled for IP. The experiences from the ATM work should not be neglected and could provide useful guidance about fruitful approaches and dangerous pitfalls

In this paper we concern ourselves with video transfers across the Internet. Many of the issues will however be valid for audio transfers as well. We consider interactive applications and live transfers for one–way video distribution. These types of transfers rarely allow the possibility of trading losses for delay, as is commonly done by retransmission for data transfers. The remaining error–correction options operate at the receiver and rely on some form of redundancy added to the signal by the sender. The most common form, of course, is the algebraic redundancy given by error–control coding. The application of error–control codes to multiplexing channels offers some problems not encountered in connection with transmission channels. The channel extents over several links that are shared statistically with other channels, and the deteriorations in terms of delay and loss of packets depend on the stochastic fluctuations in the load. This in in contrast with transmission systems for which the channel distortion is independent of the channel's utilization (*eg*, it is additive noise). As a consequence we find that providing forward–error correction for best–effort channels is complicated since the amount of redundancy cannot be determined before a session starts. It follows that the amount of redundancy should be tuned to the experienced channel state as measured by the receiver. Increase of redundancy has however to be compensated by a reduction in source rate to avoid an increased load and ensuing deterioration of the channel state. If the load level rises due to new sessions, then each session must lower the bit rate to avoid overloading the network and to provide a fair share of the resource to all sessions (Ref. 10).

Here there is a second issue to consider, namely the quality requirement. A session that was started with some expected level of quality, which had been determined necessary for carrying out the purpose of the communication, may find the load rising during the session due to competing sessions. Eventually the load may have become so large that the needed redundancy rate necessary to sustain the desired level of uncorrectable losses cannot be accepted without reducing the source rate under what is needed for the desired quality. Alternatively, the source adjusts its total rate downwards to accommodate the new sessions in an attempt to keep the load at a reasonable level. In the process of doing so, it might need to reduce the rate below what is needed for the source and redundancy rates. The consequence is of course that the session has been rendered useless to the annoyance of the users.

In this paper we will outline an architecture for dealing with this problem. We will propose a configuration with forward–error correction that is tuned to optimize the redundancy rate for the experienced channel state. However, we wish to protect ongoing sessions from new sessions in order to provide a floor for the session quality. This is done by an admission control to limit the network load. We will call our version of this controlled–load service for predictable–quality service in order to distinguish it from the service defined in RFC 2211 (Ref. 25). The outline is as follows. A short review of pertinent issues of digital video and its encoding will follow next. The subsequent section explains the tunable forward–error correction that we see as a prime function for Internet transfers. Following that we give our quality criteria and present the matching service architecture. The workings of the predictable–quality network service are explained in Section 6 and in Section 7 we outline how the service architecture may include advance reservations and multicast.

2. The video system

A video communication system is shown in Fig. 1. The digitized video is first passed to a source coder. It is often built with three system components: energy compaction, quantization and entropy coding. The energy compaction aims at putting the signal into the form most amenable to coarse quantization. Common methods for video include discrete cosine transform, subband analysis, and prediction, possibly motion estimated. The quantizer reduces the number of permissible amplitude values of the compacted signal and introduces round–off errors. The entropy coding, finally, assigns a new representation to the signal which represent the data more efficiently but there is no longer a constant number of bits per picture, and the bit rate becomes temporally varying.

2.1 Bit–rate regulation and traffic description

The bit–rate regulation is used to adapt the varying bit rate to the channel in the network. The regulator flattens the bit–rate variations by buffering and may adjust the compression to avoid overflow. The feedback reaches the quantization of the encoder and enforces a higher step–size with increased round–off error as a consequence. If the quantizer step–size is throttled frequently and heavily it may lead to visible quality fluctuations in the reconstructed signal.

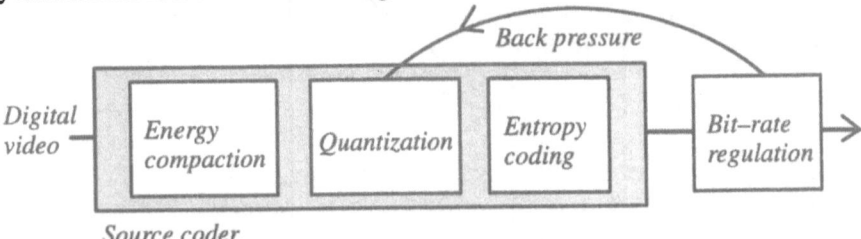

Source coder

Fig. 1   The sending side of a a video communication. The bit rate regulation consists of a smoothing buffer with back pressure to avoid overflow.

Leaky bucket descriptors (or token buckets as they also are called) have recently been studied for regulated video. The leaky–bucket parameters are the peak and sustainable rates, and the burst size. It is clear that the feedback from the regulator to the coder makes it possible to shape the bit rate to fit any choice of these leaky–bucket parameters. Whether a particular set of parameters is good or not can only be determined by subjectively evaluating the encoding quality. The leaky–bucket descriptor does not give a lower delay bound than a fixed–rate descriptor for guaranteed service (deterministic multiplexing) and does not give a lower average delay in the regulator when the two cases are allocated the same amount of capacity in the network (Ref. 12). We therefore assume that a session is restricted by a fixed–upper bound on the bitrate. This bound is for the total rate consisting of a source rate from the video coder and a redundancy rate from the channel (error–control) coder.

2.2 Layered coding

Layered (or hierarchical) coding means that the signal is separated into components with differing visual importance (Ref. 11). The layers are formed from the signal components after the energy compaction, and they are separately quantized, entropy coded and packetized. The idea is based on the realization that all quanta of rate (ΔR) allocated to a source do not yield the same amount of reduction in distortion, as illustrated in Fig. 2. However, if all allocations are within the same service class then it is reasonable to assume that the cost increases linearly with the allocated rate (denoted by curve (a)). By letting each quantum of rate being a layer and by matching service class to the importance of the layer, it would be possible to reduce the cost of the ses-

136

sion without noticeable effect on the quality (shown as curve *(b)*). Vital layers might be transferred in a class with guaranteed quality, while a signal layer that enhances the quality could be sent best effort. This reasoning assumes that a set of connections of differing capacity and quality of service is cheaper than one connection for the aggregate stream. This can be firmly established first when tariff structures for broadband networks are in place, however.

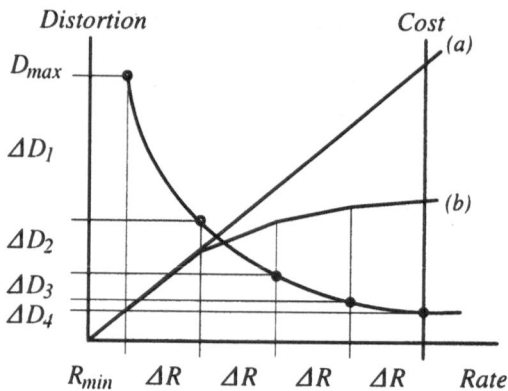

Fig. 2 Illustration of layering: increments in rate above a needed minimum (R_{min}) are treated as separate layers. The layers do not give equal reduction in distortion.

Layering is also useful within a single service class if the quality is lower than required by the more important signal components. Each layer can then be given an error protection that is commensurate with its importance. Such unequal error protection might require less total redundancy than a protection of the non–layered signal. Another application of layered coding is when a specific target bit rate or quality level cannot be stated *a priori* for the transfer. By layering, the sender can provide a range of bit rates and qualities in one and the same encoding of the information, and the particular point in that range can be chosen dynamically. It can, for instance, be beneficially used for stored programs so that rate control can be exercised when the video is being retrieved: the server can add and drop layers to fit a given channel for the transfer. For multicast, a session can be distributed over multiple multicast groups, with one layer per group. Each receiver independently decide at which quality and bitrate the reception should be by selecting the appropriate set of layers (Ref. 19).

An alternative layering method is the multiple–description technique which separates signals into layers of equal importance (Ref. 14). This means that the signal can be reconstructed as long as a single layer is received. The quality of the reconstruction depends only on the number of layers that are received, and not on which ones. So, we have two strategies for the layered source coding: The first is to separate the signal into components of different importance for which the network service quality or error protection is tailored to each layer. The second strategy is to separate the signal into layers of equal importance in order to fit a uniform error protection or service quality.

3. Tunable forward–error correction

Static forward–error correction (FEC) using algebraic error–control coding, possibly in conjunction with interleaving, is a traditional and widely used method of error control. It works well when the channel characteristics are known so that a proper code and interleaving structure can be chosen. This knowledge is usually not available for a channel with best–effort service. Feedback from the receiver about the estimated channel characteristics may be used instead to tune the error–control coding to the measured channel state (Ref. 17). For example, the simple stop–and–wait ARQ uses a trivial $(n, 1)$ repetition code, where n is incremented by one for each negative acknowledgment. More advanced codes can be employed in such schemes and the transmission can of course be continuous. Forward–error correction of packet loss requires an interleaving matrix so that the symbols of a codeword get sent in different packets. Lost packets become erasures in the codewords since the packet's sequence number indicates the position of the loss (see Fig. 3).

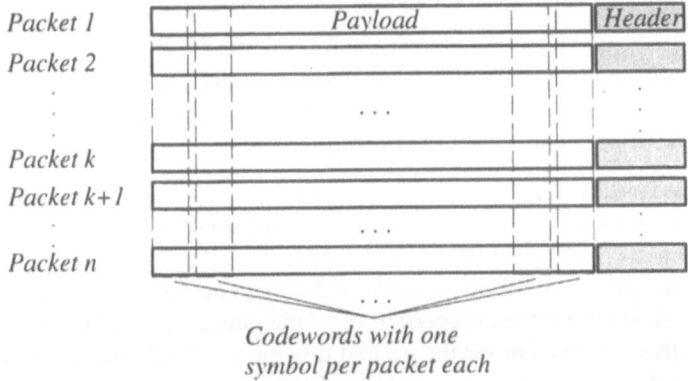

*Codewords with one
symbol per packet each*

Fig. 3 Forward–error correction for packet losses require an interleaving matrix where the rows are the packet payloads. The k upper rows hold source data and the $n–k$ lower rows hold the redundancy data for a (n, k) block code. A lost packet results in one erasure per codeword.

Algebraic redundancy is useful for all structural information such as framing information (picture and slice header in MPEG coding, for instance), and for motion vectors (in case motion–compensated prediction is used). Redundancy for video could also be perceptual which can be used to conceal losses of picture data (Ref. 24). It could consist of the signal encoded at low bitrate and quality and concealment of lost data is made by filling in the affected locations with the lower–quality redundancy data. The redundancy may also be in the form of intra–picture coded data which limits error propagation for predictive coding.

The important issue for the systems designer is to pick an appropriate error–control code. Block codes can be used and the tuning is simply accomplished by dropping parity–check or information symbols in the codewords. This is called shortening or puncturing, respectively (Ref. 2). The decoder has to know which parity–check or information symbols have been omitted in the encoding. For instance, the popular Reed–Solomon burst–error correcting codes of McAuley can be tuned in this way (Ref. 18). Also the rate–compatible punctured convolutional codes of Hagenauer can be used with code combining to reduce the code rate (the analog of shortening the block code) (Ref. 7). It is desirable to use systematic code in order not to aggravate the losses when an interleaving matrix cannot be recovered.

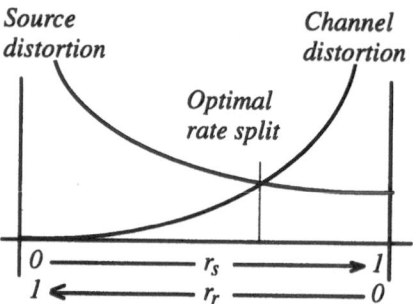

Fig. 4 Illustration of the trade–off in source and channel coding for a given total rate of 1. The source rate is r_s and the redundancy rate r_r. The goal is to minimize the total distortion. The channel–distortion curve changes with the packet–loss probability.

The combined system consists of a rate–controlled source coder and a tunable channel coder. The goal is to yield the best perceptual quality for a given loss probability and bound on the total bitrate. The actual choice of total rate is however not determined. As we shall see in Section 6, a reasonably cautious guess is needed. It should allow a fair redundancy rate, proportionate to the tuning range of the code, and a source rate that is sufficient for the desired quality. The total rate can be adjusted downwards when more experience is gathered about the network state.

Given a fixed upper bound on the rate, what is the optimal split over the source and redundancy rates? We will assign distortion–rate curves to the source coding, as usual, and to the residual loss rate at the receiver. The residual loss probability is a function of the network's loss probability and the redundancy rate. The task is therefore to find the redundancy that minimizes the sum of the source distortion, which depend on the source rate, and the channel distortion which depend on the rate of uncorrectable losses. The optimization is constrained to yield a sum of the source and redundancy rates that is below the fixed–rate bound. This is illustrated in Fig. 4. Since the loss probability is rarely known, the receiver will use estimates in the optimization. It will notify the sender of any adjustment in the source–channel rate trade off. (A theoretic study of the source–channel coding trade–off is presented in Ref. 8). Layered source coding allows this trade off to be done for each layer separately, thus offering a better tuning at the expense of complexity. Note that each layer would have its own distortion curves of the two types.

4. The desire for predictable quality

A central aspect of human quality expectations is that the quality should be uniform for the duration of the communication. For instance, a session of a high quality which during a noticeable period suffers reduced quality could be found more annoying than a session at the reduced quality without fluctuations. Different quality levels could be useful for various applications and communication contexts. The importance however is that each quality level appears to be consistent without perceptual variations. The problem with a best–effort service for transfers of audio–visual information is the tremendous variability in quality.

In telephony the grade–of–service concept has primarily included the accessibility of the network. A one–percent blocking probability during peak hour has been a typical design criterion. For ATM, the prime target has been in–call parameters such as the cell loss ratio, the total delay and the delay variations. The Internet world has mostly been concerned delay issues. The quality aspects we consider being the most important are the session blocking probability, the probability of uncorrectable packet losses and the total delay. We discuss them briefly in that order.

It is often not acceptable that a session for a scheduled meeting is blocked by the network. A network should therefore allow advanced reservations for which it is willing to offer guarantees against blocking. The guarantee could be deterministic, meaning that the session can be established at the requested time with complete certainty. It can also be statistical, giving a probability that the session cannot be set–up within a prescribed time period (*eg*, the probability of blocking a session for more than five minutes). Although advanced reservation is a reasonable request on the network services, there has been remarkably little research in this direction (see Refs. 3 and 6).

The packet–loss process in the network depend on the load variations in the statistical multiplexers (ignoring bit errors and system failures). The residual loss rate is of course the same as the network loss rate for the session if channel coding is not used. When an appropriate code is used for a given average loss rate, then the probability of residual (not correctable) losses depend on the distribution of the losses. Our aim is to make the loss rate predictable over some reasonable time window, say a few times the round–trip delay. The loss rate is measured by the receiver and reported back to the sender. The error–control code may then be tuned to the measured rate with some additional protection to account for the variations in the loss process.

Long delays lowers the interactivity of a conversation. As a rule of thumb, 150 ms can often be taken as a safe value for one–way delays (defined in ITU–T Recommendation G.114: Mean one–way propagation time; see also Refs. 13 and 15). Delay variations can be smoothed out and there is a trade–off between the buffers size and the associated buffering delay on one hand, and the probability of packet loss when the buffer over or under flows on the other hand. The delay variations could also be allowed to propagate through the system to the digital to analog conversion. The result is that the sampling interval is not maintained which in turn means a poorer rendition of motion: smooth movements could appear as a sequence of jolts. The magnitude of the jitter could be reduced by buffering to raise the lower limit. There are, however, no subjective tests reported on the jitter sizes that can be deemed acceptable for various video applications.

5. The proposed service architecture

Our design for expanding the service offering of the Internet contains three distinct classes (see Fig. 5). The best effort is similar to today's service with the possible option of specifying a preference for low delay or high throughput, as described by Ref. 16. The highest quality is given by a guaranteed service class, much along the lines proposed in RFC 2212 (Ref. 23). The differences we suggest is that the traffic descriptor consists of a fixed–rate bound and that the scheduling in the routers is not work conserving. The token–bucket traffic descriptor that is used in RFC 2212 does not offer any advantage in terms of tighter delay bound over a single–upper bound. However, using the token bucket means that all nodes must provide buffering for the bursts, while only the sender needs that buffer space when the peak–rate bound is used. Non–work conserving scheduling is used to shape the traffic down to the sum of the individual flows' declared rates (prevents packets from clumping together in the multiplexing). These two changes will limit the buffer space needed in the routers for this service class. The details can be found in Ref. 20.

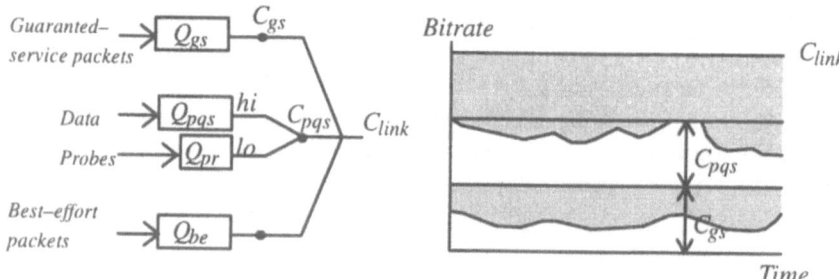

Fig. 5 The proposed service architecture: the guaranteed service is allocated a portion C_{gs} of the link capacity and the predictable–quality service a portion C_{pqs}; remaining and allocated but unused capacity is available best effort (shaded areas). The queue Q_{pqs} for data is served at higher priority than Q_{pr} for the probes within predictable–quality service class.

The third and last class we include in our architecture is a predictable–quality service. The remainder of the paper describes the design of this service class in more detail. It is an instance of the controlled–load service (Ref. 25) but we have chosen a different name for it to allow ourselves to stray away from the original proposal. Our work is also inspired by the scalable resource reservation protocol developed at EPFL (Ref. 1). The goal with the predictable–quality service is that it should provide a network state with a bounded worst case behavior (the controlled load). The novelty is that it does not rely on an explicit signalling protocol to request a channel from the network nor on a decision by the network to grant the request. Such admission control tend to be complex (Ref. 5). Instead we rely on users to probe the network to determine whether enough resources are available for conducting a sensible session. If so, the call can be established. Quality beyond that offered by the bare channel must be ensured by the end systems themselves through tunable forward–error correction, as previously discussed.

6. A predictable–quality service

A problem encountered with ATM traffic control has been the conflict between the complexity of the traffic descriptors and the calculations for admission control on the one hand, and the possible statistical multiplexing gain on the other. Already such a straightforward descriptor as the leaky bucket puts the end system in the difficult situation of estimating the parameters before the session has started. Yet the leaky bucket provides too little information to yield any statistical multiplexing gain in many cases (Refs. 12 and 21). Increasing the amount of detail provided by the traffic descriptor would make the estimation task of the end system as well as the admission control more difficult. The remedy has been to accept a simpler traffic descriptor, a fixed–rate bound, and to let the network collect the information needed for a better statistical multiplexing gain by direct measurements of the network traffic (Refs. 9 and 22). This approach is interesting and eases the role of the end system at the expense of more complexity in the network nodes for the measurements. Measurement–based admission control does not remove the need for a signaling protocol between the host and the network nodes. Furthermore, it is not clear what the network operator can state as intended goal with the service, and how the commitment to the goal is expressed.

Our goal has been to provide an admission test in order to protect ongoing sessions from new sessions. The purpose is to get a more consistent and predictable quality for a session than offered by best effort, as discussed in Section 4. Yet, we do not want to burden the routers with possibly complex measurements and analyses, nor do we want signalling more than between the host and the access gateway. There are not any guarantees given about the service from the network: the operation of the service is openly defined and it is the decision of the end system to decide whether and when to use the service, and what precautions to take in terms of forward–error correction. Although our evaluation is not yet completed, we believe we have found a design which meets our goal.

6.1 Configuration and basic principles

The predictable–quality service requires its own capacity partition on the links in the network and it gives an unsurpassable limit for the traffic. Unused capacity is however fully available to best–effort traffic so there is no waste incurred by the allocation (Fig. 5). The limit can be provided by the aforementioned non–work conservative scheduling (Ref. 20) and is needed for the estimation of available capacity: each host is required to first probe the network to see if there is capacity available for the session. The probe should only measure capacity available within the partition for the predictable–quality service class and it should not include the remaining capacity on the link, used by guaranteed and best–effort services. Routers are also required to recognize two priorities within the service class. We wish to protect ongoing sessions from the probes of coming sessions and it is easily accomplished by serving them at different priorities. This means that probe packets are only forwarded onto a link if there is capacity left over by the established sessions. The loss rate experienced by the probe is used to make a decision of whether the session can be established. Established sessions are allowed to send packets at the high priority, while blocked sessions back off and make new probes after random waiting times.

The probe consists of packet sent at the maximal bitrate that the sender wishes to use. It is a time limited stream of packets, all of the same length, sent at constant interarrival times. The packet length is equal to the longest length the sender wishes to use. The packets in the probe are carried at a low priority and will only be forwarded onto a link by a router if there is capacity allocated to the class which remains unused. The priority could be implemented in two ways, either as a threshold–based discard where low priority packets are lost whenever the queue length is above the threshold, or there could be two queues in parallel served in priority order (as in Fig. 5). The threshold based discard could of course be replaced by a more complex random early discard for which the probability of discarding a packet does not only depend on the instantaneous queue length and the packet priority, but also on the history of the queue dynamics (Ref. 4).

The probe rates from a source have to be below a given bound which is settled in a service contract with the operator. It limits the size of the sessions which can be set up. The maximally allowed rate should be only a small fraction of the capacity allocated to the service class on each link, so that statistical multiplexing works well. The maximum probe rate is policed and violating packets are simply dropped. The packet length is similarly limited in the contract to a value that is reasonable with respect to delay variations. The packets of the probe contain information about the session: the desired peak rate as well as details of the source and channel encodings. The receiver measures the probe and sends the result back to the sender. The message contains a time stamp and a certificate to prove that the measurement is recent and the process follows the requirements of the operator. The measurement could consider both the interarrival times and the loss of packets. However, we will consider only the ratio of lost to total amount of packets.

The sender is allowed to establish the session if the packet loss ratio of the probe was below a level prescribed in the service contract. The same level must be used by all parties sharing the service. The establishment is done by signalling the access gateway with a message that should include the certified and timestamped measurement report from the receiver. The access gateway will then set a peak–rate limiter to the value of the probe signal and start a meter for the call charge. The limiter monitors packet interarrival times and packet lengths and discards all packets that are too long or arrive to closely. The sender can at any time reduce the value of the established bit rate or close the connection. The peak–bit rate can only be raised by a new probe to the same receiver. The policer in the gateway enforces a limit on the sender for each host to which it has a connection.

The session establishment is blocked when the probe is unsuccessful, and the sender backs off a random time before issuing a new probe. The back–off time is drawn from a uniform distribution of some width, which is doubled for each new attempt that is blocked for the same receiver. The receiver may use the measurements to determine whether the session will start, and can alert the target application. The receiver can of course block the session if needed (for instance, by sending a reply indicating zero received packets).

6.2 Ongoing and future research

There are still many open issues regarding the service specification:

- The steadiness of the operational quality that established sessions experience is of course the main criterion of evaluation. The variance of the probing process will occasionally lead to false accepts and their likelihoods and consequences will be studied.

- Fairness issues are critical for making the service operational. We intend to study the influence of the probe durations and rates on the probability of success.

- The allocation of capacity for the predictable–quality service partly determines the blocking probability on the different paths. The background process run by the network management to allocate and de–allocate capacity to the service class is certainly important and its development has not yet been tackled.

6.3 Preliminary results

We have initiated the evaluation of the described method. The scenario is as follows. Forty identical but independent time–discrete on–off sources represent the traffic from established sessions. Each has a peak rate of one Mb/s and geometrically distributed on and off periods with average durations 2 and 3 slots, respectively, with 1 ms per slot (1000 bits). The capacity for the service is 20 Mb/s and the utilization is thereby 80 percent. All probes are 10 seconds long. Q_{pr} in Fig. 5 holds one probe packet. We have ignored the time extent of the probe packets and simply plot the probability of loss versus the number of probe packets per second in Fig. 6. The loss ratio is almost linearly proportional to the probe rate. A decision level of five percent loss would mean that 0.7 Mb/s of the available 4 Mb/s could be used by a new call.

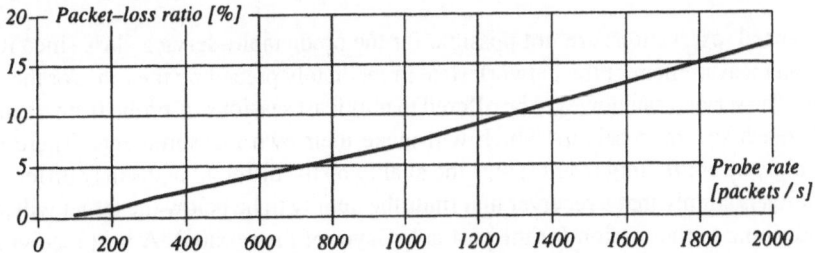

Fig. 6 The packet–loss ratio as a function of the probe rate in packets per second.

In the second simulation, we kept the probe rate constant to 1000 packets per second and varied the load for established sessions by adjusting the capacity allocation for the service class from 17 Mb/s up to 47 Mb/s (94 percent load down to 30 percent). Fig. 7 shows the packet loss ratio for the probes as a function of the load for established sessions. It is possible to limit the load by blocking sessions if the packet loss ratio is higher than a prescribed limit. Five percent loss would for instance limit the load to below 70 percent for the given traffic mode (three percent loss and a standard deviation less than one percent, measured over groups of one 25th the probe length).

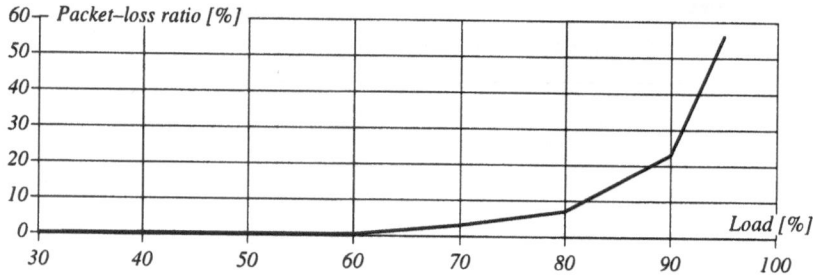

Fig. 7 The packet–loss ratio for the probe as a function of the system's load. The loss was below 0.3 percent for all loads under 60 percent.

7. Advance reservations and multicast

The service architecture should support advance reservations and multicast. The guaranteed service we briefly described above can easily incorporate advance reservations and works also for multicast sessions. The reservation request states the time when the session will begin, its duration and the desired rate. Since the guarantee is deterministic there cannot be any overbooking. Instantaneous establishment will still require a specified duration of the session and a rate; the session's start time is the same as the request time. A session can always end before the stated duration has expired, but it can only extend it by a new session establishment and will thereby run the risk of being blocked. Multicast for the guaranteed service is also doable. When the multicast session is announced it will also state the bit rate of the session (or the different layers of the session). A host that joins the multicast session reserves resources for its own branch to the closest node in the distribution tree.

Advanced reservations are not possible for the predictable–service class since it relies on measurement of the network state immediately preceding the time for the session. The service can however be offered to multicast sessions. A probe from a sender will reach several receivers which will make their own measurements. There will certainly be conflicting results since the quality on the different paths may differ. The solution is simply that a receiver that finds the quality to be below the target will stop subscribing to the session (or the particular layer of the session). A host receiving a probe successfully should report back to the sender. The number of reports could be a major concern for a session with a larger number of members. For the sender it would suffice with one report which is enough to prove to the gateway that resources are available in the network for reaching someone. The problem is to find a protocol that selects one or at most a few receivers who answers, but rarely many and never none. (This is also encountered for repeat requests in reliable multicast.)

When a new sender joins the session, some members may find that they can no longer belong to the group since their access links get overloaded. This is not a problem of the protocol but rather of the definition of the session: if all senders are promised a fair chance at sending, then there cannot be any discrimination against late arriving senders. In order to overcome this the session management has to exercise a gavel control so that not more senders are active simultaneously than what the receivers can handle. The dynamic membership of multicast sessions is handled by periodic probes from the senders. The sender times out if no one answers and backs off before sending the next probe (or terminate the session if that is acceptable). The probes should be sent on a different multicast address than the data. A receiver joining the session will first subscribe to the probe address, and can later join the main session if the probe is received successfully.

8. Summary

We have discussed the possibility of offering service quality for realtime applications by a system that combines robust source coding, tunable forward–error correction and self–admission load control. A central aspect of the architecture has been to keep functionality for the load control and quality assurance outside the network. The load control is intended to give a reasonably stable multiplex channel for which the packet loss probability can be estimated end–to–end. This estimate is used to tune the error–control coding to bring the number of uncorrectable packet losses under what is tolerable. The level of tolerable losses is in turn given by the robustness of the source coding.

References

1. W. Almesberger, T. Ferrari, and J.–Y. Le Boudec "SRP: a Scalable Resource Reservation Protocol for the Internet", March 1998. (http://lrcwww.epfl.ch/srp)

2. R. E. Blahut, *Theory and Practice of Error Control Codes*, Addison–Wesley, Reading, MA, 1983.

3. M. Degermark, T. Köhler, S. Pink, O. Schelen, "Advance Reservations for Predictive Service in the Internet" ACM–Springer J. Multimedia Syst., Vol. 5, No. 3, May 1997, pp 177 – 186.

4. S. Floyd and V. Jacobson, "Random early detection gateways for congestion avoidance," IEEE/ACM Trans. Networking, vol. 1, Aug. 1993, pp. 397–413 .

5. E. Gelenbe, X. Mang, and R. Önvural, "Bandwidth Allocation and Call Admission Control in High–Speed Networks," IEEE Comm. Mag., Vol. 35, No. 5, May 1997, pp. 122–129.

6. A. Greenberg, R. Srikant, and W. Whitt, "Resource Sharing for Book–Ahead and Instantaneous–Request Calls,"in *Proc. of ITC–15*, Elsevier, June 1997, pp. 539–548.

7. J. Hagenauer, "Rate–Compatible Punctured Convolutional Codes (RCPC Codes) and their Applications," IEEE Trans. Comm., Vol. 36, No. 4, April 1988, pp. 389–400.

8. B. Hochwald and K. Zeger, "Tradeoff Between Source and Channel Coding," IEEE Trans. Inf. Th., Vol. 43, No. 5, Sept. 1997, pp. 1412–1424.

9. S. Jamin, et al., "A Measurement–based Admission Control Algorithm for Integrated Services Packet Networks,," IEEE/ACM Trans. Networking, Feb. 1997, pp. 56–70.

10. H. Kanakia, et al., "An Adaptive Congestion Control Scheme for Real–Time Packet Video Transport," ACM Comp. Comm. Rev., Vol. 23, No. 4, Oct. 1993, pp. 20–31.

11. G. Karlsson and M. Vetterli, "Sub–band coding of video for packet networks," Optical Engineering, Vol. 27, No. 7, July 1988, pp. 574–586.

12. G. Karlsson and G. Djuknic, "On the efficiency of statistical–bitrate service for video," in *Proceedings of IFIP International Conference on Performance of Information and Communications Systems*, Lund, Sweden, May 25–28, 1998.

13. N. Kitawaki and K. Itoh, "Pure Delay Effects on Speech Quality in Telecommunications," IEEE Journal Sel. Areas Comm., Vol. 9, No. 4, May 1991, pp. 586–593.

14. J. Kovacevic, "Multiple Description Coding for Communications," in *Proceedings of the CNIT/IEEE 10th Tyrrhenian International Workshop on Digital Communications*, Ischia, Italy, Sept. 15–18, 1998.

15. T. Kurita, et al., "Effects of Transmission Delay in Audiovisual Communcations," Electronics and Commun. in Japan, Part 1, Vol. 77, No. 3, 1994, pp. 63–74.

16. J.–Y. Le Boudec, M. Hamdi, L. Blazevic, and P. Thiran, "Asymmetric Best Effort Service for Packet Networks," EPFL–DI–ICA Report No. 98/286, July 1998.

17. D. M. Mandelbaum, "An Adaptive–Feedback Coding Scheme Using Incremental Redundancy," IEEE Trans. Inf. Th, Vol. IT–20, May 1974, pp. 388–389.

18. A. J. McAuley, "Reliable broadband communication using a burst erasure correcting code,", ACM Comp. Comm Rev., Vol. 20, No. 4, Oct. 1990, pp. 297–306.

19. S. McCanne, V. Jacobson and M. Vetterli, "Receiver–driven Layered Multicast," ACM Comp. Comm. Rev, Vol. 26, No. 4, Oct. 1996, pp. 117–130.

20. M. Mowbray, G. Karlsson, and T. Köhler, "Capacity reservation for multimedia traffics," Distr. Syst. Eng., Vol. 5, 1998, pp. 12–18.

21. B. V. Patel and C. C. Bisdikian, "End–Station Performance under Leaky Bucket Traffic Shaping," IEEE Network, Sept./Oct. 1996, pp. 40–47.

22. H. Saito, "Dynamic Resource Allocation in ATM Networks," IEEE Comm. Mag., Vol. 35, No. 5, May 1997, pp. 146–153.

23. S. Shenker, C. Partridge, and R. Guerin, "Specification of Guaranteed Quality of Service" IETF RFC 2212, Sept. 1997.

24. M. Wada, "Selective Recovery of Video Packet Loss Using Error Concealment," IEEE J. Sel. Areas Comm., Vol. 7, No. 5, June 1989, pp. 807–814.

25. J. Wroclawski, "Specification of the Controlled–Load Network Element Service," IETF RFC 2211, Sept. 1997.

Video Traffic over Wireless Links: Issues and Performance

Hang Liu[1] and Magda El Zarki[2]
[1]C&C Research Laboratories, NEC USA, Inc., 4 Independence Way, Princeton, NJ 08540, USA
[2]Department of Electrical Engineering, University of Pennsylvania, Philadelphia, PA 19104, USA

Abstract

This paper presents a wireless transport system for real-time multimedia services over personal communication services (PCS) networks.The proposed system uses a hybrid ARQ scheme for error control because it can efficiently adapt to nonstationary wireless channels and yield good throughput and reliability. In particular, delay and delay jitter control related to retransmissions in the error control module is addressed. An adaptive source rate control protocol is used to deal with the fluctuation of the effective data rate due to fades and retransmissions. A novel adaptive synchronization protocol is developed to compensate for long-term delay variation due to large-scale (shadowing) fading so that synchronization is maintained and end-to-end delay is kept low. Simulation results from the performance evaluation of the system are presented

1. Introduction

In recent years, wireless personal communication services (PCS) networks based on digital technologies have been deployed and become more and more popular. Although these networks initially focus on voice and data applications, wireless users are requiring a wider range of communications services involving video and multimedia.

A mobile communication system experiences both small-scale (multipath) and large-scale (shadowing) fades [1]. Error control schemes are necessary to obtain high transmission reliability required by multimedia services. Traditionally, forward error correction (FEC) codes have been used for real-time services because they maintain a constant throughput and a bounded delay. Wireless channels are time-varying. FEC codes can be chosen to guarantee certain error rate requirements for the worst channel conditions. However, this causes unnecessary overhead and wastes bandwidth when the channel is in a good state. Recently, for wireless environments, it has been shown that hybrid ARQ schemes can significantly improve the transmission reliability and yield much higher throughput than FEC schemes because they can effectively adapt to the varying channel conditions [2]. The new video conferencing standard, H.324, will support hybrid ARQ schemes for wireless video communications [3, 4]. The MPEG-4 standard committee is also considering to adopt a hybrid ARQ scheme for video transmission in error-prone environments [5].

However, retransmissions in hybrid ARQ schemes cause delay and delay jitter. Long delays are intolerable for interactive real-time applications. In the current

PCS networks [8], the total bandwidth for a channel is constant. The wireless channel conditions change over time. Hybrid ARQ schemes adapt to the varying channel conditions by retransmitting erroneous packets. When the channel is good, no retransmissions are required and the effective data rate can be high. When the channel becomes poor, the retransmissions use up bandwidth and thus reduce the effective data rate (the effective data rate is defined as the rate of the information that is correctly transmitted). This results in a varying effective data rate from the point of view of the source.

Multimedia services require synchronization. For example, in video conferencing applications, video frames must be presented at the receiver with the same temporal relationship as that with which they were captured at the source. Video and associated audio should be played out with "lip synchronization". However the original temporal relationships may be distorted due to delay jitter introduced by retransmissions. Before presenting at the receiver, the original temporal relationships must be recovered and the synchronization must be maintained.

In this paper we investigate a wireless transport system for real-time multimedia services. We focus on low bitrate (32 kb/s) video conferencing over a PCS network. For video coding, the ITU H.263 standard [6] is considered (QCIF format and 15 frame/s) because the low bitrate makes it well suited for band limited wireless networks. The proposed system uses a hybrid ARQ scheme as error control mechanism for the transmission of compressed video data. In particular, the delay and synchronization control issues related to retransmissions in the error control module are addressed. We uses an adaptive source rate control (ASRC) scheme [7] to deal with the fluctuation of the effective data rate due to fades and retransmissions. A novel adaptive synchronization protocol is developed to compensated for long-term delay variation due to large-scale (shadowing) fading so that synchronization is maintained and end-to-end delay is kept low.

This paper is organized as follows. The next section briefly describes the system architecture. In section 3, we describe the synchronization protocol. Section 4 presents the simulation results of the performance evaluation of the wireless multimedia transport system. Finally conclusions are given in section 5.

2. Communication System

Fig. 1 shows the wireless transport system. For real-time video applications, the raw video source is compressed and passed to the error control module. The error control module prepares the bitstream for delivery by segmenting the data into packets and adding the appropriate error protection. The packets are sent over a wireless link. The video encoder uses the ITU H.263 coding standard [6]. For error control, we consider a type-II hybrid ARQ scheme with a selective repeat retransmission protocol [2, 7, 15]. The hybrid ARQ scheme employs two codes, C_0 and C_1. C_0 is a (N, K) cyclic redundancy check (CRC) code, which is used as the error detection code. C_1 is a half-rate invertible shortened RS code $(2k, k, t)$ with m bits per symbol for both error detection and correction. The initial transmission is the information packet with CRC bits. If transmission errors occur, the retransmission consists of the parity packet generated by the half-rate invertible shortened RS code. If there is no error in the parity packet, the original information data is recovered from this packet. If errors are present in it, the parity packet is combined with the previously transmitted information packet (stored in the receiver buffer) to form rate 1/2 RS codes. If the errors are correctable by the RS codes or only one retransmission is allowed due to the delay constraint of real-time services, the RS decoded message is accepted. Otherwise, if an uncorrectable error pattern is detected by the RS decoder and more retransmissions would be possible within the required delay bound, the old erroneous information packet is discarded and the retransmission of the information packet is performed. When the new information packet is received, it is used to recover the information as described before. If this fails, the new erroneous information packet and the erroneous parity packet are combined to form rate 1/2 RS codes for error correction. If the errors are still not correctable, the next retransmission will be the parity packet. This process continues, i.e. alternating transmissions of the information packet and the parity packet, until the data is successfully accepted or the allowed maximum delay is reached.

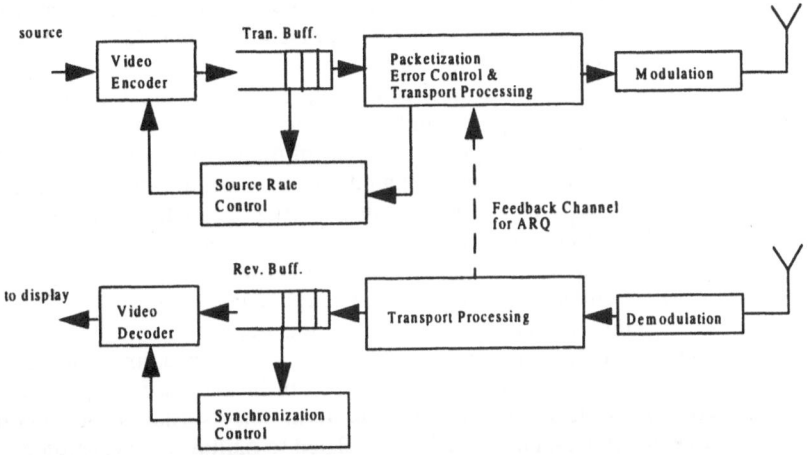

Fig.1. Diagram of wireless transport system.

An adaptive source rate control (ASRC) scheme [7] is used at the transmitter end to deal with the fluctuation of the effective data rate due to fades and retransmissions. It works in conjunction with the hybrid ARQ error control scheme to achieve the efficient transmission of real-time video with low delay and high reliability. When the channel conditions are good, the source rate is high and when the channel conditions are poor, the source rate is reduced and the extra bandwidth is used for retransmissions to reduce the channel errors. The channel conditions are estimated by the error control module of the transmitter from the outcome of recent packet transmissions based on acknowledgments fed back from the receiver. This information is passed up to the adaptive source rate control scheme. The ASRC scheme uses it to forecast the channel's effective data rate before a video frame is encoded. Once such an estimate is available, the ASRC scheme determines the target number of bits for the next encoded frame (i.e. the target source rate) based on the estimated effective data rate, the current transport buffer occupancy (i.e. the amount of data waiting for transmission before this frame), and the delay constrains. The change in source rate may result in the fluctuation of the encoded video quality. The ASRC scheme changes both the number of the intracoded macroblocks and the quantization scale used in a frame to achieve the allocated target source rate. The number of intracoded macroblocks in a frame is first adjusted. Only when the target source rate cannot be obtained by adjusting the number of intracoded macroblocks in a frame, is the quantization scale changed. This reduces the fluctuation in the value of the quantization scale during encoding so that a more uniform video quality is obtained.

In order to guarantee the end-to-end delay, a delay bound is set in the hybrid ARQ error control module. The transmitter discards any data whose delay exceeds the delay bound. An important issue is how to determine the delay bound so that good QoS (error rate, synchronization performance and end-to-end delay) can be achieved. We propose to employ a large delay bound for the error control module of the transmission system and a synchronization protocol at the application layer of the receiver to control end-to-end delay and maintain synchronization based on the channel conditions. The reason is that a large error control delay bound allows more effective recovery from transmission errors and provides the system with the capability to adapt to different channel conditions. It is noted that the compressed video data is very sensitive to transmission errors, i.e. the transmission errors greatly damage the video quality. The error control module using the hybrid ARQ scheme is adaptive. With a large delay bound, the error control module can allow more retransmissions when the channel conditions are poor so that the transmission error rate is kept low. When the channel conditions are good, the data arrives error free at the receiver in the initial transmission, then a low error rate and a low delay can still be achieved simultaneously even if the large delay bound is set.

Large-scale fading may last for a very long time, depending on the terrain, this results in long-term delay variation which causes synchronization problem. We develop an novel adaptive synchronization protocol in this paper. It piecewisely adjusts the end-to-end delay to match the long-term delay variation and smoothly recovers from synchronization errors. Issues related to multimedia synchronization in broadband wired networks have recently received significant attention, where the network delay variation is the major cause for synchronization loss [9-14]. The proposed synchronization scheme is distinct from other comparable schemes in many ways. First our synchronization control is directly based on the synchronization requirements of the application. This allows the system to achieve a desirable tradeoff between synchronization and end-to-end delay. Second we don't assume that the receiver clock is synchronized with the sender clock and that the communication delay model is previously known. It is more realistic because the channel conditions and the mobile speed are not known a priori and may change during the period of the connection. Third we combine the long-term delay equalization and the short-term synchronization error recovery in our protocol. Fourth we show that the synchronization protocol yields good performance and fits well into the wireless video conferencing architecture we proposed. We will discuss the synchronization protocol in more detail next section.

3. Synchronization

Multimedia services consist of several media streams. Each stream constitutes a succession of consecutive logical data units or media units (MUs) [9, 10]. In the case of an audio stream, MUs are individual samples or blocks of samples transferred together from the source to the sink. Similarly with video, one MU may typically correspond to a video frame. Close temporal relationships exist for the MUs both within a media stream and among the various media streams. Synchronization can be divided into intrastream synchronization and interstream synchronization. The intrastream synchronization means restoring the temporal ordering among the MUs in a single media stream before playback at the receiver. The quality of intrastream synchronization for a stream can be represented by the root mean square error (RMSE) of the inter-sample time of the MUs within a media stream [14]. For N MUs, let $G_i(n)$ and $P_i(n)$ denote the generation and the playout time of the nth MU ($n = 1, 2,...N$) of media stream i, respectively; the RMSE is defined as

$$\sigma_i = \sqrt{\frac{\sum_{n=2}^{N} [(P_i(n)-P_i(n-1))-(G_i(n)-G_i(n-1))]^2}{N-1}} \tag{1}$$

For the interstream synchronization, a MU in a stream should be played out with the corresponding MUs in the related streams. The size of MUs may not be the same in different streams. For example, in case of video conferencing, a video frame duration may be 67 ms (15 frames per second) but an audio frame duration may be 30 ms. Let $G_v(n)$ and $P_v(n)$ be the generation time and the playout time of the nth MU in the video stream, respectively, and $G_a(m)$ and $P_a(m)$ are the generation time and playout time of the mth MU in the audio stream, respectively. The nth MU in the video stream corresponds to the mth MU in the audio stream, that is, the generation time of the mth MU in the audio stream is closest to the generation time of the nth MU in the video stream. The quality of interstream synchronization between the audio and the video can be represented by the root mean square error of the playout time between the audio MUs and their corresponding video MUs. The root mean square error of interstream synchronization is defined as

$$\sigma_{av} = \sqrt{\frac{\sum_{m=1}^{M_a} [(P_a(m)-P_v(n))-(G_a(m)-G_v(n))]^2}{M_a}} \tag{2}$$

where M_a is the total number of audio MUs in the stream.

Retransmissions in the wireless transport system destroys the continuity of the media stream by introducing communication delay jitter. An approach to compensate for communication delay jitter so as to maintain intrastream synchronization is to use a buffer at the receiver. The receiver artificially delays the received data for a certain amount of control time and then plays them out with an equalized end-to-end delay. This additional artificial delay at the receiver is called the receiver buffer delay or the equalization delay [11]. If two corresponding media streams are played out by the same amount of end-to-end delay, the interstream synchronization between them is maintained.

The end-to-end delay consists of four parts. First is the collection delay. It is the time needed for the sender to collect media units and to encode them for transmission. The second part is the communication delay. Communication delay is defined here as the time from the instant that a MU enters the transport buffer to the instant that it is completely received by the receiver. This includes the transport buffer delay, the transmission delay, the propagation delay and the ARQ delay. The third part is the receiver buffer delay or the equalization delay. It is the time that an MU spends in the receiver buffer to equalize the end-to-end delay. This delay can be controlled for synchronization purpose. And the fourth is the delivery delay. Delivery delay is the time the receiver needs to process the MUs and prepare them for presentation. In this research, the collection delay and the delivery delay are not considered. They are processor related. We assume that the processors have enough power to handle the collecting/displaying and the encoding/decoding processes. The end-to-end delay is then the time from the instant that the frame enters the transport buffer to the instant that it is taken from the receiver buffer, including the communication delay and the controlled equalization delay.

Theoretically, if the maximum communication delay of a frame during a video conference session is known, the first frame is buffered and the playout starts when its end-to-end delay is greater than or equal to the maximum communication delay. All subsequent frames can then be continuously played back with their original temporal relationship and no discontinuity will occur (i.e. intrastream synchronization is maintained). However the maximum communication delay during a session depends on the channel conditions. When the channel conditions are good and there are no retransmissions, the maximum communication delay is small so that synchronization can be guaranteed with a small end-to-end delay. The channel conditions are not known a priori and may change during the connection period. Although a large equalization delay can be chosen for the worst case, this may result in unnecessary long end-to-end delays when the channel conditions are good. For interactive real-time applications, long end-to-end delays should be avoided.

It is well known that the human perception system tolerates small synchronization errors in video and audio signals. Occasional and small variations in video playout rate are tolerable. People won't notice even if the skew between the video and the associated audio is as large as 80 ms [9]. Thus an adaptive synchronization mechanism is desirable because it can provide optimal QoS for a given channel resource.

We focus on wireless real-time multimedia services such as video conferencing. Retransmissions are the major cause for communication delay jitter in the system. Without loss of generality, we consider that the session consists of a video stream and a voice stream. Voice can tolerate a much higher error rate than video and error concealment can be used effectively, therefore it is not necessary to use retransmission [4]. In this study, we assume that voice data arrives at the receiver through another physical or logical channel with less communization delay (as no retransmissions occur). The video stream is encoded using the H.263 coder. It is noted that the scheme we develop here is applicable to other video coding schemes.

Fig. 2 shows the diagram of decoding and playout mechanisms at the receiver. We assume that each MU (video or audio) carries a time stamp, indicating its generation time. For H.263 encoded video, the temporal reference field in the picture header represents the time stamp. At the receiver, a MU is instantaneously

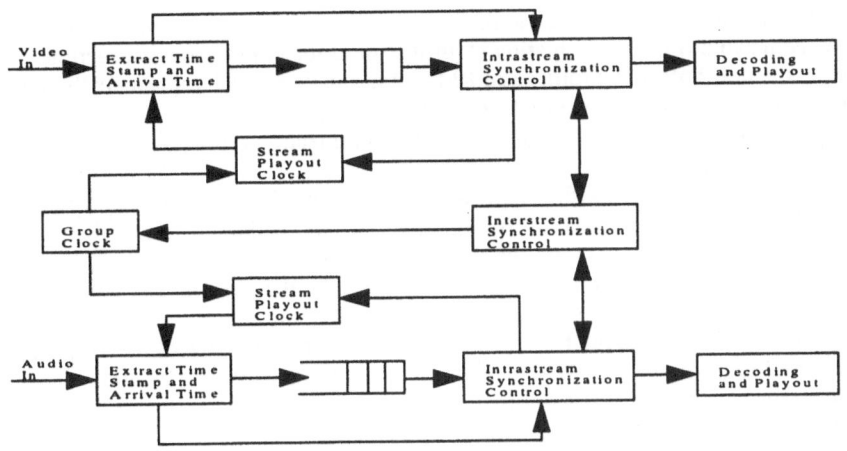

Fig.2. Diagram of playout mechanism at the receiver.

removed from the receiver buffer for playout at its playout time determined by the synchronization protocol. The playout clock is needed as a time reference. In our study, we use the stream clocks and the group clock as proposed in [14]. This structure gives flexibility in synchronization control. Each media stream has its own playout clock for intramedia synchronization. If there is no need for intermedia synchronization, the MUs are played out according to their own playout clock. If intermedia synchronization is needed, the group clock updates the stream clocks of all the related streams. The playout clock is piecewisely adjusted by the synchronization protocol based on the channel conditions which determine the communication delay characteristics. This is equivalent to changing the end-to-end delay.

The synchronization protocol achieves the following three goals: (1) during the call setup period, determining the equalization delay based on the channel conditions, (2) monitoring and estimating the communication delay characteristics and adjusting the equalization delay to match long-term delay variation due to large-scale fading, (3) when a synchronization error occurs, reducing its impact on the quality degradation. It can be decomposed into two phases: (1) playout clock adjustment, (2) playout time determination.

3.1. Playout clock adjustment

The primary function of this phase is to adjust the playout clock based on the communication delay characteristics so that the synchronization requirement is satisfied and the end-to-end delay is kept low.

Before we describe the protocol. Let us define the following items associated with the *nth* MU in stream i ($i = v$: video, $i = a$: audio).

$G_i(n)$: the generation time of MU n in stream i based on the source clock, i.e. the value of the time stamp.

$A_i(n)$: the arrival time at which MU n completely arrives at the receiver based on the playout clock,

$P_i(n)$: the playout time at which MU n is taken from the decoder buffer, decoded and played out, with reference to the playout clock. The playout time is determined by the playout time determination algorithm described in the next subsection.

$\sigma_i^{(t)}$: the threshold synchronization RMSE for stream i which represents the intrastream synchronization error that stream i can tolerate.

S_{max}: the maximum allowed intermedia synchronization error (skew) between the video and audio streams.

The stream clock is first adjusted to satisfy the intrastream synchronization requirement. The clock should be initialized in the beginning. Since the channel conditions are not known and the protocol is adaptive, the stream clock is simply set to be the time stamp carried by the first MU in the stream. The clock is then driven by the receiver system clock.

As soon as all the related stream clocks have been initialized, the group clock is initialized. The group clock is set to be equal to the slowest stream clock of all the related stream clocks. After the group clock is initialized all the related stream clocks are then resynchronized with the group clock.

We let the time carried in the time stamp of a MU be the scheduled playout time of this MU with reference to its playout clock at the receiver. Note that the scheduled playout time according to the playout clock is now the generation time of a MU (i.e. the time carried in the time stamp) according to the source clock. If there is no communication delay jitter, the arrival time of every MU (based on the playout clock) is equal to its scheduled playout time. The MU can then be played out at its scheduled time. The original temporal relationship at the source is maintained at the receiver, i.e., they are synchronized. However, in real systems, some of the MUs arrive later and some of them arrive earlier due to communication delay jitter. In our synchronization protocol, if an MU arrives earlier than its scheduled playout time, it will be played out at its scheduled playout time. If it arrives later than its scheduled playout time, a synchronization error has occurred. Its playout time will be determined by the playout time determination algorithm described in the next subsection which attempts to recover from the synchronization error smoothly.

Based on the communication delay characteristics, the playout clock is adjusted to find the optimal playout point. If the channel conditions are poor and the arrival time of many frames is beyond their scheduled time, the playout clock can be decreased so that the synchronization requirements are satisfied. This is equivalent to increasing the end-to-end delay. If the channel conditions become good and retransmissions are rare, then the arrival time of all the frames will be ahead of their scheduled time. The playout clock can then be increased. This is equivalent to decreasing the end-to-end delay.

The communication delay variation can be classified into: (1) short-term delay variations, mainly caused by the small-scale Rayleigh fading. It generally lasts for a few frames when the adaptive source rate control is used at the transmitter, and (2) long-term delay variations, caused by large-scale fading. This depends on the terrain. It may last for a long time, several minutes or tens of minutes. The adjustment mechanism for the playout clock should not react to short-term delay variations because each adjustment of the playout clock will require to pause or speed up media units, i.e., results in synchronization errors.

A sliding window is used to detect the necessity for the clock adjustment and to estimate the adjustment amount in our protocol. The window contains the MUs which were most recently received. The window size can be changed from zero to W_{max}. In the beginning or after a clock adjustment, the window size is set to be zero and then increased by one for every newly received MU until the size reaches W_{max}. The playout clock is adjusted under the following two conditions:

(1) The RMSE of intrastream synchronization is monitored for the MUs in the window. Let α ($\alpha = 0, 1, 2, 3,...W_{max}$) denote the current window size. $P_i(n)$ denotes the playout time of the nth MU ($n = 1, 2,...\alpha$) in the window when the window is not empty. As described in the next section, the $P_i(n)$ is determined by the playout time determination algorithm according to the current playout clock. Let $G_i(n)$ and $A_i(n)$ be the generation time and the arrival time of the nth MU in the window, respectively. When the window size is greater than 2, the RMSE is obtained by

$$\sigma_i = \sqrt{\frac{\displaystyle\sum_{n=2}^{\alpha} [(P_i(n) - P_i(n-1)) - (G_i(n) - G_i(n-1))]^2}{W_{max} - 1}} \tag{3}$$

If σ_i is greater than the threshold RMSE, $\sigma_i^{(t)}$, the clock needs to be adjusted. To estimate the adjustment amount, we find the maximum communication delay that the MUs in the window experienced. It is equal to

$$\Delta = Max(A_i(n) - G_i(n)), n \in Window \tag{4}$$

The playout clock is then adjusted by $-\Delta$, this is equivalent to having the end-to-end delay increased by Δ. The MU with the greatest communication delay in the window could then arrive at the receiver and be played out at its scheduled playout time according to the new playout clock, i.e., the end-to-end delay is adjusted to the maximum communication delay in the window. The window size is set to be zero every time after the playout clock is adjusted.

(2) When the window size becomes W_{max}, the arrival time of the MUs in the window is monitored. If the arrival time of every MU in the window is ahead of its scheduled playout time, that is, the arrival time of the MU is less than its scheduled playout time for W_{max} consecutive MUs, the clock is adjusted. Again $\Delta = Max(A_i(n) - G_i(n)), n \in Window$, and the playout clock is adjusted by $-\Delta$. Note that Δ is less than 0 in this case, the clock is actually increased by an amount $|\Delta|$, that is, the end-to-end delay decreases by $|\Delta|$. The window size is set to be zero after the playout clock is adjusted.

After a playout clock for a stream is updated according to the intrastream synchronization, the group clock can be adjusted to satisfy intermedia synchronization requirements. As we did in the initialization phase, the group clock is simply set equal to the slowest stream clock among all the related stream clocks. All the related stream playout clocks are then resynchronized with the group clock. After adjustment, the playout for a stream is based on its new playout clock. Note that each adjustment of the playout clock results in either pausing or speeding up the playout process.

3.2. Playout time determination

If an MU arrives at the receiver before its scheduled playout time according to the playout clock, it is played out at the scheduled playout time. However it is possible that some MUs arrives late due to retransmissions. Late arriving MUs result in synchronization errors. The second phase of the synchronization protocol is to reduce the impact of the errors. After an error occurs the protocol smoothly recovers from the error.

In our wireless transmission system, there are no retransmissions for the audio. We assume that the audio MUs arrive at the receiver on time. The audio MUs play out at their scheduled playout time. Video will experience retransmissions when the channel is in a fade. We focus on synchronization error recovery for the video stream. The playout time determination algorithm in the synchronization protocol will determine the playout time for a video frame based on its arrival time, the scheduled playout time and the intrastream and interstream synchronization requirements. In our scheme the interstream synchronization control is performed after the intrastream synchronization control. For the interstream synchronization, we let the audio stream be the master stream and the video stream be the slave stream, the slave stream is synchronized with the master stream [10, 13]. Note that the playout clock constitutes the time reference at the receiver.

After video frame n is received, its playout time is determined as follows.

3.2.1. Determine the playout time based on intrastream synchronization

(1) If frame n-1 is displayed at its scheduled playout time, frame n will be displayed at its scheduled playout time or its arrival time, depending on which is later, that is,

$$P_v(n) = Max(G_v(n), A_v(n)). \tag{5}$$

This indicates that if frame n arrives early, it is held and displayed at its scheduled playout time, otherwise, it is played out as soon as it arrives.

(2) If frame $n-1$ is displayed after its scheduled playout time (i.e., $P_v(n-1) > G_v(n-1)$), this means that frame $n-1$ is being played out behind schedule and a synchronization error has occurred. The original display period of frame $n-1$ should be $G_v(n)-G_v(n-1)$. If we display frame $n-1$ for a period of $G_v(n)-G_v(n-1)$, that is, frame n starts to be displayed at $P_v(n) = P_v(n-1) + G_v(n)-G_v(n-1)$, the end-to-end delay will go up for all subsequent frames. The delay should be as small as possible for real-time services. On the other hand, if frame n is played out at its scheduled playout time $G_v(n)$ or arrival time $A_v(n)$, the display period for frame $n-1$ will be $G_v(n)-P_v(n-1)$ or $A_v(n)-P_v(n-1)$. This value may be much smaller than the original display period of frame $n-1$ and motion jerkiness will be noticed. People are much more sensitive to a single large motion jerkiness than several consecutive small variations of the display period [12]. Therefore synchronization error should be recovered from smoothly. We set the playout time of frame n to be equal to

$$P_v(n) = Max[P_v(n-1) + G_v(n) - G_v(n-1) - \delta, G_v(n), A_v(n)] \tag{6}$$

The first term on the right side of the equation represents the smooth error recovery where δ is a parameter to ensure that the display period of frame $n-1$ is not too small. The second term means that frame n will be displayed at its scheduled playout time if the scheduled playout time results in less distortion during the display period of frame $n-1$. The third term indicates that frame n must be displayed after it is received.

3.2.2. Adjust the playout time based on interstream synchronization

The video stream is a slave stream. The next step is to adjust the playout time of video frame n so that the interstream synchronization with the master stream (audio here) can be satisfied. Let m be the mth MU in the master stream whose generation time is equal to or just ahead of video frame n. The interstream synchronization error is given as

$$\varepsilon_{inter} = [P_v(n) - P_a(m)] - [G_v(n) - G_a(m)] \tag{7}$$

If the interstream synchronization error is within the allowed range, the playout time of frame n is determined from the intrastream synchronization scheme. Otherwise the playout time is adjusted to satisfy the interstream synchronization requirement, that is,

$$if \; |\varepsilon_{inter}| \le S_{max}, P_v(n) = P_v(n)$$

$$else \; if \; \varepsilon_{inter} > S_{max}, P_v(n) = Max[P_a(m) + (G_v(n) - G_a(m)) + S_{max}, A_v(n)]$$

$$else \; if \; \varepsilon_{inter} < -S_{max}, P_v(n) = P_a(m) + (G_v(n) - G_a(m)) - S_{max} \tag{8}$$

With current common display technology, a video frame generally cannot be presented for any period at any time [9]. The display time must be synchronized with the refresh start time of the display device. We assume that the display refresh granularity is L, for example $L = 1/60$ sec. Then frame n's display time is

$$d_v(n) = d_v(n-1) + \left\lceil \frac{P_v(n) - P_v(n-1)}{L} \right\rceil L \tag{9}$$

where $\lceil a \rceil$ is the minimum integer equal to or greater than a.

4. Performance Results

In this section, we study the performance of the wireless transport system through simulations and focus on the synchronization and delay behavior of wireless video conferencing. We assume that there is a video stream and a voice stream in the video conferencing session. Voice can tolerate a much higher error rate; it is not necessary to use a retransmission scheme. In this study, we concentrate on the video stream and assume that voice data always arrives at the receiver before its scheduled playout time through another physical or logical channel.

The system model for video transmission is the same as shown in Fig. 1. We set the delay bound in the error control module to be 400 ms. This is the worst case delay that a video conferencing application can tolerate [16]. Data will be discarded by the error control module if its delay is over 400 ms. We assume that video is transmitted over a TDMA radio network with a user data rate of 32 kb/s. The total number of bits in a packet is 420, consisting of 400 data bits and 20 CRC bits. The data bits may be the original information data or the parity generated by the RS code.

In this study both small-scale fading and large-scale fading are considered. For the small-scale fading, a fading simulator based on Jack's model is used to generate the Rayleigh fading signal [17]. Large-scale fading strongly depends on the terrain. It is often based on measurements. For simplicity, we use a two-state Markov model for the large-scale fading [18]. The large-scale fading results in a change in the local mean value of the Rayleigh fading by as much as 10 dB [1]. In the simulation we assume that the local mean SNR is 30 dB for the good state and the local mean SNR 20 dB for the bad state. The QCIF "Mother and Daughter" video sequence is considered and the video frame interval is about 67 ms (15 frame/s).

In all of our simulations, the resultant frame error rate after error control is less than 10^{-4}. The picture quality degradation due to transmission errors is negligible because the large delay bound (400 ms) in the hybrid ARQ error control module allows sufficient retransmissions to correct the errors. This shows that it is beneficial to use a large delay bound in the error control module.

To get some insight, we show the typical operation of the synchronization protocol in Fig. 3. We observe that the protocol tries to keep the end-to-end delay low

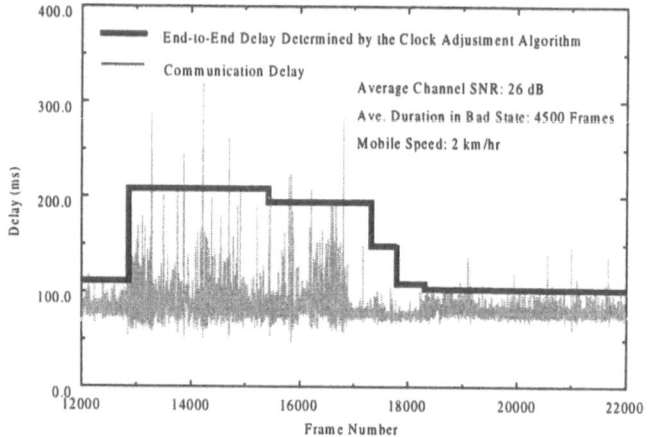

Fig.3. The communication delay and the end-to-end delay determined by the clock adjustment algorithm.

to obtain good interactivity when the channel is good and the communication delay jitter is small. When the channel becomes poor, the protocol increase the equalization delay to compensate for the large communication delay jitter introduced by retransmissions. The end-to-end delay is scarified to preserve good synchronization performance.

Next we study the impact of the channel conditions on the performance. Fig. 4 presents the resultant intrastream synchronization RMSE, interstream synchronization RMSE and the mean end-to-end delay versus the average channel SNR for a slow mobile (speed = 2 km/hr) and a fast mobile (speed =100 km/hr). The curves for the uncontrolled case are also shown in the figures for comparison. Uncontrolled means that a video frame is displayed as soon as it is received, that is, the end-to-end delay is equal to the communication delay. Audio is played with the mean communication delay of the video for the uncontrolled case.

We observe that without the synchronization control the synchronization RMSE is quite high due to the communication delay jitter. When the average channel SNR is higher, the channel is in a good state most of time and the communication delay jitter (introduced by retransmissions) becomes smaller. Thus the

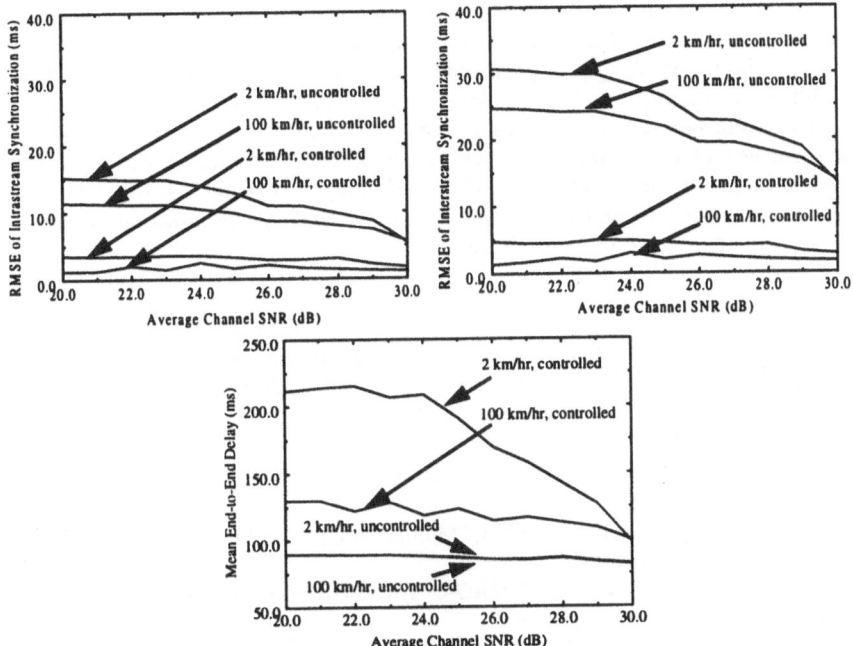

Fig.4. Intrastream synchronization RMSE, interstream synchronization RMSE and end-to-end delay for a slow mobile and a fast mobile.

synchronization RMSE decrease. When the mobile speed is high, the error pattern is more random so that a smaller synchronization RMSE is obtained due to less communication delay jitter. When synchronization control is used, the synchronization RMSE is kept low. However the mean end-to-end delay is increased compared to the uncontrolled case. This is because a controlled equalization delay is introduced to compensate for the communication delay jitter. When the average SNR and/or the mobile speed is high, the communication delay jitter decreases and less equalization delay is required to compensate for the communication delay jitter so that the resultant end-to-end delay can be lower. This shows that the adaptive synchronization protocol performs reasonably well as it can effectively adjust the equalization delay to satisfy the synchronization requirements and keep the end-to-end delay low based on the channel conditions.

5. Conclusions

Although the hybrid ARQ scheme can significantly improve transmission reliability and yield high throughput for nonstationary wireless channels, it results in variable effective data rates for current PCS networks and large delay jitter. The original temporal relationship of the real-time information streams is distorted due to delay jitter introduced by retransmissions, which causes synchronization problems in multimedia services. We studied the approach of employing a large delay bound for the hybrid ARQ error control module of the transmission system to effectively correct transmission errors and an adaptive synchronization scheme at the application layer of the receiver to maintain synchronization and to keep end-to-end delays low. A novel synchronization protocol was also presented, which adaptively changes the end-to-end delay to match the long-term delay variation due to large scale fading and smoothly recovers from the synchronization errors. The performance of the proposed transport system was evaluated. It was shown that low transmission error rates, low end-to-end delays and good synchronization are feasible when error control, source rate control and synchronization control are jointly considered and properly designed.

References

1. B. Sklar, "Rayleigh fading channels in mobile digital communication systems," IEEE Communications Magazine, Vol. 35, No. 7, PP.90-109, July 1997.
2. H. Liu and M. El Zarki, "Performance of video transport over wireless networks using hybrid ARQ," Proc. of ICUPC 96, Boston, Oct. 1996.
3. ITU-T Recommendation H.324, "Terminal for low bitrate multimedia communication," Nov. 1995.
4. ITU-T Draft Recommendation H.223, "Multiplexing protocol for low bitrate multimedia communication," Nov. 1996.
5. ISO-IEC/JTC1/SC29/WG11, "MPEG-4 syntax description language specification (MSDL), version 1.3," Sept. 1996.
6. ITU-T Recommendation H.263, "Video coding for low bitrate communication," 1996.
7. Hang Liu and Magda El Zarki, "Adaptive source rate control for real-time wireless video transmission," ACM/Baltzer J. Mobile Networks and Applications, Vol. 3, No. 1, PP. 49-60, Jan. 1998.
8. J. Padgett, C. Gunther, and T. Hattori, "Overview of wireless personal communications," IEEE Commun. Mag., Vol. 33, No. 1, PP. 28-41, Jan. 1995.
9. R. Steinmetz, "Human perception of jitter and media synchronization," IEEE JSAC, Vol. 14, No. 1, PP. 61-72, Jan. 1996.
10. R. Steinmetz, "Synchronization properties in multimedia communication systems," IEEE JSAC, Vol. 8, No. 3, PP. 401-412, Apr. 1990.
11. J. Escobar, D. Deutsch, and C. Partridge, "Flow synchronization protocol," Proc. IEEE INFOCOM 92, PP. 1381-1387.
12. M. Yuang, S. Liang, Y. Chen, and C. Shen, "Dynamic video playout smoothing method for multimedia applications," IEEE INFOCOM'96.
13. S. Yoo and D. Kim, "An unified synchronization mechanism for general multimedia services in ATM networks," IEEE INFOCOM'97.
14. C. Liu, Y. Xie, M. Lee and T. Saadawi, "Multipoint multimedia teleconference system with adaptive synchronization," IEEE JSAC, Vol. 14, No. 7, 1422-1435, Sept. 1996.
15. S. Lin and D. Costello, Error control coding: fundamentals and applications, Prentice-Hall Inc., Englewood Cliffs, NJ, 1983.
16. R. Cox and P. Kroon, "Low bitrate speech coders for multimedia communication," IEEE Comm. Magazine, Vol. 34, No. 12, PP. 34-41, Dec. 1996.
17. M. Jeruchim, P. Balaban, and K. Shanmugan, Simulation of communication systems, Plenum Press, New York, NY, 1992.
18. E.N. Gilbert, "Capacity of burst noise channels," Bell Sys. Tech. J., Vol. 39, PP. 1253-1256, 1960.

Measurement-Based Admission Control with Aggregate Traffic Envelopes

Edward W. Knightly and Jingyu Qiu

Electrical and Computer Engineering Department
Rice University
http://www.ece.rice.edu/networks

Abstract. The goal of admission control is to support the quality-of-service demands of real-time applications via resource reservation. In this paper, we introduce a new approach to Measurement-Based Admission Control (MBAC) based on adaptive and measurement-based *maximal rate envelopes* of the aggregate traffic flow. We show that such traffic envelopes provide a robust and accurate characterization of the aggregate traffic, capturing its temporal correlation as well as the available statistical multiplexing gain. In estimating applications' future performance, we introduce the notion of a schedulability confidence level which describes the uncertainty of the measurement-based "prediction" and reflects estimation errors and temporal variations in the measured envelope. We then apply principles from *extreme value theory* and devise techniques to estimate the packet loss probability for a buffered multiplexer. Our results are quite general and apply to heterogeneous and bursty traffic flows belonging to a broad class of underlying traffic distributions including Gaussian, log-normal, and Gamma.

1 Introduction

To satisfy the Quality of Service (QoS) requirements of real-time multimedia applications, networks must employ resource reservation and admission control [2, 5, 13]. However, traditional approaches to resource reservation require that an accurate characterization of each flow be specified at flow setup time, a requirement that, in practice, may be difficult for many "live" applications to meet. This is especially problematic for applications which exhibit rate variations over multiple time scales, as this behavior is not adequately characterized by standard traffic models such as the leaky bucket [8, 14].

An alternative technique for supporting applications with ill-specified traffic characteristics is a *measurement-based* service: by basing admission control decisions on measured values of traffic parameters rather than *a priori* client-specified guesses, the effects of mistaken client traffic characterizations are largely alleviated, as is the need for a traffic model which captures the exact multiple time scale behavior of each traffic flow.

This research is supported by NSF CAREER Award ANI-9733610 and NSF Grant ANI-9730104

In this paper, we propose and evaluate a new envelope-based MBAC algorithm. Our approach can be divided into two inter-related components: 1) measurement of the aggregate flow, which prescribes what to measure and how, and 2) admission control, which uses these measurements to predict future performance with admission of a new traffic flow.

To characterize the behavior of the aggregate flow, we adaptively measure its *maximal rate envelope*. Specifically, a traffic flow's *rate* is only meaningful if it is associated with a corresponding interval length. A rate envelope therefore describes the flow's behavior as a function of interval length. By measuring the *maximal* rates over the corresponding interval lengths, we directly characterize the peaks and extremes of the aggregate flow which are most likely to lead to packet loss. We apply extreme value theory [1] to obtain insights into the statistical behavior of the maximal rate envelope for a broad class of underlying traffic characteristics, including Gaussian, log-normal, and Gamma distributions, and show that this envelope is a simple and robust description of the aggregate traffic flow.

Exploiting properties of this measured aggregate traffic envelope, we devise a new admission control algorithm as follows. First, subject to the (incorrect) assumption that the past maximal rate envelope of the aggregate flow will continue to bound future packet arrivals, one could ensure that, with admission of a coarsely characterized new flow, *no* buffer overflow would occur. Conceptually, we begin with such a framework, but quantify the uncertainty of the prediction with what we term a *schedulability confidence level*, which reflects the variation in past envelope measurements and the uncertainty of the prediction of the future workload. From extreme value theory and this concept of schedulability confidence level, we estimate the quality of service parameters that applications are ultimately concerned with, such as packet loss probability and delay-bound violation probability. For example, we derive an expression for packet loss probability by determining the mean number of packets lost when the future arrivals exceed the measured envelope of the past.

To evaluate our approach, we perform an extensive set of simulation experiments using traces of compressed video. We show that our approach is robust and accurate over a wide range of scenarios and is able to extract the available statistical multiplexing gain, achieving for example utilizations of up to 79% for a 45 Mbps router with a target packet loss probability of 10^{-4} and a 20 msec delay bound. We also perform experiments with lower link rates and find that even in such cases where limited statistical multiplexing gains are available such as in a link sharing environment [7], the MBAC algorithm correctly restricts the admissible region to lower utilizations.

Previous approaches to MBAC such as [4, 6, 9, 10, 11, 17] have employed a number of measurement methodologies to characterize traffic including instantaneous peak rate, the instantaneous rate's mean and variance or moment generating function, and per-flow statistics. In addition, a number of theoretical techniques have been applied to study various aspects of MBAC including large deviations theory, Gaussian modeling, Hoeffding bounds, and decision theory.

Our approach differs first in our use of a maximal rate envelope to characterize traffic and capture the aggregate flow's interval-based behavior. Second, exploiting properties of this envelope and applying extreme value theory, we can control QoS for a buffered multiplexer servicing a broad class of underlying traffic types, including cases of moderate numbers of traffic flows in which Gaussian approximations may not be applicable. Finally, in the simulation scenario described above, we experimentally find this MBAC algorithm to significantly out-perform previous approaches.

2 Algorithm for Measurement-Based Admission Control

Here, we describe a new approach for measurement-based admission control that utilizes measured values of aggregate traffic envelopes. Our scheme consists of a measurement algorithm and an admission control algorithm. The measurement algorithm continually updates the recent empirical aggregate envelope and measures the envelope's temporal variation. The admission control algorithm, invoked upon arrival of a new flow's admission request, conceptually consists of two parts. First, we check for aggregate schedulability with an associated prediction confidence level, and second, we determine the packet loss probability. The new flow is admitted if the estimated future performance parameters satisfy the QoS requirements of all flows.

2.1 Adaptive Measurement of Aggregate Rate Envelope

In characterizing a flow's rate, an associated interval length must also be specified. For example, denoting $A[s, s+I_k]$ as a flow's arrivals in the interval $[s, s+I_k]$, $A[s, s+I_k]/I_k$ is the rate in this particular interval. Moreover, the peak rate over any interval of length I_k is given by $R_k = max_s A[s, s+I_k]/I_k$. We refer to a set of rates R_k which bound the flow's rate over intervals of length I_k as a maximal rate envelope [12].

The goal of our measurement methodology is twofold. First, by measuring an envelope of the aggregate flow, we capture the short time scale burstiness of the traffic which we will use to analyze the dynamics of a buffered multiplexer with admission of a new flow. Second, we measure the variation of the aggregate rate envelope to characterize measurement errors and longer time scale fluctuations in the traffic characteristics. Using the variance of the measured envelope, we can determine the confidence values of our schedulability condition and estimate the expected fraction of packets dropped should the schedulability condition fail to hold.

We consider time to be slotted with width $\tau = I_1$, the minimum interval of the measured rate envelope (τ may be larger than the packet transmission time). Denoting a_t as the aggregate arrivals in time slot t such that $a_t = A[t\tau, (t+1)\tau]$, we define the maximal rate envelope over the past T time slots from the current time t as

$$R_k^1 = \frac{1}{k\tau} \max_{t-T+k \leq s \leq t} \sum_{u=s-k+1}^{s} a_u \tag{1}$$

for $k = 1, \cdots, T$. Thus, the envelope $R_k^1, k = 1, \cdots T$ describes the aggregate maximal rate envelope over intervals of length $I_k = k\tau$ in the most recent $T\tau$ seconds. This envelope measures the short-time scale burstiness and autocorrelation structure of the aggregate flow.

Every T time slots,[1] the current envelope R_k^1 is measured using Equation (1) and $R_k^n \leftarrow R_k^{(n-1)}$ for $k = 1, \cdots, T$ and $n = 2, \cdots, N$. Thus, the variance of the measured envelopes over the past M windows of length T can be computed as

$$\sigma_k^2 = \frac{1}{M-1} \sum_{m=1}^{M} (R_k^m - \bar{R}_k)^2 \tag{2}$$

where $\bar{R}_k$ is the empirical mean of the R_k^m's, $\sum_m R_k^m / M$. Thus, we measure the *variability* of the aggregate envelope over $T \cdot M$ time slots to characterize the variation of the envelope itself over longer time scales.

2.2 Background on Extreme Value Theory

From Equations (1) and (2), we can adaptively measure the maximal rate envelope of the aggregate traffic. Unlike the case of deterministic service in which the envelope is bounded *a priori*, the measurement-based traffic envelope should be viewed as random for MBAC. Below, we introduce extreme value theory which we use to obtain insights into the behavior of the maximal rate envelope and serve as a theoretical framework to characterize the temporal variation of the measured peak rates. We apply this theory in Sections 2.3 and 2.4 to predict the packet loss probability with admission of a new traffic flow.

To motivate the application of extreme value theory, consider a sequence $X_1, X_2, \cdots$ of independent and identically distributed random variables with distribution $F(x)$. The maximum of n X_i has distribution

$$P(\max_{1 \leq i \leq n} X_i \leq x) = F^n(x). \tag{3}$$

Extreme value theory addresses the asymptotic distribution of $\max_{1 \leq i \leq n} X_i$: analogous to how the central limit theorem describes the distribution of sums of random variables without requiring knowledge of their exact underlying distributions, extreme value theory describes the distribution of the *extremes* of sequences of random variables for a general class of underlying distributions. In particular, for a large class of distributions $F(x)$, including Gaussian, exponential, log-normal, Gamma, Gumbel, and Raleigh distributions,

$$\lim_{n \to \infty} P(\max_{1 \leq i \leq n} X_i \leq x) = \exp[-\exp(-\frac{x-\lambda}{\delta})] \tag{4}$$

where $\exp[-\exp(-\frac{x-\lambda}{\delta})]$ is a Gumbel distribution with mean $\mu = \lambda + 0.57772\delta$, and variance $\sigma^2 = \pi^2 \delta^2 / 6$. Moreover, even if $X_1, X_2, \cdots$ are *dependent*, for most correlation structures and the same class of distributions above, the asymptotic distribution of $P(\max_{1 \leq i \leq n} X_i \leq x)$ is still Gumbel [1].

---

[1] Guidelines for setting the measurement window T, typically on the order of several seconds, are presented in Section 2.5.

2.3 Aggregate Schedulability and Confidence Level

Upon arrival of a new flow request, we first perform a test for "aggregate schedulability". This test, quite different from a strictly worst-case analysis as in [18], ensures that for a certain confidence level, no packet loss will occur if the new flow is admitted. The confidence level is required here since there is no *a priori* assurance that the past envelope will continue to bound the aggregate flow, as is the case in a deterministic approach.

Theorem 1. *Consider a new flow bounded by $r_k, k = 1, \cdots, T$ requesting admission to a first-come-first-serve server with capacity C, buffer size B, and a workload characterized by a maximal rate envelope with mean $\bar{R}_k$ with variance σ_k^2, $k = 1, \cdots, T$. With confidence level $\Phi(\alpha)$, no packet loss will occur with admission of the new flow if*

$$\max_{k=1,2,\cdots,T-1} \{k\tau(\bar{R}_k + \alpha\sigma_k + r_k - C)\} \leq B \tag{5}$$

and

$$\bar{R}_T + \alpha\sigma_T + r_T \leq C \tag{6}$$

Proof. Equation (5) considers the buffer dynamics of the multiplexer and ensures that the maximal buffer occupancy is smaller than the buffer size. Equation (6) is a *stability condition*, as it ensures that the mean rate over intervals of length T is less than link capacity with confidence level $\Phi(\alpha)$, so that the busy period is less than T also with probability $\Phi(\alpha)$.

For an interval of length T, we denote the distribution of the maximal rate R_k by $F_k(\cdot)$. With probability

$$\Phi(\alpha) = \int_{-\infty}^{\bar{R}_k + \alpha\sigma_k} dF_k$$

the future aggregate flow, denoted by $A[s, t]$, satisfies

$$P\{\max_s A[s, s+k\tau]/k\tau \leq \bar{R}_k + \alpha\sigma_k\} = \Phi(\alpha) \tag{7}$$

Since

$$\max_s\{A[s, s+t] + \hat{A}[s, s+t]\} \leq \max_s A[s, s+t] + \max_s \hat{A}[s, s+t], \tag{8}$$

the rate envelope of the past aggregate flow multiplexed with the new flow is bounded by $(\bar{R}_k + r_k + \alpha\sigma_k), k = 1, 2, \cdots, T$, also with probability $\Phi(\alpha)$. Therefore, with this bound on the envelope of the aggregate flow that includes the new flow, from [3] the maximum queue length will be bounded by Equation (5) with probability $\Phi(\alpha)$.

From extreme value theory, we approximate R_k as having a Gumbel distribution with mean $\bar{R}_k$ and variance σ_k^2 so that the parameters of the Gumbel distribution λ and δ can be obtained from the sample mean and variance of the maximal rate envelope as $\delta = \sqrt{6}\sigma_k/\pi$ and $\lambda = \bar{R}_k + 0.57772\delta$. Thus, the schedulability confidence level is

$$\Phi(\alpha) = \exp(-\exp(-\frac{\alpha - \lambda_0}{\delta_0}))$$

where $\delta_0 = \sqrt{6}/\pi$ and $\lambda_0 = 0.57772\delta$. $\square$

We make the following observations about Theorem 1. First, by applying extreme value theory, we have approximated the distribution of the maximal rate envelope for a wide class of (unknown) underlying traffic distributions. The temporal correlation structure of the aggregate flow is captured by the rate envelope for intervals up to T seconds. We address correlation at time scales greater than T in [15], but ignore it here for simplicity (again, T is typically on the order of several seconds).

Second, a new flow is not required to specify a complete rate envelope. For example, if only an estimate of a single rate is known at flow setup time, then the source may specify this value of the peak rate (r_1) for all of the r_k's. If instead the source characterizes its traffic with the standard dual leaky bucket model using parameters (P, σ, ρ), then its maximal rate over intervals of length I_k is given by

$$r_k = \frac{1}{I_k}min(PI_k, \sigma + \rho I_k) \qquad (9)$$

Third, we note that even with a modest number of multiplexed flows, $\bar{R}_k + \alpha\sigma_k >> r_k$. Consequently, if a new flow mischaracterizes its traffic parameters, the impact on the schedulability condition is relatively minor. Note further that specified traffic parameters (correct or not) for previously admitted flows are unused in the test as the impact of these flows on network performance is *measured* via the aggregate envelope.

Finally, we note that despite our use of a maximal rate envelope, our approach is able to measure and exploit the extent to which sources statistically multiplex. For example, if flows happened to synchronize in a worst-case manner, then the measured envelope would be exactly the sum of the individual envelopes of Equation (9). However, when flows are statistically independent and economies of scale are present, we expect that in practice, $\bar{R}_k + \alpha\sigma_k$, the rate envelope of the aggregate process, will be significantly less than the sum of the individual worst-case envelopes. We explore these observations experimentally in Section 3.

2.4 Packet Loss Probability

As described above, the *maximal* rate of the aggregate flow over intervals of length I_k has mean $\bar{R}_k$ and variance σ_k^2 as given by past measurements. Theorem 1 provides a no-loss schedulability condition that is satisfied with probability $\Phi(\alpha)$. However, if the future aggregate rate envelope exceeds $\bar{R}_k + \alpha\sigma_k$, then packet loss may occur, and the fraction of packets lost, or the packet loss probability, may be calculated from the following theorem.

Theorem 2. *Consider an aggregate traffic flow that satisfies the schedulability condition of Theorem 1 and has mean bounding rate $\bar{R}_k$ and variance σ_k^2 over intervals of length $k\tau$. For a link capacity C, buffer size B, and schedulability confidence level $\Phi(\alpha)$, the packet loss probability is bounded by*

$$\max_{k=1,2,\cdots,T} \frac{\sigma_k \Psi(\alpha) \cdot I_k}{\bar{R}_T \cdot T\tau} \leq P_{loss} \leq \max_{k=1,2,\cdots,T} \frac{\sigma_k \Psi(\alpha)}{\bar{R}_T} \tag{10}$$

where

$$\Psi(\alpha) = \delta_0 e^{-\frac{\alpha - \lambda_0}{\delta_0}} \tag{11}$$

Proof. Let

$$\tilde{R}_k = \bar{R}_k + \alpha \sigma_k$$

From Theorem 1, we have that no packet loss will occur unless the future bounding rate R_k exceeds $\tilde{R}_k$ for some $k = 1, \cdots T$, which occurs with probability $1 - \Phi(\alpha)$. For a particular k and exceeding rate $\tilde{R}_k$, the mean number of packets lost or dropped due to buffer overflow, denoted by L_k, satisfies

$$E(R_k - \tilde{R}_k)^+ \cdot I_k \leq E(L_k) \leq E(R_k - \tilde{R}_k)^+ \cdot T\tau$$

where

$$E(R_k - \tilde{R}_k)^+ = \int_{\tilde{R}_k}^{\infty} (r - \tilde{R}_k) dF_k$$

$$= \sigma_k \int_{\alpha}^{\infty} (x - \alpha) \frac{1}{\delta_0} exp(-\frac{x - \lambda_0}{\delta_0} - exp(-\frac{x - \lambda_0}{\delta_0})) dx$$

$$\approx \sigma_k \int_{\alpha}^{\infty} (x - \alpha) \frac{1}{\delta_0} exp(-\frac{x - \lambda_0}{\delta_0}) dx$$

$$= \sigma_k \delta_0 e^{-\frac{\alpha - \lambda_0}{\delta_0}}$$

The fraction of packets lost is the ratio of the number of violating packets to the total number of packets sent. Considering all interval lengths k, the loss probability satisfies

$$\max_{k=1,2,\cdots,T} \frac{E(R_k - \tilde{R}_k)^+ \cdot I_k}{\bar{R}_T \cdot T\tau} \leq P_{loss} \leq \max_{k=1,2,\cdots,T} \frac{E(R_k - \tilde{R}_k)^+ \cdot T\tau}{\bar{R}_T \cdot T\tau}$$

$$\max_{k=1,2,\cdots,T} \frac{\sigma_k \cdot \Psi(a) \cdot I_k}{\bar{R}_T \cdot T\tau} \leq P_{loss} \leq \max_{k=1,2,\cdots,T} \frac{\sigma_k \cdot \Psi(a)}{\bar{R}_T}$$

from Equations (11) and (12) and the fact that $Prob(R_k > \tilde{R}_k) \leq 1 - \Phi(\alpha)$. $\square$

Thus, Theorem 2 bounds the loss probability by determining the mean number of packets that are dropped due to buffer overflow when the schedulability condition of Theorem 1 is not satisfied, a condition which has probability $1 - \Phi(\alpha)$.

2.5 Discussion

Statistical Multiplexing and Maximal Envelopes In this section, we explore the use of a maximal rate envelope to capture the important multiplexing properties of the traffic flows. For two scenarios, Figure 1 depicts a flow's maximal rate envelope normalized to the flow's mean rate. In the figure, the upper curve depicts the normalized envelope of a single 30-minute trace of an MPEG-compressed action movie. Plotting the bounding rate to mean ratio vs. the interval length, the figure shows how the traffic characterization captures the maximum rates and durations of the flow's bursts. For example, for small interval lengths, the envelope approaches the source's instantaneous peak rate, which is about 10 times of the mean rate. For longer interval lengths it decreases towards the long term average rate, which is the total number of bits in the MPEG sequence divided by the length of the sequence.

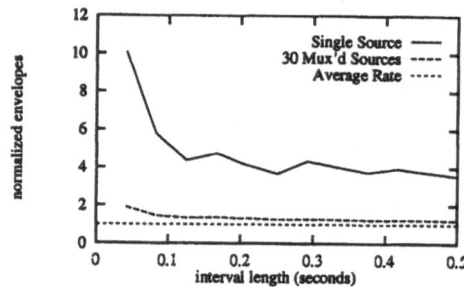

Fig. 1. Rate Envelope for Multiplexed Video Sources

The second curve of Figure 1 shows the normalized envelope of 30 multiplexed copies of this same trace. If all 30 flows are exactly synchronized, then this second curve would overlap exactly with the first. However, when the flows have statistically independent phases, the peaks of each flow do not line up exactly, and when normalized to the aggregate traffic's mean rate, the measured normalized envelope of the aggregate flow is significantly less than that of the individual flow. Hence, even with a traffic characterization which describes the flow's maximal rates, the extent to which flows statistically multiplex is evident.

We also note from Figure 1 that the autocorrelation structure of the flow is revealed from the traffic envelope. For example, with a single flow, the source's alternation between large intra-coded frames and smaller inter-coded frames is evident from the shape of the normalized rate envelopes, with its sharp drop from one to two frame times and its oscillation thereafter. However, when these video sources multiplexed, this quasi-periodicity is largely removed as evidenced by the near monotonicity of the envelope of the aggregate flow. Thus, with an increased number of flows, the normalized envelope of the aggregate traffic approaches the long-term average rate.

Finally, we note that in addition to characterizing the extreme values of the traffic flow which can be exploited for resource allocation, the maximal rate envelope has the desirable property that the variation of the *maximum* rate tends to be less than the variance of the flow itself. The following lemma shows that this is asymptotically true for uncorrelated flows with bounded rate.

Lemma 3. *Denote ϕ_t as the arrival rate at t and let the maximum rate over T time units be $\phi_T^* = \max\{\phi_1, \phi_2, \ldots, \phi_T\}$. Denoting the distribution of ϕ_t as $F(x)$, if ϕ_t is bounded by C so that $F(x) = 1$ for $x \geq C$, and ϕ_s is independent of ϕ_t, $s \neq t$, then $\lim_{T \to \infty}\{E[\phi_T^{*2}] - E[\phi_T^*]^2\} = 0$.*

Proof. The distribution of ϕ_T^* is given by $F^T(x)$ under the independence assumption so that the limiting distribution is

$$\lim_{T \to \infty} F^T(x) = U(x - C)$$

where $U(x)$ is a step function. Thus, the limiting variance of ϕ_T^* is given by

$$\lim_{T \to \infty} \text{var}(\phi_T^*) = \int_{-\infty}^{+\infty} x^2 dU(x) - \left(\int_{-\infty}^{+\infty} x dU(x)\right)^2$$

which is $C^2 - C^2 = 0$, while the variance of the rate itself asymptotically remains $\text{var}(\phi_t)$. $\square$

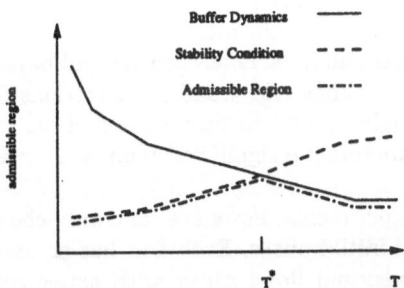

Fig. 2. Setting the Measurement Window T

Setting the Measurement Window T Here, we discuss how to set the measurement window T. With Equations (5) and (6), the applications' specified QoS will always be satisfied for any choice of the measurement window T. As we will illustrate below, a poorly set T can only under-utilize network resources. For a particular T, either the buffer test or stability condition will restrict the admission of a new flow, so that the admissible region is the smaller of the two admissible regions calculated from Equations (5) and (6). Figure 2 illustrates the relationship between the admissible region and the measurement window T. If T is set too small, σ_T, the variation of rate over intervals of length T, will be large, so that the stability condition cannot be satisfied. Alternatively, if T is set too large, the interval from which the maximum rate is measured is large, so that estimation of the maximal buffer occupancy will also be large; consequently, the buffer test becomes the bottleneck. In between these two extremes, there exists an optimal T^*, where the maximal admissible region is achieved. From our experiments, we find that our proposed MBAC algorithm is quite robust to different choices of T. Regardless, we note that setting of the measurement time scale is a fundamental one to MBAC, and further discussions of its proper setting can be found in [17] for example.

3 Experimental Results

In this section, we evaluate the performance of the new MBAC algorithm. The workload consists of a set of twenty 30-minute traces of MPEG and JPEG compressed video from [16]. With this collection of traces and an implementation of our MBAC algorithm, we perform a set of trace-driven simulation and admission control experiments for a wide variety of traffic mixes and network capacities.

3.1 Workload, Experimental Scenario, and Performance Metrics

The MPEG traces exhibit both fast-time-scale rate variations primarily due to the coder's alternation between large intra-coded frames and smaller inter-coded frames, and slow-time-scale rate variations due to scene changes. The twenty traces exhibit considerable heterogeneity and burstiness with mean rates ranging from 175 to 960 kbps and standard deviations from 269 to 892 kbps. In addition to the variability and autocorrelation of the traffic for an individual trace, these traffic flows retain a significant temporal correlation structure when they are multiplexed.

In the following experiments, flows are randomly chosen from the 20 traces with randomly shifted initial phase. Each flow has an exponential holding time with mean 900 seconds, and flows arrive with exponential inter-arrival times with mean 5 seconds. A new flow provides the MBAC algorithm with its leaky bucket parameters and QoS requirements (delay, P_{loss}). Our MBAC algorithm is then invoked to measure the aggregate traffic and perform admission control as described in Section 2. We consider a first-come-first-serve multiplexer with buffer size B, the video traces as the workload, and a link capacity of C Mbps. We set the buffer size to be C times the required delay bound so that the delay-bound violation probability is the same as the packet loss probability. We measure the fraction of packets that are dropped due to buffer overflow as the empirical loss probability, and measure the fraction of time that the multiplexer is busy as the average utilization of the multiplexer. We perform repeated simulations for each scenario and report average results along with 95% confidence intervals.

3.2 Performance of MBAC Algorithm

The goal of an admission control algorithm is to admit the maximum number of traffic flows possible without violating any of their QoS requirements, i.e., to maximize resource utilization subject to some QoS constraints. Consequently, we evaluate the effectiveness of an admission control algorithm by comparing the admissible region achieved by the MBAC algorithm with the true admissible region obtained experimentally under the same set of QoS constraints.

In addition to evaluating the proposed MBAC algorithm, we also compare with two algorithms from the literature. First, we consider the approach of Jamin et al. [11] which uses an equivalent token bucket model to characterize the aggregate traffic flow of each class. Admission control is performed by measuring and controlling the average bandwidth utilization $\hat{\nu}$ and the experienced maximal queueing delay $\hat{D}$ to target values. In our experiments, we use the recommended performance tuning parameters of [11] with v set to 0.9, $\lambda = 2$, $S = 1/24$ sec, and $T = 3$ sec.

We also compare with Floyd's approach [6], which, like the algorithm proposed in this paper, addresses scenarios with moderate numbers of multiplexed traffic flows such as in a link-sharing environment in which the available capacity is partitioned to support a number of services and traffic classes. The approach is based on the Hoeffding bound which utilizes the measured mean rate of the aggregate flow together with the individual peak rates as specified and policed for each flow. In the simulations, we smooth the video source over 12 frames to reduce the peak rate and maximize [6]'s admissible region.

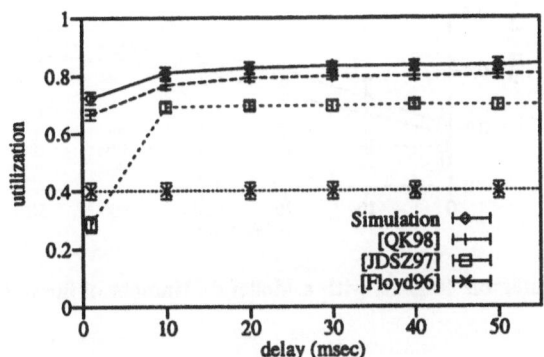

Fig. 3. Admissible Regions from Simulations and MBAC Algorithms

Figure 3 depicts the results of our experiments for target $P_{loss} = 10^{-4}$, and capacity $C = 45$ Mbps. In the figure, a point on the "QK98" curve represents the average utilization achieved by our MBAC algorithm subject to the QoS constraints as given by the delay bound depicted on the horizontal axis and a loss probability of 10^{-4}. The "Simulation" curve depicts the experimental admissible region, or the maximum utilization achievable in simulation subject to these same QoS constraints. We observe that the proposed MBAC algorithm is quite accurate in controlling the admissible region subject to the QoS constraints. It captures most of the available statistical multiplexing gain with the difference between the utilizations of the "Simulation" and "QK98" curves less than 5%. Moreover, the target QoS objectives are met and measured loss probabilities are within one or two orders of magnitude below the target value. For example, for 20 msec delay, the mean measured loss probability is 0.8×10^{-5}. We also observe that the algorithm distinguishes among different QoS requirements, properly increasing the admissible region as the delay requirement becomes less restrictive.

We observe from Figure 3 that our algorithm outperforms both "JDSZ97" [11] and "Floyd96" [6] in that a higher utilization is achieved while still satisfying the target QoS. In particular, for small delay requirements below 10 msec, [11] is conservative, rejecting admission of new flows because the delay test fails (Equation (7) in [11]). For larger delay bound requirements, its utilizations are approximately 70% compared to simulated utilizations above 80%. A further advantage of the proposed approach is our control of QoS parameters directly via the delay and loss probability, rather than indirectly through the "target utilization" parameter. Finally, the admissible region of [6] is approximately 40% for all delays.

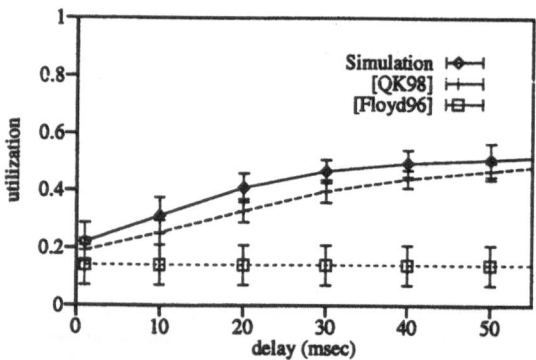

Fig. 4. Admissible Regions with a Moderate Number of flows, C = 5 Mbps

As discussed above, the capacity allocated to a particular class of traffic can be quite small in a link sharing environment, and a Gaussian approximation of the aggregate traffic distribution may be inaccurate. Both our proposed algorithm and [6] address this scenario, respectively with application of extreme value theory and Hoeffding bounds. To compare these two MBAC algorithms with a moderate number of flows, we perform experiments with a link capacity of 5 Mbps, which allows only 5 to 10 active video sources. Figure 4 depicts the resulting admissible regions for target $P_{loss} = 10^{-4}$ and new users arriving with a mean interval of 30 seconds. Observe that at this small capacity, the simulated utilization is much lower than for a 45 Mbps link, as significant statistical multiplexing gains are not possible with so few sources. Regardless, the "QK98" curve indicates that the proposed algorithm is still able to control the admissible region, closely following the simulated curve and incorporating the region's increasing trend with larger delay bounds; moreover, measured loss probabilities are within an order of magnitude below the target value.

4 Conclusions

In this paper, we introduced a novel algorithm for Measurement-Based Admission Control (MBAC) that exploits measured maximal rate envelopes of the aggregate traffic flow to allocate network resources. Our approach uses the ability of the aggregate flow's envelope to reveal the critical characteristics of the traffic for admission control, such as the extent to which flows are statistically sharing network resources and the autocorrelation structure of the aggregate flow. Since there is no assurance that the aggregate flow will continue to be bounded by its past behavior, we developed new theory to quantify the confidence level of a schedulability condition that considers estimation errors and long-time-scale fluctuations of the aggregate envelope. Moreover, employing extreme value theory, we devised new techniques to estimate the packet loss probability by deriving an expression for the expected number of packets lost when the schedulability condition is violated. Our goal is to design an accurate and robust MBAC algorithm that encompass a wide range of traffic types, buffer sizes, and link capacities, including situations with a relatively modest number of multiplexed flows per traffic class.

References

1. E. Castillo. *Extreme Value Theory in Engineering*. Academic Press, 1988.
2. D. Clark, S. Shenker, and L. Zhang. Supporting real-time applications in an integrated services packet network: Architecture and mechanism. In *Proceedings of ACM SIGCOMM'92*, pages 14–26, Baltimore, Maryland, August 1992.
3. R. Cruz. A calculus for network delay, part I : Network elements in isolation. *IEEE Transactions on Information Theory*, 37(1):114–121, January 1991.
4. Z. Dziong, M. Juda, and L. Mason. A framework for bandwidth management in ATM networks – aggregate equivalent bandwidth estimation approach. *IEEE/ACM Transactions on Networking*, 5(1):134–147, February 1997.
5. D. Ferrari and D. Verma. A scheme for real-time channel establishment in wide-area networks. *IEEE Journal on Selected Areas in Communications*, 8(3):368–379, April 1990.
6. S. Floyd. Comments on measurement-based admissions control for controlled-load services, July 1996. Lawrence Berkeley Laboratory Technical Report.
7. S. Floyd and V. Jacobson. Link-sharing and resource management models for packet network. *IEEE/ACM Transactions on Networking*, 3(4):365–386, August 1995.

8. M. Garret and W. Willinger. Analysis, modeling and generation of self-similar VBR video traffic. In *Proceedings of ACM SIGCOMM '94*, pages 269–280, London, UK, August 1994.

9. R. Gibbens, F. Kelly, and P. Key. A decision-theoretic approach to call admission control in ATM networks. *IEEE Journal on Selected Areas in Communications*, 13(6):1101–1114, August 1995.

10. I. Hsu and J. Walrand. Dynamic bandwidth allocation for ATM switches. *Journal of Applied Probability*, 33(3):758–771, September 1996.

11. S. Jamin, P. Danzig, S. Shenker, and L. Zhang. A measurement-based admission control algorithm for integrated services packet networks. *IEEE/ACM Transactions on Networking*, 5(1):56–70, February 1997.

12. E. Knightly and H. Zhang. D-BIND: An accurate traffic model for providing QoS guarantees to VBR traffic. *IEEE/ACM Transactions on Networking*, 5(2):219–231, April 1997.

13. A. Lazar and G. Pacifici. Control of resources in broadband networks with quality of service guarantees. *IEEE Communications*, pages 66–73, October 1991.

14. A. Lazar, G. Pacifici, and D. Pendarakis. Modeling video sources for real time scheduling. *ACM Multimedia Systems Journal*, 1(6):253–266, April 1994.

15. J. Qiu. *Measurement-Based Admission Control in Integrated Services Networks*. M.S. Thesis, Rice University, May 1998.

16. O. Rose. Statistical properties of MPEG video traffic and their impact on traffic modeling in ATM systems. In *Proceedings of IEEE Conference on Local Computer Networks*, pages 397–406, Minneapolis, MN, October 1995.

17. D. Tse and M. Grossglauser. Measurement-based call admission control: Analysis and simulation. In *Proceedings of IEEE INFOCOM '97*, Kobe, Japan, April 1997.

18. D. Wrege, E. Knightly, H. Zhang, and J. Liebeherr. Deterministic delay bounds for VBR video in packet-switching networks: Fundamental limits and practical tradeoffs. *IEEE/ACM Transactions on Networking*, 4(3):352–362, June 1996.

Coexistence of QoS and Best-Effort Flows *

Klara Nahrstedt, Shigang Chen
Department of Computer Science
University of Illinois at Urbana-Champaign
{klara,s-chen5}@cs.uiuc.edu

Abstract. The future high-speed networks will need to support diverse traffic and provide services to flows with Quality of Service (QoS) requirements as well as to best effort flows. In this paper we analyze the coexistence of the QoS and best effort flows from the routing and scheduling point of view. We concentrate in our routing and scheduling analysis on the *network bandwidth* resource. We present two sets of source routing algorithms: (1) the *bandwidth-constrained routing with imprecise state information* for QoS flows, and (2) the *maxmin fair routing* for best effort flows. Furthermore, we discuss *two level hierarchical scheduling* which is tailored towards the needs raised by the coexistence of QoS and best effort flows.

1 Introduction

The future high-speed networks will carry many concurrent flows with diverse requirements. Hence, it is crucial that the network bandwidth and other network resources are shared effectively and fairly among all competing flows. In this paper we will consider two sets of flows: *QoS flows* which post QoS requirements on the established end-to-end path, and *best effort flows* which do not have any specific QoS requirements on the established end-to-end path. There exist numerous network services which need to be examined and revisited when coexistence of these flows is studied. In this paper we will analyze two network services: *Routing* and *Scheduling*. In our analysis we will concentrate on the *network bandwidth* resource.

Routing: In general, the problem of routing is difficult due to a number of reasons. First, distributed multimedia applications have diverse QoS constraints on delay, delay jitter, loss ratio, bandwidth, etc. Multiple constraints often make the QoS routing problem intractable. In particular, finding the least-cost path with one path constraint or

---

* This work was supported by the ARPA grant F30602-97-2-0121 and the NSF Career grant NSF CCR 96-23867.

finding a feasible path with two independent path constraints is NP-complete [10]. Second, any future integrated-service network will carry both QoS traffic and best-effort traffic, which makes the issue of performance optimization complicated. A primary task of routing is to maximize the *resource efficiency*, which is measured by two goals. One goal is to maximize the number of QoS flows that are admitted into the network, which is equivalent to minimize the *call-blocking ratio*. The other goal is to optimize the throughput and responsiveness of best-effort traffic. The two goals may contradict each other.

In this paper we present two sets of source routing algorithms: (1) the *bandwidth-constrained routing with imprecise state information* for QoS flows, and (2) the *maxmin fair routing* for best effort flows.

Scheduling of flows with diverse network requirements can be done by a single-level weighed fair queuing algorithms. However, the complexity of this algorithm increases with the number of flows and the book-keeping of flow states becomes more complex. Hence, a hierarchical scheduling is more appropriate to deploy as it scales better towards the increasing number of flows with different network requirements. We modified the hierarchical fair queuing scheduling algorithm proposed by Bennett and Zhang [2] and tuned it towards specific needs raised by the co-existence of QoS flows and best effort flows.

The paper is outlined as follows: The Section 2 presents the network model and specifies the notation for the bandwidth partition between the QoS and best effort flows. Section 3 describes bandwidth-constrained routing algorithms using the imprecise state model. Section 4 analyses the best effort maxmin routing algorithm which uses the fairness-throughput relation for route finding and allocation. Section 5 proposes an integrated hierarchical packet scheduling which enforces network bandwidth for both QoS and best effort flows. The paper concludes with Section 6.

2 Model

A network is modeled as a set V of nodes that are interconnected by a set E of full-duplex, directed communication links. Let F be the set of flows in the network. We study connection-oriented networks where each flow has a fixed source (destination) and is assigned a fixed route through which all packets of that flow are transmitted in the FIFO order. For a flow $f \in F$, the set of links on its route is denoted as $L(f)$. The set of flows through a link l is denoted as $F(l)$.

We study two types of flows in this paper. A *QoS flow* f has a bandwidth requirement $B(f)$, which must be guaranteed (reserved for f on each link in $L(f)$) in order to ensure an acceptable quality. An example is an audio session between two remote users. The set of QoS flows in the network is denoted as F_{qos}. The set of QoS flows through a link l is denoted as $F_{qos}(l)$. A *best-effort flow* can operate at any bandwidth level, and hence the reservation of bandwidth is not needed. Examples are file transmission (*ftp*), web-page download and database retrieval. The set of best-effort flows in the network is denoted as F_{best}. The set of best-effort flows through a link l is denoted as $F_{best}(l)$. We have $F_{qos} + F_{best} = F$ and $F_{qos}(l) + F_{best}(l) = F(l)$.

Each link l has a bandwidth capacity $C(l)$, among which the part reserved by the QoS flows is denoted as $C_{qos}(l)$ and the part available to the best-effort flows is denoted as $C_{best}(l)$. $C_{qos}(l) = \Sigma_{f \in F_{qos}(l)} B(f)$, and $C_{best}(l) = C(l) - C_{qos}(l)$. It is clear that the values of $C_{qos}(l)$ and $C_{best}(l)$ are not fixed. When a new QoS flow f is routed through l, $C_{qos}(l)$ is increased by $B(f)$ and $C_{best}(l)$ is decreased by the same amount. In order to prevent the best-effort flows from being starved, $C_{qos}(l)$ is upper-bounded by $\lambda C(l)$, where λ is a system parameter less than 1. For every link l, the condition $C_{qos}(l) \leq \lambda C(l)$ must *always* be satisfied. Hence, the bandwidth for best-effort flows is at least $(1 - \lambda)C(l)$.

A new QoS flow f with $B(f)$ requirement can be accepted by a link l only if $\lambda C(l) - C_{qos}(l) \geq B(f)$. $\lambda C(l) - C_{qos}(l)$ is called the *residual* bandwidth of l and is denoted as $bandwidth(l)$. A simple path p consists of a set of connected links. The residual bandwidth of a path is defined as

$$bandwidth(p) = \min_{l \in p}\{bandwidth(l)\}$$

3 QoS Routing

We study the *bandwidth-constrained routing* problem, i.e., finding a path p from s to t such that $bandwidth(p) \geq B$, where s, t and B are the source node, the destination node and the bandwidth requirement [2] of a new QoS flow, respectively. Any path from s to t satisfying the above constraint is called a *feasible* path.

---

[2] We use B as an abbreviation of $B(f)$ when only a single QoS flow is in discussion.

We use the source routing strategy [5, 11, 12], in which every node maintains an image of the *global network state*, based on that a routing path is computed at the source node. However, the global state, which is typically maintained by a link-sate (or distance-vector) protocol, is inherently *imprecise* in a dynamic network where the traffic load changes constantly. The imprecision is especially noticeable in large wide-area networks due to the following reasons. First, it takes non-negligible propagation delay for a local state change to be broadcasted to other nodes. Second, a link-state (or distance-vector) protocol updates the state information periodically or upon triggering when significant state change is detected. There exists a tradeoff between the update frequency and the overhead involved. For large scale networks, the excessive communication overhead often makes it impractical for the update frequency to be high enough to cope with the dynamics of network parameters such the bandwidth availability on every link. Third, the hierarchical approach is likely to be used to solve the scalability problem of routing in large networks [9]. However, the state aggregation in hierarchical routing increases the level of imprecision [11].

The imprecision of state information directly affects the routing performance. The goal of our algorithm is to maximize the probability of success in finding a feasible path in a dynamic network environment where the available information is imprecise. In the following, we shall first propose an imprecise state model, based on which the routing algorithm is described.

Imprecise State Model : Every node maintains a state vector which has an entry, denoted as B_l, for every link $l \in E$.

1) Bandwidth: B_l keeps the residual bandwidth available on link l. B_l is updated periodically by a link-state protocol. It is inherently *imprecise* in a dynamic network. We propose a simple imprecise state model which can be easily implemented. An additional state variable ΔB_l is required.

2) Bandwidth Variation: ΔB_l keeps the *estimated* maximum change of B_l before the next update. The *actual* residual bandwidth of link l, denoted as $bandwidth(l)$, is expected to be between $B_l - \Delta B_l$ and $B_l + \Delta B_l$ in the next period.

ΔB_l is updated periodically together with B_l. We consider an arbitrary update of ΔB_l and B_l. Let ΔB_l^{old} and ΔB_l^{new} be the values of ΔB_l before and after the update, respectively. Similarly, let B_l^{old} and B_l^{new} be the values of B_l before and after the update, respectively. B_l^{new} is provided by a link-state protocol. ΔB_l^{new} is calculated as follows.

$$\Delta B_l^{new} = \alpha \times \Delta B_l^{old} + (1 - \alpha) \times |B_l^{new} - B_l^{old}|$$

The above formula is similar to the one used by TCP to estimate the round-trip delay. The factor α (< 1) determines how fast the *history* information (ΔB_l^{old}) is forgotten, and $(1 - \alpha)$ determines how fast ΔB_l^{new} converges to $|B_l^{new} - B_l^{old}|$.

By the above formula, it is still possible for the actual residual bandwidth to be out of the range $[B_l - \Delta B_l, B_l + \Delta B_l]$. One way to make such probability negligible small is to enlarge ΔB_l.

Routing Algorithm : The purpose of our routing algorithm is to maximize the probability of success in finding a feasible path, given B_l and ΔB_l, $\forall l \in E$. An important part of the algorithm is to calculate the probability of a link l satisfying a given bandwidth requirement. Such a probability is determined by the probability distribution function of $bandwidth(l)$. Let us consider the simple case. Assume that $bandwidth(l)$ is a random variable which value is uniformly distributed in $[B_l - \Delta B_l, B_l + \Delta B_l]$. The probability density function of $bandwidth(l)$ is

$$f(x) = \begin{cases} \frac{1}{2\Delta B_l} & x \in [B_l - \Delta B_l, B_l + \Delta B_l] \\ 0 & x \notin [B_l - \Delta B_l, B_l + \Delta B_l] \end{cases}$$

Let us consider a new flow which source node, destination node and bandwidth requirement are s, t and B, respectively. The probability of a link l satisfying the bandwidth requirement is

$$Pr(bandwidth(l) \geq B)$$

$$= \int_{B}^{+\infty} f(x)\, dx$$

$$= \begin{cases} \frac{B_l + \Delta B_l - B}{2\Delta B_l} & B \in [B_l - \Delta B_l, B_l + \Delta B_l] \\ 1 & B < B_l - \Delta B_l \\ 0 & B > B_l + \Delta B_l \end{cases}$$

The probability of a path p satisfying the requirement is

$$Pr(bandwidth(p) \geq B)$$

$$= \prod_{l \in p} Pr(bandwidth(l) \geq B)$$

$$= \begin{cases} \prod_{l \in p} \frac{min\{B_l + \Delta B_l - B, 2\Delta B_l\}}{2\Delta B_l} & \forall l \in p, B < B_l + \Delta B_l \\ 0 & \exists l \in p, B \geq B_l + \Delta B_l \end{cases}$$

The routing algorithm is designed to find a path p from s to t which maximizes $Pr(bandwidth(p) \geq B)$ as follows:

1. Let $E' = \{l \mid B < B_l + \Delta B_l, l \in E\}$. Remove all links in $E - E'$, and the resulting graph is denoted as $\langle V, E' \rangle$. If there does not exist a path from s to t in $\langle V, E' \rangle$, reject the flow and return.

2. $\forall l \in E'$, a weight $w_l = -log \frac{min\{B_l + \Delta B_l - B, 2\Delta B_l\}}{2\Delta B_l}$ is assigned.

3. Use the Dijstra's shortest path algorithm to find the *least-weighted* [3] path p from s to t in $\langle V, E' \rangle$.

4. Select p as the routing path. If the flow is successfully established through p, return. Otherwise, reject the flow and return.

4 Best-Effort Routing

For the best-effort flows, we shall first review the *maxmin bandwidth allocation*, based on which a new routing policy, called the *maxmin fair routing*, is defined. Throughout this section, we call $C_{best}(l)$ the *available bandwidth (capacity)* for best effort flows of link l.

---

[3] The least-weighted path p is the one which minimizes $\sum_{l \in p} w_l$.

Maxmin Bandwidth Allocation : The maxmin allocation was first proposed by Jaffe [8] as a flow control technique which distributes the network bandwidth fairly among the best-effort flows. Much further research [1, 3, 14] has been done since then. It has been accepted by the ATM Forum as one of the traffic management approaches for the ABR(Available Bit Rate) service.

Its name comes from the fact that the maxmin allocation always *maximizes* the bandwidth allocated to those flows that receive the *minimum* bandwidth among all flows. The maxmin allocation has two basic properties: *Fairness property, Maximum throughput property*.

The maxmin bandwidth of each best-effort flow is determined by the bottleneck link on its route. A global bottleneck based algorithm which assigns the maxmin bandwidth to each best-effort flow was described in [13] and is briefly summarized below. A distributed algorithm was given in [8].

Fairness-Throughput Optimality : We formalize an important property for the maxmin allocation. A *feasible bandwidth allocation* $\Psi : F_{best} \to R^+$ is a function which satisfies the following condition

$$\forall l \in E, \sum_{f \in F_{best}(l)} \Psi(f) \leq C_{best}(l)$$

Let $\Psi(F_{best})$ be the *list* of values $(\Psi(f) \mid \forall f \in F_{best})$ in the increasing order. Note that in mathematics $\Psi(F_{best})$ normally represents a *set* of values $\{\Psi(f) | f \in F_{best}\}$. In this paper we make a different interpretation by introducing an increasing order to $\Psi(F_{best})$ and using it as an ordered *list*. A link l is said to be *saturated* by Ψ if

$$\sum_{f \in F_{best}(l)} \Psi(f) = C_{best}(l)$$

Definition 1 *Given two feasible bandwidth allocations Ψ and Ψ' on F_{best}, we define the fairness-throughput relations: (1) $\Psi(F_{best}) = \Psi'(F_{best})$ if the two lists are identical, and (2) $\Psi(F_{best}) > \Psi'(F_{best})$ if there exists a prefix of $\Psi(F_{best})$, $(b_1, b_2, ..., b_i)$, and a prefix of $\Psi'(F_{best})$, $(b'_1, b'_2, ..., b'_i)$, such that $b_i > b'_i$ and $b_j = b'_j, 1 \leq j \leq i - 1$.*

The above relations place a total order on the set of all feasible allocations. The ordering is based on two performance measurements, *fairness* and *throughput*. Fairness and throughput are often conflicting measurements and the fairness-throughput relations defined in Definition 1 evaluate an allocation based on a measurement which provides a tradeoff between the fairness and the overall throughput. We proved a theorem showing that the maxmin allocation maximizes the fairness-throughput performance [6].

Definition of Maxmin Routing : Let $\langle V, E \rangle$ be a network where bandwidth is alway allocated to flows by maxmin. Let F_{best} be the set of existing best-effort flows, each of which has a fixed route. Consider a new best-effort flow f_0. The task of routing is to assign a route r to f_0. Each different route for f_0 results in a different maxmin bandwidth allocation $\Psi_{m,r}$ on $F_{best} \bigcup \{f_0\}$. The purpose of *maxmin routing* is to find a route r such that $\Psi_{m,r}(F_{best} \bigcup \{f_0\}) \geq \Psi_{m,r'}(F_{best} \bigcup \{f_0\})$, for any feasible route r' of f_0.

The maxmin routing is a new problem which is different from the maxmin allocation studied by the previous publications. The problem solved by the latter is as follows: given a network and a set of flows with fixed routes, how to assign the network bandwidth to the flows such that the network performance is optimized. The maxmin routing, however, assumes the network bandwidth is assigned based on the maxmin allocation. It then introduces another dimension, new flows. The problem to be solved is how to assign routes to new flows such that the performance of the maxmin allocation can be maximized.

Maxmin Routing Algorithm : first adds f_0 to every link in the network [4] and then iteratively removes f_0 from the links which have the smallest per-flow bandwidth until the route for f_0 is found. By removing f_0 from the links with the smallest bandwidth per flow, the algorithm effectively routes f_0 around the most congested links and therefore maximizes the bandwidth allocated to the congested flows, which equals maximizing the low end of $\Psi_{m,r}$ and thus equals maximizing $\Psi_{m,r}$ because the low end of $\Psi_{m,r}$ is of more significance by definition.

The algorithm below consists of two phases. In the first phase, the bottleneck link of f_0 is found, which determines the maxmin bandwidth for f_0; in the second phase, the algorithm finds the rest of the route which maximizes $\Psi_{m,r}$. We mark links either green or red. Green links are candidates to form a route for f_0; red links are either not on any

---

[4] Note that we are not adding the flow to the links of the *real* network but to the data structure representing the network at a node doing the source routing.

paths from the source to the destination or considered to be congested and thus rejected by the algorithm. A path consisting of only green links is called a *green path*. A detailed description of both phases, further discussion and approximation algorithms of best effort routing can be found in [6].

5 Integrated Packet Scheduling

Design Goals: When there exist many concurrent flows in the network, it is crucial that the limited bandwidth and other resources are shared effectively and fairly among all competing flows. We have two design goals for the scheduling of data packets: *Guaranteeing QoS* and *Ensuring fairness*.

Hierarchical Scheduling: We present a modified two-level hierarchical scheduling algorithm which achieves the two design goals. A packet scheduling algorithm *operates on each individual link l*. The algorithm is a two-level hierarchy. On the first level, the link capacity is divided between two logical scheduling servers: the *QoS server* and the *best-effort server*. The capacity of the QoS server is $C_{qos}(l) = \sum_{f \in F_{qos}(l)} B(f)$, and the capacity of the best-effort server is $C_{best}(l) = C(l) - \sum_{f \in F_{qos}(l)} B(f)$. The values of $C_{qos}(l)$ and $C_{best}(l)$ change when the flow set $F_{qos}(l)$ changes. On the second level, the QoS server schedules the flows in $F_{qos}(l)$ and the best-effort server schedules the flows in $F_{best}(l)$. The QoS server guarantees that every flow f in $F_{qos}(l)$ receives a bandwidth of $B(f)$. The best-effort server makes sure that every flow in F_{best} receives an equal share of $C_{best}(l)$ according to maxmin bandwidth allocation.

QoS Server: The QoS server must maintain two invariants.

I1) The capacity of the QoS server, $C_{qos}(l)$, must be $\sum_{f \in F_{qos}(l)} B(f)$ at any time. Whenever a new QoS flow f joins in $F_{qos}(l)$, $C_{qos}(l)$ must be increased by $B(f)$ immediately; whenever an existing QoS flow f leaves $F_{qos}(l)$, $C_{qos}(l)$ must be decreased by $B(f)$.

I2) $\forall f \in F_{qos}(l)$, the QoS server assigns a bandwidth no less than $B(f)$ to f, regardless the dynamics of the network state.

Best-effort Server: The best-effort server has two properties.

P1) The capacity of the best-effort server, $C_{best}(l)$, is always equal to the link bandwidth left over by the QoS server. When a new QoS flow joins and thus $C_{qos}(l)$ increases, $C_{best}(l)$ must decrease accordingly; when an existing QoS flow leaves and thus $C_{qos}(l)$ decreases, $C_{best}(l)$ must increase accordingly[5]

P2) The best-effort server distributes its capacity $C_{best}(l)$ fairly among all flows in $F_{best}(l)$. Any two flows whose packet queues remain back-logged should receive the same share of bandwidth. The flows whose queues are not back-logged receive less bandwidth which is equal to the incoming data rate.

Scheduling within QoS Server: We assume that the invariant I1 always holds. We implement the scheduling within the QoS server by the weighted fair queuing as follows.

1) A packet queue is maintained for each flow $f \in F_{qos}(l)$. The arrival packets are inserted into the queue in the FIFO order. A timestamp t_{qos}^i is calculated for the ith arrival packet.

$$t_{qos}^i \leftarrow max\{V_{qos}, t_{qos}^{i-1}\} + \frac{sp_i}{B(f)}$$

where V_{qos} is the reference *virtual time* [2] of the QoS server, sp_i is the length of the packet, t_{qos}^{i-1} is the timestamp of the $(i-1)$th packet and $B(f)$ is used as the *weight*. V_{qos} is a variable maintained by the QoS server, keeping track of the timestamp of the last transmitted packet from $F_{qos}(l)$. It is used to determine where the timestamp of a new or resumed QoS flow should start. Note that there is a single variable V_{qos} used by all flows in $F_{qos}(l)$.

2) The scheduling among flows in $F_{qos}(l)$ is based on the timestamps. Whenever the QoS server becomes idle, the packet with the smallest timestamp among all queues is selected for transmission.

The above weighted fair queuing assigns bandwidth to flows based on their weights. The bandwidth received by $f \in F_{qos}(l)$ is equal to $\frac{B(f)}{\sum\limits_{f' \in F_{qos}(l)} B(f')} \times C_{qos}(l) = \frac{B(f)}{C_{qos}(l)} \times C_{qos}(l) = B(f)$, if all flows are back-logged. Hence, the invariant I2 holds. Readers are referred to [7] for the detailed study of fair queuing.

---

[5] Note that $C_{qos}(l)$ has an upper bound ($\lambda C(l)$, where $\lambda < 1$ and $C(l)$ is the overall link capacity) as specified in Section 2.

Scheduling within Best-effort Server: Assume the property P1 always holds. How to achieve this will be discussed shortly. We implement the scheduling within the best-effort server by the weighted fair queuing as follows. The property P2 is achieved by assigning an equal weight to every flow.

1) A packet queue is maintained for each flow $f \in F_{best}(l)$. The arrival packets are inserted into the queue in the FIFO order. The weight of each flow is 1. A timestamp t^i_{best} is calculated for the ith arrival packet of f.

$$t^i_{best} \leftarrow max\{V_{best}, t^{i-1}_{best}\} + \frac{sp_i}{C_{best}(l)/|F_{best}(l)|}$$

V_{best} is a variable maintained by the best-effort server, keeping track of the timestamp of the last transmitted packet from $F_{best}(l)$. V_{best} is used as a reference *virtual time* of the server to determine where the timestamp of a new or resumed best-effort flow should start.

2) The scheduling among flows in $F_{best}(l)$ is based on the timestamps. Whenever the best-effort server becomes idle, the packet with the smallest timestamp among all non-empty queues is selected for transmission.

We have three observations about the above scheduling.

O1) Those flows whose queues remain back-logged receive the same share of bandwidth from the best-effort server because they have the same weight of 1.

O2) There may exist flows with empty queues due to insufficient incoming data packet rate, which may result from an upstream bottleneck link. These flows consume less bandwidth than the others simply because at times there are no packets in the queues for scheduling. Because the queue is not back-logged, the outgoing data rate, which is the actual bandwidth consumed, must be equal to the incoming data rate.

O3) Our scheduling is work-conserving, which means the capacity of the best-effort server will be fully utilized as long as there are back-logged queues.

Additional flexibility may be achieved by assigning different weights to different types of flows. Some interactive flows demand relatively small bandwidth. However, the instant bandwidth availability is critical to their performance. Examples are distributed games such as playing chess or cards over the Internet. Some other flows are relatively bandwidth-insensitive. Examples are non-interactive video retrieval and large file transmission working in the background. We can modify the scheduling of the best-effort server by classifying the flows into different categories, to each of which a different weight w is assigned. The timestamp calculation becomes $t^i_{best} \leftarrow max\{V_{best}, t^{i-1}_{best}\} + \frac{sp_i}{w}$. The flows with larger weights receive more prompt service and/or larger bandwidth shares. For the most critical flows, a special timestamp of -1 is assigned to every of their packets so that the packets will always be transmitted before those of other flows.

Scheduling between Two Servers: The QoS server and the best-effort server are *logical servers* using the same physical link. When both servers have packets to send, we must select one of them for the actual transmission. We want the scheduling between the two servers satisfies the invariant I1 and the property P1, i.e., the QoS server receives a capacity of $\sum\limits_{f \in F_{qos}(l)} B(f)$ and the best-effort server receives a capacity of $C(l) - \sum\limits_{f \in F_{qos}(l)} B(f)$.

The weighted fair queuing is used again, where the two servers are modeled as two logical flows, whose packets are from the physical flows in $F_{qos}(l)$ ($F_{best}(l)$) sorted by the timestamps. Let the weight of the QoS server be $W_{qos} = \sum\limits_{f \in F_{qos}(l)} B(f)$ and that of the best-effort server be $W_{best} = C(l) - \sum\limits_{f \in F_{qos}(l)} B(f)$. W_{qos} and W_{best} are not fixed in the run-time; they change when $F_{qos}(l)$ changes.

1) The ith packet selected by the QoS server is assigned a timestamp

$$T^i_{qos} \leftarrow max\{V_{link}, T^{i-1}_{qos}\} + \frac{sp_i}{W_{qos}}$$

where T^{i-1}_{qos} is the timestamp assigned to the $(i-1)$th packet selected by the QoS server. V_{link} will be explained shortly. The ith packet selected by the best-effort server is assigned a timestamp

$$T^i_{best} \leftarrow max\{V_{link}, T^{i-1}_{best}\} + \frac{sp_i}{W_{best}}$$

where T^{i-1}_{best} is the timestamp assigned to the $(i-1)$th packet selected by the best-effort server.

V_{link} is a variable maintained by the physical link,[6] keeping track

---

[6] In more precise words, by the node in charge of the link.

of the timestamp — T_{qos}^i or T_{best}^i depending which server the packet is from — of the last packet transmitted by the physical link. V_{link} is used as a reference *virtual time* of the link to determine where the timestamp should start when a packet arrives at an empty QoS or best-effort server,

2) When both servers select packets, the packet with the smaller timestamp will be transmitted.

The bandwidth received by the QoS server is $\frac{W_{qos}}{W_{qos}+W_{best}} \times C(l) = \frac{W_{qos}}{C(l)} \times C(l) = W_{qos} = \sum_{f \in F_{qos}(l)} B(f)$, and the bandwidth received by the best-effort server is $\frac{W_{best}}{W_{qos}+W_{best}} \times C(l) = W_{best} = C(l) - \sum_{f \in F_{qos}(l)} B(f)$.

Overhead: We study the per-packet computational overhead of our algorithm. For scheduling within the QoS server, finding the smallest timestamp among all flows in $F_{qos}(l)$ takes $O(log|F_{qos}(l)|)$, if a balanced binary tree such as a heap tree is maintained. For scheduling within the best-effort server, finding the smallest timestamp takes $O(log|F_{best}(l)|)$. For scheduling between the QoS server and the best-effort server, finding the smaller timestamp takes $O(1)$. Two timestamps, t_{qos}^i and T_{qos}^i or t_{best}^i and T_{best}^i, are calculated for each packet, which takes a small constant time. Therefore, the total overhead for scheduling a single packet is $O(log|F_{qos}(l)| + log|F_{best}(l)|)$, which is reasonably small and comparable to the time complexity $O(log|F(l)|)$ of the single-level fair queuing scheduling.

6 Conclusion

We presented possible algorithms for routing and scheduling which allow coexistence of QoS and best effort flows in future high-speed networks. Our network routing algorithms took into account state imprecision in routers, maxmin bandwidth allocation, and existing link state information. Our scheduling algorithms enforced effective and guaranteed bandwidth allocation for QoS flows, and fair sharing of bandwidth for best effort flows.

In summary, our integrated routing and scheduling framework allows for (1) bandwidth QoS routing when intermediate nodes carry imprecise state information which is a realistic assumption in current and future networks; (2) finding a best-effort route according to fairness-throughput performance relation. This type of relation optimizes the maxmin bandwidth allocation, hence with our maxmin routing algorithm we find a route which will be optimized according to the maxmin bandwidth allocation; (3) approximation algorithm to find best-effort routes according to maxmin bandwidth allocation using link states only; (4) starvation avoidance in case of best effort flows because we maintain an upper bound on the bandwidth allocation for QoS flows, which never encompasses the entire link bandwidth; (5) maximal throughput/link utilization because we allow for sharing of bandwidth by best-effort flows which is not utilized by QoS flows. This means that if only few QoS flows are routed and scheduled through a link, the remaining bandwidth can be shared among best-effort flows.

References

1. B. Awerbuch and Y. Shavitt. Converging to approximated max-min flow fairness in logarithmic time. *accepted for Infocom'98, San-Francisco, CA*, April 1998.

2. J. Bennett and H. Zhang. Hierarchical packet fair queueing algorithms. *ACM SIGCOMM'96*, 1996.

3. D. Bertsekas and R. Gallager. Data networks. *Prentice Hall, Englwood Cliffs, N.J.*, 1992.

4. S. Chen and K. Nahrstedt. Distributed qos routing with imprecise state information. *IEEE Seventh International Conference on Computer, Communication and Networks*, 1998.

5. S. Chen and K. Nahrstedt. Maxmin fair routing in connection-oriented networks. *Proceedings of IASTED European Conference on Parallel and Distributed Systems, Vienna, Austria*, July 1998.

6. A. Demers, S. Keshav, and S. Shenker. Analysis and simulation of a fair queueing algorithm. *ACM SIGCOMM'89*, pages 3–12, 1989.

7. J. M. Faffe. Bottleneck flow control. *IEEE Transactions on Communications*, COM-29(7):954–962, July 1981.

8. ATM Forum. Private network network interface (pnni) v1.0 specifications. May 1996.

9. M. Garey and D. Johnson. *Computers and Intractability: A Guide to the Theory of NP-Completeness.* New York: W.H. Freeman and Co., 1979.

10. R. Guerin and A. Orda. Qos-based routing in networks with inaccurate information: Theory and algorithms. *Infocom'97, Japan*, April 1997.

11. D. H. Lorenz and A. Orda. Qos routing in networks with uncertain parameters. *Infocom'98*, March 1998.

12. Q. Ma, P. Steenkiste, and H. Zhang. Routing high-bandwidth traffic in maxmin fair share networks. *Proceedings of SIGCOMM'96*, August 1996.

13. J. Mosley. Asynchronous distributed flow control algorithms. *Ph.D. thesis, MIT, Dept. of Electrical Engineering and Computer Science, Cambridge, MA*, 1984.

Impact, Characterization and Modeling of Intermedia Relationships in a Multimedia-System Performance Evaluation Framework

Alfio Lombardo, Giovanni Schembra
Istituto di Informatica e Telecomunicazioni, Università di Catania
V.le A. Doria 6 - 95125 Catania - ITALY
phone: +39 095 7382376 - fax: +39 095 338280

Abstract
The time relationships which characterize intermedia synchronization in a multimedia stream affect network performance. In this paper an analytical framework is presented in order to model the multimedia traffic and to evaluate the performance of a multiplexer loaded by multimedia sources. The emission process of each multimedia source is defined as the superposition of heterogeneous correlated emission processes, each of which models one monomedia source as a switched batch Bernoulli process (SBBP). In order to characterize the intermedia relationships a new parameter, here referred to as the Intermedia Synchronization Index (ISI), is defined in terms of the cross-covariance between the monomedia streams in the multimedia flow, and it is calculated in terms of the parameters of the emission process modeling the multimedia source. The influence of the ISI on loss performance is shown by means of examples in a case study.

1. Introduction

The advent of multimedia services has created a new performance requirement: intermedia synchronization. In a multimedia traffic flow, in fact, temporal relationships exist which characterize the mutual evolution of the component monomedia traffic streams.

For example, in a slide show session in which the voice silence period detection is used, the slide emission rate varies according to the changes in the voice activity which alternates high emission and low emission periods. Again, in an MPEG source consisting of video and voice VBR coded sources, variations occurring in the statistical parameters which characterize each of the component traffic streams are correlated with each other because the activity changes in the voice and video sources are related to the same user behavior. This is the case, for example, of a video-telephone application, where the user gesticulates more when he is speaking than when he is listening, so the activity of both the audio and video sources alternates between periods during which the average bit rate is high, and periods during which it is low; because these activity changes in both voice and video are due to the same user behavior the changes in the voice and video traffic statistics have to be considered as correlated events. These intermedia relationships clearly affect the second-order statistics of the emission process modeling a multimedia source and so, when the performance of a network loaded by multimedia traffic is to be evaluated, a multimedia source cannot be defined as the superposition of independent monomedia emission processes; intermedia correlation, that is, correlation among the component processes, has to be modeled [1-2].

Monomedia traffic models have been widely studied in the last years. In particular intramedia correlation, that is, correlation in a monomedia stream, has already been modeled in the literature for both video traffic [3-8] and voice traffic [8-11] by using Markov-based processes constituted by N sub-processes, one for each activity state of the source, where transitions between the above sub-processes model activity changes in the source. In this paper we derive a paradigm of a multimedia source taking into account both intramedia and intermedia correlation, that is, considering the activity changes of the component monomedia sources to be correlated with each other. To this end we will use the most general Markov-based process in the discrete-time domain: the switched batch Bernoulli process (SBBP) [12]. It is an arrival process representing the number of packets emitted in the slot n, modulated by an underlying Markov chain with a state space $\mathfrak{S}$.

In Section 2 we describe how to design an SBBP capturing the activity changes of a monomedia source, and then in Sections 3 and 4 we introduce the paradigm we propose for modeling the intermedia relationships of a multimedia stream. After examples of multimedia source modeling, introduced in Section 5, in Section 6 we outline how to address performance evaluation of a multiplexer loaded by a number of multimedia sources and in Section 7 we discuss the impact of intermedia relationships in a system performance evaluation framework. Section 8 concludes the paper.

2. Modeling monomedia sources

A monomedia flow is usually represented as the superposition of sequences of Information Units (IUs). The exact IU definition is media-dependent: for example, an IU corresponds to a talkspurt for a voice source, a frame for a video source, or a picture for a still-picture source.

Variations in the statistics of the IU emission process may occur due to sudden changes of the source behavior; in order to model the above variations different source activities have to be defined. Source activities in a process may differ in:

- the statistics of the IU interarrival time;
- the statistics of the IU duration;
- the statistics of the bit rate emission in an IU.

Source activity changes have already been modeled in literature through Markov-based processes constituted by N sub-processes, one for each activity state of the source [3-11]. According to this approach, in this section we will model each monomedia source by using an SBBP process , $V(n)$, whose underlying Markov chain is constituted by a number of macrostates each modeling one of the different activity states of the source[1]. Each macrostate contains the states relating to the activity to be modeled. So a monomedia source will consist of a number of SBBP sub-processes, one for each activity state of the source, and the transitions between the above sub-processes will model activity changes in the source. In particular, as an SBBP is characterized by a parameter set (Q, B), where Q is the transition probability matrix of the state of the underlying Markov chain, and B is the matrix describing the emission process in each state, in the following sections we will exploit the definition of the transition matrix Q to capture the statistics of both IU interarrival time and IU duration, and the definition of the B matrix to capture the statistics of the bit rate emission in an IU.

As an example, in the following sub-sections we will characterize the SBBP processes modeling a VBR voice source, a video MPEG source and a still picture source, respectively.

2.1 VBR voice source model

In this section we refer to a voice source where coding is applied to the output of a Voice Activity Detector (VAD). The source emits a sequence of talkspurts; if we assume each talkspurt is coded with fixed rate, the bit rate during an IU is constant. In this case the voice emission process is therefore characterized by the statistics of the talkspurt interarrival time and the statistics of the talkspurt duration. Therefore the activity in a voice source changes when the above statistics change and a voice source can be modeled by means of an SBBP, whose underlying Markov chain is constituted by a number of macrostates each modeling one of the different activities of the voice source. Each macrostate contains the states OFF and ON of the related activity. A transition between two of these macrostates models an activity change in the source. In this way the SBBP process modeling a voice source, $V^{(Voice)}(n)$, is the one shown in Fig. 1 where three different activities are envisaged. Referring to this figure, let $\Psi^{(Voice)}$ be the set of the activity macrostates and $\Phi^{(Voice)} = \{OFF, ON\}$ the set of the OFF and the ON states within each activity macrostate. Therefore, the whole state space of the underlying Markov chain of the SBBP modeling a voice source as a whole, $\mathfrak{I}^{(Voice)}$, is given by the Cartesian product of the two component subspaces, that is:

$$\mathfrak{I}^{(Voice)} = \Psi^{(Voice)} \times \Phi^{(Voice)} \tag{1}$$

and the state of the underlying Markov chain at the slot n, $S^{(Voice)}(n) \in \mathfrak{I}^{(Voice)}$, is:

$$S^{(Voice)}(n) = \left(A^{(Voice)}(n), Z^{(Voice)}(n) \right) \tag{2}$$

where $A^{(Voice)}(n) \in \Psi^{(Voice)}$ is the activity macrostate and $Z^{(Voice)}(n) \in \Phi^{(Voice)}$ is the state within the activity macrostate. A voice source is characterized by:

---

[1] The number of macrostates as well as the number of states in each of them of course depends on the peculiar features of the monomedia source to be modeled.

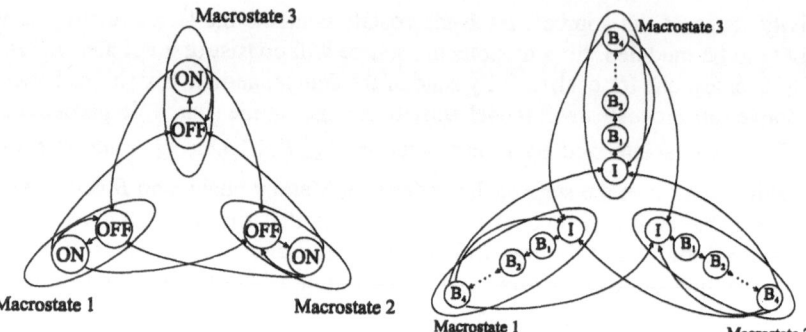

Figure 1: SBBP model for voice an still picture sources

Figure 2: SBBP model for video sources

- $\overline{T}_{(a,OFF)}$ and $\overline{T}_{(a,ON)}$, $\forall a \in \Psi^{(Voice)}$: the average duration of the states OFF and ON, respectively, during the activity macrostate a;

- $\overline{T}_a^{(Voice)}$, $\forall a \in \Psi^{(Voice)}$: the average duration of the activity macrostate a;

- $q_{a',a''}^{(Voice)} = \lim_{n \to \infty} \text{Prob}\left\{ A^{(Voice)}(n+1) = a'' \middle/ \left(A^{(Voice)}(n) = a', a' \neq a'' \right) \right\}$

  $\forall a', a'' \in \Psi^{(Voice)}$: the probability that the activity of the voice source moves from the macrostate a' to the macrostate a'', provided that the activity leaves the macrostate a';

- $P^{(Voice)}$: the packet-rate during the ON periods.

From these source parameters, the parameter set of the SBBP $V^{(Voice)}(n)$, $\left(Q^{(Voice)}, B^{(Voice)} \right)$, can be calculated as follows [11]:

$$Q_{[(a',z'),(a'',z'')]}^{(Voice)} = \lim_{n \to \infty} \text{Prob}\left\{ \begin{matrix} A^{(Voice)}(n+1) = a'', & / A^{(Voice)}(n) = a', \\ Z^{(Voice)}(n+1) = z'' & / Z^{(Voice)}(n) = z' \end{matrix} \right\} \qquad (3)$$

$$= \begin{cases} \dfrac{\Delta_{Voice}}{\overline{T}_{(a',ON)}} \cdot \left(1 - \dfrac{\overline{T}_{(a',OFF)} + \overline{T}_{(a',ON)}}{\overline{T}_{a'}^{(Voice)}} \right) & \text{if } a' = a'', \ z' = ON \text{ and } z'' = OFF \\[4mm] \dfrac{\Delta_{Voice}}{\overline{T}_{(a',OFF)}} & \text{if } a' = a'', \ z' = OFF \text{ and } z'' = ON \\[4mm] 1 - \dfrac{\Delta_{Voice}}{\overline{T}_{(a',z')}} & \text{if } a' = a'' \text{ and } z' = z'' \\[4mm] \dfrac{\Delta_{Voice}}{\overline{T}_{(a',ON)}} \cdot \dfrac{\overline{T}_{(a',OFF)} + \overline{T}_{(a',ON)}}{\overline{T}_{a'}^{(Voice)}} \cdot q_{a',a''}^{(Voice)} & \text{if } a' \neq a'', \ z' = ON \text{ and } z'' = OFF \\[4mm] 0 & \text{otherwise} \end{cases}$$

$$B_{[(a,z),r]}^{(Voice)} = \lim_{n \to \infty} \text{Prob} \left\{ V^{(Voice)}(n) = r / S^{(Voice)}(n) = (a,z) \right\} = \tag{4}$$

$$\begin{cases} 1 & \text{if } z = OFF \text{ and } r = 0 \\ 0 & \text{if } z = OFF \text{ and } r = 1 \\ P \cdot \Delta_{Voice} & \text{if } z = ON \text{ and } r = 0 \\ 1 - P \cdot \Delta_{Voice} & \text{if } z = ON \text{ and } r = 1 \end{cases}$$

where Δ_{Voice} is the chosen slot duration.

2.2 MPEG video source model

An MPEG video source emits frames organized in Groups of Pictures (GoPs); as frames are emitted at fixed time instants an activity change in an MPEG video source determines a variation in the probability density function (pdf) of the emission process of each frame of the GoP.

As the I-frame value in each GoP mainly determines the activity in the whole GoP, we assume activity changes cannot occur within a GoP, that is, activity changes can only occur when source is moving from one GoP to another. So an MPEG video source can be modeled by an SBBP with an underlying Markov chain constituted by a number of macrostates, each characterizing an activity state, grouping a number of states, one for each frame in a GoP. A transition between two of these macrostates models an activity change in the source. In this way the underlying Markov chain of the SBBP process modeling a video source, $V^{(Video)}(n)$, is the one shown in Fig. 2, where three different activities are envisaged. As in the voice case, referring to this figure, let $\Psi^{(Video)}$ be the set of the activity macrostates and $\Phi^{(Video)}$ the set of states within each activity macrostate. If we consider, for example, a GoP constituted by $l_{GoP} = 6$ frames and a subGoP of 3 frames, we have $\Phi^{(Video)} = \{I, B_1, B_2, P_1, B_3, B_4\}$. Let $\mathfrak{S}^{(Video)} = \Psi^{(Video)} \times \Phi^{(Video)}$ be the state space of the underlying Markov chain of the SBBP modeling the video source as a whole, $V^{(Video)}(n)$; the state of the underlying Markov chain $V^{(Video)}(n)$ at the slot n, $S^{(Video)}(n) \in \mathfrak{S}^{(Video)}$, is:

$$S^{(Video)}(n) = \left(A^{(Video)}(n), Z^{(Video)}(n) \right) \tag{5}$$

where $A^{(Video)}(n) \in \Psi^{(Video)}$ is the activity macrostate and $Z^{(Video)}(n) \in \Phi^{(Video)}$ is the state within the activity macrostate. If we take a Gamma distribution for the emission pdf of each kind of frame [14], an MPEG video source is characterized by:

- $\mu_I(a)$ and $\sigma_I^2(a)$, $\forall a \in \Psi^{(Video)}$, mean value and variance of the Gamma function $f_a^{(I)}(r)$, pdf of the I-frames when the activity macrostate is a;

- $\mu_P(a)$ and $\sigma_P^2(a)$, $\forall a \in \Psi^{(Video)}$, mean value and variance of the Gamma function $f_a^{(P)}(r)$, pdf of the P-frames when the activity macrostate is a;

- $\mu_B(a)$ and $\sigma_B^2(a)$, $\forall a \in \Psi^{(Video)}$, mean value and variance of the Gamma function $f_a^{(B)}(r)$, pdf of the B-frames when the activity macrostate is a;

- $\overline{T}_a^{(Video)}$, $\forall a \in \Psi^{(Video)}$: the average duration of the activity macrostate a;

- $q_{a',a''}^{(Video)} = \lim_{n\to\infty} \text{Prob}\left\{ A^{(Video)}(n+1) = a'' / \left(A^{(Video)}(n) = a', a' \neq a'' \right) \right\}$,

   $\forall a', a'' \in \Psi^{(Video)}$: the probability that the activity moves from the macrostate a' to the macrostate a'', provided that the activity leaves the macrostate a'.

Let us give an ordering relation to the elements of the set $\Phi^{(Video)}$. In this way, $\forall z', z'' \in \Phi^{(Video)}$, $z'' = z'+1$ holds if z'' is the frame following to z' in the sequence $I, B_1, B_2, P_1, B_3, B_4$; thus, for example, $B_3 = P_1 + 1$. If we use a slot duration, Δ_{Video}, equal to the time duration of one frame, the parameter set of the SBBP $V^{(Video)}(n)$, $\left(Q^{(Video)}, B^{(Video)} \right)$, can be calculated from the previous source parameters as follows [6]:

$$Q_{[(a',z'),(a'',z'')]}^{(Video)} = \lim_{n\to\infty} \text{Prob}\left\{ \begin{matrix} A^{(Video)}(n+1) = a'', \\ Z^{(Video)}(n+1) = z'' \end{matrix} \middle/ \begin{matrix} A^{(Video)}(n) = a', \\ Z^{(Video)}(n) = z' \end{matrix} \right\} \tag{6}$$

$$= \begin{cases} 1 & \text{if } a' = a'' \text{ and } z'' = z'+1 \\ \dfrac{l_{GoP}}{\overline{T}_{a'}^{(Video)}/\Delta_{Video}} & \text{if } a' = a'' \text{ and } z' = B_4, \ z'' = I \\ \dfrac{l_{GoP}}{\overline{T}_{a'}^{(Video)}/\Delta_{Video}} \cdot q_{a',a''}^{(Video)} & \text{if } a' \neq a'' \text{ and } z' = B_4, \ z'' = I \\ 0 & \text{otherwise} \end{cases}$$

and

$$B_{[(a,z),r]}^{(Video)} = \lim_{n\to\infty} \text{Prob}\left\{ V^{(Video)}(n) = r / S^{(Video)}(n) = (a,z) \right\} = \tag{7}$$

$$= \begin{cases} f_a^{(B)}(r) & \text{if } z = B_i, \ \forall i \in \{1, K, 4\} \\ f_a^{(P)}(r) & \text{if } z = P_i, \ \forall i \in \{1\} \\ f_a^{(I)}(r) & \text{if } z = I \end{cases}$$

2.3 Still Picture source model

A still picture source emits files each consisting of the coded information of one slide. As each of the above IUs is transmitted by the source at a constant bit rate, the still picture source is characterized by the statistics of the interarrival time of the files and by the statistics of the file length. So, like a the voice source, a still picture source is characterized by:

- $\overline{T}_{(a,OFF)}$ and $\overline{T}_{(a,ON)}$, $\forall a \in \Psi^{(StPic)}$: the average duration of the states OFF and ON, respectively, during the activity macrostate a;

- $\overline{T}^{(StPic)}$, $\forall a \in \Psi^{(StPic)}$: the average duration of the activity macrostate a;

- $q_{a',a''}^{(StPic)} = \lim_{n\to\infty} \text{Prob}\left\{ A^{(StPic)}(n+1) = a'' / \left(A^{(StPic)}(n) = a', a' \neq a'' \right) \right\}$,

   $\forall a', a'' \in \Psi^{(StPic)}$: the probability that the activity of the still picture source moves from the macrostate a' to the macrostate a'', provided that the activity leaves the macrostate a';

- $P^{(StPic)}$: the packet-rate during the ON periods;

Therefore, a still picture source can be modeled by an SBBP process $V^{(StPic)}(n)$, as in Fig. 1, characterized by the sets $\Psi^{(StPic)}$, $\Phi^{(StPic)}$, $\mathcal{S}^{(StPic)}$ and by $Q^{(StPic)}$ and $B^{(StPic)}$ matrices calculated according to (3) and (4).

3. Modeling multimedia source as mutually correlated SBBP processes

In a multimedia traffic flow the monomedia traffic streams coming from a multimedia source are driven by a single application according to the user needs. For this reason, when system performance is to be evaluated, a multimedia source cannot be modeled as the superposition of independent emission processes, each modeling a monomedia source, but correlation among the above processes has to be taken into account [15]. So, a multimedia source is to be considered as an aggregate of two or more monomedia sources correlated with each other: the correlation among the component monomedia sources lies in the fact that the behavior of each source depends on that of the other sources. For this purpose we correlate the monomedia SBBP processes with each other by defining, for each monomedia process, the average time the process spends in the activity macrostates as a function of the activity macrostate of the other monomedia source.

Since in the majority of multimedia applications there is one monomedia source (the master) which is completely independent of the others (the slaves), and these are dependent only on the first [1-2][16], the activity macrostate duration of the slave sources only will be defined as a function of the master activity state, that is, $\overline{T}_a^{(Sl)} = \overline{T}_a^{(Sl)}\left(A^{(M)}(n)\right)$ $\forall a \in \Psi^{(Sl)}$ where we refer to the master and slave emission processes as $M(n)$ and $Sl(n)$, respectively. Of course, the transition probability matrix of the slave source, $Q^{(Sl)}$, also depends on the activity macrostate of the master source, that is, $Q^{(Sl)}\left(A^{(M)}(n)\right)$. For the sake of simplicity, in this paper we will consider each multimedia source as being constituted by two monomedia sources, the master and one slave. Once this case has been studied, extension to multimedia sources with more than one slave is straightforward. Let $S^{(M)}(n) = \left(A^{(M)}(n), Z^{(M)}(n)\right)$ and $S^{(Sl)}(n) = \left(A^{(Sl)}(n), Z^{(Sl)}(n)\right)$ be the states of the underlying Markov chains of $M(n)$ and $Sl(n)$ respectively. A multimedia source as a whole can now be modeled as the superposition of the above emission processes, that is, as an SBBP process $W(n)$ whose states are the elements of the Cartesian product of the states of the component sources; the state of the multimedia source is therefore $S^{(W)}(n) = \left(S^{(M)}(n), S^{(Sl)}(n)\right)$, and its state space, $\mathcal{S}^{(W)}$, is the Cartesian product of the state space of the master, $\mathcal{S}^{(M)}$, and of the slave, $\mathcal{S}^{(Sl)}$. So the generic element of the transition matrix for the multimedia source from the state $s'_W = \left(s'_M, s'_{SI}\right)$ to the state $s''_W = \left(s''_M, s''_{SI}\right)$, where $s'_M = \left(a'_M, z'_M\right)$, $s'_{SI} = \left(a'_{SI}, z'_{SI}\right)$, $s''_M = \left(a''_M, z''_M\right)$, and $s''_{SI} = \left(a''_{SI}, z''_{SI}\right)$, is:

$$
Q^{(W)}_{[(s'_M, s'_{SI}),(s''_M, s''_{SI})]} = \lim_{n \to \infty} \mathrm{Prob}\left\{S^{(W)}(n+1) = \left(s''_M, s''_{SI}\right) \middle/ S^{(W)}(n) = \left(s'_M, s'_{SI}\right)\right\} =
$$
$$
= Q^{(M)}_{[s'_M, s''_M]} \cdot Q^{(SI)}\left(a''_M\right)_{[s'_{SI}, s''_{SI}]}
\tag{8}
$$

and the generic element of the emission matrix is:

$$
B^{(W)}_{[(s'_M, s'_{SI}), r_W]} = \lim_{n \to \infty} \mathrm{Prob}\left\{W(n) = r_W \middle/ S^{(W)}(n) = \left(s'_M, s'_{SI}\right)\right\} =
$$
$$
= \sum_{\substack{r_M = 0 \\ r_M + r_{SI} = r_W}}^{r^{(M)}_{MAX}} \sum_{r_{SI} = 0}^{r^{(SI)}_{MAX}} B^{(M)}_{[s'_M, r_M]} \cdot B^{(SI)}_{[s'_{SI}, r_{SI}]}
\tag{9}
$$

where $r_W \in \left\{0, \mathrm{K}, r^{(W)}_{MAX}\right\}$ and $r^{(W)}_{MAX} = r^{(M)}_{MAX} + r^{(SI)}_{MAX}$ is the maximum number of packets a multimedia source can emit in the slot n, the sum of the maximum number of packets the master and the slave can emit in one slot.

Using the matrix obtained in (8), it is possible to calculate the steady-state probabilities of the multimedia source as the solution of the following system:

$$
\begin{cases}
\underline{\pi}^{(W)} \cdot Q^{(W)} = \underline{\pi}^{(W)} \\
\sum_{s_W \in \mathfrak{I}^{(W)}} \pi^{(W)}_{[s_W]} = 1
\end{cases}
\tag{10}
$$

where $\pi^{(W)}_{[s_W]} = \lim_{n \to \infty} \mathrm{Prob}\left\{S^{(W)}(n) = s_W\right\}$ is the generic element of the steady-state probability array $\underline{\pi}^{(W)}$. Moreover, we can easily calculate the marginal steady-state probabilities for both the master and the slave sources, which are:

$$
p^{(M)}_{s_M} = \sum_{s_{SI} \in \mathfrak{I}^{(SI)}} \pi^{(W)}_{[(s_M, s_{SI})]} \qquad\qquad p^{(SI)}_{s_{SI}} = \sum_{s_M \in \mathfrak{I}^{(M)}} \pi^{(W)}_{[(s_M, s_{SI})]}
\tag{11}
$$

Let us denote by $A^{(W)}_{r_W}$ the diagonal matrix having the r_W-th column of $B^{(W)}$ on its main diagonal and the other elements all equal to zero. Thus, the multimedia source transition matrix, including the probability of r_W packets being emitted in one slot, is:

$$
C^{(W)}(r_W) = Q^{(W)} \cdot A^{(W)}_{r_W}
\tag{12}
$$

The matrices $Q^{(W)}$ and $B^{(W)}$ or, equivalently, $C^{(W)}(r_W)$, completely describe the emission process of the multimedia source.

4. Characterization of intermedia relationships

Although the correlation and cross-correlation functions of multimedia traffic can be derived as in [1], in this paper, in order to quantify the synchronization between the master and the slave sources belonging to the same multimedia source, we define the Intermedia Synchronization Index (ISI) as the cross-correlation coefficient [17] between the master traffic stream, $M(n)$, and the slave traffic stream, $Sl(n)$. If we denote the cross-covariance between the two streams as $\sigma_{M\,SI}$

and their standard deviations as σ_M and σ_{Sl}, the ISI can be defined as follows:

$$\text{ISI} = \frac{\sigma_{M\,Sl}}{\sigma_M \cdot \sigma_{Sl}} \tag{13}$$

where:

$$\sigma_{M\,Sl} = \text{E}\{M(n) \cdot Sl(n)\} - \text{E}\{M(n)\} \cdot \text{E}\{Sl(n)\} \tag{14}$$

$$\sigma_M = \sqrt{\text{E}\{[M(n)]^2\} - \text{E}^2\{M(n)\}} \qquad \sigma_{Sl} = \sqrt{\text{E}\{[Sl(n)]^2\} - \text{E}^2\{Sl(n)\}} \tag{15}$$

In order to calculate the ISI in terms of the parameters of the multimedia source model introduced in this paper, we derive the first- and second-order moments of the master and slave processes in (14) and (15) as follows:

$$E\{M(n)\} = \sum_{r_M=0}^{r_{MAX}^{(M)}} r_M \sum_{s_M \in \Im^{(M)}} B_{[s_M, r_M]}^{(M)} \cdot \underline{\pi}_{[s_M]}^{(M)} \qquad E\{Sl(n)\} = \sum_{r_{Sl}=0}^{r_{MAX}^{(Sl)}} r_{Sl} \sum_{s_{Sl} \in \Im^{(Sl)}} B_{[s_{Sl}, r_{Sl}]}^{(Sl)} \cdot \underline{\pi}_{[s_{Sl}]}^{(Sl)}$$

$$E\{M(n) \cdot Sl(n)\} = \sum_{r_M=0}^{r_{MAX}^{(M)}} \sum_{r_{Sl}=0}^{r_{MAX}^{(Sl)}} r_M \cdot r_{Sl} \cdot \sum_{s_M \in \Im^{(M)}} \sum_{s_{Sl} \in \Im^{(Sl)}} B_{[s_M, r_M]}^{(M)} \cdot B_{[s_{Sl}, r_{Sl}]}^{(Sl)} \cdot \underline{\pi}_{[(s_M, s_{Sl})]}^{(W)} \tag{16}$$

$$E\{M^2(n)\} = \sum_{r_M=0}^{r_{MAX}^{(M)}} r_M^2 \sum_{s_M \in \Im^{(M)}} B_{[s_M, r_M]}^{(M)} \cdot \underline{\pi}_{[s_M]}^{(M)} \qquad E\{Sl^2(n)\} = \sum_{r_{Sl}=0}^{r_{MAX}^{(Sl)}} r_{Sl}^2 \sum_{s_{Sl} \in \Im^{(Sl)}} B_{[s_{Sl}, r_{Sl}]}^{(Sl)} \cdot \underline{\pi}_{[s_{Sl}]}^{(Sl)}$$

It is easy to prove that $-1 \le \text{ISI} \le 1$, and that, in particular, $\text{ISI} = 0$ means that the slave source is independent of the master. Let us note that a multimedia source has a maximum positive correlation ($\text{ISI} = +1$) when the difference between the two monomedia processes multiplied by two integer positive constants, h_1, h_2, is equal to an integer constant, l:

$$h_1 M(n) - h_2 Sl(n) = l \tag{17}$$

On the contrary, a multimedia source has a maximum negative correlation ($\text{ISI} = -1$) when the sum of the two monomedia processes multiplied by two integer positive constants, h_1, h_2, is equal to a non-negative integer constant, l':

$$h_1 M(n) + h_2 Sl(n) = l' \tag{18}$$

As we will see in Section 7, the ISI characterizes the intermedia correlation relationships; so it has to be taken into account to estimate the performance of a multiplexer loaded by multimedia sources.

5. Multimedia source modeling examples

In this section we apply the multimedia source model previously defined to the measured output traffic flows coming from two real multimedia applications: a remote teaching application and a slide show with a voice commentary.

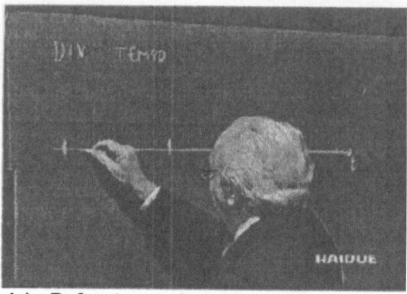

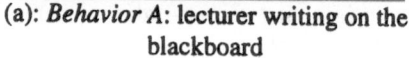

(a): *Behavior A*: lecturer writing on the blackboard   (b): *Behavior B*: lecturer speaking to the students

Figure 3: Different behaviors

5.1 MPEG coded teaching session

As a first example, let us consider a distance learning session in which the lecturer is filmed and both voice and video are encoded using the MPEG compression standard. More specifically, for the voice encoding a 64 kbit/s PCM encoder with a VAD is used; video is encoded with a GoP of six frames and a subGoP of three frames, like the one taken as an example in Section 2.

From empirical observation of a 2-hour transmission, we noted two different kinds of behavior on the part of the lecturer: one, a typical image of which is depicted in Fig. 3a, when the lecturer is writing on the blackboard (behavior A), and another, a typical image of which is depicted in Fig. 3b, when the lecturer speaks to the students and is facing them (behavior B). These two kinds of behavior determine two different statistics in the IU emission processes of both the voice and video sources and, for this reason, we use two different activity macrostates in modeling each monomedia traffic stream. Let L' and H' be the two activity macrostates modeling the voice emission process; L' presents a low emission activity, while H' presents a high emission activity. Analogously, let L'' and H'' be the two activity macrostates modeling the video emission process; macrostate L'' corresponds to the behavior A of the lecturer and is characterized by a high intra- and inter-frame correlation, that is, by a low emission activity, given that the blackboard takes up most of the image and the lecturer's movements are slow; macrostate H'' corresponds to the behavior B of the lecturer and is characterized by a higher emission activity because of the lower values of both the previous correlations. Of course, given that both the voice and the video sources behave according to only two activity macrostates, the macrostate transition probabilities, provided that the initial macrostate is left, are $q_{L',H'}^{(Voice)}(a) = q_{H',L'}^{(Voice)}(a) = 1$, $\forall a \in \Psi^{(Video)}$, and $q_{L'',H''}^{(Video)} = q_{H'',L''}^{(Video)} = 1$.

VIDEO ACTIVITY	Activity average duration	I-frame	B-frame	P-frame
H''	$\overline{\overline{T}}_{L''}^{(Video)} = 3.65$ sec	MEAN (bits)		
		158810	22688	49630
		VARIANCE (bits ^{2})		
		$4.24 \cdot 10^5$	$55.3 \cdot 10^3$	$2.00 \cdot 10^5$
L''	$\overline{\overline{T}}_{H''}^{(Video)} = 6.16$ sec	MEAN (bits)		
		141470	20209	44208
		VARIANCE (bits ^{2})		
		$3.07 \cdot 10^5$	$50.1 \cdot 10^3$	$1.21 \cdot 10^5$

a) MPEG video source parameters

VIDEO ACTIVITY	VOICE ACTIVITY	Average activity duration
H''	H'	$\overline{\overline{T}}_{H'}^{(Voice)} (H'') = 1.64$ sec
	L'	$\overline{\overline{T}}_{L'}^{(Voice)} (H'') = 0.91$ sec
L''	H'	$\overline{\overline{T}}_{H'}^{(Voice)} (L'') = 0.96$ sec
	L'	$\overline{\overline{T}}_{L'}^{(Voice)} (L'') = 1.05$ sec

VOICE ACTIVITY	Average duration state OFF	Average duration state ON
H'	$\overline{\overline{T}}_{(H',OFF)} = 39.4$ sec	$\overline{\overline{T}}_{(H',ON)} = 2.77$ sec
L'	$\overline{\overline{T}}_{(L',OFF)} = 3.27$ sec	$\overline{\overline{T}}_{(L',ON)} = 2.70$ sec
Peak Bit Rate $P^{(Voice)} = 64$ kbit/s		

b) PCM voice source parameters
Table 1: Audio/Video MPEG source parameters

From measurement of the traffic statistics it was observed that, on average, the macrostates L'' and H'' present a longer duration than L' and H', that is, the video stream shows a slower dynamic than the voice stream; for this reason, in this case, we take the video as the master source and calculate the duration of the voice source macrostates, L' and H', depending on which macrostate the video source is in. The durations measured for the macrostates L'' and H'' for the video source are $\overline{\overline{T}}_{L''}^{(Video)} = 3.65$ sec and $\overline{\overline{T}}_{H''}^{(Video)} = 6.16$ sec, while the durations measured for the macrostates L' and H' for the voice source are $\overline{\overline{T}}_{L'}^{(Voice)} (L'') = 1.05$ sec and $\overline{\overline{T}}_{H'}^{(Voice)} (L'') = 0.96$ sec when the video source is in the macrostate L'', and

$\overline{T}_{L'}^{(Voice)}(H'') = 0.91$ sec and $\overline{T}_{H'}^{(Voice)}(H'') = 1.64$ sec when the video source is in the macrostate B''. This determines a positive intermedia correlation which, as discussed in Section 4, can be represented by the ISI parameter whose measured value for this voice/video trace is $ISI = 1.129 \cdot 10^{-1}$. Table 1 summarizes the values of the parameters characterizing both the voice and the video source.

Using the SBBP parameters calculated by replacing the above values in equations (3), (4), (6) and (7), we have computed, according to (13), the same ISI parameter value as the measured one.

5.2 JPEG/LD-CELP commented slide show session

As a second example, let us consider a slide show with a voice commentary generated by a tourist information guide for the "Uffizi Gallery" in Florence. In this case, a JPEG coding technique was applied to each slide while, for the voice encoding a ITU-T Rec. G. 728 (16 kbit/s LD-CELP) encoder with a VAD, is used. Here again we observed two different forms of behavior; in fact, the speaker only presents some of the slides during high voice-emission periods, mainly those coinciding with the presentation of the painter (behavior A), whereas most of the slides, mainly concerning the paintings, are presented during low voice-emission periods, in order to allow the tourist to enjoy the contents of the still pictures in silence (behavior B).

Again, these two kinds of behavior determine different statistics in the IU emission processes of both the voice and the still picture source and we use two different activity macrostates in modeling each monomedia traffic stream. Let H' and L' be the two activity macrostates modeling the high-emission activity of the voice source when the painter is introduced, and the low-emission activity of the voice source when the paintings are shown, respectively. Analogously, let H'' and L'' be the two activity macrostates modeling the still picture emission process; macrostate H'' is characterized by a low IU interarrival time and, therefore, by a high emission activity; macrostate L'', on the contrary, is characterized by high IU interarrival time. Of course, given that in this case both the voice and the still picture sources behave according to two only activity macrostates, the macrostate transition probabilities, provided that the initial macrostate is left, are $q_{H'',L''}^{(StPic)}(a) = q_{L'',H''}^{(StPic)}(a) = 1$, $\forall a \in \Psi^{(Voice)}$, and $q_{H',L'}^{(Voice)} = q_{L',H'}^{(Voice)} = 1$.

From measurement of the traffic statistics it was observed that, on average, the macrostates H' and L' present a longer duration than H'' and L'', that is, the voice stream shows a slower dynamic than the still picture stream; for this reason, in this case, we take the voice as the master source and calculate the duration of H'' and L'' depending on which macrostate the voice source is in.

More specifically, the durations measured for the macrostates H' and L' for the voice source are $\overline{T}_{H'}^{(Voice)} = 102$ sec and $\overline{T}_{L'}^{(Voice)} = 78$ sec, while the durations measured for the macrostates H'' and L'' for the still picture source are $\overline{T}_{H''}^{(StPic)}(H') = 39$ sec and $\overline{T}_{L''}^{(StPic)}(H') = 12$ sec when the voice source is in the macrostate H', and $\overline{T}_{H''}^{(StPic)}(L') = 21$ sec and $\overline{T}_{L''}^{(StPic)}(L') = 18$ sec when the voice

ACTIVITY	Activity average duration	Average duration state OFF	Average duration state ON
H'	$\overline{\overline{T}}_{H'}^{(Voice)} = 102.0$ sec	$\overline{\overline{T}}_{(H',OFF)}^{(Voice)} = 40$ msec	$\overline{\overline{T}}_{(H',ON)}^{(Voice)} = 2.75$ sec
L'	$\overline{\overline{T}}_{L'}^{(Voice)} = 78.0$ sec	$\overline{\overline{T}}_{(L',OFF)}^{(Voice)} = 5.21$ sec	$\overline{\overline{T}}_{(L',ON)}^{(Voice)} = 2.75$ sec
Peak Bit Rate $P^{(Voice)} = 16$ kbit/s			

a) LD-CELP voice source parameters

VOICE ACTIVITY	STILL PICTURE ACTIVITY	Average activity duration
H'	H''	$\overline{\overline{T}}_{H''}^{(StPic)}(H') = 27$ sec
	L''	$\overline{\overline{T}}_{L''}^{(StPic)}(H') = 24$ sec
L'	H''	$\overline{\overline{T}}_{H''}^{(StPic)}(L') = 66$ sec
	L''	$\overline{\overline{T}}_{L''}^{(StPic)}(L') = 12$ sec

STILL PICTURE ACTIVITY	Average duration state OFF	Average duration state ON
H''	$\overline{\overline{T}}_{(H'',OFF)} = 2.9$ sec	$\overline{\overline{T}}_{(H'',ON)} = 0.1$ sec
L''	$\overline{\overline{T}}_{(L'',OFF)} = 11.9$ sec	$\overline{\overline{T}}_{(B'',ON)} = 0.1$ sec
Peak Bit Rate $P^{(StPic)} = 280$ kbit/s		

b) Still-picture source parameters
Table 2: JPEG/LD-CELP source parameters

source is in the macrostate L'.

Table 2 summarizes the values of the parameters characterizing both the voice and the still picture sources. Let us note that, in the latter table, the still picture source presents the same average ON-period duration within both macrostate H'' and macrostate L'', given that the quality and the dimension of the presented slides do not depend on the speaker's behavior. Moreover, as we have considered JPEG files characterized by an average dimension of 3.5 Kbytes and transmitted at a rate of $P^{(St)} = 280$ kbit/s, the still picture source is characterized by average ON-period durations of $\overline{\overline{T}}_{(H'',ON)} = \overline{\overline{T}}_{(L'',ON)} = (3.5 \cdot 10^3 \cdot 8)/(280 \cdot 10^3) = 0.1$ sec.

This second example is characterized by a negative intermedia correlation which, as discussed in Section 4, is represented by an ISI $= -1.67 \cdot 10^{-3}$, when a slot duration of $\Delta = 1.41$ msec is considered.

6. Multiplexer system model

Let us now apply the multimedia source model introduced in the previous section to define a finite-buffer model for a multimedia multiplexer, Σ. Let c be the capacity of the output link, expressed in bit/sec. We will use a discrete-time approach and we will assume as the slot duration, Δ, the time needed to transmit a packet on the multiplexer output link. We assume a *late arrival system with immediate access* model [18]: packets arrive in batches, and a batch of packets potentially arrives immediately before the end of a time slot. An arriving packet can enter the server facility if it is free, with the possibility of it being ejected almost instantaneously. Note that in this model, a packet service time is counted as the number of slot boundaries between the point of entry to the service facility and the packet departure point. Therefore, even though we allow an arriving packet to be ejected almost instantaneously, its service time is counted as 1, not 0. Let us assume that the buffer can accommodate at most K packets, including the one being served, if any; thus, if r packets arrive when k packets $(k \geq K - r)$ are in the buffer, only $K - k$ packets are accommodated, and the remaining $r - K + k$ packets are discarded. As we are interested in calculating the performance for one particular multimedia source like that modeled in Section 3, we model the system as loaded by this source, here referred to as the Tagged Source (TS), and by external traffic generated by the aggregate of the other sources, here referred to as the External Source (ES). To this purpose let us model ES with an SBBP, $V^{(ES)}(n)$, whose state in the slot n is $S^{(ES)}(n)$, and let $\left(Q^{(ES)}, B^{(ES)}\right)$ be its parameter set.

A complete description of the multiplexer at the nth slot requires a two-dimensional state, $S^{(\Sigma)}(n) = \left(S^{(Q)}(n), S^{(A)}(n)\right)$, where $S^{(Q)}(n) \in \{0,1,\ldots,K\}$ is the buffer state in the nth slot, i.e. the number of packets in the queue and in the server at the observation instant, and $S^{(A)}(n) = \left(S^{(TS)}(n), S^{(ES)}(n)\right)$ is the state of the aggregated input traffic in the same slot. Due to the assumptions of the *late arrival system with immediate access* model, given that the observation instant is preceded by the departure, the buffer will never be totally full at the observation instants and we therefore have $S^{(Q)}(n) \in \{0,1,\ldots,K-1\}$. Now, the state transition probability for the Markov chain modeling the multiplexer Σ can be defined as:

$$\lim_{n \to \infty} \text{Prob}\left\{S^{(Q)}(n+1) = k, S^{(A)}(n+1) = s''_A / S^{(Q)}(n) = h, S^{(A)}(n) = s'_A\right\} =$$

$$= \begin{cases} \left(C^{(A)}(0) + C^{(A)}(1)\right)_{[s_A, s'_A]} & \text{if } h = 0, k = 0 \\ \left(C^{(A)}(k+1)\right)_{[s_A, s'_A]} & \text{if } h = 0, k \leq K-2, k \leq h + r_{MAX}^{(A)} \\ \left(C^{(A)}(k-h+1)\right)_{[s_A, s'_A]} & \text{if } h > 0, h-1 \leq k \leq K-2, k < h + r_{MAX}^{(A)} \quad (19) \\ \left(\sum_{r=k-h+1}^{r_{MAX}^{(A)}} C^{(A)}\right)_{[s_A, s'_A]} & \text{if } k = K-1, k < h + r_{MAX}^{(A)} \\ 0 & \text{otherwise} \end{cases}$$

where $r_{MAX}^{(A)}$ is the maximum number of arrivals to the multiplexer in one slot, and $C^{(A)}(r)$, transition probability matrix of the aggregated sources including the probability of r cells being emitted in one slot, can be calculated as follows:

$$C^{(A)}(r) = \sum_{r_1=0}^{r_{MAX}^{(TS)}} \sum_{r_2=0}^{r_{MAX}^{(ES)}} C^{(TS)}(r_1) \otimes C^{(ES)}(r_2) \qquad \forall r \le r_{MAX}^{(A)} \qquad (20)$$
$$\scriptstyle r_1 + r_2 = r$$

where $C^{(ES)}(r)$ is the transition probability matrix of the ES source including the probability of r cells being emitted in one slot, and can be calculated like in (12). Of course, $C^{(A)}(r)$ is a null matrix if $r > r_{MAX}^{(A)}$. From (19) we can easily calculate the state transition probability matrix $Q^{(\Sigma)}$ characterizing the system as a whole as in [1-2] and therefore the steady-state probability array of the system Σ as the solution of the following linear system:

$$\begin{cases} \underline{\pi}^{(\Sigma)} \cdot Q^{(\Sigma)} = \underline{\pi}^{(\Sigma)} \\ \underline{\pi}^{(\Sigma)} \cdot \underline{1} = 1 \end{cases} \qquad (21)$$

where $\underline{1}$ is a column array all of whose elements are equal to 1, and $\underline{\pi}^{(\Sigma)} = \left[\underline{\pi}_{[0]}^{(\Sigma)}, \underline{\pi}_{[1]}^{(\Sigma)}, K, \underline{\pi}_{[K-1]}^{(\Sigma)} \right]$ is the steady-state probability array, whose generic element, $\underline{\pi}_{[k]}^{(\Sigma)}$, is the array containing the steady-state probabilities of the multimedia source when the buffer is in the state k, that is, the generic term of $\underline{\pi}_{[k]}^{(\Sigma)}$ is:

$$\pi_{[k,w]}^{(\Sigma)} = \lim_{n \to \infty} \mathrm{Prob}\left\{ S^{(Q)}(n) = k, S^{(A)}(n) = s_A \right\} \qquad (22)$$

When the queueing system is solved, it is possible to evaluate the loss, jitter and skew performance of the multimedia multiplexer as in [1-2] [6]

7. Impact of intermedia relationships on traffic performance
In this section we will analyze how intermedia relationships affect performance of an ATM multiplexer loaded by multimedia traffic. To this purpose we consider an ATM multiplexer with an output link with a capacity $c = 300$ kbit/s, loaded by $N = 2$ multimedia traffic streams. More specifically, as an example, we use sources relating to a tourist information guide, like that one introduced in Section 5.2. In order to highlight the influence of the ISI on the multiplexer performance, we consider sources consisting each of a voice source and a still picture source characterized by the same emission pdf of that one in Section 5.2 with a different intermedia correlation (sources S_0, S_1, S_2, S_3, S_4). We obtain these traffic streams from the parameters listed in Table 2 by moving the presentation of a suitable number of slides from a voice activity period, to the other one. In order that these new sources maintain the same emission pdf, the still picture source in all the multimedia sources S_0, S_1, S_2, S_3, S_4 emits the same number of slides during a voice activity cycle, that is, during a period with duration equal to the sum of the average durations of one high activity period and one low activity period of

the voice source in Table 2 (180 s). The maximum and the minimum number of slides emitted by the slave source during each master activity period can be achieved by changing only the slave macrostate duration, or the slave state duration. too. Of course, the second choice gives the maximum variation range of the ISI because it allows us to move a higher number of slides. In any case the range is lower than $[-1,+1]$. Let S_1 and S_2 be the multimedia sources obtained from Table 2 by changing only the slave macrostate durations during the master high-emission periods, $\overline{T}_{H''}(H')$ and $\overline{T}_{L'}(H')$, and during the master low-emission periods, $\overline{T}_{H''}(L')$ and $\overline{T}_{L'}(L')$. So doing S_1 is characterized by 19 slides during the master high activity periods and 26 during the master low activity periods ($ISI = -3.50 \cdot 10^{-3}$), and S_2 is characterized by 34 slides during the master high activity periods and 11 during the master low activity periods ($ISI = +4.65 \cdot 10^{-3}$). Analogously, let S_3 and S_4 be the multimedia sources obtained by changing also the duration of the OFF periods between two successive slides, $\overline{T}_{(H'',OFF)}$ and $\overline{T}_{(L'',OFF)}$: S_3 is characterized by all the 45 slides during the master low activity periods ($ISI = -1.57 \cdot 10^{-2}$), and S_4 is characterized by all the 45 slides during the master high activity periods ($ISI = +1.20 \cdot 10^{-2}$). Finally, source S_0 is achieved in such a way the voice and the still picture sources are uncorrelated, that is, ISI=0.

The loss probabilities curves obtained when two of these source are multiplexed together, are shown in Fig. 4. The curves are obtained by using the analysis introduced in [1]. From these figures it is evident the dependence of the loss probability on the intermedia correlation and, therefore, the importance of the ISI parameter in performance evaluation.

8. Conclusions

In this paper an analytical framework is presented in order to model the multimedia traffic and to evaluate how time relationships which characterize intermedia synchronization in a multimedia stream affect network performance.

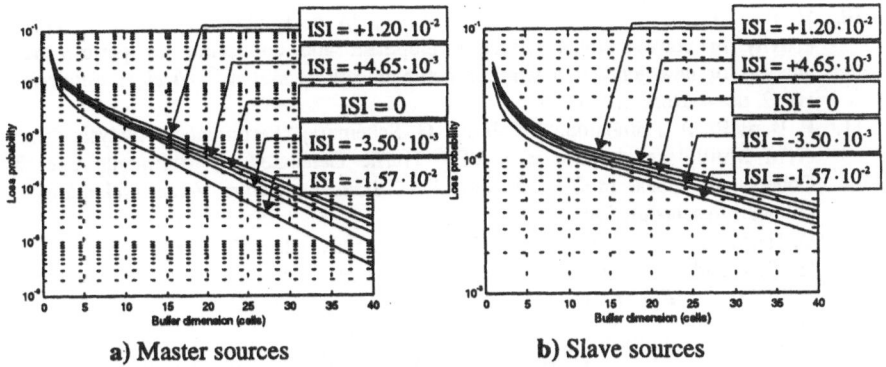

a) Master sources   **b)** Slave sources

Figure 4: Loss probability

The emission process of each multimedia source has been modeled as the superposition of a number of SBBP processes correlated with each other, each modeling one monomedia source. In order to characterize the intermedia relationships a new parameter, here referred to as the Intermedia Synchronization Index (ISI), has been defined and calculated in terms of the parameters of the emission process modeling the multimedia source introduced.

References

1. A. La Corte, A. Lombardo, G. Schembra, "An Analytical Paradigm to Calculate Multiplexer Performance in an ATM Multimedia Environment", *Computer Networks and ISDN Systems*, vol. 29, no. 16, December 1997.

2. A. Lombardo, G. Schembra, "An Analytical Paradigm to Compare Routing Strategies in an ATM Multimedia Environment", *IEEE/ACM Transactions on Networking*, vol. 5, no. 6, December 1997.

3. B. Maglaris, D. Anastassiou, P.Sen, G. Karlsson and J, D. Robbins, "Performance models of statistical multiplexing in packet video communication", *IEEE Transaction On Communications*, vol. 36, no. 7, July 1988

4. P. Sen, B. Maglaris, N. Rikli and D. Anastassiou, "Models for Packet switching of variable-bit-rate video sources", *IEEE Journal On Selected Areas in Communications*, vol. 7, no. 6, June 1989.

5. P. R. Jelenkovic, A. A. Lazar, "The effect of multiple time scales and subexponentiality in MPEG video streams on queueing behavior", *IEEE Journal on Selected Areas in Communications*, vol. 15, no. 6, August 1997.

6. A. Lombardo, G. Morabito, G. Schembra, "An Accurate and Treatable Markov Model of MPEG-Video Traffic", *IEEE Proc. Infocom '98*, San Francisco, USA, April 1998.

7. A. Lombardo, G. Morabito, S. Palazzo, G. Schembra , "Intra-GoP Modeling of MPEG Video Traffic", *IEEE Proc. ICC '98*, Atlanta, USA, June 1998.

8. A. Lombardo, G. Morabito, S. Palazzo, G. Schembra, "A Fast Simulation Model of MPEG Video Traffic", *IEEE Proc. GLOBECOM '98*, Sidney, Australia, November 1998.

9. P. T. Brady, "A model for generating on-off speech patterns in two-way conversation", *Bell Systems Technical Journal*, pp. 2445-2472, September 1969.

10. H. P. Stern, S. A. Mahmoud, K. K. Wong, "A comprehensive model for voice activity in conversational speech-development and application to performance analysis of new-generation wireless communications system", *Wireless Networks*, vol. 2, no. 4, December 1996.

11. F. Beritelli, A. Lombardo, S. Palazzo, G. Schembra, "Performance Analysis of an ATM Multiplexer Loaded with VBR Traffic Generated by Multimode Speech Coders", to appear in *IEEE Special JSAC Issue on Future Voice Technologies*, publication scheduled for 4Q98.

12. O. Hashida, Y. Takahashi, and S. Shimogawa, "Switched Batch Bernoulli Process (SBBP) and the discrete-time SBBP/G/1 queue with application to statistical multiplexer", *IEEE J. Selected Areas in Communications*, volume 9, no. 3, pp. 394-401, 1991.

13. R. Steinmetz, "Multimedia encoding standards", *ACM/Springer Multimedia Systems*, vol. 1, no. 5, 1994.

14. C. Huang, M. Devetsikiotis, I. Lambadaris, A. R. Kaye, "Modeling and Simulation of Self-Similar Variable Bit Rate Compressed Video: A Unified Approach", *ACM SIGCOMM '95*, Cambridge, Massachusetts, January 1995.

15. P. Venkat Rangan, S. S. Kumar, S. Rajan, "Continuity and Synchronization in MPEG", *IEEE Journal on Selected Areas in Communications*, vol. 14, no. 1, January 1996.

16. P. Venkat Rangan, H. M. Vin, S. Ramanathan, "Designing an On-Demand Multimedia Service", *IEEE Communication Magazine*, July 1992.

17. W. Feller, An Introduction to Probability Theory, vol. I, vol II, Wiley, New York 1966.

18. J. J. Bae, T. Suda, R. Simha: "Analysis of Individual Packet Loss in a Finite Buffer Queue with Heterogeneous Markov Modulated Arrival Processes: A study of Traffic Burstiness and a Priority Packet Discarding", *Proc. IEEE Infocom '92*, Florence (Italy).

Multimedia Network Management as Global Control

Giovanni Pacifici[1], Rolf Stadler[2], Mun Choon Chan[3]

[1] IBM T.J. Watson Research Center, Hawthorne, NY 10532,
giovanni@watson.ibm.com
[2] Center for Telecommunications Research, Columbia University,
New York, NY 10027, stadler@ctr.columbia.edu
[3] Bell Labs, Lucent Technologies, Holmdel NJ 07733,
munchoon@dnrc.bell-labs.com

Abstract. Global control refers to influencing the global state of a
system on a slow time scale, in order to achieve management objectives. In multimedia networks, global control operations can be
realized by tuning the behavior of network controllers in a coordinated way. We distinguish between four classes of global control
operations, namely, managing protocols and algorithms, managing
adaptivity, managing robustness, and managing (indirect) interactions among controllers. This paper focuses on managing indirect
interactions among controllers. Such interactions occur, because controllers adapt to changes in the network state and—by applying
control—influence the network state and thus other controllers in
a complex way. We present experiments in which different resource
controllers of a multimedia network interact. We show that the performance of the system is strongly affected by indirect interactions
among these controllers. We demonstrate how the management system can control the effect of such interactions, by modifying specific
control parameters associated with the control system. These results
were obtained using an emulation platform which runs an executable
model of the traffic control system and management agents, emulating in real–time the behavior of routers, connection managers, link
admission controllers, etc.

1 Introduction

Global control refers to influencing the global state of a system on a slow
time scale, in order to achieve management objectives. From a performance
point of view, the state of a multimedia network is determined by interacting
resource controllers, such as schedulers, routers, flow controllers, etc. Each
of these controllers contributes to allocating a specific resource (e.g., link
bandwidth) in response to service requests, by taking into account the current
state of the resource, the resource capacity, and the request intensities. The
difficulty in managing multimedia networks stems from the large number of
controllers running concurrently and on different time scales. Although the
behavior of a single controller taken in isolation has been extensively studied

will interact and influence each other in a real network, and how to manage such interactions.

Aspects of global control have been addressed within the standardized frameworks for network management, such as those developed jointly by the ISO and ITU committees [9] or by the Internet community [1, 16]. These frameworks provide models for defining the structure of management information, and they specify protocols for exchanging this data between functional entities known as managers and agents. Unified modeling of performance–related management information [14] and the definition of generic interfaces for monitoring [7] belongto this category. With few notable exceptions, such as [6, 5], most of the work in this area has focused on monitoring.

In this paper, we concentrate on the control aspect. We categorize global control operations into four classes. These are: managing protocols and algorithms, managing adaptivity, managing robustness, and managing indirect interactions among controllers. Indirect interactions among controllers result from the fact that controllers adapt to changes in the network state and—by adapting their control policies—influence the state and thus other controllers.

We focus specifically on indirect interactions and report on experiments which show that this type of interaction among the different resource controllers can lead to dramatically different states (in term of network performance) as the load pattern in the network changes. We show how the management system can control the effect of such interactions by modifying specific control parameters associated with the control system. The desired value of these control parameters depends on the environment and, therefore, cannot be fixed during the design/provisioning phase of the network. We conclude that, in order to achieve the objectives of service effectiveness and efficiency (e.g., guaranteeing QOS to different applications while achieving a high degree of network utilization), the management system must be designed in such a way that control parameters can be changed during the operational phase of the network.

The experimental results presented in this paper were obtained using a prototype of a management and control architecture running on a high–performance emulation platform [2]. Our experimental environment is composed of two subsystems: a manager station and an emulation platform. The manager station allows a human operator to interact with the multimedia network through high–level control primitives, supported by a 3D interface. The emulation platform runs an executable model of the traffic control system and management agents, emulating in real–time the behavior of routers, connection managers, link admission controllers, etc.

2 Global Control

2.1 A Management Framework

Fig.1 outlines an architecture we have developed for managing multimedia network services, which focuses on providing global control and dynamic abstractions of the network state [15]. This architecture identifies three separate entities that operate on network resources: a real-time traffic control system, a management system, and a management application. The traffic control system directly regulates the competition for network resources and operates in real-time. The management system controls the operations of the traffic control system, while the management application supervises these activities, pursuing management objectives. These subsystems interact asynchronously and run on different time scales. To cope with the high-speed and dynamic nature of real-time traffic, the traffic control system works on a time scale of microsecond to millisecond, while the management system and application act on a time scale of seconds or minutes.

The task of the traffic control system is to provide end-to-end quality of service to distributed applications, while utilizing resources in an efficient way. This system can be seen as a collection of mechanisms, each of which operates asynchronously and solves a specific resource control problem. Examples of real-time control mechanisms are buffer management and scheduling, flow control, routing and admission control. The operations of the traffic control system can be tuned by changing control parameters associated with each mechanism. Changing the parameters of a single controller results in a different resource control policy for that controller and, in turn, may result in a different operating point for other controllers. The network state is the result of the interaction of these real-time control mechanisms.

The task of the management system is to provide the functionality for pursuing management objectives. The management system executes its task by interacting with the traffic control system, following the monitor/control paradigm. This means that it monitors the network state and takes control actions in order to influence this state. Control actions result in changing specific parameters in the traffic control system.

Management applications monitor the network state and perform global control by acting upon management parameters. A detailed example, describing the management parameters used for controlling the resources available to different traffic classes and for adjusting the adaptivity of the traffic control system, is presented in [15].

2.2 A Model of the Control System

The general approach to system management is to model the system to be managed as a set of managed objects which are organized into a (hierarchically structured) Management Information Base (MIB). In [12] functional

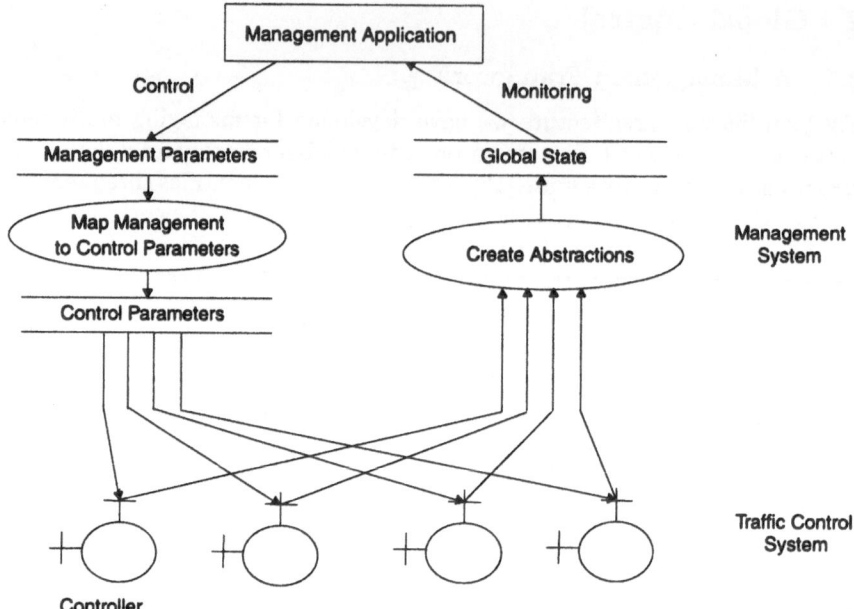

Fig. 1. The management architecture.

components, such as network controllers, are integrated into the MIB framework as part of the Integrated Network Information Model. This approach focuses on the data aspect of the system to be managed and on the relationships between objects.

For our purpose, however, the functional aspect (including algorithms, etc.) and the way network components interact play a key role. Therefore, dealing with data abstractions alone is not sufficient. Our approach is to start from a software abstraction of the service delivery system as the model of the system to be managed.

Our model of the control system contains a set of cooperating and interacting controllers, which perform the tasks of service creation and service delivery in a distributed fashion. They exhibit the following characteristics (1) to (4).

(1) A control system consists of *a large number of controllers*, with many instances of the same controller class spread throughout the system, in order to support distributed operation and increase fault tolerance. A limited number of controller classes exist, each of which encapsulates a specific functionality, such as running a network control algorithm (admission control, routing, etc.).

(2) *Controllers are adaptive.* They adapt their behavior to changes in the environment in which they operate. A good example is a router which periodically adapts to the load pattern in the network. Also, controllers adapt or change their behavior according to operations executed by the manage-

ment system. Note that, by changing their behavior, controllers themselves influence (or perturb) the environment in which they run.

(3) *Controllers interact* in two fundamentally different ways. First, they interact via direct cooperation. Distributed algorithms, such as routing, are realized as a set of cooperating controllers. Also, controllers run protocols among themselves for service creation, including connection setup and resource negation/renegotiation, and for information transfer. Second, as mentioned above, controllers influence each other indirectly via perturbation of the environment. This means that the change in the behavior of one controller affects that of other controllers via a change in the system state, e.g., through a change in traffic statistics.

(4) Most *controllers are realized in software.* Using a software paradigm, we model them as communicating active objects. They are multi-threaded, i.e., they have separate address spaces, and they communicate by exchanging messages, which allows them to run on different time scales. Furthermore, they are designed as persistent objects, which are created during the initialization phase of the control system.

2.3 Categorizing Global Control Operations

We understand the system state as an abstraction of the states of the controllers in the control system. Therefore, modifying the behavior of one or more of these controllers changes the system state. The task of management in the sense of global control thus refers to changing the behavior of the controllers in a coordinated way in order to achieve management objectives.

We distinguish between four basic operations related to global control, which the management system executes on the control system.

(1) *Managing protocols and algorithms*: A (possibly distributed) algorithm or a protocol can be managed by modifying control parameters associated with it. In [3] we report on an experiment in which blocking constraints of call admission controllers are changed, in order to influence the traffic mix in a multimedia network.

(2) *Managing adaptivity*: This operation refers to how sensitively the controllers react to changes in the environment, e.g., to traffic fluctuations. In [3] we provide measurements that show the influence of adaptivity parameters, such as the time scale to recompute control policies, on the network utilization.

(3) *Managing robustness*: Control mechanisms generally are based upon assumptions about the traffic model, the user behavior, etc. Using these assumptions, control algorithms are developed that provide statistical QOS guarantees. Managing the robustness of a set of controllers refers to defining a safety margin within which the original assumptions can vary. A similar concept is applied in civil engineering for building bridges and constructing aircrafts. In the case of networking, managing robustness means striking (and possibly changing) the balance between aggressive control policies that allow

for high system utilization and conservative control policies that provide high confidence that the QOS objectives can be met.

(4) *Managing the (indirect) interaction among controllers*: This is the most difficult and challenging task with respect to global control. Unlike protocol interaction, this type of communication is not intended by the designer of the control system. It results from the fact that controllers are adaptive and interact with the environment.

3   Experiments

In this section we present results of simulation experiments which illustrate the behavior of interacting controllers and their impact on the network state. We focus our study on three different classes of controllers: *routers, admission controllers* and *schedulers*.

We start by describing the controllers and how their behavior can be changed by tuning parameters associated with each of them (Sec 3.1). Without loss of generality, we refer to specific resource control algorithms. Sec 3.2 presents experimental results that illustrate both the impact of the interaction among controllers and the quantitative effect of management operations. The experiments are conducted on a prototyping platform described in [2].

3.1   The Resource Controllers

We focus on three different levels of control: scheduling, admission control and routing. Scheduling mediates the low level contention for service between cells of different classes of network traffic. Admission control regulates the acceptance or blocking of incoming traffic on a network link. Routing selects an end-to-end path from source to destination.

The task of the admission controller is to accept or reject calls so as to maximize some utility function. The admission controller must guarantee the required QOS at both the cell level (delay, cell loss, etc.) and the call level (call blocking constraints) to every call admitted into service on the link. From the point of view of the admission controller, the link capacity can be expressed by its *schedulable region S* [8], which defines how many calls of a given class the link can support, while guaranteeing the appropriate cell–level QOS to each class. The schedulable region S is an N–dimensional space, where N is the number of classes (such as voice, video, facsimile, data) recognized by the link controller. The resource state is defined by the occupancy vector $\mathbf{x}$, which represents the number of calls of each class currently active in the link. In order to guarantee the QOS to each class of traffic, the occupancy vector can assume only values that are inside the schedulable region, i.e., $\mathbf{x} \in S$.

The information available to the admission controller includes the boundaries of the schedulable region S, the call arrival and departure rates λ^i and μ^i associated with each class, the utility C^i generated by a call for each class and the call blocking constraints associated with each class κ^i. In [8], a linear programming formulation is given to find an admission control policy that maximizes the expected system utility, while providing QOS guarantees to the various classes of traffic. An *intensity estimator* calculates the traffic arrival and departure intensities, λ^i and μ^i, by averaging the number of arrivals and departures over a given time interval, while a *capacity estimator* determines the boundaries of the schedulable region S by estimating the cell–level QOS for each possible value of the link state $\mathbf{x}$. We assume that the quality of service required by the application traffic is expressed in terms of *upper bound* on the maximum link delay. The parameter Q^i_j identifies the value of the upper bound for calls of class i at link j.

The router is responsible for selecting an end–to–end path for establishing a connection between a source and a destination. The router must guarantee the required end–to–end cell level QOS for each established connection. The end–to–end delay experienced by a class i connection along path r is always bounded by,

$$X^i_r = \sum_{j \in L_r} (Q^i_j + D_j) . \tag{1}$$

where Q^i_j is the upper bound on the class i queuing delay for link j, and D_j is the propagation delay for link j. When establishing a class i connection between two source destination pairs (n, m) the selected route is the one that minimizes the sum of the weight on each link, i.e.,

$$\min_{r \in R_{m,n}} J(r) = \sum_{j \in L_r} w^i_j \tag{2}$$

subject to

$$X^i_r \leq S^i \tag{3}$$

where S^i is the maximum end–to–end delay allowed for class i, and L_r is the set of links belonging to route r.

The weight w^i_j for link j are a function of the link utilization *estimated* over a specified time interval. In particular, if $\rho^i_{U,j}(n)$ is the average value of the link utilization in the n-th estimation interval, the weight $w^i_j(n)$ is expressed by,

$$w^i_j(n) = \begin{cases} \rho^i_{U,j}(n) & \text{if } n = 0 \\ w^i_j(n-1) & \text{if } \mid w^i_j(n-1) - \rho^i_{U,j}(n) \mid \leq \alpha^i_j \\ \rho^i_{U,j}(n) & \text{otherwise} \end{cases} \tag{4}$$

where α^i_j is a constant (usually called *bias factor*) used to damp possible oscillations of this algorithm [11].

Each one of the above controllers monitors the network state and takes control actions in order to influence the allocation of a resource. Every time the network state changes, a different resource allocation policy is adopted.

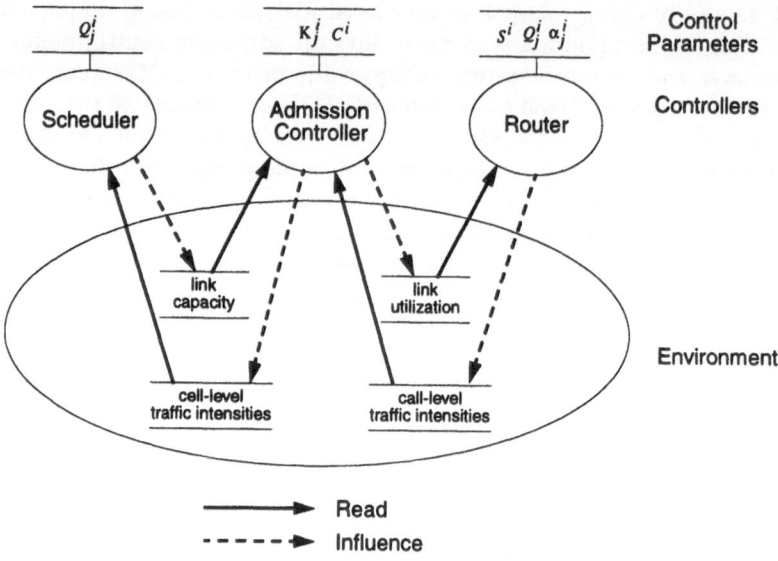

Fig. 2. Indirect interaction among controllers.

For example, when the intensity estimator detects a change in the traffic intensity λ^i, it triggers the computation of a new admission control policy.

Fig. 2 illustrates how the three controllers described above interact and influence each other by perturbating the environment that they control. For example, the scheduler decisions will influence the size of the schedulable region [8], which in turn is used by the admission controller to compute the admission policy. The actions of the admission controller impact the link utilization[HYM93], which is used by the router to determine the link weights and select end-to-end routes. The admission controller can also effect the cell-level traffic intensities and therefore indirectly influence the scheduler operations. Since the router determines the path of each call, the call-level traffic intensities for each link are influenced by the routing policy. Since the scheduler impacts the decisions of the admission controller, which in turns influences schedulers and routers and is influenced by the router, when a perturbation is introduced in the environment (e.g., network load pattern is changed), a state changed is rippled through each controller until a new steady state is reached (sometimes a steady state may never be reached; in these cases the system state may oscillate until a new perturbation is introduced).

The behavior of each controller can be tuned by changing control parameters. Some control parameters influence only a single controller, while other parameters may impact more than one controller. One such parameter is the upper bound on the maximum link delay Q_j^i. If Q_j^i is increased, the size of the schedulable region, i.e. the link capacity, increases, since the scheduler on link j is allowed to delay transmission of cells for a longer period. At the same time, when Q_j^i is increased, the number of end–to–end routes that satisfy the constraint in eq. 3 decreases, since only routes with fewer number of hops (that satisfy the same end–to–end delay bound S^i) can be used.

In the next section we show how the parameter Q_j^i can be used to manage the interactions among controllers.

3.2 Measurements Results

In this section, we will present the results of simulation experiments which illustrate the effect of the (indirect) interaction among controllers in a multimedia network. We also demonstrate the ability to control the global state of the system by tuning control parameters. The simulation experiments were carried out using the experimental environment described in [2].

All experiments were conducted using a $N_N = 16$ node network, organized in a 4 X 4 grid topology, giving a total of $N_L = 24$ links as shown in Fig. 3. The network traffic was composed of two different classes (video and voice). Each traffic class i originating from node j with destination node k, consisted of a stream of statistically identical calls with Poisson arrivals at rate $\lambda^i(j, k)$ and i.i.d. exponentially distributed holding times with mean $1/\mu^i(j, k)$. In all the experiments, $\mu^I(j, k) = \mu^{II}(j, k) = 0.01 \ \forall j, k$. All connections were bidirectional and required the same amount of resources in both directions.

Each traffic class was distinguished by a unique set of performance constraints and per–call traffic characteristics. For all links, the size of the schedulable region and the associated value for Q_j^i were the same. The size of the schedulable region was determined through simulation. Finally, for every simulation, run a transient period of 10,000 seconds was allowed before any measurements were taken, and the simulations were run for 25,000 seconds. The 95 % confidence bounds for the measured performance criteria were well within 5 % of the observed values.

We used two different traffic load scenarios. The first scenario simulated the traffic load resulting from a video–on–demand type of service in a *cluster pattern*. In particular, we assumed that two video servers were located on node 3 and 6 of the grid in Fig. 3. The clients for the server on node 3 resided on nodes 14 and 16. The clients for the server on node 6 resided on nodes 4, 12, 15 and 16. The aggregated traffic among clients and server on node 3 was 0.025 calls/sec for class I call, and 0.25 calls/sec for class II call. The aggregated traffic among clients and server on node 6 was 0.03 calls/sec for class I call, and 0.30 calls/sec for class II call. On all other nodes, a server sent out background traffic of ($\lambda_N^I = 0.001$ calls/sec and $\lambda_N^{II} = 0.01$ calls/sec) to clients uniformly distributed on all other nodes. Aggregated network traffic intensity was 6.9 for class I calls and 69 for class II calls.

The second load scenario simulated the traffic load from the same service in a *uniform pattern*. In this case, the servers were located at node 6 and

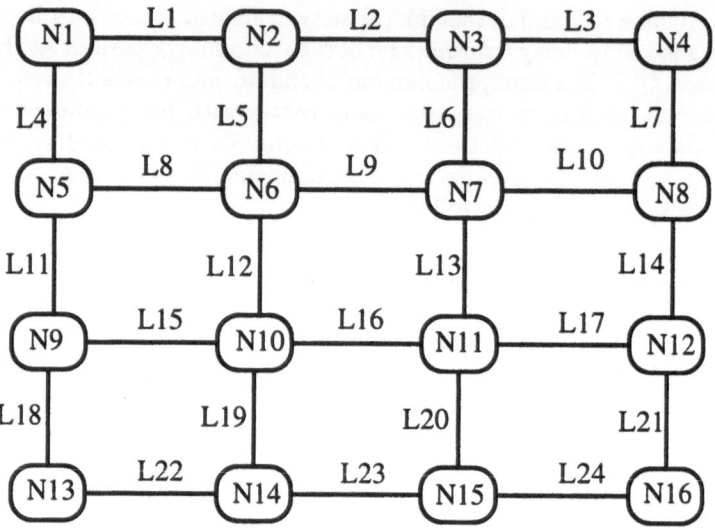

Fig. 3. Topology of the network used in the experiments.

11, while the clients for both servers were uniformly distributed on all other nodes. Each of the servers sent an aggregated traffic load of 0.16 calls/sec for class I call and 1.6 calls/sec for class II call to the clients. On all other nodes, a server sent out background traffic of ($\lambda_N^I = 0.001$ calls/sec and $\lambda_N^{II} = 0.01$ calls/sec) to clients uniformly distributed on all other nodes. Aggregated network traffic intensity was 33.4 for class I calls and 334 for class II calls.

The cluster pattern represented a situation in which most of the network load was concentrated around a few network nodes. On the other hand, the uniform pattern represented a case in which the network load was more evenly spread over the network.

In the first experiment, we set the value of the upper bound on the maximum link delay $Q=(Q^i=Q^i_j) \; \forall i, j$) to 5ms and we started loading the network with a uniform pattern. After the network had reached its steady state, we switched the traffic load to the cluster pattern and we measured the changes in the network blocking probability. The changes in the traffic load pattern resulted in an increase in the network blocking probability, as shown in Fig. 4. In this figure, the x-axis represents different values of Q, while the y-axis shows the call blocking probability for the entire network. State A is the initial state (i.e, uniform pattern and Q=5ms), while state B is the steady state after the change in the load pattern. By changing the pattern from uniform to cluster (whereby the system state moves from A to B), the blocking probability increases, even though the network load decreases by a factor of 5).

The same measurements were performed for a value of Q=4ms. We obtained state C for the cluster case and state D for the uniform case. Fig. 5 gives more complete measurements.

Most important, Fig. 4 shows how the parameter Q can be used for management purposes: The system in a "bad" state B can be brought back into a "good" state C, by changing Q from 5ms to 4ms.

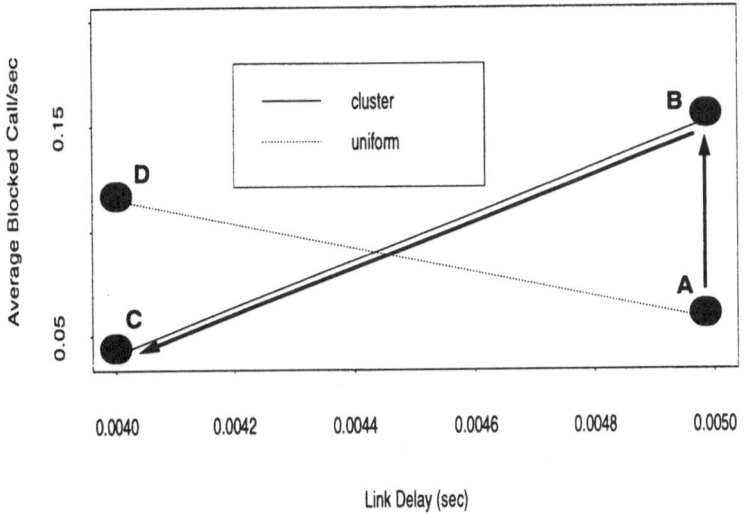

Fig. 4. The influence of link delay constraint on the network performance for a 16–node grid network with two different traffic patterns. Arrow BC shows the effect of a global control operation.

The impact on the network state of different settings of the parameter Q cannot be easily predicted for two reasons. First, the parameter changes the behavior of more than one controller at a time; second, while increasing Q increases the link capacities, larger values of Q reduce the number of end–to–end routes that a call can use. As Q increases, the size of the schedulable region increases, since cells can sustain longer delays. Increasing the link capacity triggers a change in the admission control policy, which allows the admission controller to admit more calls onto the link. However, the set of end–to–end routes that satisfy the constraint in eq. 3 decreases, since only routes with fewer hops (that satisfy the same end–to–end delay bound S^i) can be used.

A better understanding of the impact of the parameter Q on the network behavior can be derived from our next set of experiments. In these experiments we measured the call blocking probabilities for values of Q in the range [1ms, 5ms]. Our measurements indicate that, in the cluster case (Fig. 5(a)), the gain in link capacity seems to outweigh the loss in route options until

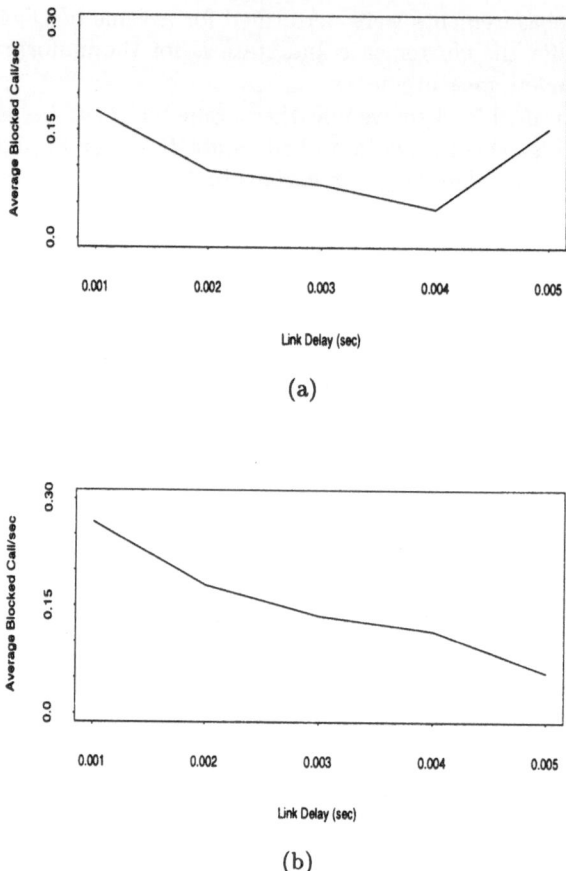

Fig. 5. Influence of link delay constraint on network performance for a 16–node grid network under different load: (a) cluster pattern (b) uniform pattern.

Q=4ms. However, by increasing Q further to 5ms, the loss of route options becomes dominant. As a result, the network blocking probability increases dramatically. On the other hand, in the uniform case (Fig. 5(b)), the gain in link capacity always seems to outweigh the loss in route options for Q=1ms to Q=5ms.

This observation is further clarified in Figs. 6 and 7, which show the distribution of the average number of calls blocked per second for each source-destination (SD) pair. The support of the distribution is the average hop count. Each point in the plots corresponds to a SD pair. (There is a total of 16 x 15 = 240 SD pairs per plot.)

The envelopes of the blocked call distribution for the uniform case (Fig.

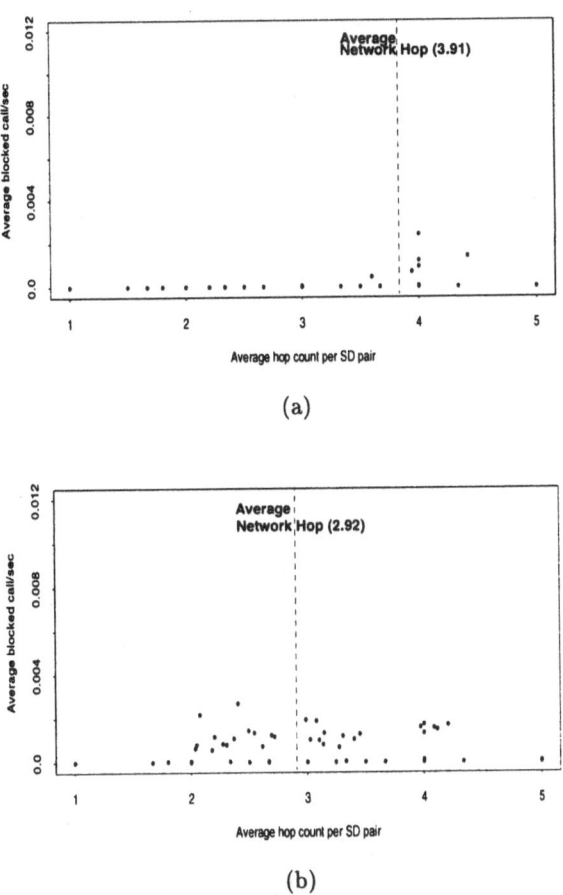

(a)

(b)

Fig. 6. Distribution of calls blocked per second for each source-destination pair on a 16–node grid network (Q=4ms) under different load: (a) Cluster pattern (b) Uniform pattern.

6(b) and 7(b)) is more uniform than those for the cluster case (Figs. 6(a) and 7(a)). The shapes of the these envelopes are influenced by how well the routers balance the load in the network. For each SD pair, the associated router has a number of possible routes to chose from, and the choice depends on the link load, since we are using a minimum weight routing strategy. When the number of route options decreases (due to increasing Q from 4ms to 5ms), the available routes might contain congested links. For example, in state B (Fig. 7(a)), most of the blocking occurs for traffic over the SD pairs (6,14), (6,12) and (6,16), while the other three active SD pairs experience little blocking. The congested links are links 9 and 10, which contribute significantly to the

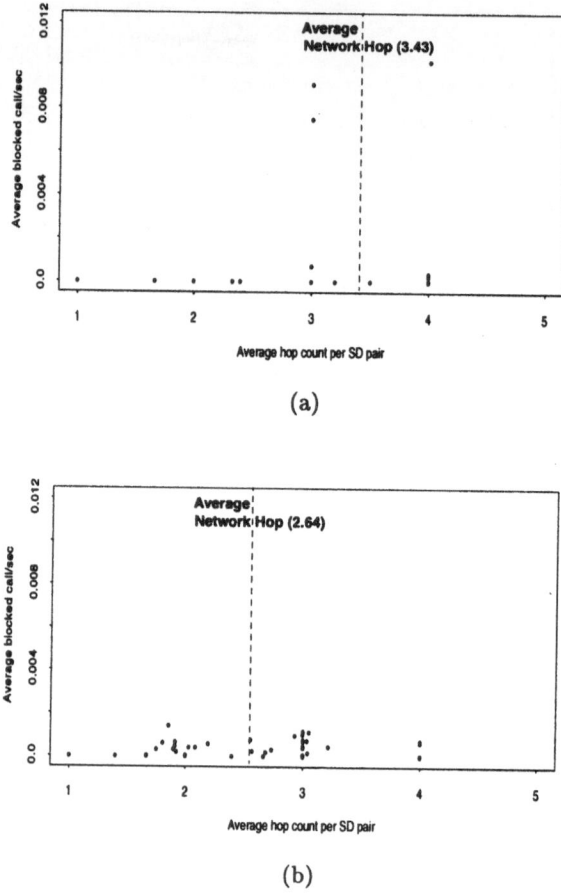

(a)

(b)

Fig. 7. Distribution of calls blocked per second for each source-destination pair on a 16–node grid network (Q=5ms) under different load: (a) Cluster pattern (b) Uniform pattern.

network blocking.

3.3 Discussion of the Results

We have presented experiments in which three different classes of network controllers interacted by reading and writing the system state. We showed that the performance of the system is strongly affected by indirect interactions among these controllers. These interactions are complex in nature and their effects are difficult to predict.

Our results demonstrate that interactions can be managed by tuning the controllers. In particular, we used the parameter Q, which impacts the three classes of controllers, to recover from the performance degradation that occurs under certain load pattaerns. The effect of changing Q can be significant (for example, by reducing blocking from 0.16 to 0.05). We believe that this type of global control can be exploited for performance management purposes.

Our results show that the impact of the parameter Q cannot be easily predicted. There is no intrinsic optimal value for Q; rather, the best choice of Q depends on the traffic load.

Acknowledgement

The work reported here was supported by the Department of the Air Force, Rome Laboratory, under contract F30602-95-R-0143. It was conducted using the resources of the Cornell Theory Center.

References

1. Jeffry D. Case, Keith M. McCloghrie, Marshall T. Rose, and Steven Waldbusser. *Introduction to Version 2 of the Internet–Standard Network Management Framework.* RFC–1441, April 1993.

2. M.C. Chan, G. Pacifici, and R. Stadler. Prototyping network architectures on a supercomputer. In *Fifth International Symposium on High Performance Distributed Computing*, Syracuse, NY, August 1996.

3. Mun Choon Chan, Giovanni Pacifici, and Rolf Stadler. Managing multimedia network services. *Journal of Network and Systems Management*, 5(3), 1997.

4. Domenico Ferrari and Dinesh C. Verma. A scheme for real-time channel establishment in wide-area networks. *IEEE Journal on Selected Areas in Communications*, SAC-8(3):368–379, April 1990.

5. P. Georgatsos and D.P. Griffin. A general framework for routing management in multi-service atm networks. In *IFIP/IEEE International Symposium on Integrated Network Management (IM '97)*, San Diego, CA, May 1997.

6. David Griffin and P. Georgatsos. A TMN system for VPC and routing management in ATM networks. In *Proceedings of the IFIP/IEEE International Symposium on Integrated Network Management*, pages 356–369, San Barbara, CA, May 1995. Elsevier Science (North–Holland).

7. Stephen Hayes. Analyzing network performance management. *IEEE Communications Magazine*, 31(5):52–58, May 1993.

8. Jay M. Hyman, Aurel A. Lazar, and Giovanni Pacifici. A separation principle between scheduling and admission control for broadband switching. *IEEE Journal on Selected Areas in Communications*, 11(4):605–616, May 1993.

9. ISO. *Information Processing Systems — Open System Interconnection — Systems Management Overview.* ISO/IEC, IS 10040, 1991.

10. Van Jacobson. Congestion avoidance and control. In *Proceedings of the ACM SIGCOMM*, pages 316–329, Stanford, CA, August 1988.

11. Atul Khanna and John Zinky. The revised ARPANET routing metric. In *Proceedings of the ACM SIGCOMM*, September 1989.

12. S. Kheradpir, W. Stinson, J. Vucetic, and A. Gersht. Real-time management of telephone operating company networks: Issues and approach. *IEEE Journal on Selected Areas in Communications*, 11(9):1385–1403, December 1993.

13. Suk Lee and Asok Ray. Performance management of multiple access communications networks. *IEEE Journal on Selected Areas in Communications*, 11(9):1426–1437, December 1993.

14. B. Neumair. Modeling resources for integrated performance management. In H.G. Hegering and Y. Yemini, editors, *Proceedings of the IFIP/IEEE International Symposium on Integrated Network Management*, pages 109–121. Elsevier Science (North–Holland), Amsterdam, The Netherlands, 1993.

15. Giovanni Pacifici and Rolf Stadler. An architecture for performance management of multimedia networks. In *Proceedings of the IFIP/IEEE International Symposium on Integrated Network Management*, pages 174–186, Santa Barbara, California, May 1995. Elsevier Science (North–Holland).

16. Marshall T. Rose and Keith M. McCloghrie. *Structure and Identification of Management Information for TCP/IP based Internets*. RFC–1155, May 1990.

Malleable Multimedia Presentations: Adaptive Streaming Tradeoffs for Best-Quality Fast-Response Systems

Wei Zhao[1], Marc Willebeek-LeMair[2] and Prasoon Tiwari[2]

[1]Department of Computer Science, University of Maryland, College Park, MD 20742, zw@cs.umd.edu

[2]IBM T.J. Watson Research Center, Yorktown Heights, NY 10598, mwlm@watson.ibm.com, ptiwari@watson.ibm.com

Abstract
The evolution of today's computer networks to higher bandwidths and increased connectivity to a wider range of end-user devices will further exacerbate the need for adaptive streaming solutions. Motivated by these adaptive streaming techniques, we introduce the notion of a malleable presentation and outline a framework for the transmission of such presentations across heterogeneous networks to different end-user devices. We devise a method for partitioning a schedule into independent segments. Our adaptive scheduling scheme uses the segmentation scheme to efficiently make schedule changes for delivering the highest quality presentation based on specified network bandwidth and device constraints.

1 Introduction

We are witnessing a fundamental transition period in the history of communications as traditional analog communications such as telephony and television are being digitized and delivery of these media is moving to packet-based IP networks. With the emergence of higher backbone speeds and faster access connectivity, high quality digital television and stereo audio will proliferate the Web. This network convergence will also help spawn new computing and communications devices with different functions and capabilities. Such a transition does not occur overnight. During the transition period, and for a long time to come, a broad range of network connection bandwidths and device capabilities will exist.

Streaming multimedia content over such networks presents a unique set of challenges. Due to the continuous nature of multimedia streams, the sensitivity of multimedia quality to network conditions, and relatively large time needed to measure network conditions it is difficult to adapt the (stored or live) multimedia

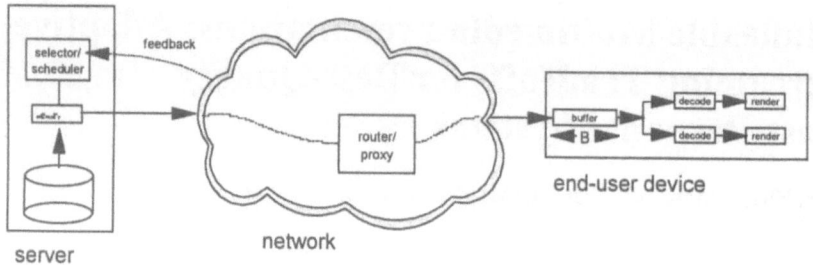

Figure 1. *Components of a multimedia streaming system include a server, network and end-user device.*

stream in mid-session. In order to deal with these difficulties in a graceful manner, it is essential for a multimedia presentation to be *malleable*, as defined in Section 2. An important component of a malleable presentation is a transform method used to adapt the presentation to varying network and client conditions.

Streamed multimedia delivery solutions include server, network and client components which coordinate and manage the transformation and transmission of malleable multimedia presentation streams across a network (Figure 1). The server is typically where scheduling and transmission of the original presentation occurs. Proxies or routers (multicast) may intercept the stream and modify it before delivering it to the end-user device where it is eventually decoded and rendered. We anticipate that these multimedia streams will be composed of a combination of old and new media data types including, images, 3D graphics, animation, audio and video. The multimedia presentations will specify the choreography of the multimedia objects to indicate when each is to be presented and in what location on the end-user rendering device.

Previous Results

A critical component in a multimedia delivery solution is the scheduling algorithm that dictates when and at what quality each multimedia object is transmitted. There has been considerable research on this topic. One common strategy has been a feedback-based mechanism. Information about the current network condition is sent back to the sender, which in turn adjusts its sending rate [4,6,11,12]. Rate adjustment can also be carried out by the receiver, such as in [5] where a receiver adjusts its receiving rate by dynamically changing its subscription among a set of concurrent multicast groups. Typically, feedback-based mechanisms are associated with rate-adjustable or multi-layered codecs for real-time interactive audio and video applications. In general, these techniques are not particularly efficient with variable-bit-rate multimedia streams, nor did they consider the scheduling for synchronized composite multimedia (e.g. SMIL [10]) that is being standardized.

For the purpose of our discussion we will refer to the multimedia presentation illustrated in Figure 2(a). This example depicts a presentation consisting of eight objects (o_1-o_8) which are to be presented at times, t_1-t_8, respectively. These objects may represent a variety of multimedia types including a graphic image such as a viewgraph, an audio clip or segment, or a video clip. Two of the most important criteria measuring a transmission schedule are its bandwidth requirement and the initial startup delay. Typically, the bandwidth requirements is a function of the initial startup delay and the client buffer size. The initial startup delay is the amount of time the client must wait while buffering the data sent by the server before starting to render the presentation. The client buffer size is the amount of data that the client can use at any given time to store the received multimedia stream in preparation for rendering.

In order to analyze the nature of the transmission processes of such a presentation we stack the objects with their respective sizes as illustrated in Figure 2(b). The profile of the presentation, $L(t)$, represents the amount of data that must be sent by any time point, t, in the presentation. Given a buffer size limit B at the receiving node, the boundary $U(t)$ represents the maximum amount of data that can be sent to the client up to time t. The region bounded by the *lower bound curve $L(t)$* and the *upper bound curve $U(t)$* determines the *feasible region* [1,3] within which any valid transmission schedule curve must lie.

Among all valid transmission schedules, we are most interested in the ones with the least possible bandwidth requirement. Note that the transmission rate of a transmission schedule at any time instant is the slope of the schedule curve at that point. An optimal smoothing algorithm was given in [1]. The algorithm produces a piecewise linear transmission schedule that is as smooth as possible, with the minimum peak rate and minimum rate variance. The algorithm is quite straightforward: starting from $t=0$, the algorithm looks for the longest linear segment within the feasible region (Figure 2(b)). If the segment terminates on the

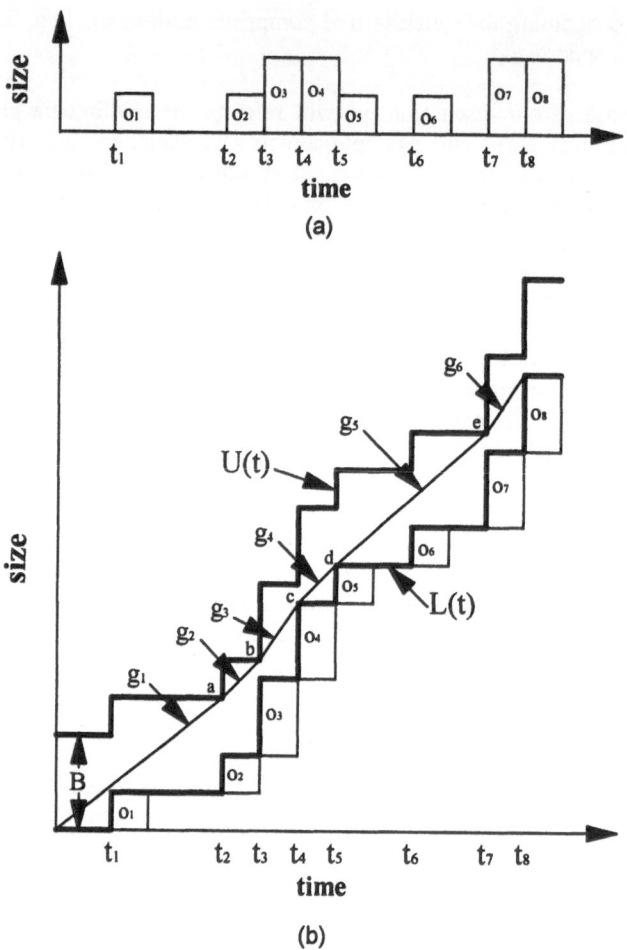

Figure 2. *Illustration of the scheduling Algorithm A. (a) Given a presentation schedule consisting of objects o_1-o_8 with presentation times t_1-t_8. (b) For a client with buffer size B, the feasible region is bounded by the lower, L(t), and upper, U(t) data transmission limits. A schedule S, consisting of change points a-e and segments g_1-g_6, is created* using Algorithm A.

lower bound curve, it generates a schedule segment ending at the latest upper bound point it touches, and vice versa. The procedure is then repeated from the new start point. The algorithm has a running time of $O(N)$, where N is the number of objects in the presentation. We refer to the algorithm as algorithm A throughout the paper. We call the points on the schedule where the slope changes as change points. There are two types of change points, *buffer empty points* on the lower bound curve where the slope decreases and *buffer full points* on the upper bound curve where the slope increases.

Zhao, *et.al.*[2] demonstrated the relationship between startup delay and bandwidth in deriving optimal transmission schedules based on the same feasible region technique and showed that the bandwidth/delay relationship can be calculated in $O(N)$ time. Based on these results, some preliminary scheduling guidelines related to the notion of scaleable multimedia were introduced.

This paper extends that work and introduces an adaptive scheduling technique for multimedia presentations that are layer-structured and may be streamed under dynamic and heterogeneous network and end-system conditions. Section 2 introduces the notion of malleable media. Section 3 presents a technique for partitioning a schedule into independent segments for efficient handling of object changes. Our adaptive scheduling approach is described in Section 4 making use of the malleable presentation concept and the efficient schedule segmentation principle. We state our conclusions and directions of future work in Section 5.

2 Malleable Media: Adaptation and Scalability

In this section, we describe a framework for working with media in heterogeneous environments. In a given context, a piece of media is called a *media element* if either it can not be further decomposed, or such a decomposition is not desirable[1]. A *media collection* is a set of one or more media elements.

Heterogeneous environments and varying and scarce network conditions, imply that a given media collection may not be usable in certain conditions; e.g., an audio stream can not be used because it requires more than the available bandwidth. This brings out the need for adapting a media collection to fit into the available resources. Traditionally, this has been done for media elements by defining scalable codecs (e.g. MPEG-2 etc. [8,9,13]) or media transcoding (e.g. in [7]). We extend this notion of scalability to media collections.

Our approach is based on a *layered media element* model that defines malleability through the concept of layering. The concept is inspired by the layered (or progressive) encoding techniques of digital audio and video. In the layered model, media consists of a number of versions, each having a different size and different perceptive quality. Each version is associated with a *layer index* indicating the quality it represents. A higher layer index indicates a higher quality and presumably a larger size.

A *static malleable media* object consists of one or more stored media elements and other static malleable media objects, and a *transform* method that takes as

---

[1] Examples of media elements are audio stream, graphics, video stream etc. In some cases, one may wish to decompose a video stream into frames; then individual frames would be media elements.

input a QoS^2 (Quality of Service) object and an *end-user device capabilities*[3] object, and returns a sequence of media elements called a *playback*. A *static malleable presentation* consists of one or more temporally and spatially related static malleable media objects or presentations, and a *transform* method that takes as input a QoS object and an end-user capabilities object, and returns a *playback*.

A playback is considered to have achieved an overall perceptive quality of layer *l* if it contains a version at layer *l* or above for every media element in the presentation. We call such a playback a *layer l playback*. We further define a *uniform layer l playback* as having all media elements at exactly layer *l*. In the remainder of this paper, we assume that there are N elements $o_1, o_2, ..., o_N$, in a malleable presentation ordered by their presentation time. There are L layers for each element. Media element o_i's version at layer *l* is denoted by o_i^l.

As one example, consider a user browsing a video archive across a network. In particular, this user is browsing a video sequence consisting of frames $o_1, o_2, ...,$ o_N. Suppose that L versions of each frame are available. A *playback from i to j* is a sequence f_i. The *version sequence* associated with this playback is $l_i, l_{i+1}, l_{i+2}, ...,$ l_j. The playback is *monotonic* if its version sequence is monotonically non-decreasing. Monotonic playbacks are of special interest because, much like progressive GIFs, its quality improves with time. Let F_1 and F_2 be two playbacks from *i* to *j*. F_1 *dominates* F_2 if every number in the version sequence of F_1 is at least as large as the corresponding number in the version sequence of F_2. A playback is *feasible* if it satisfies all constraints specified in the problem. A monotonic playback is *best quality* if it is not dominated by another monotonic feasible playback. Our user makes a request to the video archive to play from frame i to j. Our objective is to deliver the best quality playback to the user.

As another example, consider the authoring of a malleable presentation consisting of a speaker making a technical presentation. The base layer can be chosen to include only the slides outlining the key points. The next layer can add a mono-audio stream. Then there could be additional layers which can include a synchronized video stream showing the speaker, and a stereo audio stream, etc.

3 Efficient Scheduling by Segmentation

By partitioning a schedule into independent segments we are able to efficiently reconstruct portions of a schedule to accommodate changes in the size or synchronization of multimedia objects or adapt to variations in network bandwidth. This is particularly useful for adaptive scheduling schemes in dynamic network environments as well as for efficient static scheduling

---

[2] The QoS object, not defined here, is an abstraction of the quality of service available from the network. Typically, it contains the available bandwidth and delay and loss properties.

[3] The client capabilities object, also not defined here, is an abstraction of the capabilities of the client.

potentially incorporated into multimedia authoring tools. Given a multimedia presentation schedule S created by Algorithm A, we are able to show that the optimal schedule can be partitioned into independent segments based on the change points defined in the schedule.

Theorem 3.1 [Segment Independence Theorem]: Let (L_1,U_1) and (L_2,U_2) be the feasible regions of a presentation before and after an object o with deadline c changes size, let S_1, S_2 be the optimal transmission schedules in (L_1,U_1) and (L_2,U_2), respectively. Then:

(a) If object o decreases in size, the change points of S_2 can only differ from those of S_1 in interval (E,F), where E,F are the last buffer empty point before c and the first buffer full point after (and including) c in S_1, respectively. Furthermore, E,F still remain buffer empty and full points in S_2, respectively (see Figure 3).

(b) If object o increases in size, the change points of S_2 can only differ from those of S_1 in interval (F',E'), where F',E' are the last buffer full point before c and the first buffer empty point after (and including) c in S_1, respectively. Furthermore, F',E' still remain buffer full and empty points in S_2, respectively.

We will prove *Theorem* 3.1 using *Lemma* 3.1 below.

Lemma 3.1 Let c_1 be a change point of both S_1 and S_2, and c_2 be the next change point of S_1, then c_2 must be the next change point of S_2 of the same type (full/empty), if one of the following is true:
(a) c_2 is a buffer empty point of S_1 before c.
(b) c_2 is a buffer full point of S_1, but c_2 is before E.

Proof of Lemma 3.1: Observe that if object o's size decreases by d, then $L_2 = L_1$ before c, $L_2 = L_1 - d < L_1$ after (and including) c. Similarly, $U_2 = U_1$ before c and $U_2 = U_1 - d < U_1$ after (and including) c.

First consider part (a). From algorithm A, the extension of the line from c_1 to c_2 must penetrate U_1 first (before penetrating L_1), since $L_2<L_1$, $U_2<U_1$, it must also penetrate U_2 first (before penetrating L_2). So c_2 is the next buffer empty point of S_2. For part (b), let E_1 be the first buffer empty point after c_2 (E_1 is before c), we show that the extension of the line from c_1 to c_2 must penetrate L_1 first, before or at E_1. This is because there can only be buffer full points between c_2 and E_1, so the schedule can only increases in slope from c_2 to E_1. Since E_1 is on L_1, the extension of the line from c_1 to c_2 must penetrate L_1 first, before or at E_1. Since E_1

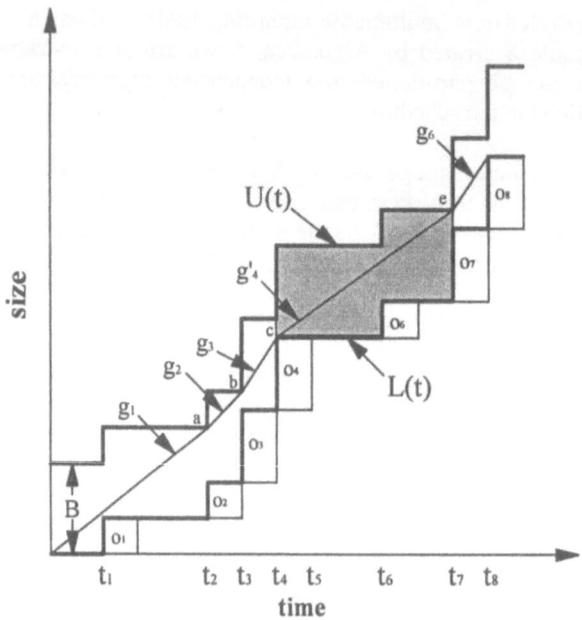

Figure 3. *Illustration of the Segment Independence Theorem. In this example, object o_5 scheduled at time t_5 is removed from the original presentation shown in Figure 2. As a consequence, we see that the new schedule S' differs from the original schedule S only in the shaded region between the last buffer empty point ($E=c$) before t_5 and the first buffer full point ($F=e$) after t_5.*

is before c, L_2 and U_2 are identical before E_1, so c_2 is on U_2 and the extension of the line from c_1 to c_2 must also penetrate L_2 first, therefore c_2 is the next buffer full point of S_2. End of proof.

Proof of Theorem 3.1: We prove claim (a) first; the proof of claim (b) is symmetric.

Start from $t=0$, we use *Lemma* 3.1 (a) and (b) iteratively depending on the type of change point of S_1. The procedure continues until we reach E, which by *Lemma* 3.1 must also be a buffer empty point of S_2. So the first part of claim (a) holds.

We proceed to the second part of the claim. To show this, we rotate the feasible region (L,U) one hundred and eighty degrees on the surface. Now the region is both upside-down and left-right swapped. We have a new feasible region (L',U') with $L'(t) = U(N) - U(N-t)$ and $U'(t) = U(N) - L(N-t)$, N is the deadline of the last object. For valid schedule S in (L,U), the rotated version is $S'(t) = U(N) - S(N-t)$. It's easy to see S' is a valid schedule in (L',U'). We show the schedule calculated by algorithm A in (L',U') is identical to the rotated version of the schedule calculated in (L,U). From the properties of the optimal schedules, both of them are the shortest path between the same two end-points. But the shortest path is unique [1], so they must be the same.

Now we rotate both (L_1,U_1) and (L_2,U_2), and get (L_1',U_1'), (L_2', U_2'). By the above rotation formulas, it is not difficult to see that $L_1'=L_2'$ and $U_1'=U_2'$ up to (and including) N-c, and $L_2'=L_1'$-d and $U_2'=U_1'$-d after N-c. Let S_1',S_2' be the optimal schedules of the rotated regions, then by the same argument, S_2' can only differ from S_1' after E', where E' is the last buffer empty point of S_1' before (and including) N-c. Since rotation swaps U and L, and also swaps left and right, E' is actually the first buffer full point of S_1 after (and including) c, which is F. Thus S_2 (with rotated version S_2') can only differ from S_1 after (and including) F, and F is also a buffer full point of S_2. End of proof.

4 Adaptive Scheduling Approach

The key problem in scheduling malleable multimedia is the adaptive handling of the tradeoffs between the quality of the presentation and the available network bandwidth and end-user device capabilities. In particular, we are interested in finding the highest quality presentation under a given bandwidth constraint. With the layered multimedia model, we are given some guidelines of how the quality of a presentation can be gauged, and through these guidelines, principles and algorithms can be devised to handle the tradeoff issues.

4.1 Maximum Uniform Presentation

From a perceptual standpoint, it is not desirable to have a presentation whose quality varies frequently and drastically over time. An ideal choice is thus to find a *maximum uniform presentation* (*i.e.* a uniform presentation with the maximum layer index under the given bandwidth constraint). There are exactly L different uniform presentations, each with a unique layer index from 1 to L, where L is the maximum layer index. Since the required bandwidth of a presentation grows monotonically with object size, a uniform presentation at a higher layer requires a higher bandwidth. Therefore, a binary search on the set of layers can be used to find the highest layer at which the uniform presentation satisfies a given bandwidth constraint. During each iteration, the optimal schedule for the uniform presentation being inspected is calculated in $O(N)$ time and tested against the bandwidth bound. Depending on the test result, the search interval is cut in half as in a binary search. The algorithm has a running time $O(NlogL)$, where N is the number of objects.

The following function *Find_Max_Uniform* returns the layer index of the maximum uniform presentation under the bandwidth constraint BW. Function *Binary_Search* finds the index recursively using a binary search. Both functions call *Optimal_Schedule* to calculate the optimal transmission schedule for a uniform presentation at a certain layer *l*, which is denoted by *Layer[l]*.

Algorithm: Finding Maximum Uniform Presentation

Function *Find_Max_Uniform*
    $S = Optimal\_Schedule(Layer[1])$;
    If ($Peak(S) > BW$) Return 0;
    $S = Optimal\_Schedule(Layer[L])$;
    If ($Peak(S) <= BW$) Return L;
    Return *Binary_Search*(1, L - 1);
End

Function *Binary_Search(low, high)*
    If (*low* >= *high* - 1) Return *high*;
    mid = floor((*low* + *high*) / 2);
    $S = Optimal\_Schedule(Layer[mid])$;
    If ($Peak(S) <= BW$) Return *Binary_Search(mid, high)*;
    Else Return *Binary_Search(low, mid* - 1);
End

4.2 Enhancing Maximum Uniform Presentation

In many cases, a maximum uniform presentation at layer *l* may not fully utilize the available bandwidth *BW*. We already know it is not possible to raise the entire presentation to layer *l*+1 by its maximality. However, portions of the presentation could potentially be enhanced to layered *l*+1 utilizing the residual bandwidth. From a user's perspective, for example, if listening to an entire speech is not feasible, listening to part of the audio stream is often more desirable than missing it entirely.

Starting with the uniform presentation at layer *l*+1, we calculate its optimal schedule S_{l+1} consisting of a series of schedule segments. Based on the slopes of these segments, we divide the time axis of the entire presentation into two types of interleaved intervals. The intervals containing segments with slopes not exceeding *BW* are defined as *enhanceable intervals*, while those with slopes greater than *BW* are called *breaking intervals*. The two types of intervals are delimited by boundary points at which the slope changes from under *BW* to above *BW* or vice versa. We further define the boundary points between two types of intervals as belonging to the breaking interval they delimit.

Lemma 4.1 Every breaking interval starts at a buffer full point and ends at a buffer empty point of S_{l+1} in the feasible region.

<Proof>: A breaking interval starts when the segment slope of S_{l+1} increases from no higher than *BW* to above *BW*. According to the optimal schedule algorithm, a

schedule slope increase only occurs at a buffer full point. Similarly, a breaking interval ends at a buffer empty point where the slope decreases. End of proof.

Theorem 4.1 The presentation with all objects within the enhanceable intervals enhanced to layer $l+1$, while keeping all objects in the breaking intervals at layer l, is a feasible presentation under bandwidth constraint BW.

<Proof>: We establish the claim by constructing a valid transmission schedule S for the above presentation with a peak bandwidth not exceeding BW. It is constructed as follows: within each enhanceable interval where objects have been enhanced to layer $l+1$, S follows the trace of S_{l+1}. Clearly, these segments of S have slopes not exceeding BW by the definition of enhanceable intervals.

Within each breaking interval where objects remain at layer l, S is constructed as the optimal schedule of the uniform layer l *subpresentation*, from the *buffer full point* at the start of the breaking interval to the *buffer empty point* at the end of the breaking interval. From the essential bandwidth algorithm [2], it is clear that the peak rate R of this portion of S is the essential bandwidth of this uniform layer l subpresentation. By the essential bandwidth formula [2], the essential bandwidth of a subpresentation can not exceed the essential bandwidth of the whole presentation, which in turn is upper bounded by the peak rate ($<=BW$) of the optimal schedule of the whole presentation. Hence, R can not exceed BW.

The only issue left is whether these two types of separately constructed schedule pieces of S do concatenate vertically with each other at their boundaries. This is indeed ensured by *Lemma* 4.1, which states that breaking intervals do start at a buffer full point and end with a buffer empty point of S_{l+1}. End of proof.

Theorem 4.2 A feasible layer l presentation under bandwidth constraint BW can not contain a uniform layer $l+1$ subpresentation spanning a breaking interval.

<Proof>: Recall that a layer l presentation is the one with all objects at layer l or higher. Suppose it contains a uniform layer $l+1$ subpresentation spanning a breaking interval, which starts at a buffer full point and ends at a buffer empty point of S_{l+1}. the peak rate of this portion of S_{l+1}, which is higher than BW, is the essential bandwidth of the uniform layer $l+1$ subpresentation. Consequently, the essential bandwidth of the entire presentation must exceed BW, making it infeasible. Contradiction. End of proof.

The following $O(N)$ algorithm enhances the uniform layer l presentation by classifying intervals and enhancing objects in the enhanceable intervals.

Algorithm: Enhancing the Maximum Uniform Presentation
Function *Enhance_Uniform(l)*
    Classify_Intervals();
    $P = \text{Layer}[l]$;

```
    For each object o in an enhanceable interval
        Enhance o to layer l+1 in P;
    Return P;
End


Function Classify_Intervals
    S_{l+1} = Optimal_Schedule(Layer[l+1]);
    Current = empty breaking interval;
    For each Segment s in S_{l+1}
        If (Slope(s) <= BW and Current is a breaking interval)
            Terminate Current;
            Current = empty enhanceable interval;
        If (Slope(s) <= BW and Current is an enhanceable interval)
            Terminate Current;
            Current = empty breaking interval;
        Append s to Current;
End
```

We need to point out that the enhanced presentation in *Theorem* 4.1 is not necessarily "*maximal*". Some of the layer l objects could still be enhanced to layer $l+1$ while maintaining the presentation's feasibility. The "ceiling line" approach in [2] can be used to construct a maximal enhancement. On the other hand, maximality is not the only concern. Minimizing quality transitions and maximizing uniform quality piece lengths are some of the other desirable characteristics. Thus the number of quality transitions can be controlled by not enhancing some of the enhanceable intervals.

Based on the segment independence results, it is not necessary to recalculate the entire schedule for the enhanced presentation. We can identify portions of S_l (*i.e.* optimal schedule of uniform layer l presentation, from which only increments are made) and S_{l+1} (from which only decrements are made) where the schedule is not affected. Only remaining portions of the presentation need to be rescheduled.

There is an interesting correlation between uniform presentation enhancement and layering refinement. If the layering is defined on a very fine scale with many layers, the bandwidth distance between layers are small and enhancements are not as crucial. However, this puts more effort on the authoring. We can incorporate tools into the authoring system that provide instant feedback of the bandwidth distance and layering refinements that can be built on by the author.

5 Conclusion

Driven by the need to support a wide range of connectivities as well as end-user devices, adaptive streaming solutions motivate the idea of malleable multimedia presentations. We have introduced the notion of malleable presentations and outlined a framework for the transmission of such presentations across heterogeneous networks to different end-user devices. We devise a method for partitioning a schedule into independent segments. Our adaptive scheduling method uses the segmentation scheme to efficiently make schedule changes for delivering the highest quality presentation based on specified network bandwidth and device constraints. Our approach adaptively selects between different versions (L) of a multimedia presentation (of N objects) and computes the presentation schedule in $O(N*\log L)$ time. Our notion of malleable multimedia and the associated scheduling techniques can be incorporated into existing multimedia authoring languages such as SMIL [10], providing them with efficient scheduling solutions for adaptive multimedia streaming.

We have begun to extend these ideas to a dynamic adaptive streaming scheme which adjusts to fluctuations in the available bandwidth during an ongoing presentation playback. In many cases network conditions will vary during the life of a presentation transmission. Without stringent quality of service mechanisms to guarantee bandwidth and control the amount of traffic over a given path through the network, best effort techniques will persist. Hence, given a feedback mechanism which will permit a server to monitor the network bandwidth, new parameters can be fed into the scheduling algorithm to dynamically adjust a presentations quality level and determine a corresponding schedule. Using the segment independence properties, transitions in quality can be made efficiently at segment boundaries.

A *dynamic malleable media object* is similar to a static malleable media object except that its transform method takes as input a sequence of QoS objects and a sequence of end-user device capability objects, and produces a sequence of media elements. A *live (static or dynamic) malleable* object is one where the transform method also receives one or more (sequences of) media objects as input.

Malleable presentations provide a way to automatically trade off the current available network and client resources and the best possible presentation quality. In fact, it makes this tradeoff an integral part of the presentation itself. Therefore, malleable presentation authoring tools would allow the author to specify these tradeoffs, and even to tailor the presentation so it yields better tradeoffs.

One last issue we would like to mention is client buffer effects. In the multimedia streaming process, the client buffer size imposes a limit on the amount of data that can be *pre-delivered* by the server, thus having a significant effect on the transmission schedule and bandwidth requirement of a presentation. In order to minimize bandwidth usage, the client is expected to advertise a *buffer size as large as possible* within its operating limit. With the increasingly heterogeneous receiver devices, it is difficult to predict a universal or expected system metric such as the client buffer size, which can differ by a few orders of magnitude. On the other hand, it is desirable for the client to use *as little buffer as necessary* whenever possible, for example, when the client buffer space is scarce or is shared by multiple concurrent processes. In situations where there is extra bandwidth in the network, it would be desirable to utilized it to minimize the client buffer size usage. We studied the buffer effect on transmission schedules and the tradeoffs between bandwidth and client buffer size. Our results show that with a small precomputation time of $O(N\log N)$, the on-demand scheduling given an arbitrary client buffer size takes only $O(\log N)$ time. The two-way mapping between bandwidth and client buffer takes the same instantaneous $O(\log N)$ time, enabling efficient bandwidth/client-buffer tradeoff decisions.

References

1. J. Salehi, Z. Zhang, J. Kurose and D. Towsley, "Supporting Stored Video: Reducing Rate Variability and End-to-End Resource Requirements through Optimal Smoothing", IEEE/ACM Transaction on Networking, 6(4), August 1998.
2. W. Zhao, T. Seth, M. Kim and M. Willebeek-LeMair, "Optimal Bandwidth/Delay Tradeoff for Feasible-Region-Based Scaleable Multimedia Scheduling", Proceedings of the IEEE INFOCOM'98, March 1998.
3. W. Feng and S. Sechrest, "Critical Bandwidth Allocation for Delivery of Compressed Video", Computer Communications, 18, October 1995.
4. H. Kanakia, P. Mishra, and A. Reibman, "An Adaptive Congestion Control Scheme for Real-time Packet Video Transport", Proceedings of the ACM SIGCOMM'93, August 1993.
5. S. McCanne, V. Jacobson and M. Vetterli, "Receiver-driven Layered Multicast", Proceedings of the ACM SIGCOMM'96, August 1996.
6. P. Pancha and M. El Zarki, "MPEG Coding for Variable Bit Rate Video Transmission", IEEE Communications Magazine, May 1994.
7. E. Amir, S. McCanne and H. Zhang, "An Application Level Video Gateway", Proceedings of the ACM Multimedia '95, November 1995.
8. L. Delgrossi, C. Halstrick, D. Hehmann, R. Herrtwich, O. Krone, J. Sandvoss and C. Vogt, "Media Scaling in a Multimedia Communication System", Multimedia Systems Journal, Springer-Verlag, 2, 1994.
9. D. LeGall, "MPEG: A video compression standard for multimedia applications", Communications of the ACM, 34(4), April 1991.
10. World-Wide-Web Consortium, "Synchronized Multimedia Integration Language (SMIL) 1.0 Specification", W3C Recommendation, (http://www.w3.org/TR/REC-smil), June 1998.
11. N. Manohar, M. Willebeek-LeMair, A. Prakash, "Applying Statistical Process Controls to the Adaptive Rate Control Problem: A Framework for the Streaming of Heterogeneous Streams", SPIE and IS&T Multimedia Computing and Networking 1998 Conference, January 1998.
12. M. Naghshineh and M. Willebeek-LeMair, "End-to-End QoS Provisioning in Multimedia Wireless/Mobile Networks Using an Adaptive Framework", IEEE Communications Magazine, November 1997.
13. L. Chiariglione, "MPEG and Multimedia Communications - A Guided Tour to MPEG-1 and MPEG-2", (http://www.cselt.stet.it/ufv/leonardo/paper/mpeg.htm), August 1996.

Network Traffic Issues for Interactive Multipoint Multimedia Applications

George C. Polyzos*

Center for Wireless Communications
and
Computer Systems Laboratory
Department of Computer Science and Engineering
University of California, San Diego
La Jolla, CA 92093-0114, U.S.A.

Abstract

We consider interactive multipoint multimedia communications over heterogeneous networks. In order to support such applications in an effective and efficient manner, we advocate the use of Multi-resolution Layered Coding (MLC) and present architectural alternatives that can support MLC within the scope of existing standards and technologies. In particular, in order to support real-time video transmission over ATM networks, we have proposed the use of multiple virtual-channels with different Quality of Service (QoS) levels, instead of the standard two-level priority scheme based on the Cell Loss Priority bit. Similarly, implicit (for IPv4) or explicit (for IPv6) flow IDs can be used in the Internet environment. We have also considered the combined use of both guaranteed QoS streams, for the critical but highly compressible components of continuous media, and "best-effort" service, for the enhancement layers, as a means to provide a statistical multiplexing gain and increase the number of flows admitted by networks supporting QoS guarantees. Finally, we provide some illustrative examples of the impact of congestion or transmission errors leading to cell or packet loss on images coded with MLC.

1   Introduction

The role of images in human communication is central. In the last few years digital images and video have been a driving force for the design and deployment of high-speed communication networks and distributed multimedia applications. The direction is towards full integration of transmission and processing of the various media, with a Broadband Integrated Services Digital Network (B-ISDN) one of the best examples of the goals to be achieved.

The success of the World Wide Web, following the availability of point-and-click browsers and improvements in the image and multimedia capabilities of browsers and distributed multimedia applications, provides additional

---

* Email: `polyzos@cs.ucsd.edu`, Tel.: +1-619-534-3508, FAX: +1-619-534-7029

incentives for the full integration of digital images and video in everyday computing. It has also fueled increasing pressure for user customization of media presentation. Note that differentiation in data handling based on (media) content is important mainly when real-time presentation (or processing) of the media is required. Otherwise, images and video can (and probably should) be treated as any other data transported by communication networks or processed by computer systems.

Due to the very high data volumes of images, audio, and video it is important to provide efficient network mechanisms for these types of traffic even when high-speed fiber-optic networks are available, but more so if heterogeneous networks with possibly severely bandwidth constrained links or subnets (e.g., wireless) are considered. In addition to very high throughput requirements, images, audio and video are usually associated with real-time interactive applications and thus they also present rather stringent delay constraints. More precisely, in order for audio and video to be effectively used in interactive communication, i.e., without forcing the communicating subjects to modify their behavior from that of face-to-face communication, the delay between transmitter and receiver is expected to be minimal. Various guidelines set this end-to-end delay tolerance in the tenths or few hundreds of ms [3].

Traditionally, transmission of Continuous Media (CM), such as real-time audio and video, has been based on Constant Bit Rate (CBR) circuit-switching channels. Therefore, source coding schemes for CM were designed specifically to produce output at a specific, constant, bit rate, typically that of the channel to be used, independently of the instantaneous information content of the signal to be transmitted. This coding typically results in variable signal quality, even though when bandwidth is not severely limited, potential temporary quality degradation is usually engineered to be imperceptible.

For transmission over packet-switching networks, the problem of CM coding changes. There is typically no a priori constraint on the maximum instantaneous bit rate produced by the coding scheme, but instead, an effort is made to produce constant quality signal at a given average bit rate (or level of quality) by using Variable Bit Rate (VBR) schemes. Peak-to-mean bit-rate ratios for VBR codecs is usually high; a 4.7 value is reported for one particular codec [4].

The use of VBR coding is based on the capability of packet-switching networks to use statistical multiplexing for efficient utilization of network resources. One can then economize on resources when the information content of the signal is low and expend them at a higher rate when it is high, achieving the best overall (constant) quality at a given cost. The potential for significant economies from statistical multiplexing was one of the main arguments for the selection of the Asynchronous Transfer Mode (ATM), the transmission and multiplexing mode standardized for B-ISDN, over the circuit-switching alternative.

Of course, there will be instances where peak demands are presented to the system (actually, its "best-effort" component) simultaneously. In most of these cases the system will be unable to immediately satisfy these demands. In traditional packet-switching networks these situations have been handled through queueing of the requests and congestion control. With real-time services, however, queueing might introduce unacceptable delays. Therefore, one of the techniques considered for traffic smoothing and controlling user perceived latency in times of high congestion is the dropping of parts of the signal that might be of secondary importance. This approach is made possible by the use of hierarchical or layered coding and appropriate traffic labeling and prioritization.

1.1 Layered Coding

Hierarchical or Layered Coding (LC) techniques split signals into components of varying importance [8]. The aggregation of these components reconstructs the original data, but subsets of the data can also provide various degrees of approximation to the original signal. Signal subsets are coded separately (possibly with different coding schemes), and can therefore be decoded separately. By careful design, the first (or first few) components in the hierarchy can be a good approximation to the overall signal, providing a good first impression of the information without requiring all the components to be received (and decoded) first (see Fig. 1). LC has advantages in many facets of image and video transmission [11, 18].

LC with a variable spatial resolution [1], has received less attention than LC with a fixed resolution, despite its adoption in coding standards. Two-layer coding has been mostly considered in the literature [4, 11, 18, 5, 10], with the first, coarse or low frequency, subsignal usually referred to as the *base layer*, and the second, higher resolution, subsignal termed the *enhancement layer*.

We use the term Multi-resolution Layered Coding (MLC) to refer to LC with more than two layers of spatial resolution. One well-known form of MLC is pyramidal coding, with variable spatial resolution across layers [1]. Our intention here is to differentiate MLC from LC with a spatial fixed resolution hierarchy (e.g., subband coding) or even from single resolution LC using partitioning of the (transform space) coefficients into two or more layers for transmission (e.g., see [12]). We focus on MLC because schemes with variable spatial resolution provide better image quality at multiple and lower resolution levels[16](p. 96). This is a very desirable feature for some important applications which require good quality using the lower layers or tight control over the coding error. Actually, with MLC the quality can be optimal at the specific resolutions selected to define the layers in the hierarchy. The main potential disadvantage of LC, and MLC in particular, is the increased overhead; this is discussed below.

Fig. 1. Hierarchical progression up to the (a) first (80x60, 0.048 bpp), (b) second (160x120, 0.093 bpp), (c) third (320x240, 0.291 bpp), and (d) fourth layer (640x480, 1.029 bpp) of 4-layer HJPEG image (bridge). Images (a)-(c) have been expanded to the size of the original image (640x480) for comparison.

The existing JPEG image compression standard has a provision for MLC, referred to as the hierarchical mode [16] of JPEG (HJPEG). This mode of JPEG is based on pyramidal coding and has different spatial resolution at each layer.[2] Briefly, the image is first low pass filtered and then subsampled by the desired number of multiples of 2 in either or both dimensions and encoded using either the *sequential* or the *progressive* JPEG mode [16]. Then, the encoded reduced-size image is decoded, interpolated and upsampled by the same number of multiples of 2 horizontally and/or vertically. This upsampled image is then used as a prediction of the original image at this resolution and the difference image is computed, encoded, and transmitted. Finally, the last two steps are repeated until the original image at full resolution has been encoded.

We have implemented in software the HJPEG codec [6] and have used it to obtain experimental results investigating the properties of MLC when used for image and video transport. Note that even though JPEG has been designed for still images, it is also widely used for video transmission, pro-

---

[2] Layers are called "frames" in JPEG terminology, but we prefer to use the term "layer" here to avoid confusion with video frames.

viding (only) intraframe compression. Furthermore, because this scheme uses only intraframe compression it is much more robust in the presence of errors than coding methods exploiting both temporal and spatial redundancy, such as MPEG.[3]

1.2 Overview and Contributions of this Paper

The remainder of this paper is structured as follows. Section 2 discusses architectural alternatives for various network technologies that can effectively support and exploit the features of LC and Multi-resolution Layered Coding (MLC) in particular. Section 3 provides an overview of the impact and the approaches for handling transmission errors, focusing on concealment techniques and the implications for LC and network architectures. Finally, we present our conclusions in section 4.

The main contributions of this paper are the following. First, the proposal and arguments for using MLC for image and continuous media transmission over packet networks, especially in order to support multicast and heterogeneity. Second, the proposal to use multiple virtual-channels with different Quality of Service (QoS) levels, instead of the standard two-level priority scheme based on the Cell Loss Priority bit, for supporting real-time video transmission over ATM networks. Similarly, implicit (for IPv4) or explicit (for IPv6) flow IDs can be used in the Internet environment. Third, the proposal to use both guaranteed QoS connections for the critical, but highly compressible parts of CM signals, and "best-effort" connections for the enhancement layers as the means to provide a statistical multiplexing gain and increase the number of connections admitted by networks supporting QoS. And finally, an illustration of the impact of transmission errors, leading to cell or packet loss, on MLC schemes.

2 Transport Architecture for MLC Media

The advantages of LC are best exploited when the networking architectures support differential treatment of traffic. Even though priorities and labeling for different types of traffic have been a part of the Internetworking Protocol (IP) since its original design, they have seldom been implemented in network system software or used in production mode on the Internet. The situation has been similar for most other wide-area networks. Local Area Network (LAN) specifications, notably FDDI, have been at the forefront of supporting priorities and Quality-of-Service (QoS) guarantees. However, in this case the impact of QoS guarantees for real-time imaging is probably less critical due to

---

[3] MPEG uses intraframe coded only frames periodically to achieve a desired level of redundancy; at the extreme, intraframe coded frames could be used exclusively, obtaining similar levels of robustness and compression with motion JPEG.

the relatively high bandwidth available on LANs and the small propagation delay.

Productive use of priorities and service differentiation still seem to be in the future of networking, with both ATM and the next generation of IP (IPv6) expected to use traffic differentiation in order to provide improved QoS and/or to guarantee it. A general framework for the dissemination of real-time CM through packet switching networks is the Multimedia Multicast Channel (MMC) [13, 14]. The MMC design is independent of specific technologies, such as IP or ATM, but compatible with MLC and network architectures that provide services with different QoS guarantees.

2.1 ATM

One of the main reasons why considering more than two layers for LC has been avoided in the past is a lack of explicit support for MLC in ATM and the existence of an explicit mechanism supporting two-layer coding, namely the use of the Cell Loss Priority (CLP) bit in ATM cells. However, with the increasing interest in wireless ATM and heterogeneous networks, it is appropriate to consider more general mappings between the requirements and capabilities of various technologies.

The traditional mapping of LC to ATM service has been through the use of the Cell Loss Priority (CLP) bit in the ATM cell header, as defined in the ATM specification. ATM switches that experience congestion are expected to first drop cells that have the CLP bit set. This action should relieve congestion and thus protect cells that have not been designated as candidates for dropping. Obviously, this technique is only effective if a significant amount of traffic with the CLP bit set can be found at a switch at times of congestion.

Two mechanisms have been suggested for setting the CLP bit. First, the network at its entry points can set the CLP bit when incoming traffic violates its traffic contract with the network, i.e., as a soft implementation of a policing function. This mode of operation is content independent and either irrelevant or possibly detrimental to LC of images and continuous media. The second mechanism depends on a source designating some cells as less important and candidates for dropping, based on their information content. This is exactly where a match between LC and the ATM specification exists, because designating some cells as useful but non-critical is fully compatible with the definition of LC.

The CLP bit approach has some drawbacks and is not flexible enough given the various applications of LC that one can envision. First, there is direct support (differentiation) for only two layers. Second, without further service specification the QoS provided to the two layers, and particularly to the layer using cells with the CLP bit set, is not known. Therefore, we propose the following alternative architecture.

Without necessarily involving the CLP bit, multiple priorities can be implemented through the use of multiple virtual circuits (e.g., ATM virtual channels) with different QoS specifications. Fig. 2 illustrates this idea. In this figure, S is the source, A, L, H, and W are receivers (adaptive, low quality, high quality and wireless), S1 – S5 (ATM) switches, N1 – N5 (ATM) networks with N5 a wireless network, and B a base station. The number of concentric circles for the source and receivers represent terminal capabilities or user quality settings. Lines between devices depict separate Virtual Channels (VCs), with differences in thickness symbolizing differences in traffic characteristics and QoS. In particular, the dashed lines stand for a VC with lower QoS, possibly with the CLP bit set, that could be used for fast reaction to local congestion by dropping traffic at will.

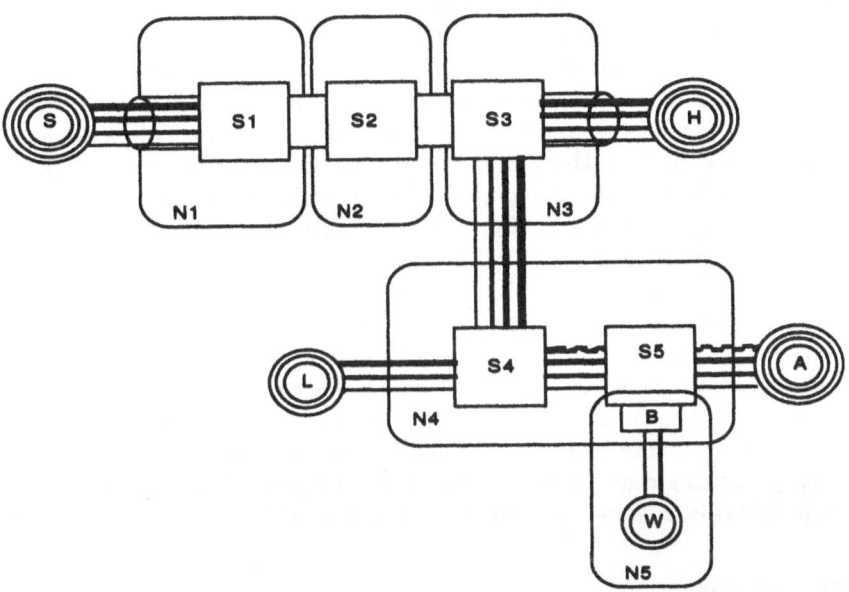

Fig. 2. Transport of a MLC signal over multiple virtual channels in a heterogeneous ATM network. S is the source, A, L, H, and W are receivers (adaptive, low quality, high quality and wireless), S1 – S5 (ATM) switches, N1 – N5 (ATM) networks with N5 a wireless network, and B a base station. The number of concentric circles for the source and receivers represent terminal capabilities or user quality settings.

This proposal has some specific implications for ATM switch designs, which we discuss next, but we believe it is in the mainstream of the rationale for the adoption of ATM and the implementation of B-ISDN. First, it relies on switches using per-VC queueing and forwarding. However, this seems to be the trend in new ATM switch offerings. Second, it increases, by a factor equal to the number of layers, the number of VCs that switches and network controllers need to service and manage. This is a potential drawback of the approach, but we believe it is minor compared with the advantages it offers, particularly in the multipoint case where separate connections might have to be maintained for receivers with incompatible settings. Furthermore, ATM provides a mechanism to address this problem in core switches where it is more critical because of the traffic aggregation: the Virtual Path (VP) mechanism. A VP provides a way for ATM switches that are not the final destination of traffic to treat and switch traffic from multiple VCs collectively. For example, in Fig. 2, S2 "sees" a VP going through it where the VCs in question are contained, but is unaware of their detailed set-up. The only constraint is that the QoS offered to the VP must be able to satisfy the QoS of the most stringent VC it carries. Alternative designs, e.g., having more than one VP with different QoS levels, in order to decrease the "cost" of less demanding traffic at core switches, are easy to arrive at.

A major advantage of this approach is that a receiver does not need to *receive* all the data transmitted by the source, but instead it can select to "subscribe" (connect) to only the VC that it is interested in. This is extremely significant for bandwidth limited wireless mobile terminals uninterested in, for example, high-resolution components of the signals, participating in multicast sessions, as shown in Fig. 2. In that case, the high-resolution signal does not need to be transmitted over the air and the source needs not be aware of it. Furthermore, the base station, B, can be simple and completely unaware of the media content it carries (instead of a protocol/media converter that would otherwise be required). Finally, depending on the available signaling protocols, adjustments to signal quality could be made easily and dynamically by adding or dropping connections. Note that the demultiplexing cost for receiving multiple VCs arriving at a receiver (as opposed to the signal processing cost for synthesizing the final presentation signal from its components) is minimal based on our experimental findings reported in [17].

2.2 Internet

Similar techniques can be applied in IP networks. The current IP specification (IPv4) has two (related) mechanisms that can be used for traffic labeling and prioritization: the Type-of-Service field and an explicit priority field that allows eight levels of priority to be specified.

In addition, even if the sources themselves do not explicitly specify the type of traffic, some service differentiation could be achieved for applications that use the UDP or TCP transport layer protocols by having routers snoop at the UDP/TCP port numbers. This assumes no fragmentation, but fragmentation is now considered a bad idea anyway. A method for specifying the service equivalent of a connection in a connectionless network such as the Internet, without support or involvement of the applications, is the use of (implicit) flows, as described in [2].

Explicit flows, on the other hand, are part of the specification for the next generation of IP (IPv6), through a new IP packet format [7]. In addition, a new priority structure, potential support for reservations and QoS guarantees, etc., are introduced with IPv6 and related technologies such as RSVP [19] and others.

3 Error Control and Concealment

Despite considerable efforts to perfect concealment schemes exploiting spatial and temporal redundancies for non-layered coding, only a few efforts have been reported specifically for LC, most concentrating on video. Often the need for concealment is ignored due to the inherent error resilience of LC with a protected base layer. Indeed, most approaches assume that the base layer can be adequately protected. However, errors in the base layer cannot be effectively concealed in the absence of a reference picture and accurate motion vectors. A simple zero or mean substitution is usually not effective in the case of a damaged base layer. Further, in the case of hierarchical coding with variable spatial resolutions, these techniques may aggravate the situation.

3.1 Decoder Resynchronization

With variable length coding, loss of even a single packet can lead to loss of decoder synchronization, which can damage an image catastrophically. Synchronization flags, usually preceded by a definition header, can be inserted around entropy coded segments in order to identify them as such without the need to decode the compressed data. The definition header contains the specification of a restart interval, which is an integer multiple of the minimum coded unit, e.g., the block. Various error conditions such as missing markers or out of range values trigger error recovery procedures at the decoder. When the decoder detects an error condition, it scans for the next flag in order to resynchronize by resetting the decoder. The relative frequency of restart flags can be increased leading to increased robustness at the expense of coding efficiency; this is a basic design trade-off. Note, however, that recovery is not possible from all error conditions; in particular, header information loss is catastropic [16].

The impact of the loss of synchronization with various coding schemes is depicted in Fig. 3. A restart flag every 8 lines is used in this case. We observe that with proper use of resynchronization schemes, the decoding process can maintain synchronization, despite errors leading to block loss. Note also that use of LC is advantageous.

Alternatively, the header field of a packet can convey direct addressing information, thereby reducing the chance of catastrophic synchronization error, particularly since the number of pixel blocks within a packet is rather small. Furthermore, in the case of interframe coded video using conditional

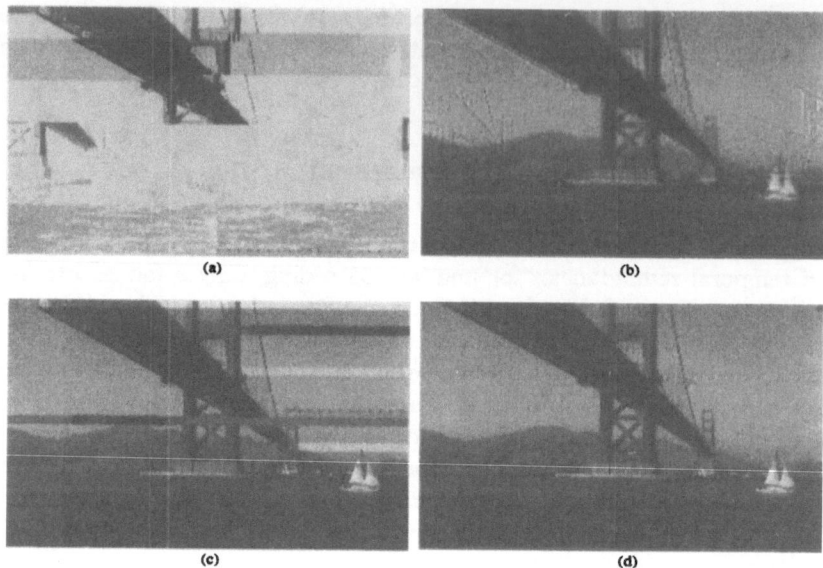

Fig. 3. 2.5% loss in (a) BJPEG without restart interval (b) BJPEG with restart marker every 8 lines, (c) 4-layer HJPEG without restart interval, (d) 4-layer HJPEG with restart marker every 8 lines. The HJPEG images are fully protected up to the third layer.

replenishment, as packets update blocks at specific locations excluding stationary regions, a short restart interval is naturally attained. Hence, in these cases the impact of synchronization errors is usually confined.

3.2   Error Resilience of MLC

Here we illustrate the error resilience of MLC considering the possibility of errors at the base layer or, more generally, the lower layers of a MLC image and potentially multiple levels of protection, adjusted to the importance of each layer. Depending on the situation, the different protection levels can be achieved through various techniques. For example, multiple QoS levels or priorities could be used if the problem is switch congestion leading to cell or packet loss, or different degrees of FEC can be applied if the loss is due to transmission errors.

First, note that by the definition of MLC, even complete loss of the last layer is secondary or can even be insignificant or unnoticeable. This is illustrated in Fig. 1. The original image, shown in Fig. 1 (d), is coded and transmitted as a 4-layer MLC image. Any errors or packet loss occurring in layer 4 data can be masked by displaying the decoded image including only layers 1–3. The result is a lower resolution, but otherwise perfect image, as shown in Fig. 1 (c).

This example demonstrates the basic error resiliency of LC. Note that up to roughly 2/3 of the size of the 4-layer coded image can be lost with little impact. Furthermore, if errors occurred even in layer 3, that layer can be dropped too, resulting in the image shown in Fig. 1 (b). There is now noticeable degradation in quality from the loss of the low spatial frequency signal, but still the image is probably preferable to images with much less data in error, but no LC or sophisticated concealment, as shown in Fig. 3.

Thus, terminating the decoding early, i.e., disregarding all received information from the last layer (or last few layers), is the simplest, but also a very effective error control strategy when errors occur in the higher layers. Of course, refined versions of this strategy can be used when the errors are concentrated on parts of the image, leading to an image that is identical to the original, except for a reduced resolution in parts of the picture.

Block loss in intermediate layers can be tolerable depending on their levels in the image hierarchy and the loss patterns. In general, the impact of errors and loss in the lower layers is far greater than that at higher layers (see Fig. 4) and packet loss crossing multiple layers is more detrimental. Given these observations and a fixed budget for error correction overhead, it becomes apparent that multiple protection levels, with higher protection for the smaller but more significant lower layers, would be beneficial.

Fig. 4. Impact of errors at the lower layers. From left to right: original *Lena* image, and 6-layer MLC image with all blocks lost up to the first (0.07% of total size), second (0.37%), and the third (1.54%) layer. Each MLC image is obtained by aggregating differential layers with neutral gray background.

An interesting case demonstrating the power of MLC is full-layer MLC, i.e., MLC where the number of layers employed is such that the base layer is a single block (assuming an image of appropriate dimensions). Fig. 5 illustrates the effect of evenly spaced (non-random) errors leading to loss on two different coding techniques: non-layered JPEG and full-layer HJPEG (MLC). Note that no special protection or different treatment is applied to any of the layers. For non-layered JPEG concealment using mean pixel values from the closest pixels in undamaged neighboring blocks was applied.

In Fig. 5 (a) the block loss rate is 10.9%. Black blocks in the left image

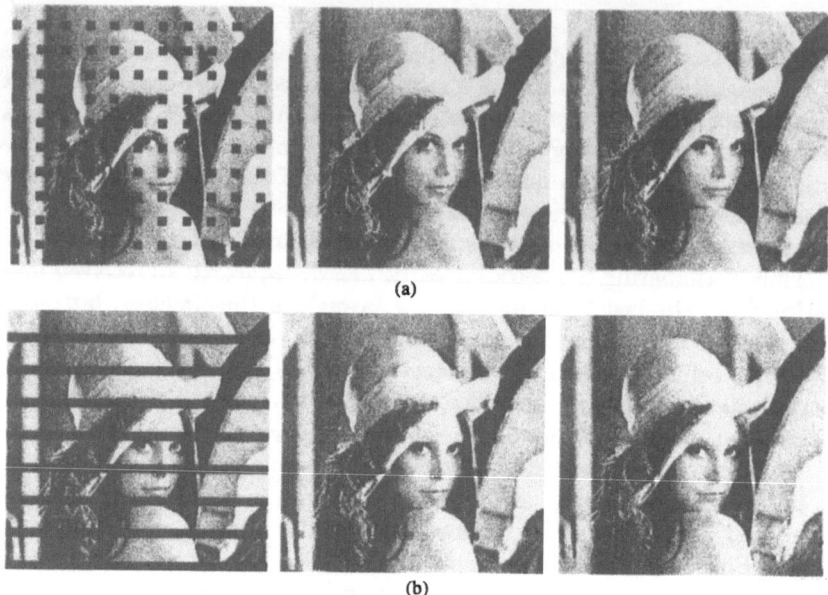

Fig. 5. Images with (a) 10.9% evenly distributed block loss and (b) 25.1% strip block loss. From left to right: Loss marked as black, non-layered coding with mean substitution, and MLC image with zero substitution. Note that there is no block loss in the base layer of the MLC image.

show the positions of block loss. The second image shows the reconstructed non-layered JPEG image after mean substitution. Artifacts are clearly visible in this case. The last image uses full-layer MLC with a simple zero substitution. It shows slight blurring, which is less annoying in subjective evaluation. Due to the impact of considerable block loss in the lower layers of MLC, full-layer MLC images often display dithering artifacts and blurring, which decrease the PSNR to a level just comparable to, or sometimes and particularly at low loss rates even lower than, non-layered images.

With block loss stretched through multiple blocks at the increased rate of 25% shown in Fig. 5 (b), a better image quality is obtained with full-layer MLC. In this case full-layer MLC shows less artifacts than the non-layered image with mean substitution, albeit with more blurring. The objective measure, PSNR, did not indicate a much improved image fidelity. Observe that in the case of block loss in and/or near the bottom layer, the image loses much of its low spatial frequency signals and texture richness.

4 Conclusions

We have discussed how various features of Layered Coding (LC) can be applied in order to improve the responsiveness of image-based applications and the efficiency of packet switching networks transporting images and video in real-time. Congestion control, multicasting of continuous media (CM), terminal and network heterogeneity, and error control for wireless networks are key areas for the application of LC techniques in order to solve critical networking problems. For all but the first application, we argued that Multi-resolution Layered Coding (MLC) has significant advantages over the traditional two-layer coding schemes. The adoption of an open-loop control approach, suggested by the real-time character and the stringent delay constraints of interactive CM and made possible by MLC, also solves the multipoint feedback control problem.

We have also presented architectural alternatives in the context of various packet switching networking technologies, that can effectively support and exploit the features of LC, and MLC in particular. For transport over ATM networks, we proposed to adopt multiple ATM virtual channels with different QoS levels, instead of the standard two-level priority scheme based on the CLP bit. For the next generation of the Internetworking Protocol, IPv6, flow IDs can be used to efficiently achieve the same goal through IP routers.

Finally, examining the performance of MLC in the presence of errors (leading to packet loss) possibly impacting all layers in the image hierarchy, we showed the importance of appropriate, in general different, degrees of protection for the lower layers. The error resiliency of MLC is particularly important when feedback-based error control is not feasible or cost-effective. Note that MLC can provide a reasonable error concealing effect through full-layering at low error rates.

References

1. P. J. Burt and E. H. Adelson, "The Laplacian Pyramid as a Compact Image Code," *IEEE Transactions on Communications*, vol. 31, no. 4, pp. 532–540, April 1983.
2. K.C. Claffy, H.-W. Braun, and G.C. Polyzos, "A Parameterizable Methodology for Internet Traffic Flow Profiling," *IEEE Journal on Selected Areas in Communications*, vol. 13, no. 8, pp. 1481–1494, October 1995.
3. D. Ferrari, "Client Requirements for Real-Time Communication Services," *IEEE Communications Magazine*, vol. 28, no. 11, pp. 65–72, November 1990.
4. M. Ghanbari, "Two-layer Coding of Video Signals for VBR Networks," *IEEE Journal on Selected Areas in Communications*, vol. 7, no. 5, pp. 771–781, June 1989.
5. M. Ghanbari and V. Seferidis, "Cell-loss Concealment in ATM Video Codecs," *IEEE Transactions on Circuits and Systems For Video Technology*, vol. 3, no. 3, pp. 238–247, June 1993.

250

6. J.K. Han and G.C. Polyzos, "Networking Applications of the Hierarchical Mode of the JPEG Standard," Proc. IEEE International Phoenix Conference on Computers and Communications (IPCCC'96), Phoenix, AZ, pp. 58–64, March 1996.

7. C. Huitema, *IPv6: The New Internet Protocol*, Prentice Hall PTR, Upper Saddle River, NJ, 1996.

8. G. Karlsson and M. Vetterli, "Packet Video and its Integration into the Network Architecture," *IEEE Journal on Selected Areas in Communications*, vol. 7, no. 7, pp. 739–751, June 1989.

9. M. Khansari, A. Jalali, E. Dubois, and P. Mermelstein, "Robust Low Bit-Rate Video Transmission over Wireless Access System," *Proc. ICC*, pp. 571–575, September 1994.

10. L. Kieu and K. Ngan, "Cell-loss Concealment Techniques for Layered Video Codecs in an ATM Network," *IEEE Transactions on Image Processing*, vol. 3, no. 5, pp. 666–677, September 1994.

11. D.S. Lee and K. H. Tzou, "Hierarchical DCT Coding of HDTV for ATM Networks," *Proc. ICAASP'90*, Albuquerque, NM, pp. 2249–2252, April 1990.

12. P. Pancha and M. El Zarki, "MPEG Coding for Variable Bit Rate Video Transmission," *IEEE Communications Magazine*, vol. 32, no. 5, pp. 54–66, 1994.

13. J.C. Pasquale, G.C. Polyzos, E.W. Anderson, and V.P. Kompella, "The Multimedia Multicast Channel," *Internetworking: Research and Experience*, vol. 5, no. 4, pp. 151–162, December 1994.

14. J.C. Pasquale, G.C. Polyzos, E.W. Anderson, and V.P. Kompella, "Filter Propagation in Dissemination Trees: Trading Off Bandwidth and Processing in Continuous Media Networks," Proc. Fourth International Workshop on Network and Operating System Support for Digital Audio and Video, pp. 259–268, November 1993.

15. J.C. Pasquale, G.C. Polyzos, and G. Xylomenos, "The Multimedia Multicast Problem," *Multimedia Systems*, ACM/Springer-Verlag, vol. 6, no. 1, pp. 43–59, January 1998.

16. W.B. Pennebaker and J.L. Mitchell, *JPEG Still Image Data Compression Standard*, Van Nostrand Reinhold, 1993.

17. G.C. Polyzos and K. Taylor, "A Prototype Video Dissemination Application over ATM," Proc. of IEEE ICC'95, Seattle, WA, pp. 1262-1266, June 1995.

18. W. Verbiest, L. Pinnoo, and B. Voeten, "The Impact of the ATM Concept on Video Coding," *IEEE Journal on Selected Areas in Communications*, vol. 6, no. 9, pp. 1623–1632, December 1988.

19. L. Zhang, S. Deering, D. Estrin, S. Shenker, and D. Zappala "RSVP: A New Resource Reservation Protocol," *IEEE Network*, vol. 7, no. 5, pp. 8–18, 1993.

A Multi-Channel Approach to Voice Activity Detection in Noisy Environments based on Time Delay Estimation

Francesco Beritelli, Salvatore Casale, Alfredo Cavallaro
Istituto di Informatica e Telecomunicazioni - University of Catania
V.le A. Doria 6, 95125 Catania - Italy
e-mail: {beritelli, casale, acavallaro}@iit.unict.it

Abstract

In the field of mobile communications correct Voice Activity Detection (VAD) is a crucial point for the perceived speech quality, the reduction of co-channel interference, and power consumption in portable equipment. The Fuzzy Voice Activity Detector (FVAD) recently proposed in [1], shows that a valid alternative to deal with the problem of activity decision is to use methodologies like fuzzy logic. In this paper we propose a multichannel approach to activity detection using both fuzzy logic and time delay estimation. Objective tests confirm a significant improvement over traditional methods, above all in terms of a reduction in activity increase for non stationary noise.

1. Introduction

A Voice Activity Detector (VAD) aims to distinguish between speech and several types of acoustic background noise even with low signal-to-noise ratios (SNRs). Therefore, in a typical telephone conversation, a VAD, together with a comfort noise generator (CNG), achieves a silence compression. In the field of multimedia communications, silence compression allows the speech channel to be shared with other information, thus guaranteeing simultaneous voice and data applications. In a cellular radio system that uses the Discontinuous Transmission (DTX) mode, such as the Global System for Mobile communications (GSM), a VAD reduces co-channel interference (increasing the number of radio channels) and power consumption in portable equipment. Moreover, a VAD is vital to reduce the average bit rate in future generations of digital cellular networks, such as the Universal Mobile Telecommunication Systems (UMTS), which provide for variable bit-rate (VBR) speech coding. Most of the capacity gain is due to the distinction between speech activity and inactivity [2].

The performance of a speech coding approach based on phonetic classification, however, strongly depends on the classifier, which must be robust to every type of background noise [2]. As is well known, for example the performance of a VAD is critical for the overall speech quality, above all with low SNRs. When some of speech frames are detected as noise, intelligibility is seriously impaired due to speech clipping in the conversation. If, on the other hand, the percentage of noise detected as speech is high, the potential advantages of silence compression are not obtained. In the presence of background noise it may be difficult to distinguish between speech and silence, so for voice activity detection in wireless environments more efficient algorithms are needed [1][2][3].

Though the Fuzzy Voice Activity Detector (FVAD) recently proposed in [1] performs better than the solutions presented in literature [4][5], it exhibits an activity increase, above all in the presence of non-stationary noise. This paper therefore deals with the problem of speech activity detection in noisy environments, using parameters available only in a multichannel system, such as delay estimation and the difference in the power level between the channels. Referring to systems which use arrays of microphones for various purposes, such as privileging a certain incoming signal direction or detecting and locating a signal source to be found in the reception area [6][7][8][9][10], we thought that a multichannel system could be useful to obtain new information to be used in the detection of speech activity in noisy environments. This idea is also confirmed by recent studies on the central nervous system. It has been observed, in fact, that the *medial nucleus of the superior olive*, the structure whose neurons receive signals from both ears, is related with the location of sounds, which are said to be distinguished according to the differences in their arrival times. These time differences, in fact, are one of the elements used to locate sounds in space: a sound coming from a certain side reaches the ear on that side first and then, a few tens of milliseconds later, the other ear. Starting from this consideration, VADs have been developed which use Time Delay Estimation (TDE) as their only parameter, correlating the delay between the signals in the two channels on the basis of a hypothetical position for the source generating the signal. A further structure uses a fuzzy system whose input is the delay estimate, the difference in the levels of the signals on the two microphones, and the continuous output of the FVAD, thus obtaining greater precision in detecting between activity and non-activity.

2. The Fuzzy Vad (FVAD)

The functional scheme of the Fuzzy Voice Activity Detector (FVAD) is based on a traditional pattern recognition approach. The four differential parameters used for speech activity/inactivity classification are the same as those used in G.729 Annex B [4] and are: the full-band energy difference, the low-band energy difference, the zero-crossing difference and the spectral distortion. The matching phase is performed by a set of fuzzy rules obtained automatically by means of a new hybrid learning tool [11]. As is well known, a fuzzy system allows a gradual, continuous transition rather a sharp change between two values. So, the Fuzzy VAD proposed returns a continuous output ranging from 0 (Non-Activity) to 1 (Activity), which does not depend on whether the single inputs have exceeded a threshold or not, but on an overall evaluation of the values they have assumed (defuzzyfication process). The FVAD translates several individual parameters into a single continuous value which, in our case, indicates the degree of membership in the Activity class and the complement of the degree of membership in the Non-Activity class. The final decision is made by comparing the output of the fuzzy system, which varies in a range between 0 and 1, with a fixed threshold experimentally chosen by minimizing the sum of Front End Clipping (FEC), Mid Speech Clipping (MSC), OVER, Noise Detected as Speech (NDS) [12] and the standard deviation of the MSC and NDS parameters. In this way we found an appropriate value for the hangover module that satisfies the MSC and NDS statistics, reducing the total error. The hangover mechanism chosen is similar to that adopted by the GSM [5].

3. The new approach

The use of more than one microphone in a speech activity detection system gives the opportunity to exploit the parameters obtained by two registrations of the same signal, which would not be available if only one microphone were used. Below we present some new solutions for activity/non-activity detection, either using a single parameter (delay estimate between the two signals) or combining it with others such as the difference in the signal level between the two microphones.

3.1 VADs base on TDE

The delay between two signals is generally determined by plotting the x-axis on which the maximum cross-correlation function between the two signals is obtained, or by pre-filtering the signals before calculating the cross-correlation, thus obtaining the so-called generalised cross-correlation [13][13][14][15][16]. One such technique is PHAT (PHAse Transform): if a particular filter is chosen, its output is a delta function centred on delay. An algorithm we used for delay estimation is based on calculation of the Euclidean DIstance (EDI) between the two displaced signals, using 80-sample windows. If we call the signals on the two channels $x_1(k)$ and $x_2(k)$, the distance between them is calculated as follows:

$$d = \frac{1}{M} \sqrt{\sum_{K=1}^{M} \left(x_1(k) - x_2(k)\right)^2}$$

with $M = L - |S|$ where S indicates the shift and L indicates the length of the window. The windows of the two signals are made to shift between -3T and 3T, where T indicates the sampling time. This choice was dictated by the distance between the two microphones, which we assumed to be about 10 cm: at this distance the maximum delay between the two signals with a sample rate of 8000 Hz, is almost three samples in length. The shift with which the minimum distance is obtained is taken as an estimate of the delay between the two signals in that sampling window. In this paper we propose speech detectors which use an estimate of the delay between the two signals as their decision parameter.

Considering that the speech source is close to the main microphone, a hypothesis can be formulated as to the delay between the signals recorded in periods featuring speech activity. Once it has been estimated, if it is close to the delay expected in the event of speech activity, the frame is very likely to be an activity one, whereas if the delayed estimated is far from that expected in speech activity periods, the relative frame is considered to be a noise frame. For all the methodologies proposed, we used a VAD hangover to eliminate mid-burst clipping of low levels of speech. The mechanism is similar to the one used by the GSM VAD [5], choosing suitable hangconst and burstconst values [1]. Table 1 gives the values used for each algorithm. With these values speech clipping was reduced while activity increased remained virtually unaltered.

	PHAT	EDI
Burstconst	3	3
Hangconst	10	8

TABLE 1: Burstconst and hangconst values

3.2 The Delay Fuzzy VAD (D_FVAD)

Besides considering a single parameter for speech activity detection, we thought of using a system with several inputs. More specifically, we trained a fuzzy network inputting the continuous output of the FVAD with no hangover, the delay between the two channels, and the power level on the two microphones. Bearing in mind that the voice source is closer to the main microphone, it is plausible that in speech activity periods the power level of the signal recorded on the main microphone will be higher than that on the other microphone. A scheme of how the D_FVAD works is shown in Fig. 1. Of course an approach of this kind means a slight increase in the computational load. The output of the fuzzy system is a continuous value which ranges from 0 to 1. In order to establish an optimal threshold value with which to compare the fuzzy system output, we analyzed the total misclassification error with respect to a threshold value , F_{th}, ranging between 0 and 1. The threshold was chosen in such a way as to achieve a trade-off between the values of the four parameters FEC, MSC, OVER and NDS. Although some of them (specifically MSC and FEC) can be improved by introducing a successive hangover mechanism, which delays the transitions from 0 to 1, the presence of a hangover block makes the values of the OVER and NDS parameters worse. The latter were therefore given priority over MSC and FEC in choosing the threshold. The minimum total error is achieved with about F_{th} =0.26. We chose F_{th}=0.3 , so as to reduce the value of OVER and NDS; as mentioned previously, the corresponding increase in FEC and MSC can be solved by introducing a hangover mechanism. The threshold F_{th} was also chosen so as to minimize the variance of the parameters affected by the hangover: this then allows us to design a suitable hangover for our VAD. We used a VAD hangover to eliminate mid-burst clipping of low levels of speech. The mechanism is similar to the one used by the GSM VAD.

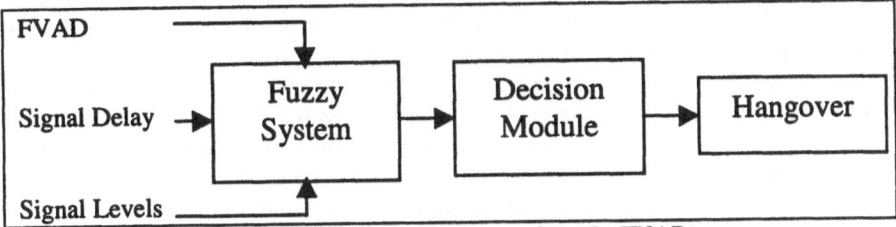

Fig. 1 Functional scheme of the D_FVAD

4. Speech databases

In order to test the proposed algorithms, two different databases were used, one simulated and one real. The first was obtained by simulating a real situation in which 2 spatially separated microphones pick up two displaced signals in the presence of various types of noise, and the second was made up of telephone conversation recordings made in different environments, sampled at a frequency of 8000 Hz, linearly quantized at 16 bit/sample and recorded using a telephone with two microphones.

4.1 Simulated Database

The simulation system has to reproduce a real telephone conversation on a telephone with two microphones placed at a distance of about 10 cm from each other. The two signals recorded on the two microphones will be the sum of the desired speech signal and background noise, with various suitable delays. Fig. 2 shows a geometrical model of the system on the plane passing between the positions of the two microphones and the noise source.

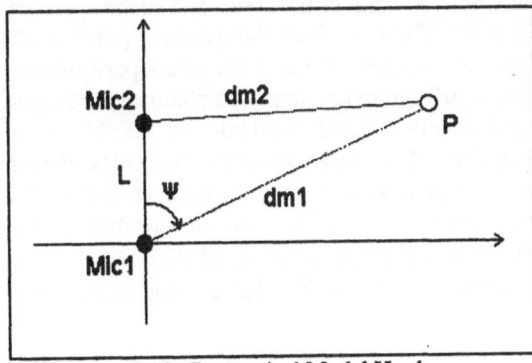

Fig.2 Geometrical Model Used

It is hypothesised that the location of the voice source is the same as that of microphone 1 (mic1), while the noise source may be located at any point P. L indicates the distance between the two microphones, dm1 the distance between mic1 and the noise source and dm2 that between the second microphone (mic2) and the noise source. On mic1 the voice source has a zero delay, whereas the noise signal undergoes a delay with respect to the instant at which it is generated, due to the distance dm1. On mic2 the voice signal arrives with a delay due to the distance L, whereas the delay for the noise signal is due to the distance dm2. The distance L between the two microphones was set to 10 cm, and dm1 was set to 10m. The scheme used to obtain the simulated database is shown in Fig. 3

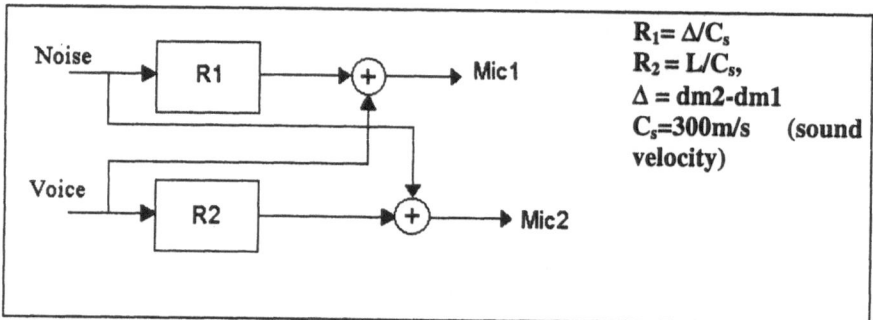

Fig. 3 Scheme used to obtain the simulated database.

The speech phrases used to obtain the simulated database contains sequences recorded in a non-noisy environment (Clean sequences, SNR=60 dB), sampled at 8000 Hz and linearly quantized at 16 bits per sample. The database was marked manually as active and non-active speech segments. In order to have satisfactory statistics as regards the languages and the speakers, the male and female speakers and the languages were equally distributed in the database. Further, to respect the statistics of a normal telephone conversation (about 40% of activity and 60% of non-activity), we introduced random pause segments, extracting from an exponential population the length of talkspurt and silence periods. We considered speech sequences at 22 dBovl i.e. from the overload point of 16 bit word length, whereas the effects of background noise on VAD performance was tested by adding various types of stationary and non-stationary background noise (Car, White, Traffic and Babble), made available by CSELT, to the clean testing sequence at different signal-to-noise ratios (20, 10, 0 dB). Movement of the noise source is simulated by generating a random number every 1000 frames, on the basis of which the position of the noise source may or may not vary. Any variation causes an increase in the angle ψ by steps of 30°, while the distance from the origin of the axes remains unaltered. The initial value of the angle ψ is set to 0°.

4.2 Database Obtained

The database obtained comprises speech sequences sampled at 8000 Hz and linearly quantized at 16 bit/sample, recorded by means a telephone with 2 microphones. One of the two microphones is located in the classical position, while the other is placed at a distance of about 10 cm, on top of the telephone. By means of the two microphones, the telephone picks up two signals that are a combination of a speech signal and environmental noise. Each channel is then amplified and then sent as input to the audio board. The signals of the two channels at the system output are two recordings of the same signal which are different due to the different distances between the sources and the microphones. The database contains 18 telephone conversations made by 6 different speakers, 3 males and 3 females. Each speaker recorded 3 conversations in different environments; more specifically, a noiseless environment, one with traffic noise and one with babble noise. Each conversation lasted for about two minutes. The marking needed for VAD performance evaluation was performed manually, detecting the active speech and non-active speech segments. These were divided into 10-ms frames and were given a flag (0 or 1) according to the class they belonged to (Activity/Non-Activity). Table 2 shows the structure of the database. The database was then subdivided into a learning and testing database, the latter naturally containing different phrases and speakers from the former. The learning database, lasting about 2 minutes, contains speech conversation in a noiseless environment.

Speakers	Language	# Frame (10 ms)
Male	Italian	97380
Female	Italian	95580
Total Activity		46224
Total Non-Activity		146736

TABLE 2: Structure of the Database Obtained

5. Experimental results

In this Section we compare the performance of the standard VAD ITU-T G.729 [4], the FVAD [1] and the new solutions proposed in this paper, distinguishing between applications using the simulated database and those using the one based on recordings of real conversations. The results were analyzed considering the percentage of FEC and MSC in active voice frames, to calculate the amount of clipping introduced, and the percentage of OVER and NDS in non-active voice frames, to calculate the increase in activity. The tables 3 and 4 give the results obtained with the simulated database using various SNRs, in terms of FEC, MSC, OVER, NDS, FEC+MSC, OVER+NDS and total error (FEC+MSC+OVER+NDS), comparing the performance of the algorithms using PHAT and EDI with the performance of the G.729 Annex B and FVAD. All results were averaged on the six types of background noise: white, car, street, restaurant, office and train noise. In terms of FEC it can be seen that the FVAD is comparable with G.729, and performed better than the other two, even if the performance obtained using the EDI is only slightly worse. In terms of MSC, the FVAD still outperforms G.729, PHAT and EDI. In this case, however, PHAT performs better than the EDI at low SNRs. In terms of OVER, on the other hand, the performance of PHAT and EDI is very similar and better than that of the FVAD and G.729, whatever the SNR value. The same applies to NDS, where PHAT and the EDI gave a virtually null error for all SNR values. Considering activity clipping in terms of FEC+MSC, the FVAD performed better than the other techniques at all SNR values. As regards activity increase, PHAT and EDI performance was similar and better than that of the FVAD Finally, in terms of total error (FEC+MSC+OVER+NDS), the EDI gave the best performance, followed by PHAT, the FVAD and the G.729, at all SNR values. This is accounted for by the fact that EDI and PHAT perform very well in terms of OVER and NDS.

Fig. 4 gives the results obtained applying the G.729, FVAD and the D_FVAD to the acquired (real) database. In this case, all results were averaged on the three types of background noise, present in the acquire database. In terms of FEC and OVER the G.279 is similar to FVAD, but is worse in the other cases. In terms of FEC and MSC, it can be seen that FVAD and D_FVAD perform in a comparable fashion. In terms of OVER and NDS the D_FVAD performs better than the FVAD. As far as activity clipping is concerned, FVAD and D_FVAD performance is comparable, as was to be expected. In terms of activity increase, on the other hand, the D_FVAD performs decidedly better than the FVAD. Finally, considering the total misclassification error, the D_FVAD gave the lowest values. So all in all a system like the D_FVAD improves on the performance of the FVAD.

SNR (dB)	FEC (%)				MSC (%)				OVER (%)				NDS (%)			
	G.729	FVAD	PHAT	EDI	G.729	FVAD	PHAT	EDI	G.729	FVAD	PHAT	EDI	G.729	FVAD	PHAT	EDI
00	4,37	4,53	17,93	7,74	25,87	18,58	26,35	32,33	9,20	10,54	0,61	0,88	27,57	12,90	0,05	0,01
10	1,70	0,80	4,73	1,68	13,69	3,73	20,35	9,65	5,01	9,40	1,27	1,99	30,89	12,82	0,09	0,05
20	0,61	0,22	1,41	0,34	6,71	1,02	12,25	2,16	1,83	6,76	2,18	3,19	31,67	13,76	0,27	0,10

TABLE 3: Objective comparisons in terms of FEC, MSC, OVER and NDS

SNR (dB)	FEC + MSC (%)				OVER + NDS (%)				TOTAL ERROR (%)			
	G.729	FVAD	PHAT	EDI	G.729	FVAD	PHAT	EDI	G.729	FVAD	PHAT	EDI
00	30,23	23,11	44,28	40,07	36,78	23,44	0,66	0,89	67,01	46,55	44,94	40,96
10	15,38	4,53	25,08	11,33	35,91	22,23	1,36	2,04	51,29	26,75	26,44	13,37
20	7,32	1,24	13,66	2,50	33,50	20,53	2,45	3,29	40,82	21,77	16,11	5,79

TABLE 4: Objective comparisons in terms of FEC+MSC, OVER+NDS, Total Error

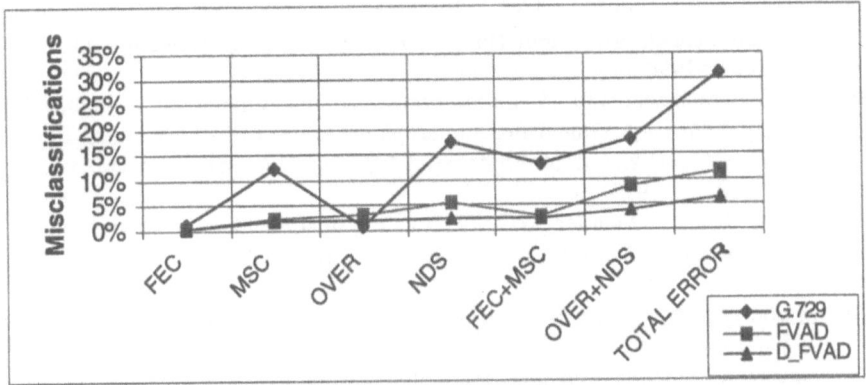

Fig. 4 Objective comparisons between G.729, FVAD and D_FVAD

6. Conclusion

In conclusion, we have presented a new voice activity detector based on time delay estimation and fuzzy logic. The new approach is more efficient than the traditional threshold method since it exploits all the information in the pattern of input parameters due to the fuzzy logic approach and the multi-channel information due to the presence of two microphones. As the results show, using a single parameter (delay estimate) the FVAD performs better in terms of OVER and NDS but worse in terms of activity clipping. When a fuzzy network trained with input given by the FVAD output and delay estimates is used, however, the results obtained reflect the positive features of the FVAD and delay estimate techniques, giving a great reduction in activity increase which is fundamental for more efficient use of the communication channel.

Acknowledgements
The authors wish to thank the CSELT audio coding group for providing the noise database

References
[1] F. Beritelli, S. Casale, A. Cavallaro, "Improved VAD G.729 Annex B for Mobile Communications Using Soft Computing", Contribution ITU-T, Study Group 16, question 19/16, Washington, 2-5 September 1997.

[2] K. Srinivasan, A. Gersho, "Voice Activity Detection for Cellular Networks", IEEE Workshop on Speech Coding for Telecommunications, Oct. 1993, pp. 85-86.

[3] J. Stegmann, G. Schroeder, "Robust Voice Activity Detection Based on the Wavelet Transform", Proc. IEEE Workshop on Speech Coding, Pocono Manor, Pennsylvania, USA, September 7-10, 1997, pp. 99-100.

[4] Rec. ITU-T G.729 Annex B, 1996 .

[5] ETSI GSM 06.32 (ETS 300-580-6) "European digital cellular telecommunications system (Phase 2); Voice Activity Detection (VAD)", September 1994.

[6] M. S. Brandstein and H. F. Silverman. A new time-delay estimator for finding source locations using a microphone array. LEMS Technical Report 116, LEMS, Division of Engineering, Brown University, Providence, RI 02912, March 1993.

[7] Knapp, C.H., and Carter, G.C., "Estimation of time delay in the presence of source or receiver motion," J. Acoust. Soc. Amer., v.61, n.6, pp. 1545-1549, 1977.

[8] Martin Drews. Time Delay Estimation For Microphone Array Speech Enhancement Systems. Esca. Eurospeech 95. 4° European Conference on Speech Communication and Technology. Madrid, September 1995. ISSN 1018-4074

[9] M. Omologo, P. Svaizer, "Talker Localization and Speech Enhancement in a Noisy Environment using a Microphone Array based Acquisition System", Proceedings Eurospeech, Berlin, September 1993, pp. 605-609.

[10] M. Omologo, P. Svaizer "Acoustic Event Localization using a Crosspower-Spectrum Phase based Technique", Proc. ICASSP, Adelaide 1994, pp. 11273-11276.

[11] M. Russo, "FuGeNeSys: Fuzzy Genetic Neural System for Fuzzy modelling", to appear in IEEE Transaction on Fuzzy Systems, Aug. 1998.

[12] C.B. Southcott et al. "Voice Control of the Pan-European Digital Mobile Radio System" ICC '89, pp. 1070-1074.

[13] IEEE Trans. Acoust., Speech, Signal Processing. Special Issue on Time-Delay Estimation, volume ASSP-29, June 1981.

[14] Knapp, C.H., and Carter, G.C., "The generalized correlation method for estimation of time delay," IEEE Trans. Acoustics, Speech and Signal Processing, v.ASSP-24, n. 4, pp. 320-327, 1976.

[15] G. C. Carter, Coherence and Time Delay Estimation. IEEE Press, 1993.

[16] Segal, M., Weinstein, E., and Musicus, BR., "Estimate-maximize algorithms for multichannel time delay and signal estimation," IEEE Trans. Signal Processing, v.39, n.1, pp. 1-15, 1991.

Queueing performance evaluation by means of traffic sampling methods

Irene Cozzani, Stefano Giordano, Michele Pagano

Department of Information Engineering
University of Pisa, ITALY
Phone +39 50 568539, Fax +39 50 568522

Abstract

We present an application of sampling procedures in a distributed test and measurement system, which is able to provide real time and reliable end-to-end performance evaluations for packet broadband networks. The novel idea introduced in this work is the attempt to reduce the quantity of control data required to carry out the further management activities. Sampling is hence introduced for relieving the complexity of management functions, typically very complex in the current heterogeneous broadband networks. A simulation model of a "Passive QoS 2-point measurement system" working on an ATM network has been built and tested with a synthetic traffic as well as with a real ATM trace. Our results show the effectiveness of sampling in terms of the trade-off between the sampling rate, the attained estimation accuracy and the cost of taking measures (assumed to be proportional to the sample size).
They also give a theoretical/quantitative method usable by network operators to design control instrumentation.
Area of interest: analysis, measurements and performance evaluations.

1 Introduction

Reliability of broadband integrated telecommunication networks relies on complex management functions that must be properly designed to allow the provision of a wide range of services with different quality requirements. Broadband network aim nowadays to support different kinds of traffic and applications like digital TV, video on demand, cooperative working, videoconferencing on top of the classical services (voice, data, video, images). This complex traffic mix has to be transferred in a reliable way; therefore beside traffic modeling, already quite extensively considered in the framework of B-ISDN networks [1], proper monitoring and performance evaluation techniques have to be developed in order to make network management effective and timely [2]. Our approach to network management has a twofold basis: a measurement-based philosophy and the sampling theory. Sampling aims to make monitoring and control functions less invasive and more efficient. In fact by using the less possible information, the cost of making measurements and the presence of the control messages will be reduced. Our work is concerned with the application of sampling survey into real measurement devices. A test and measurement system is a set of instruments aimed to provide feedback information to network operators about the real functioning of the tested network; in order to allow them to carry out policy and management functions. Current measurement instruments are primary standalone devices performing extensive measurements often only at one-point, that is they can not inter-work to compare data from different points and therefore are not able to control the whole network, but only specifics islands. Furthermore they often provide only "out-of-service testing" and not "in-service testing" dealing with real user traffic. On the contrary we have used a *distributed* test and measurement system that allows the evaluation of the *end-to-end* performances effectively perceived by the user. It can be applied to complex network scenarios like those approaching due to the even more spread use of broadband networks [1] as well as of the Internet [2]. In this work we use the proposed monitoring system for estimation of quality of service (QoS) parameters of an ATM network. QoS parameters are given as functions of the sampling fraction. It has turned out to be the external parameter that will allow network administrators to speed up the management procedures, but on the other hand it also fixes the accuracy level of the measurements. The shown relationship among sampling rate, attained accuracy and implementation details gives a methodology for assessing the reliability of the proposed evaluation method, moreover it clearly states that application of sampling relieves heaviness of standard monitoring and management actions. The analyzed test & measurement environment is described in section 2, along with a brief outline about the meaning of sampling application into network functions. Sections 3 presents the results obtained from simulations, for Poisson and real ATM traffic and emphasizes the effectiveness of the sampling approach. Section 4 concludes the paper.

2 Application of sampling to the distributed passive monitoring system: network performance evaluation

The considered test & measurement system carries out passive measures thanks to its non-intrusiveness feature. Standard testing methods usually insert a sort of known traffic into the monitored connection and measure it expecting that it experiences the same effects of the user flows, even if the presence of this intrusive testing traffic modifies the performance that it is supposed to measure. On the contrary passive measurements are obtained through a complete separation between the monitored network and the monitoring system (Fig. 1), which does not inject any traffic into the network. The testing system is composed by measurement devices located at different measurement points (MP) throughout the network and analysis consoles. They are elaboration points running control and/or estimation functions, which turn the data received by all of the measurement devices into network-wide information, allowing end-to-end performance evaluation [3]. Probes are the simplest measurement devices of this type; they have only passive monitoring capability, because do not modify the user traffic and perform "data reduction" in three successive steps: capture of cells belonging to the connection of interest (e.g. *filtering* on VCI or VPI), identification of the occurrence of particular events (*sampling*) and transmission of proper data ("*event data*") to the elaboration points (*control message generation*).

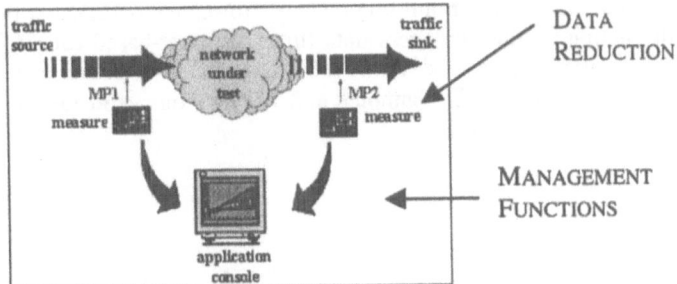

Fig. 1 Tested network and monitoring system

Our activity has followed two stages: studies of suitable sampling techniques usable in the measurement devices and implementation of the related estimation procedures in the elaboration points. We seek to answer the following questions:

I.   Is sampling a smart solution for simplifying network monitoring?
II. Does statistical inference theory prove the effectiveness of the sampling mechanism already implemented into real measurement instruments?
III. Are there few and simple external parameters, settable by the network operators, able to control the level of drawn samples and at the same time the obtained accuracy (trade-off between them)?

Question I is addressed in this section, whereas results of section 3 answer questions II and III.

2.1 Outlines about sampling

Sampling is used whenever global characteristics of the population under study are required but the analysis of every element would be too expensive or too long. A *sample selection plane* is used to select a fraction of elements (*sample*) and then *estimation procedures* compute the unknown population parameters from the sample data (*estimates*)[4]. In packet networks sampling is a way for choosing some particular packets among those crossing the network for obtaining useful data to feed in policy functions: performance parameters computed from it will be considered valid for the overall connection. Sampling methods have been used in [5] for the evaluation of traffic parameters (packet length, interarrival time distributions etc), whereas we addressed sampling use to the evaluation of performances perceived by the end users.

2.2 Implementation of sampling into the monitoring system

Figure 2 shows the analysed 2-point measurement system testing an ATM network modelled as a single queue. We have designed the monitoring system for QoS evaluation, considering cell transfer delay and cell loss ratio, as they are defined by the ITU-T and ATM Forum. The two probes located at both network edges perform Simple Random Sampling (random choice of n elements out of a population of N) by means of event occurrence identification. An event is a situation that triggers the generation of control information (event-data). The definition of events and the structure of event-data are tightly correlated and depend on the target applications they are used for. Currently probes pick out user packets when a particular function of the bit patterns in the cells payload is met.

Then the elaboration point computes performance parameters through estimation rules acting on the two event-data flows. Content-based rules for the selection of packets allow measure devices, located at different points of interest, to capture exactly the same traffic samples and to carry out valid QoS evaluation through correlation procedures.

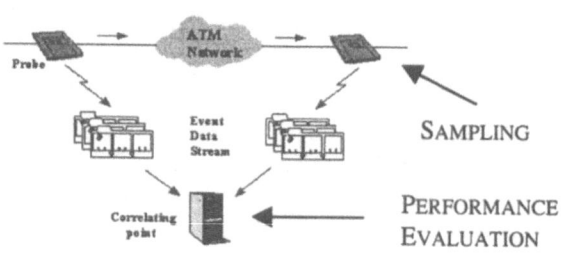

Fig. 2 The case studies: 2-point measurement system on an **ATM** network

The Simple Random Sampling estimator of a population mean $\overline{Y}$ is the sample mean: $\overline{y} = \dfrac{1}{n}(\sum\limits_{i=1}^{n} y_i)$ (y_i is the generic observation, i^{th} element of the sample) [6].

Sampling based methods inevitably yield estimation errors due because global validity conclusions are drawn on the basis of a reduced part of the data. Accuracy is usually expressed by the variance of the estimated parameter $MSE = E\left[(\overline{y} - E[\overline{y}])^2\right]$, which gives an average value of the deviation between the estimation and the real value. It also allows to compare the precision obtained with different sampling techniques, or to determine the size of the sample needed for obtaining a predefined error. Our aim has been to investigate the relationship between the two main contrasting effects of sampling: accuracy and implementation costs (assumed to be proportional to number of drawn samples).

2.3 ATM-QoS evaluation: standard methodologies

The "ATM Forum Traffic management specification, version 4.0" [7] and the "ITU-T Recommendation I.371 and I.356" [8,9] specify QoS parameters for B-ISDN and also give rules to compute them. We have dealt with Cell Loss Ratio (CLR), and Mean Cell Transfer Delay (CTD) shown in fig.3:

$$CLR = \frac{total\ lost\ cells}{tot\ transmitted\ cells\ (observed\ population)} \quad (1)$$

$$CTD = \frac{sum\ of\ (t_2 - t_1)\ for\ a\ number\ of\ samples}{number\ of\ samples} \quad (2)$$

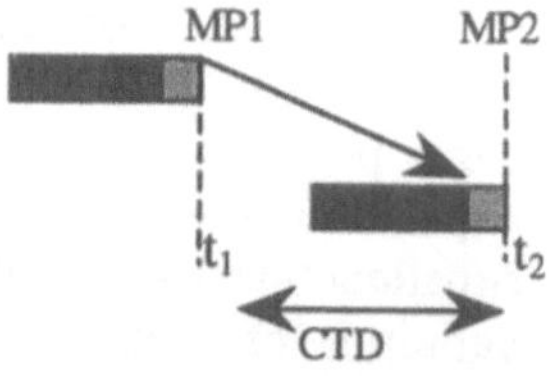

Fig. 3 QoS parameters definition

In order to use these rules, it is necessary to have the right data available: the aim of a measurement system is just to provide this data in the proper format and quantity. ITU-T recommendations I.356 and I.610 [9,10] standardise the use of Operation and Maintenance (OAM) streams for providing "*in-service*" testing capability. OAM cells are inserted into the monitored connection between user cells, in order to experience the same conditions. Their payloads carry data suitable for management purpose (number of cells since the previous OAM cell and timestamps) to allow loss and delay evaluations in the downstream switch. The major disadvantage of this solution is that OAM cells are "intrusive traffic" and therefore their presence affects the performance of the network that they are attempting to measure as explained before.

2.4 ATM-QoS evaluation: the proposed method

The system of figures 2 allows "*in-service*" testing too, still in agreement with the ITU-T specifications, but it uses a "*non-intrusive*" technology not affecting the user traffic. QoS estimations are made using event-data (packet counters and timestamps). Each of them is univocally identified by a label that allows the processing point to detect the correspondent couples of data in the event-data streams coming from the two probes. Events triggered by the same cell in the two measurement points have in fact the same value of the identifier field. After that the correlator computes CLR through (1) by making a difference of the numbers of cells passing the two measurement points, whereas timestamps provide values for t_1 and t_2 in expression of CTD (2).

3   Simulation results

An event-driven simulation model of the system in fig. 2 has been built in order to estimate QoS values as functions of the sampling fraction $f=n/N$. It is the ratio of the size of the extracted sample "n" to the size of the population "N" (total number of transmitted cells). The accuracy of the estimate is compared with the one expected from the theory of simple random sampling.

3.1 Poisson input traffic: methodology and model validation

Poisson traffic has been generated with exponential distributed interarrival times and random payloads. It has been used for investigating the behaviour of the sampling mechanism and its effects on the estimation of performance parameters. The network is represented by a M/D/1/L queue running at 155Mbps with FIFO service discipline. The following plots show a comparison between the QoS parameters computed by the correlating point and the theoretical ones. They are given by the reporting statistics blocks of the used simulator [11] through a statistical analysis of the ATM network status as it is modified by the presence of each transmitted cell (as if it was f=1). Figures 4 show CLR due to buffer overflow as a function of time, with a first transient period where the queue starts to fill. These graphs are related to the limit cases of our investigation: $f\sim10^{-4}$ and $f\sim10^{-1}$, anyway for each value of f in between, even with the lowest ($f\sim10^{-4}$), the estimated CLR lies exactly on the reference line: CLR is a cumulative quantity not affected by a reduction of data used to compute it. Network operators have to choose f depending on the required timeliness. The higher the f value the more timely are results available, but more resources are necessary.

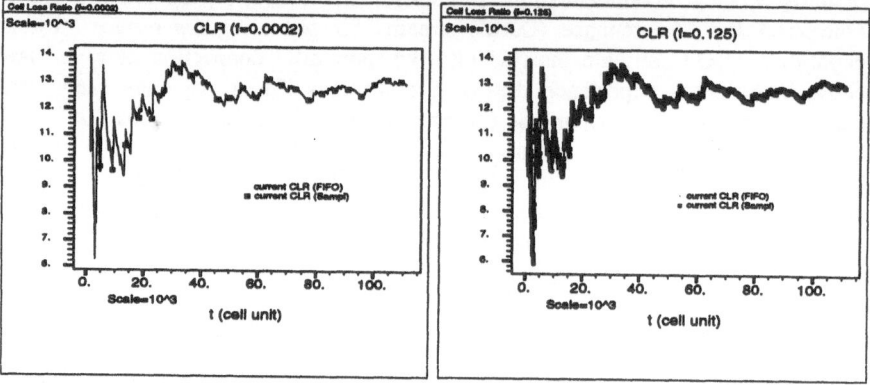

Fig. 4 Cell Loss Ratio vs time, sampled data and FIFO statistics ($f\sim10^{-4}$- 10^{-1}) - Poisson.

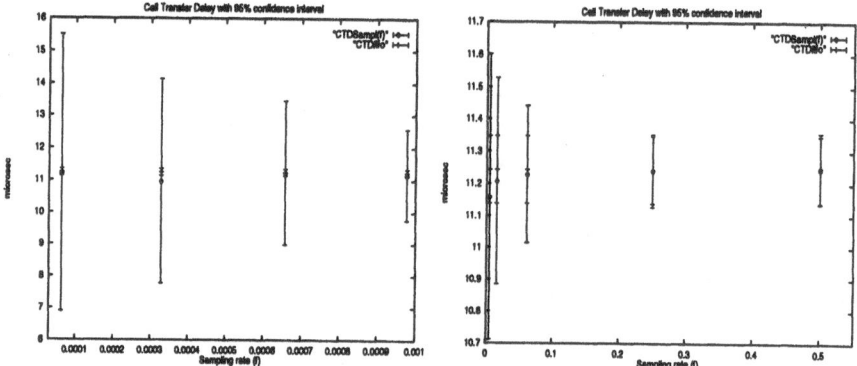

Fig. 5 Mean Cell Transfer Delay - Poisson traffic

The waiting time in the FIFO queue and the fixed service time form the cell delay CTD. Theoretical values come from the statistics blocks of the FIFO and the estimated values are given by the sample mean of the observations. In figures 5 both values are plotted versus the sampling rate f (10^{-4} up to 10^{-1}), along with their respective 95% confidence intervals. Values of f in the range (10^{-2},1) give a good fit between the theoretical and estimated points, but impairments appear when f decreases below 10^{-3}. With low values of f the small fraction of sampled cells (one in 10^{3} or 10^{4} in average) seems to be a poor representation of the tested traffic

The different conclusions we had for the two analysed QoS parameters are due to their different nature. Cell loss evaluation requires the number of cells lost since the beginning of monitoring and that is why for each value of f the CLR estimates are always in agreement with the theoretical values. Instead delay measures are instantaneous values of the random variable "cell delay" that have to be averaged altogether. A trade-off between sampling rate and desired accuracy becomes in this case essential. Network operators have to choose a value of f able to guarantee an acceptable accuracy level, and at the same time the desired reduction of the cost of taking measures (the lower f is, the less expensive and invasive the capture technique is, but less accuracy is achieved). CTD probability density functions (Pdf) in figures 6 give a deeper insight of the effects of f on accuracy. For each value of f the two histograms refer to the Pdf derived by the sampled data and to the theoretical one. As f increases the fitting between the two histograms becomes better.

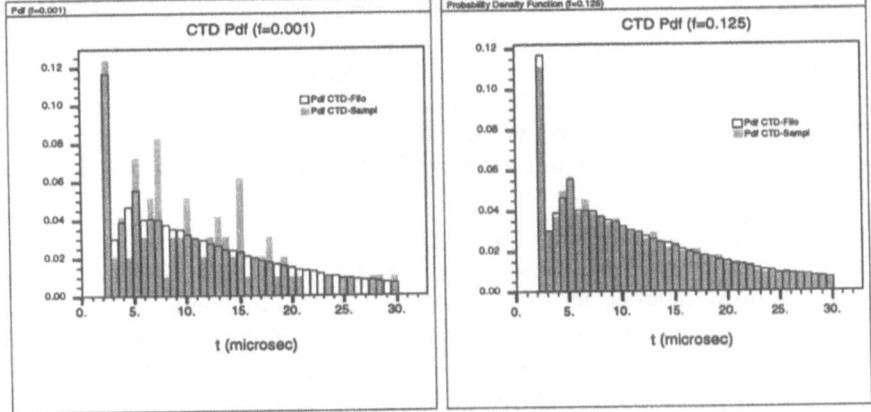

Fig. 6 CTD Probability density functions - Poisson traffic

The Mean Square Error is an indicator of the inaccuracy concerned with the sampling/estimation procedures; figure 7 shows the errors predicted by the theory of Simple Random Sampling Std Dev_Theor (CTD) = $\sigma_{\overline{y}} = \sqrt{S^2 \dfrac{1-f}{n}}$ (S^2 is the population variance) and the simulated one Std Dev_Est (CTD) = $\sqrt{MSE}$). The good fitting represents one of the main findings of this analysis: it states that the sampling mechanism implemented into the measurement devices is reportable to the simple random scheme.

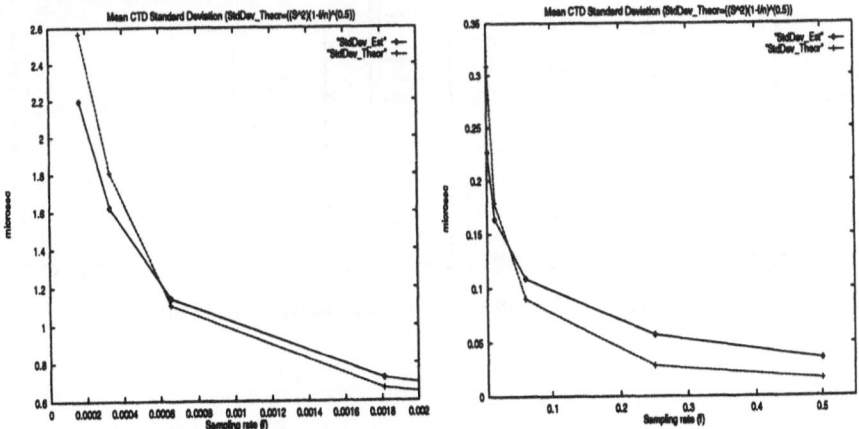

Fig. 7 Theoretical and simulated variances of the Mean Delay - Poisson traffic

This theoretical basis allows the interpretation of the QoS plots and above all is a reading key for the accuracy/cost trade-off. Quantitative assessment can be drawn for quantifying drawbacks and advantages of the sampling application. For mean cell transfer delay the value of $f \sim 10^{-3}$ divides the range of sampling rate in two groups: for f up to 10^{-3}, the estimates have 95% confidence intervals less than 0.4µsec around a mean value of 11.25µsec; afterwards confidence intervals start to increase quickly up to ~4µsec for $f \sim 10^{-4}$, that is around 38% of the estimated value (fig.5). In the same way by decreasing the sampling rate the standard deviation increases: in fig.7 we can assess standard errors lower than 0.3µsec (accuracy tighter that 3%) with sampling rate higher than 10^{-3}, that allows reductions of the cost of taking measure by 10^3 times. Lower sampling rates give even better gains (reduction of processed data: 1cell in 10^4 or 10^5 cells), but they yield higher standard errors.

3.2 A real scenario: trace of ATM traffic

The first theoretical part provides a working methodology allowing us to consider a more realistic environment. Simulations has been run feeding the queue with a trace of real ATM traffic captured by the monitoring system explained in section 2, working over an ATM Transatlantic Link at 34Mbps. The trace captured on 30th September 1997 refers to data traffic exchanged between US and UK and carries timestamps (resolution of 100nsec) and the cells contents. We show the mean cell delay against the sampling rate (fig. 8) and its accuracy (fig. 9). As it was in fig.5 we have a good matching between the theoretical and estimated CTD values and also the accuracy is still quite close to the one expected for the Simple Random Sampling theory. Standard errors are in the range (0..7.6)µsec with sampling rate up to 10^{-3}. They are even better values than the ones obtained with Poisson traffic, since the reference transfer delay is in the order of msec instead of µsec.

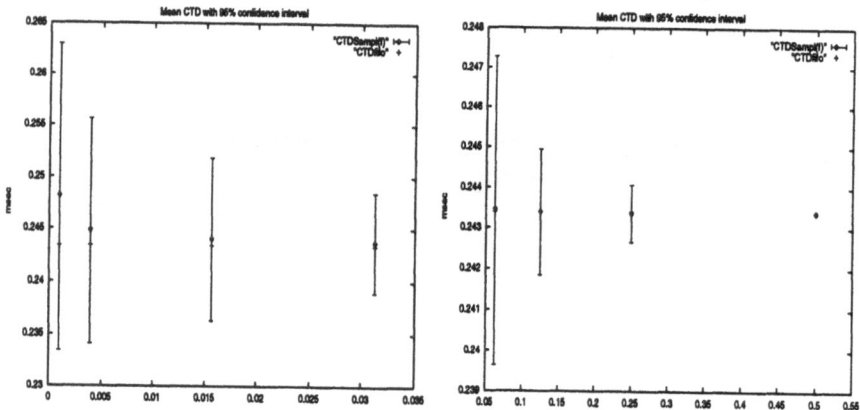

Fig. 8 Mean Cell Transfer Delay vs sampling rate - real traffic

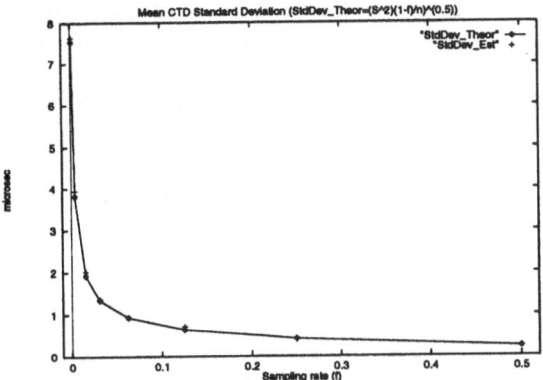

Fig. 9 Theoretical and simulated variances of the Mean Delay - real traffic

4 Conclusions

We have evaluated end-to-end QoS parameters of an ATM network by means of measurements of live traffic, obtained using a distributed test & monitoring system that does not affect the tested traffic. Our contribution is concerned with a theoretical and simulative analysis of the interaction among sampling technique, estimation rules and the functioning of the monitoring system. We have provided first of all the theoretical rationale of the testing/monitoring system and then quantitative results usable to choose the proper sampling rate. Comparisons of the QoS parameters (CLR and CTD), estimated by sampling, against the ones obtained without sampling give assessment for the accuracy of the proposed method and the inherit simplification of the control functions. This statistical analysis let us to answers to the initial questions:

I.   Capturing of user traffic is essential to perform monitoring of the network status and the further management functions. Every kind of measurement device should base its functioning on specific strategies for capturing information to use in policy and/or quality assessment functions. Sampling theory gives rules for selecting sets of cells/packets and for computing the target metrics. Furthermore sampling reduces the amount of resources (bandwidth, time, memory) required to deal with the processed data in order to make monitoring and management functions less heavy and faster.

II. We have simulated the functioning of real measurement devices (probes) and compared it with the one expected from the statistical inference theory. Standard deviations (graphs 7 and 9) of the estimated target parameters are good matching of the theoretical reference lines. This proves that the capturing algorithm, based on the cells contents, is a reliable implementation of simple random sampling and it works with Poisson input traffic as well as with real traces. That represents the rationale underpinning the functioning of this real monitoring system. Better accuracy levels could be obtained with other methods, like the Stratified or Systematic Sampling [6]. However their implementation will be more complex, because certainly requires hardware and software changes into the production of the instruments, since two successive steps of grouping and selection of data are involved. These aspects will be investigated in the future.

III. The sampling rate f is the key external parameter usable by network operators to balance implementation costs and attained accuracy. By using graphs like 7 or 9 requirements on the accuracy can be turned into specifications on the implementation details of the sampling mechanism and viceversa. If a minimum accuracy is required ($Std \leq Std_{Max}$) then the reported graphs give the minimum sampling rate to use in order to obtain an acceptable accuracy. On the contrary if a limit on the bandwidth (or other resources) available to carry control data fixes an upper threshold for f ($f \leq f_{max}$) then network operators can read on the graphs the attainable accuracy ($Std \geq Std(f_{max})$).

The analyzed measurement system monitors the network status in real time and provides management centres with the right data to carry out control and design routines. We have use it for performance evaluation purpose, but each of the management functions can be addressed with a proper design of the cell capturing mechanism (sampling), the control data production and the elaboration procedures (estimation in this case). The theoretical study concerning sampling effects on management decision has been confirmed by simulation results and, along with indications about obtainable gains, it is the main outcome of this work.

References

[1] J. Roberts "*Traffic modeling and traffic engineering for B-ISDN*", ITC15 Tutorial Washington DC, 22 June 1997

[2] S. Floyd and V. Paxson "*An overview of Internet Engineering, Measurements and Modeling*", ITC15 Tutorial Washington DC, 22 June 1997

[3] M. Cociglio, R. Cole, "*Discovering the secrets of ATM Networks*", ISS97

[4] P.R. Krishnaiah, C.R. Rao, "*Handbook of Statistics, Volume 6: Sampling*", North-Holland, 1988

[5] K.C. Claffy, G.C. Polyzos, H.W. Braun, "*Application of Sampling Methodologies to Network Traffic Characterization*", ACM Computer Communications Review, vol. 23, n. 4, pp.194-203, October 1993

[6] W.Cochran, "*Sampling Techniques*", John Wiley & Sons, 1987

[7] ATM Forum Technical Committee, "*Traffic Management Specification Version 4.0*", 4/1996

[8] ITU-T Recommendation I.371 "*Traffic Control and Congestion in B-ISDN*", 11/1995

[9] ITU-T Recommendation I.356 "*B-ISDN ATM layer cell transfer performance*", 2/1996

[10] ITU-T Recommendation I.610 "*B-ISDN Operation and Maintenance Principles and Functions*", 11/1995

[11] "BONeS DESIGNER Core Library Guide" and "User's Guide", Alta Group, Comdisco System, Inc.

Modelling DQDB multimedia sources at cell/segment level

Rosario G. Garroppo, Stefano Giordano, Michele Pagano

Department of Information Engineering
University of Pisa, ITALY
Phone +39 50 568539, Fax +39 50 568522

Abstract

Measurement analysis over real test-beds highlights the bursty nature of the traffic offered by multimedia sources to broadband networks providing a LAN-to-LAN interconnection service. This paper focuses on a quite neglected issue, i.e. the link between the bursty behaviour of traffic processes at frame level (namely IEEE 802.3/Ethernet frames measurements) and the distribution of busy/idle periods in ON/OFF models adopted at the segment/cell level. The starting point for this work is represented by a set of measurements carried out over the DQDB "Tuscan MAN" by means of an ad-hoc hardware prototype. The analysis of the acquired data pointed out the heavy tailed nature of the inactive period length distribution which motivated our choice to model the offered traffic using a simple ON/OFF model, characterised by a Pareto distribution of the time permanence in the OFF state. The relevance of this characteristic has been emphasised by comparing the complementary probability of the queue occupancy in an infinite buffer queueing system loaded by three kinds of traffics: actual data, synthetic traces generated by the proposed model and traditional ON/OFF sequences.

1. Introduction

LAN-to-LAN interconnection will represent (at least in the short and medium term) the most relevant service to be provided by broadband networks. Its relevance derives not only from traditional data applications (file transfer, remote login, data base access) in a wide area network environment, but also as local access networks to new multimedia multicast communication services [1]. This is possible due to the traditional broadcast capability of LAN architectures that stimulated multicast voice and video communications in a broadband integrated environment putting the basis for national and global multicast information infrastructures. Performance evaluation of networks providing multimedia services requires an accurate modelling of the traffic offered by each source. Since April 1993 a Metropolitan Area Network, providing asynchronous connectionless service and permanent isochronous channels, is running in Tuscany. Our interest is directed to the analysis of the traffic offered by a set of departmental LANs to the LAN Bridging interface provided by the MAN and offering connectionless service. From a teletraffic viewpoint, the relevance of the considered scenario derives from the possibility to test a working broadband infrastructure in presence of both traditional best effort services (i.e. file transfer, remote login, access to hypermedia databases, etc.) and multimedia real time applications (voice and video).

The analysis of the traffic offered at frame level confirmed the presence of Long Range Dependence (LRD) [2], previously highlighted in [3] considering the Bellcore LAN network. In this paper we extend such an analysis starting from our experimental trial which gives us the possibility to analyse traffic behaviour at DQDB (Distributed Queue Dual Bus) slot level. The highly variable and bursty nature of the DQDB data sources is efficiently captured by means of a two state model with heavy tailed distribution (namely a discrete version of the Pareto function) of the OFF periods length. Moreover, this feature has also a deep impact on network performance as confirmed by the different behaviour of queueing systems loaded by actual traffic streams and synthetic traces generated by classical Markovian ON/OFF models (with geometric distribution of the sojourn time). Simulation studies have emphasised that, in order to provide realistic queueing performance, it is necessary to take into account the heavy tailed distribution of the OFF periods length.

The cell/segment based architecture of the DQDB MAN permits to have some insight in the effects of segmentation procedures on the statistical features of the LAN traffic, which is also extremely relevant for LAN-to-LAN interconnection services over ATM networks. Indeed, it is reasonable to assume that the traffic characteristics at frame level will likewise determine a clustered pattern when considering ATM cells streams. In accordance with our analysis at DQDB level this element has to be taken into account when evaluating the actual statistical multiplexing gain reachable in an ATM environment.

Starting from the measurement scenario described in Section 2, the DQDB level model of acquired data is presented in Section 3, while Section 4 is devoted to traffic traces analysis and model parameters estimation. The validation of the model in terms of queueing performance is then carried out in Section 5 and the Conclusions Section highlights the relevance of proper traffic modelling in a cell-based network scenario.

2. Measurement Scenario

The traffic measurements were carried out in Pisa (Italy) over a broadband network infrastructure represented by a DQDB MAN (known as "Tuscan MAN") linking the towns of Pisa, Florence and Siena.

The MAN is a 140 Mbps dual folded bus based on the former QPSX (Queued Packet Synchronous Switch) proposal of the IEEE 802.6 (DQDB) standard [4] and uses an optical transmission system based on the Plesiochronous Digital Hierarchy (PDH). According to the QPSX protocol, the overall information (i.e. user data plus higher level protocols overhead) is segmented in fixed-size cells of 64 bytes plus a 5 bytes header. The MAN infrastructure [1] was realised by means of five functional network elements:

- Customer Gateways (CGW);

- Edge Gateways (EGW);

- Subnetwork Routers (SR);

- Customer Network Interface Units (CNIU);

- SMDS Edge Gateways (SEG).

As shown in figure 2.1 the measurement test-bed involves only two network elements: CGW and EGW. The CGW is located at the customer premises and provides for the interconnection of customer equipment to the MAN (LANs equipped with routers and/or an audio/video codec if the nodes support an isochronous access). The EGW (located inside the Central Office of the national provider Telecom Italia) is directly connected to the dual folded bus running at 140 Mbps, while an optical point-to-point link at 34 Mbps (E3) represents the connection to the CGW, offering the LAN bridge service.

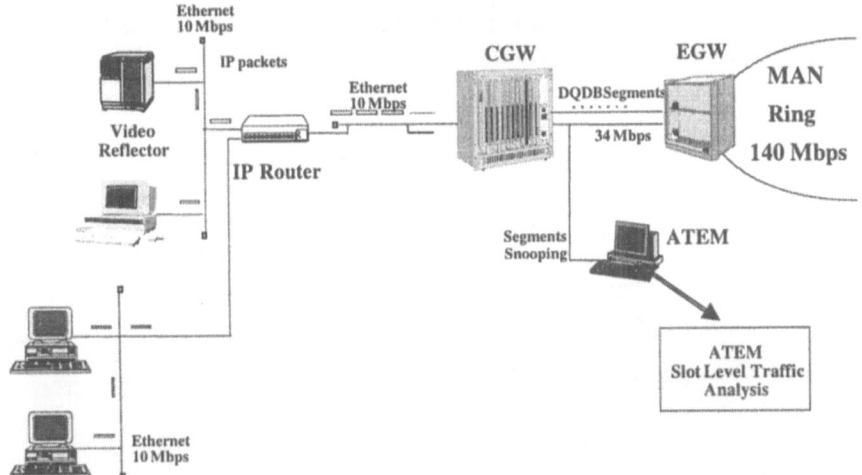

Fig 2.1. Measurement test-bed

The main novelty of the measurements described in this paper is that they are not carried out at the Ethernet level [2][5] by means of a public domain software analyser (such as TCPDump), but they involve the direct monitoring of DQDB slots by means of an hardware prototype (ATEM) connected to the CGW at the Department of Information Engineering of the University of Pisa as shown in Figure 2.1. ATEM essentially consists of a hardware interface, developed by Siemens Telecomunicazioni, now Italtel, that enables the connection of a standard PC to a DQDB node. This simple monitoring tool permits to measure both the average traffic level (with a 1 second time resolution) and the DQDB slot status (though for a limited period of time, that determines the length of the acquired sequences). Data analysed in this paper have been captured according to the latter option. It is relevant to point out how these data could not be obtained by simply segmenting the corresponding Ethernet traces since they also take into account DQDB segmentation/reassembly procedures and buffering at CGW level.

3. Modelling Background

The nature of the acquired data suggests the use of a simple ON/OFF model, whose two states correspond to a busy or idle slot respectively. To fully describe the model it is sufficient to characterise the distribution of the sojourn time in each state and to estimate the relevant parameters from the experimental data. The underlying assumption that greatly simplifies the model is represented by the independence between ON and OFF periods; its correctness will be confirmed by queueing simulations described in Section 5.

The characterisation of the source behaviour during its active periods is essentially determined by the distribution of the Ethernet frame size and by the DQDB segmentation procedure (see figure 3.1). Thus the length of the ON distribution assumes only a limited number of relatively small values (the maximum is 25): in accordance to the Ethernet analysis carried out in [6] a trimodal distribution can be assumed.

MAC PDU

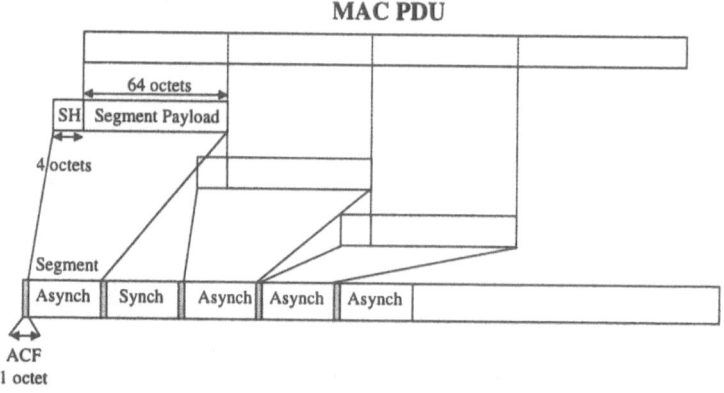

Figure 3.1. – Segmentation procedure

The situation dramatically changes for the OFF periods because they are not characterised by a proper time scale. In fact the bursty nature observed for the frame level traffic and the very short duration of each time unit (i.e. a DQDB slot) determine a great variability of the OFF periods distribution, characterised by the presence of very long silence intervals. To quantify this qualitative behaviour it is sufficient to remember that during each E3 frame (i.e. 125 μsec.) 5 asynchronous slots are sent corresponding to a logical slot duration of 25 μsec.: this means that a 25 msec. silence (1/4 of the measurement interval considered in [2][6]) corresponds to 1000 idle slots.

The observation of traffic at DQDB slot level after frame segmentation in the CGW highlight the following conclusion: the traffic burstiness at Ethernet level leads to an ON/OFF model in which only the OFF periods distribution is heavy tailed. In particular, we have considered the ceiling of the random values generated according to the Pareto function:

$$P(X > x) = \left(\frac{\theta}{x+\theta}\right)^{\alpha} \quad x \geq 0$$

where θ is commonly used to shift the mean value of the distribution, while α is related to the weight of the distribution tail. The resulting discrete time distribution will be referred in the following as Modified Pareto.

Estimation techniques for the parameters α and θ presented in [7] and [8] need a large amount of data to achieve significative results. Otherwise we can simply estimate α considering the complementary probability (i.e. P(X>x)) for the OFF periods distribution and evaluating the asymptotic slope (in log-log scale) of this function with a MMSE technique. Moreover, for values of α belonging to the interval (1, 2) (that corresponds to the case of interest with a finite mean and theoretically infinite variance), the θ parameter can be easily determined considering that the mean value of the distribution is:

$$E[X] = \frac{\theta}{\alpha - 1}$$

4. Data Analysis

A post-processing of the acquired data was needed to eliminate samples concerning the isochronous channel (which corresponds to a deterministic pattern). Then, we obtained a sequence of 0s and 1s corresponding to idle and busy slots respectively.

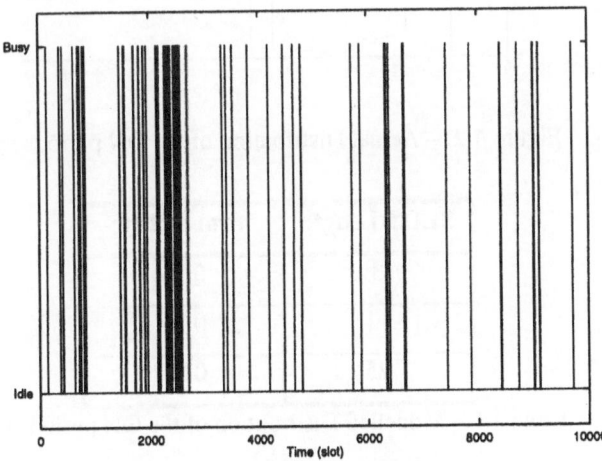

Figure 4.1. – Acquired DQDB slots

As shown in figure 4.1, the traffic pattern is characterised by the presence of clustered arrivals alternated to inactivity periods with a highly variable duration in accordance to the modified Pareto distribution hypothesis introduced in the previous Section. The analysis of the measured traffic was directed to the validation of the proposed model and to the estimation of its parameters. As mentioned above, the ON state distribution is related to the packet lengths at LAN Bridging Interface of the CGW that segments the Ethernet frames in DQDB slots.

The distribution, presented in figure 4.2, highlights the presence of a limited number of possible lengths. In particular we can note that a relevant number of ON periods last only three or four slots (around 75%), while the maximum duration (25 slots) is assumed only for a little percentage of ON periods (around 1%). Busy slot sequences of intermediate lengths are also observed with a prominent peak of the presentation frequency around 11. These experimental evidences have lead us to approximate the actual distribution considering only three possible values of the ON period length as described in table 4.1; the associated presentation probabilities have been chosen in order to match the actual mean value of the observed ON periods.

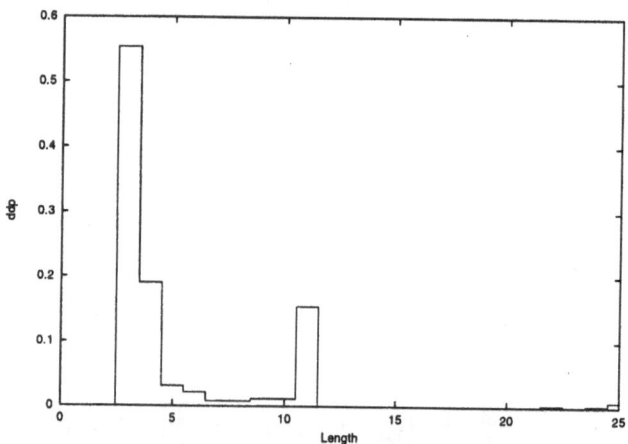

Figure 4.2. – Actual Distribution of the ON periods length

SLOT Length	Probability
3	0.75
11	0.24
25	0.01

Table 4.1 – Modelled Distribution of the ON periods length

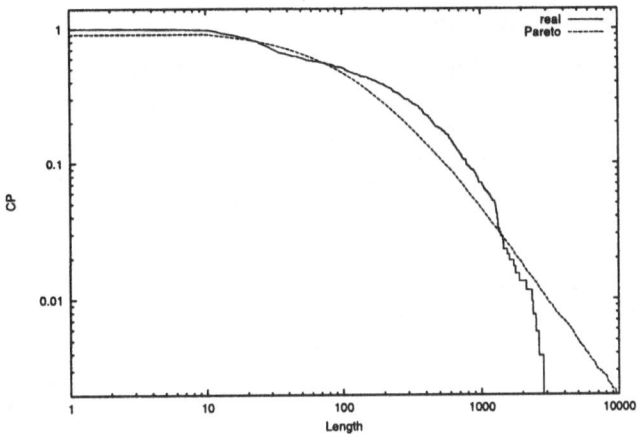

Figure 4.3. – Distribution of the OFF periods length

The analysis of the OFF periods is shown in figure 4.3, which highlights the slow decay of the tail for the complementary probability distribution; the slope estimation leads to a value of the α parameter equal to 1.53.

Trace	Busy slots percentage	Variance/ mean ratio	Mean ON periods length	Mean OFF periods length	OFF Period Std. Dev/ Mean
Acquired data	1.72%	7.41	5.01	287.24	1.53
Modif. Pareto	1.79%	7.56	5.12	280.90	4.25
Geometric	1.77%	7.45	5.13	285.24	1

Table 4.2 – Relevant Statistics of the considered traces

To highlight the effectiveness of the model, the same figure also presents the complementary probability for a longer (10^7 samples) synthetic trace generated according to the modified Pareto distribution. The two curves are in good agreement for estimated probabilities greater than 0.01, while the limited number of available samples (the measured trace is around 160 000 slots long) makes meaningless the comparison over a wider range of values. Indeed, it is necessary to take into account that, considering the parameters values summarised in table 4.2, a time period of 160.000 slots corresponds to few hundreds state transitions. The last line in the table refers to the classical ON/OFF model considered in the next Section.

5. Simulation Results

The proposed generator differs from classical ON/OFF models because of the heavy tailed distribution of the OFF periods, which appears to be a relevant statistical feature of actual broadband traffic. To assess the effectiveness of the model, it is important to evaluate the impact of that traffic feature on the network behaviour. To this aim we have analysed the queueing performances of an infinite buffer queue with deterministic service time. The choice of a suitable model for traffic engineering purposes is generally a trade-off solution between two different goals: the effectiveness of the considered model in capturing the actual system behaviour and its analytical simplicity. Mathematical tractability represents one of the most appealing features of ON/OFF models with geometric sojourn time in each state in sharp contrast with several experimental measurements like those described in Section 4. This drawback determines incorrect predictions of the actual network behaviour: indeed the neglecting of busy slot clustering phenomena leads to an optimistic estimation of queueing performance and to the related underdesigning of network infrastructures. The introduction of a modified Pareto model for the inactive periods distribution is aimed to overcome such drawback without making the model too cumbersome.

The performance analysis refers to the complementary probability behaviour for different values of the normalised load. It involves the comparison between the simulation results obtained by feeding the queue with real and synthetic traffic traces. The latter are generated considering the Modified Pareto and the geometric distribution for the inactivity periods. Moreover, all the synthetic traces refer to the ON periods distribution summarised in table 4.1 and present the same mean as the original data (see table 4.2).

The queueing results can be seen in figures 5.1 and 5.2 for two different values of the Normalised Traffic Load. Both simulation scenarios highlight the inadequacy of the two-state Markovian model in capturing the actual behaviour of real broadband traffic and show a good agreement between experimental traces and the proposed model. The comparison of these experimental results emphasises how the heavy tailed distribution of the OFF periods represents the key element in determining the queueing performance. The high variability of the Modified Pareto distribution and the related presence of long OFF periods determine the clustering of busy slots over relatively short time intervals as it happens for the real trace, while the geometric distribution leads to a more regular spreading of the activity periods. The different queueing behaviour is essentially related to the clustered nature of busy slots patterns, a consequence at the DQDB level of the bursty nature of Ethernet traffic, which can be taken into account by means of the heavy tailness of the OFF distribution.

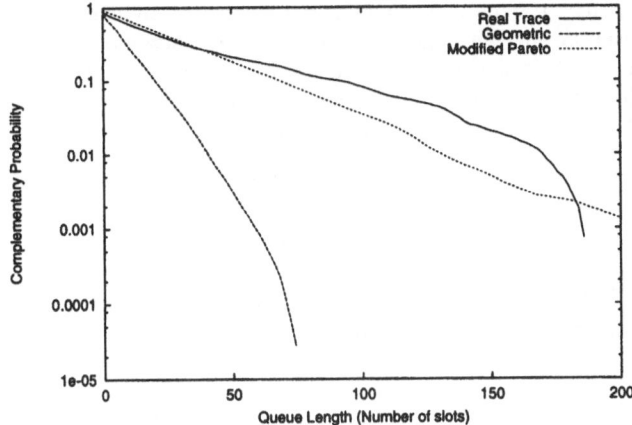

Figure 5.1 – Complementary Probability for Normalised Traffic Load 0.6

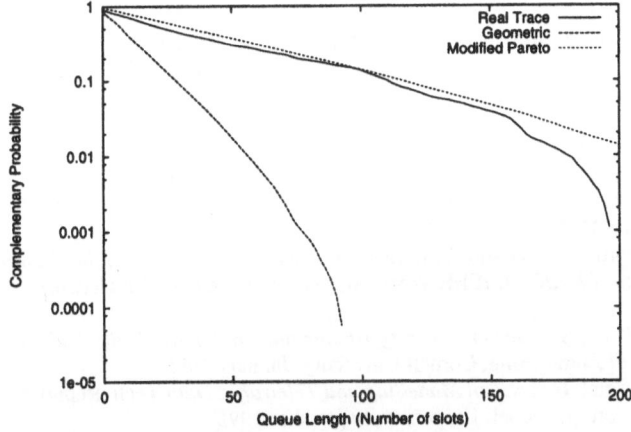

Figure 5.2 – Complementary Probability for Normalised Traffic Load 0.7

6. Conclusions

The analysis of DQDB slot level traffic highlighted the heavy tailness of the OFF periods length. Its impact on the queueing performance was evaluated by means of discrete event simulations. These studies have raised up the inadequacy of traffic models with light tailed distribution of the sojourn times. Hence, the main result of this paper is that elegant approaches based on traditional ON/OFF models leads to optimistic prediction of the actual system behaviour. Thus for the engineering purpose (i.e. the dimensioning of a network structure according to user needs), this elegant theory could be useless since it often carries to underdimensioning of network structures.

This paper provides a simple, but realistic, model for traffic sources characterisation in an operating cell-based network, focusing on LAN interconnection service, i.e. on the asynchronous service offered by the DQDB MAN. The main goal of the proposed model is to achieve an effective fitting of measured traffic statistics with a limited number of parameters.

Moreover the paper raises up the problem of developing new mathematical tools for realistic performance evaluation. In spite of this future theoretical developments, the proposed model is useful by itself in the field of discrete event simulations, since it permits to generate very long sequences and can be easily integrated in CAMAD simulations tools (like Alta Group's BONeS Designer [9]) for performance estimation of complex network systems.

References

[1] S. Giordano, G. Pierazzini, F. Russo *Multimedia experiments at the University of Pisa: from videoconferencing to random fractals* Proc. INET '95, Honolulu, June 1995

[2] M. Cinotti, E. Dalle Mese, S. Giordano, F. Russo *Long Range Dependence in Ethernet traffic offered to interconnected DQDB MANs* Proc. IEEE ICCS '94, Singapore, November 1994

[3] W. E. Leland, M. S. Taqqu, W. Willinger, D. V. Wilson *On the Self-Similar Nature of Ethernet Traffic (extended version)* IEEE Transaction on Networking, Vol. 2, n. 1, February 1994, pp.1-15

[4] IEEE 802.6 Standard *Distributed Queue Dual Bus Metropolitan Area Networks* December 1990

[5] R.G. Garroppo, S. Giordano, S. Miduri, M. Pagano, F. Russo *Statistical Multiplexing of Self-Similar VBR videoconferencing traffic* Proc. of IEEE GLOBECOM '97, Phoenix, November 1997

[6] M. Cinotti, S. Giordano, F. Romani, F. Russo *On the Self Similar Nature of the traffic offered to a MAN* 7th IEEE Workshop on LANs and MANs Networks, Florida, March 1995

[7] M.F. Kratz, S. I. Resnick *The QQ-Estimator and Heavy Tails* Tech. Report N. 1122, College of Engineering, Cornell University, January 1995

[8] S. I. Resnick *Heavy Tail Modelling and Teletraffic Data* Tech. Report N. 1134, College of Engineering, Cornell University, September 1995

[9] BONeS DESIGNER *Core Library Guide* and *User's Guide*, Alta Group, Comdisco System, Inc.

A Novel Control Scheme for VBR Video Transmission over ATM Networks

C. Perissinotto[1] G. Parladori[2] and G. A. Mian[3]

[1] CEFRIEL, via Fucini 2, 20133 Milano, Italy

[2] Alcatel-Telecom, via Trento 30, 20059 Vimercate (MI), Italy

[3] Dipartimento di Elettronica e Informatica, via Gradenigo 6, 35100 Padova, Italy

Abstract

The Asynchronous Transfer Mode (ATM) technology is reaching a considerable level of acceptance in many contexts. One of the key characteristics of ATM is the ability to provide statistical guarantees on performance. In other words, each user has to agree with the network on a set of parameters identifying the connection characteristics (e.g. Traffic Type, Rate, Quality of Service -QoS- parameters, etc.) which will be guaranteed by the network in case the call is accepted.

Real-time video applications are foreseen as one of the major users of B-ISDN. In particular the ability of the network to manage variable bit rate (VBR) traffic can be conveniently exploited by compressed video transmission systems. More precisely, with respect to the constant bit rate (CBR) video-codec, the quality of decoded signal can be kept more stable and statistical multiplexing scheme can be used in multi-channel video distribution systems for a more efficient bandwidth utilisation.

In this paper we propose a new scheme of VBR video coder, based on MPEG-2 technology, which has the ability to work at given target quality, under the traffic characteristics constraints agreed with the network.

1. Introduction

The great interest towards VBR video coding is based on two important facts. First the video quality can be kept quite constant compared with CBR transmission where the necessity of using a fixed number of bits for a constant number of frames may produce considerable variations of the quality; second, in multi-channel transmission, it is possible to take advantage of statistical multiplexing feature saving a considerable amount of bandwidth.

The ATM network offers VBR services providing at same time Quality of Service guarantees. In particular end-to-end delay and cell jitter characteristics are very critical for real-time services. In order to be able to guarantee the agreed quality of service, the ATM network implements two functions on the incoming traffic: the Connection Admission Control (CAC) and the Usage Parameter Control (UPC). The first is performed in order to ensure that the admission of a call will not compromise the QoS of existing connections. During the call setup the source establishes a service contract with the network, specifying the QoS required and a set of traffic descriptors, more specifically the Sustained Cell Rate (SCR), the Peak Cell Rate (PCR) and the Burst Length (BL), defined as the amount of time the source can transmit at the PCR. The second is performed during the connection's lifetime in order to verify the respect of the agreement.

In this paper we propose a VBR coding system that produces an MPEG-2 compatible bit stream and assures the respect of the agreement established with the ATM network, besides the best quality achievable with the available resources. The paper is organised as follows: in section 2 we study in depth the problem of VBR coding with constraints imposed by the ATM network. Section 3 contains a description of the system proposed, whereas the simulation results are presented in Section 4.

2. VBR Video Coding With Network Constraints

The problem of respecting the agreement established with the network is particularly relevant when the source is a VBR video coder: in this case the traffic has a very bursty character and is strictly dependent on input statistics, so that its characteristics change with time, whereas the traffic descriptors are static.

Another requirement imposed by real-time video applications is the proper scheduling of the output data that ensures a given end-to-end delay: the data for a video sequence are generated with a certain periodicity and have to be transmitted to the destination within a given time constraint to ensure that the decoder will have the information needed to decode and generate the pictures with the same periodicity as in the coder.

For a given network contract it is desiderable for the user to obtain the requested quality if the generated traffic meets the allocated network resources. On the contrary the system should provide the best possible quality in case the available resources are not enough. It is worth to note that in this work we have considered the PSNR as quality measure.

In order to satisfy all these constraints, it is necessary to control the traffic characteristics and the quality of decoded pictures directly at source. The proposed coding system produces an MPEG-2 compatible bit stream respecting the network and real-time constraints as well as maintaining the advantages of VBR coding.

The user specifies a PSNR target, $PSNR_t^o$, the maximum end-to-end delay and traffic descriptors like SCR, PCR and BL: if the resources allocated by the network allow to support the quality and the delay imposed, the scheme permits to obtain decoded pictures with PSNR very close to the target value. Otherwise, with the available resources the system is able to generate a traffic that respects the network and real-time constraints, assuring the best sustainable quality.

3. System Description

The block diagram of the system is shown in Figure 1: it is based on an MPEG-2 coder, whose output is fed into the transmission buffer. The Traffic Shaper

Module (TSM) takes data from the buffer and transmits them to the ATM network: the amount of data to be transmitted is determined by the Bit Rate Control Module (BRCM), which also controls the number of bits generated for each frame. The PSNR Control Module (PSNR-CM) controls the quality of the decoded pictures. The activity of the PSNR-CM and BRCM modules is co-ordinated by the Target Quality Control Module (TQCM), which ensures a consistent behaviour of the system.

The most important components of the scheme of Fig. 1 are the PSNR Control Module, the Bit Rate Control Module and the Target Quality Control Module that are described later on.

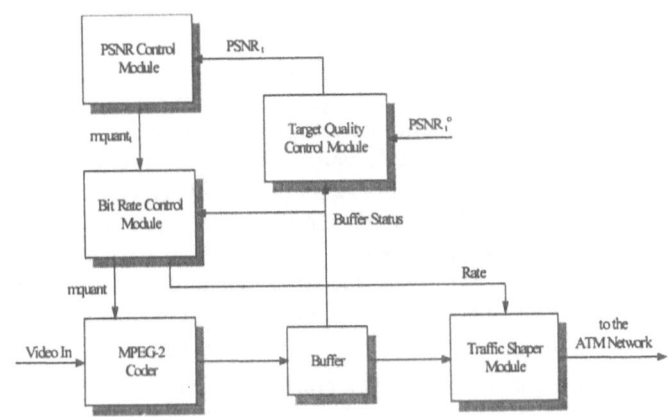

Fig. 1 Block diagram of the proposed coding system.

3.1 PSNR Control Module (PSNR-CM)

This module uses a statistical model of DCT coefficients, formerly applied to the bit rate control [Ref 5], in which these coefficients are modelled as vectors of independent random variables. The possibility of using this model in order to control the quality derives from the observation that the PSNR depends mainly on the quantisation step (hence on the parameter mquant that selects the set of quantisers used in the MPEG-2 coder).

By keeping mquant constant for all macroblocks of a frame, the achievable PSNR is easy predictable. In fact, if we know the probability density function (pdf) of the random variables that model the DCT coefficients, the quantisation error variance for each coefficient is given by

$$\sigma_i^2 = \sum_k \int_{I_k} (a - \hat{a}_k)^2 f_i(a) \, da , \quad (3.1)$$

where $f_i(a)$ is the pdf of the i-th coefficient, I_k the k-th quantisation interval and $\hat{a}_k = Q(a)$ is the associated quantisation level. Exploiting the DCT orthogonality, it's possible to estimate the reconstruction error variance

$$\sigma_e^2 = \frac{1}{M} \sum_{i=0}^{M-1} \sigma_i^2 , \qquad (3.2)$$

where M is the number of DCT coefficients, and then the PSNR

$$PSNR = 10 \log_{10}\left(\frac{255^2}{\sigma_e^2}\right). \qquad (3.3)$$

The DCT coefficients pdfs are estimated from the histograms of rounded DCT coefficients (correspondingly the integrals of (3.1) reduce to sums): the PSNR Control Module makes use of these histograms in order to select the quantiser which assures the PSNR closest to the target value. This computation may be done efficiently by means of a binary search on the mquant range, exploiting the fact that the PSNR decreases when mquant increases.

The PSNR is an objective measure of picture quality: its main drawback is the fact that not always its indications agree with the quality perceived by the observer. In order to improve the subjective quality of the pictures, we have introduced an adaptive quantisation, in which a given value of mquant is changed at each macroblock according to the local activity, as proposed in [Ref 1]. In this case the task of the PSNR-CM is to provide a reference mquant that, modulated in function of the local activity, allows one to obtain a PSNR close to the target value. If mquant varies at macroblock level, the model is not able to maintain a low prediction error, because the adaptive quantisation changes locally the difference $a - \hat{a}_k$ for the same a in equation (3.1). The problem has been solved modelling the effect of the adaptive quantisation with a corrective factor η that "weighs" the prediction given by the model: the factor η is computed from the last picture, and depends on the coding type (I, P or B picture) and on the mquant used as reference.

3.2 Bit Rate Control Module (BRCM)

The Bit Rate Control Module carries out two tasks. First it calculates the number of bits to be transmitted to the network in a time interval equal to the frame period, second it estimates an upper bound for the quantity of data that can be generated for each frame and guarantees that this limit is not exceeded.

For the first task has been adopted the strategy of transmitting the maximum number of bits compatible with the constraints: such strategy results in a more efficient use of the transmission channel, as shown in [Ref 4].

With regard to the second function, the upper bound to the number of data that can be generated for each frame is calculated as described in [Ref 2]: this limit is necessary in order to respect the constraints on traffic characteristics and on maximum end-to-end delay. We assume that the network uses the Leaky Bucket algorithm as traffic control mechanism, with the Bucket parameters chosen depending on the traffic descriptors.

The BRCM must also ensure that the upper bound is not exceeded: the rate control is performed at level of frame, slice and macroblock. At frame level the statistical model of DCT coefficients is exploited like described in [Ref 5], in order to predict the bit rate achievable with the mquant given by the PSNR-CM. If this prediction is compatible with the upper bound, the coding process begins with the mquant given by the PSNR-CM; on the contrary the BRCM searches the minimum mquant which is associated to a prediction lower than the limit and this new mquant is used like reference. At slice and macroblock level the mquant is updated depending on the number of bits still available for the current picture.

3.3 Target Quality Control Module (TQCM)

The PSNR-CM and the BRCM both act on mquant with opposite aims: to achieve a given quality level the first, to limit the number of generated bits the second. Without co-ordination there is the risk of loosing the possibility of maintaining more constant the picture quality.

For this reason has been introduced the TQCM, which task is to estimate the **best sustainable quality**, that is the best quality that the network resources allow to sustain with certain stability.

In order to verify if the available resources are adequate to the imposed PSNR, a Virtual Buffer has been defined: its contents, B^{virt}, is given by the sum of the transmission buffer contents and the Bucket contents, and its size is equal to

$$B^{virt}_{max} = SCR \times delay + LB_{max},$$

where $delay$ is the end-to-end delay and LB_{max} is the Bucket size. The term $SCR \times delay$ represents the Effective Buffer Size (EBS) [Ref 8], that is the maximum number of bits that can be stored in the transmission buffer without violating the real-time constraint. The EBS may be approximated with this product only if the Bucket is full: this approximation is allowed because in calculating the target we consider the Virtual Buffer just when the Bucket is full. When the Virtual Buffer is almost full, the present target is judged incompatible with the network agreement and must be updated. In order to avoid occasional variations of the target, the change of the target quality is based on the average Virtual Buffer contents, calculated with a first order IIR smoothing filter.

The PSNR obtained in the last frame is used to evaluate the actual average quality of the previous frame, $PSNR_{avg}$, given by another IIR smoother.

The Target Quality Control Module is a finite state machine with four states, each characterised by a different control law for the PSNR target. Fig. 2 shows the transition state diagram of the TQCM.

At the beginning it lies in the PSNR_CTRL status and the target is maintained at the level imposed by the user, $PSNR_t = PSNR_t^o$. The transition from the PSNR_CTRL status to the CONGESTION status occurs when the average Virtual Buffer contents exceeds the 65% of its capacity (this condition always forces the TQCM into the CONGESTION status). In the CONGESTION status the target PSNR is given by the equation

$$PSNR_t = \left(\frac{B^{virt}}{B^{virt}_{max}} \right) \times PSNR_{avg} + \left(1 - \frac{B^{virt}}{B^{virt}_{max}} \right) \times PSNR_t^o. \quad (3.6)$$

With a relation of this kind the contribution of the user target increases when the average contents of virtual buffer decreases, whereas, when B^{virt} approaches

B_{max}^{virt} , increases the contribution of $PSNR_{avg}$.

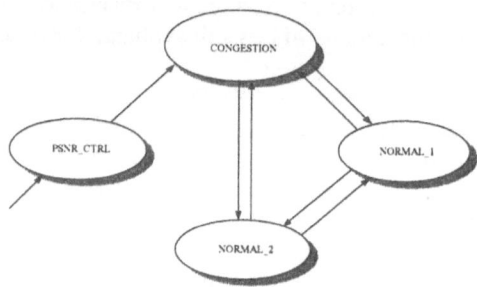

Fig. 2 Transition diagram for the Target Quality Control Module.

In order to make easier the convergence towards the best sustainable quality, two other states, NORMAL_1 and NORMAL_2, have been introduced: the transition to the NORMAL_1 status occurs when the transmission buffer becomes empty and the Bucket is not full, while the transition to the NORMAL_2 status occurs if the transmission buffer empties and the Bucket remains full. When the TQCM is in the NORMAL_1 status, the target PSNR is obtained adding to $PSNR_{avg}$ another term, positive or negative depending on the Bucket dynamics: in particular, if the Bucket is emptying, the term is positive and equal to

$$\left(1 - \frac{LB}{LB_{max}}\right) \times PSNR_{trsh}, \quad (3.7)$$

where LB are the Bucket contents and $PSNR_{trsh}$ is a constant fixed to 0.04 dB. If the Bucket is filling, the term is negative and equal in module to

$$\left(\frac{LB}{LB_{max}}\right) \times PSNR_{trsh}. \quad (3.8)$$

Also in the NORMAL_2 status the PSNR target is obtained adding to $PSNR_{avg}$ another term, positive or negative depending on the transmission buffer dynamics. The absolute value of this term is calculated with equations analogous to the (3.7) and (3.8) with the only difference that the transmission buffer contents (and not the Bucket contents) is considered and a constant factor of 0.02 dB is used.

4. Simulation Results

Fig. 3 shows the average absolute PSNR prediction error vs the target PSNR for the test sequence Basket when only the PSNR Control Module is active: from this graph we may appreciate the great accuracy of the estimate, being the error lower than 0.12 dB both with adaptive both with non adaptive quantisation. Fig. 4 gives the PSNR as a function of the frame number, the number of bits generated and that of the bits transmitted for the same sequence without the TQCM: in this case the PSNR is highly variable and the number of generated bits seems to oscillate around the SCR. Fig. 5 shows the same data when the TQCM too is active: the target quality is adjusted and the PSNR results to be more constant.

The advantages achievable with the TQCM are evident from Fig. 6 and Fig. 7: when this module is active, the average PSNR is higher and the PSNR standard deviation lower than the one obtained without target control. Finally Fig. 8 shows the PSNR obtained when the network resources are adequate to support the quality desired by the user: in this case the BRCM is idle and the PSNR-CM is able to maintain the PSNR close to the target quality for most of the sequence.

4. Conclusions

In this paper the problem of the VBR video coding in an ATM context has been considered and a coding system which is able to generate a traffic that respects the network and real-time constraints has been proposed. The system is able to assure a given quality level if the network resources are sufficient; on the contrary it is able to achieve the best quality sustainable with the available resources.

Given the described characteristics, the system is suitable for high demanding professional real-time video applications.

References

1. Standard Draft, MPEG Video Committee Draft, MPEG 92/328, Nov. 1992.
2. A. R. Reibman, B. G. Haskell, 'Constraints on Variable Bit-Rate Video for ATM Networks', IEEE Trans. On Circuits and System for Video Technology, Vol. 2, No. 4, pp. 361-371, Dec. 1992.
3. P. Pancha, M. El Zarki, 'MPEG Coding For Variable Bit Rate Video Transmission', IEEE Communications Magazine, pp. 54-66, May 1994.
4. K. Joseph, D. Reininger, 'Source Traffic Smoothing and ATM Network Interfaces for VBR MPEG Video Encoders', Proc GLOBECOM 95, Singapore, Nov. 1995, pp. 1761-1767, 1995.
5. S. Bilato, G. Calvagno, G. A. Mian, R. Rinaldo, 'Accurate Bit-Rate and Quality Control for the MPEG Video Coder', Proc ICIP 97, Santa Barbara, Oct. 1997, pp. 571-574, 1997.
6. J. Chen, D. W. Lin, 'Optimal Bit Allocation for Coding of Video Signals over ATM Networks', IEEE Journal on Selected Areas in Communications, Vol. 15, No. 6, pp. 1002-1015, Aug. 1997.
7. C. Hsu, A. Ortega, A. R. Reibman, 'Joint Selection of Source and Channel Rate for VBR Video Transmission Under ATM Policing Constraints', IEEE Journal on Selected Areas in Communications, Vol. 15, No. 6, pp. 1016-1028, Aug. 1997.
8. W. Luo, M. El Zarki, 'Quality Control for VBR Video over ATM Networks', IEEE Journal on Selected Areas in Communications, Vol. 15, No. 6, pp. 1029-1039, Aug. 1997.

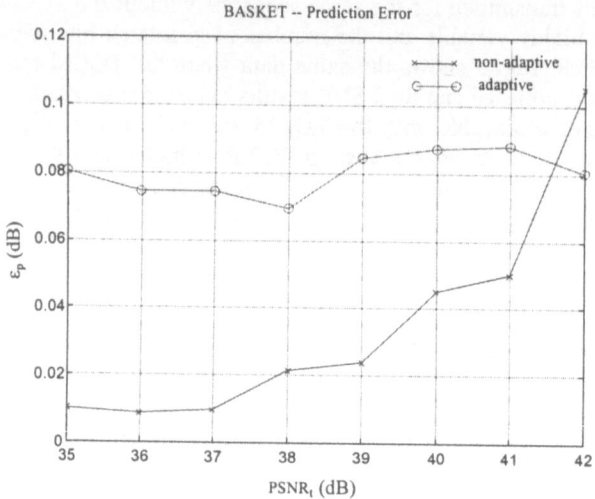

Fig. 3 Average absolute PSNR prediction error as function of the target PSNR.

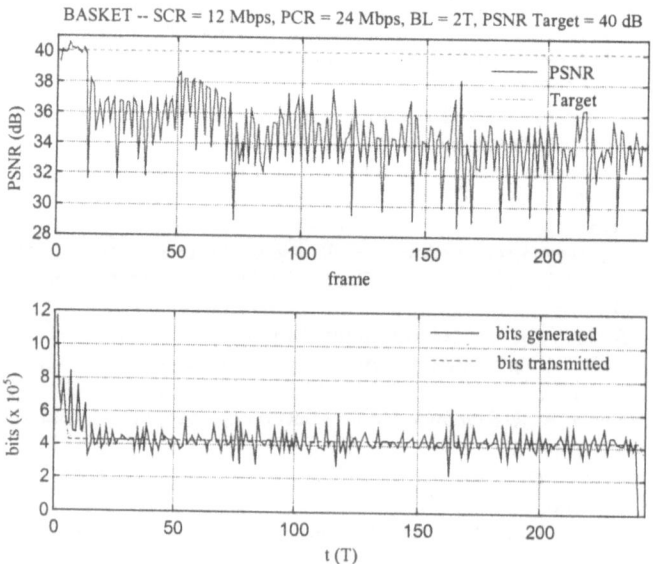

Fig. 4 PSNR and rate traces without Target Quality Control Module.

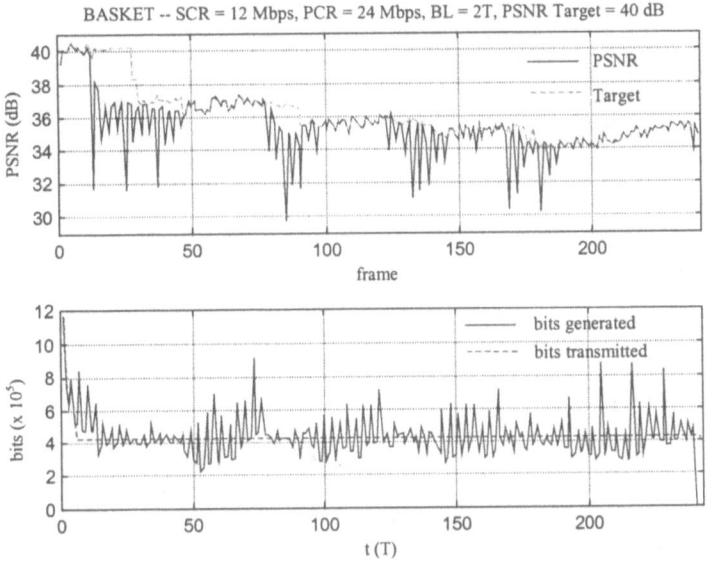

Fig. 5 PSNR and rate traces with Target Quality Control Module.

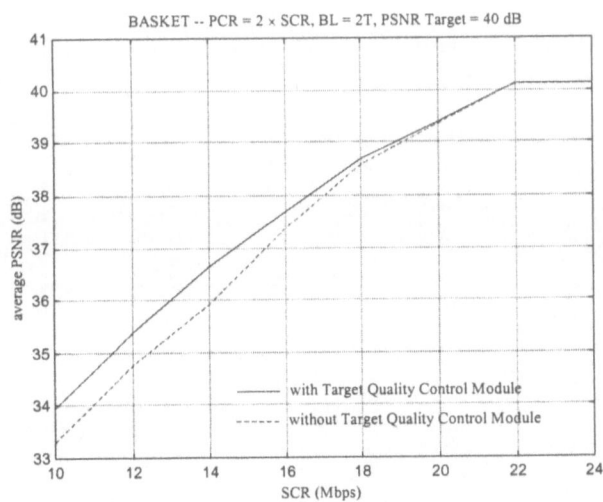

Fig. 6 Average PSNR with and without Target Quality Control Module.

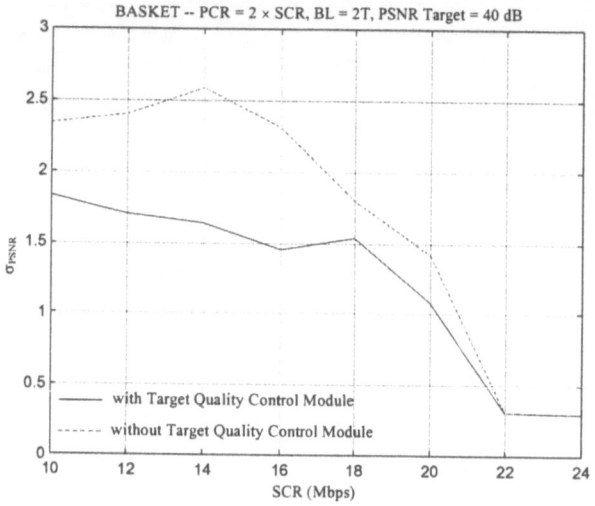

Fig. 7 PSNR standard deviation with and without Target Quality Control Module.

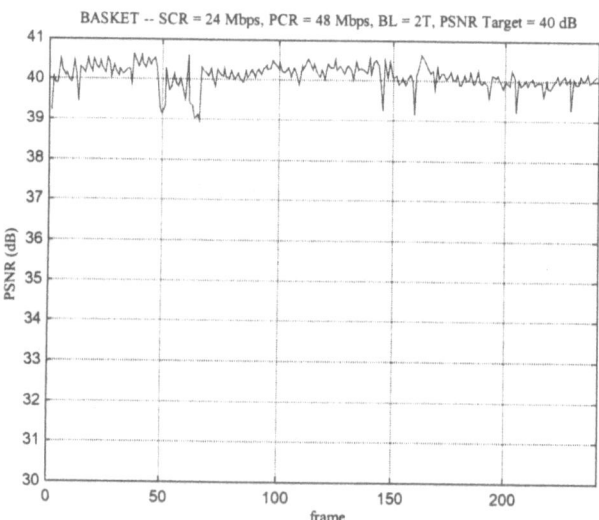

Fig. 8 PSNR trace with network resources consistent with the desired PSNR.

Part 3

Access Techniques

Passive Optical Networks: Results from project FSAN

OKADA Kenji
NTT Access Network Systems Laboratories

1. FSAN Initiative

Each telecom operator has different network architecture, system configuration, or deployment plan in access networks because each of them has different geographical conditions, development strategy, or service requirements. However, every telecom operators target the cost reduction of optical access system. In 1995, FSAN was established to have common system specifications in order to reduce the cost of optical access system for early deployment.

2. FTTx

FSAN selected ATM-PON system as the study item of Access. FSAN aims at finding cost-effective solutions by defining a multi-vendor interface in the ATM-PON system so as to obtain maximum world-wide volume as products. Service requirements by operators and technologies by suppliers have been harmonized to specify the interface between the ONU and the OLT of the ATM-PON system. The concept of FTTx is produced to overcome the diversity in geographical circumstances and the existing infrastructure.

3. ATM-PON System

There are three levels to have common specifications of the ATM-PON system; those are the physical media dependent layer, the Transmission Convergence layer, and the management channel layer. The physical media dependent layer is to specify optical characteristics so as to have commonality in optical devices. The Transmission Convergence layer including ranging protocol is to specify transmission protocol so as to have commonality in transmission processing devices, and the management channel layer is to specify management and control between the ONU and the OLT. At this stage OAN-WG has achieved the specifications up to TC layer, and the management layer is under discussion at present.

4. Physical Layer

Bi-directional transmission is economically accomplished by use of the WDM technique of 1.3 um region and 1.5 um region wavelengths on a single fiber. There are also 2-fiber solution by using 1310 nm region wavelength, however 1-fiber WDM transmission is strongly recommend. The ATM-PON systems have two options. One is the nominal symmetric line rates of 155.52 Mbit/s and another is asymmetrical line rates of 155.52 Mbit/s upstream and of 622.08 Mbit/s downstream.

5. ODN Requirements

Classification of optical path loss is consistent with Class B and Class C specified in Rec. G.982. Two classes of attenuation ranges are selected for the ATM-PON system.

The operators prefer to the optical path loss of 10-30 dB to cover both ranges of Class B and C by a single system; however, the ranges were split into two classes due to cost-effective solution of optical devices. The maximum differential optical path loss means the difference between the highest optical path loss and the lowest one in the same ODN. The maximum differential optical path loss is 15 dB to meet the requirement defined in ITU-T Rec G.982.

6. TC Layer

The upstream cell contains 3 byte overhead besides the ATM cell or the PLOAM cell.

"Guard time" provides enough distance between two consecutive cells or mini-slots to avoid collisions, which length is 4 bits at least. "Preamble" extracts the phase of the arriving cell or mini-slot relative to the local timing of the OLT, and/or acquire bit synchronization and amplitude recovery. "Delimiter" is a unique pattern indicating the start of the ATM cell or mini-slot, which can be used to perform byte synchronization. The boundaries of the overhead are not fixed to allow the vendors to use their proprietary receiver implementation. The OLT will set up the length of guard time, preamble and delimiter within 3 bytes through the TC layer OAM channel when the communication was initiated by the OLT.

7. Standardization Activities

All FSAN specifications discussed in the OAN-WG have been input to ITU-T, ETSI and ATM Forum by FSAN members on behalf of the FSAN group and discussed in these bodies. SG15 of ITU-T has a responsibility to study optical access systems for local access networks and studies the ATM-PON systems as a new draft Recommendation G.PONB. In the last SG15 meeting in February, the proposed ATM-PON specifications by the FSAN have been accepted as it is and agreed to be determined as a new Recommendation G.983, and will be made decision in the next SG15 meetings in October as a formal ITU-T Recommendation. The FSAN agreed ATM based SNIs are being studied in the SG13, and G.967.1 and G967.2 on VB5 interface are expected to be decided as Recommendation by June in 1998 and be April in 1999 respectively. TM3 of ETSI studies the ATM-PON system and its functional architecture. Every specification has been input by one of OAN-WG member as same as ITU-T, and the discussion or questions to FSAN OAN-WG from ETSI have been feed back. FSAN specifications have been accepted in ETSI as it is. Residential Broad-Band (RBB) group of ATM Forum studies the ATM-PON system as one of the access system. Formerly FSAN specifications were proposed to the RBB and agreed to include its baseline document. Now the ATM Forum is going to only refer to ITU-T Recommendation G.PONB (G.983) in its specification on the ATM-PON systems to avoid duplicated maintenance of the specifications.

8. Multi-Vendor Interface

After completing the unique solution for ATM-PON, Multi-vendor interface will be completed in a few month because OAN-WG are studying specifications on management channel and have a plan to input it to ITU-T as soon as possible. Standardized ATM-PON system will be available for any operators from any vendors in a year.

9. Workshop

FSAN management committee decided to have FSAN conference to inform the study result and gain the comment or feedback from the market. The first conference was held in 1996. The second one was in 1997. The third one was in last March. Next conference will be held as a session of Globecom access network mini-conference in November in Sydney.

296

FSAN Initiative

Motivation

(1) Each telecom operator has different network architecture, system configuration, or deployment plan in access networks due to different geographical conditions, development strategy, or service requirements.
(2) Every telecom operators target the cost reduction of optical access systems.

Establishment

(1) In 1995, consortium named G7 was established to have common system specifications in order to reduce the cost of optical access system for early deployment.
(2) ATM-PON system has been studied as common requirement.

 NTT

Members of FSAN Gx

(1) Members has their own access networks as operators.
(2) Members has to exhibit their requirements to access networks.

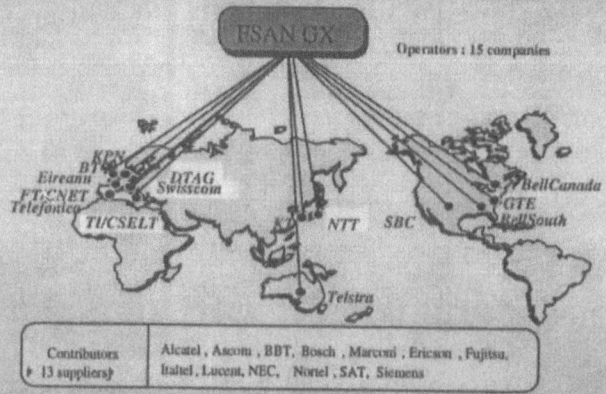

| Contributors (13 suppliers) | Alcatel , Ascom , BBT, Bosch , Marconi , Ericson , Fujitsu, Italtel, Lucent, NEC, Nortel , SAT, Siemens |

NTT

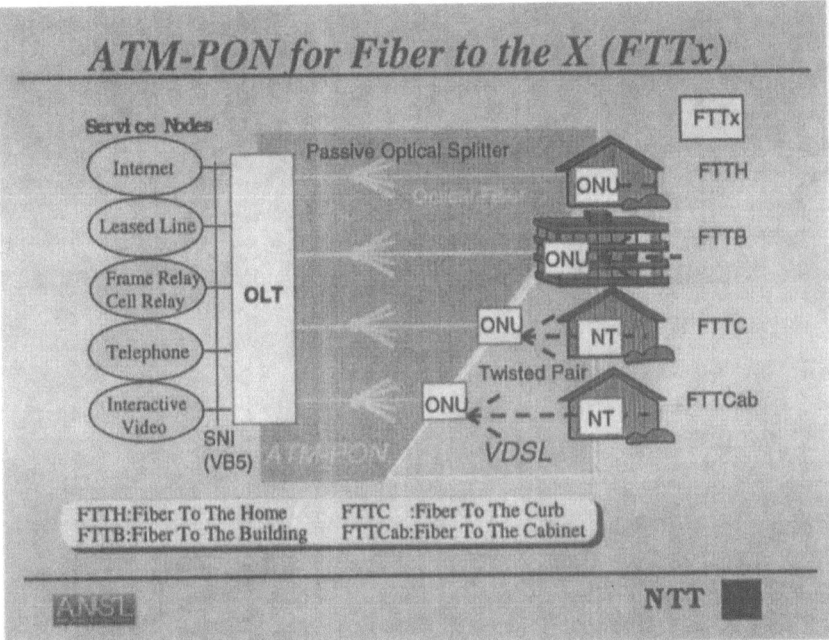

ATM-PON for Fiber to the X (FTTx)

FTTH:Fiber To The Home   FTTC :Fiber To The Curb
FTTB:Fiber To The Building   FTTCab:Fiber To The Cabinet

NTT

Existing Standard Bodies

Defect of Existing Standard Bodies

(1) There are many stamp standards.
(2) It takes long term to agree to standardize.

Defact Standard Activities in OAN-WG

(1) OAN-WG aims at unique and best solution.
(2) OAN-WG completes specifications in a short time.

NTT

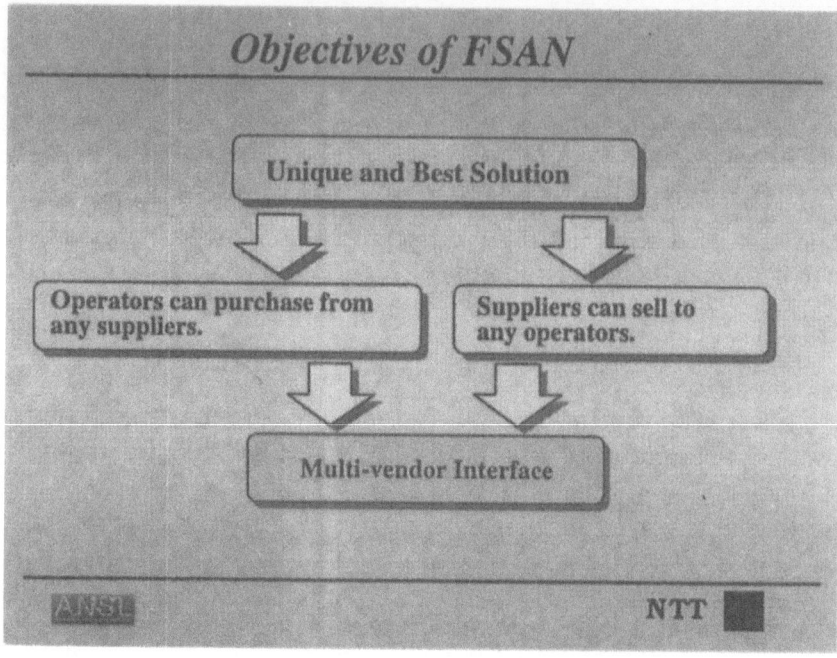

Objectives of FSAN

Unique and Best Solution

Operators can purchase from any suppliers.

Suppliers can sell to any operators.

Multi-vendor Interface

ANSL   NTT

Activity Policy in FSAN

(1) Open to world operators

- Access network operator

- Development activity in access network area

(2) Defacto standard

- Input to ITU-T, ETSI, ATM-F

(3) Global design center for the access network

- To design the best and unique solution

ANSL   NTT

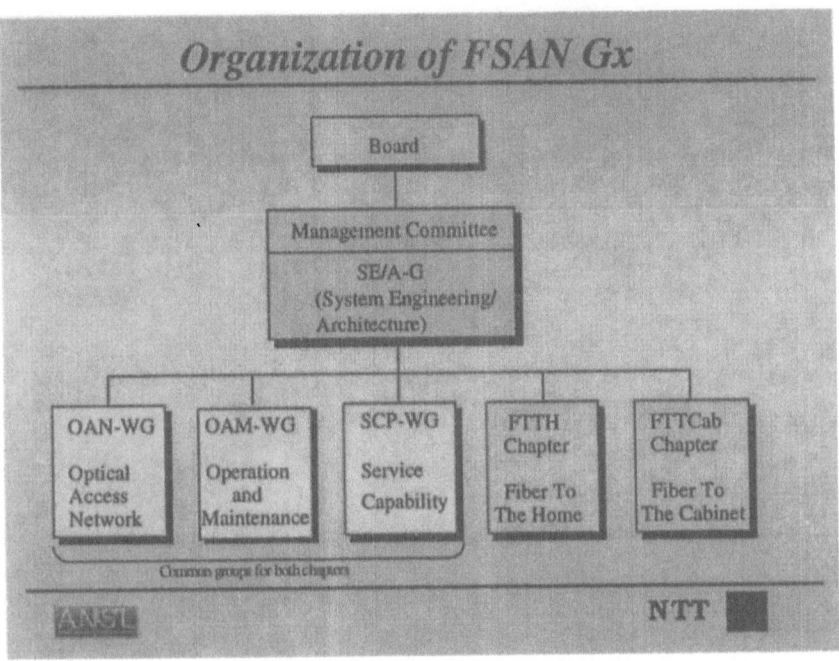

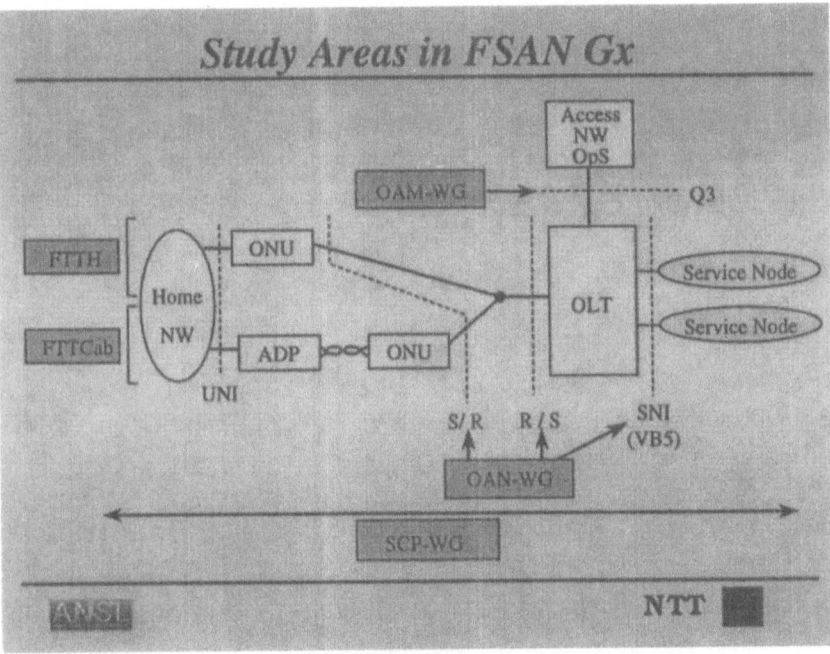

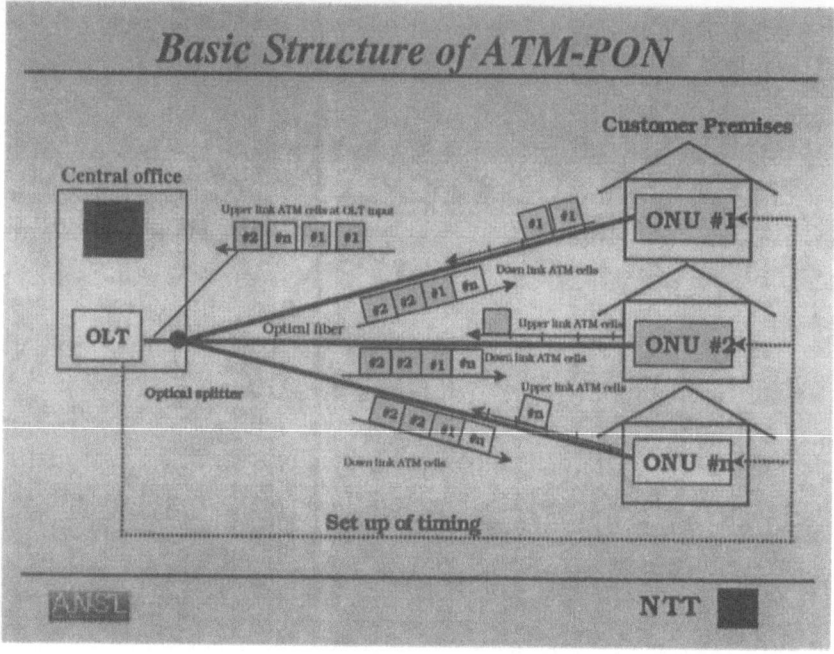

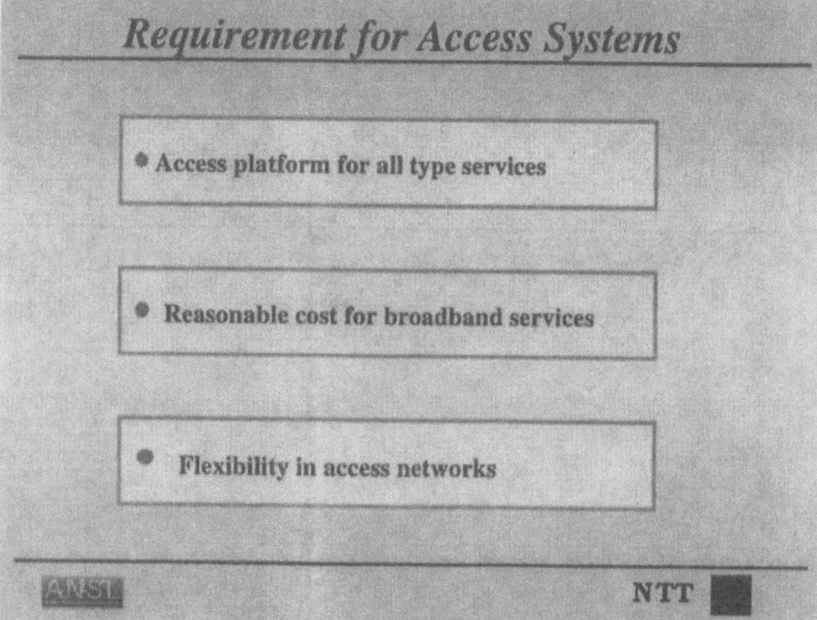

Specific interfaces for UNI

Service	UNI point to point	UNI point to multipoint
IP router	Ethernet 10BASE-T, ATM(25M), ATM(155M)	To be defined
ATM VC switch	ATM(25M)	To be defined
VOD	ATM(25M)	To be defined
Switched Video Broadcasting	ATM(25M)	To be defined
VP leased line	ATM(155M, ..., 25M)	
ISDN	I.430, I.431	To be defined

  NTT

Specific interfaces for SNI

Service Node	SNI
IP router	VB5.1
ATM VC switch	VB5.1, VB5.2
VOD	VB5.1, VB5.2
Switched Video Broadcasting	VB5.1
VP leased line	VB5.1
ISDN	V5, TR303 etc., VB5.1, VB5.2

 NTT

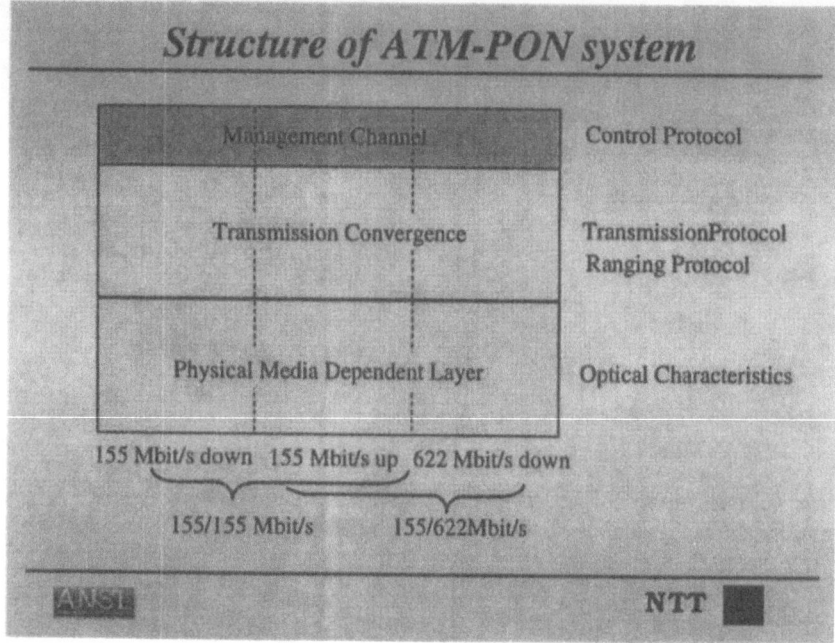

Structure of ATM-PON system

Management Channel	Control Protocol
Transmission Convergence	TransmissionProtocol Ranging Protocol
Physical Media Dependent Layer	Optical Characteristics

155 Mbit/s down   155 Mbit/s up   622 Mbit/s down

155/155 Mbit/s   155/622Mbit/s

ANSI   NTT

The ODN requirements

Items	Unit	Specification
Fiber type	-	ITU-T Rec. G.652
Attenuation range (G.982)	dB	Class B: 10 - 25 Class C: 15 - 30
Differential optical path loss	dB	15
Maximum optical path penalty	dB	1
Maximum differential logical reach	km	20
Maximum fiber distance between S/R and R/S points	km	20
Minimum supported split ratio	-	Restricted by path loss and ONU addressing limits. PON with passive splitters (16 or 32 way split)
Bi-directional transmission	-	1-fiber WDM or 2-fiber

ANSI   NTT

Requirements to physical media dependent layer

Items	Specification
Transmission line bit rate	Symetric 155.52 Mbit/s Asymetric 155.52 Mbit/s downstream 622.08 Mbist/s upstream
Transmission methodology Operating wavelength	WDM 1.3 £ m / 1.5 £ m Two fiber 1.3 £ m
Optical path loss	10 - 25 dB : class B 15 - 30 dB : class C
Differential optical path loss	15 dB
Optical level	see "example"

NTT

Structure of upstream cell

3 byte overhead

Guard time	Preamble	Delimiter	ATM cell or PLOAM cell

          NTT

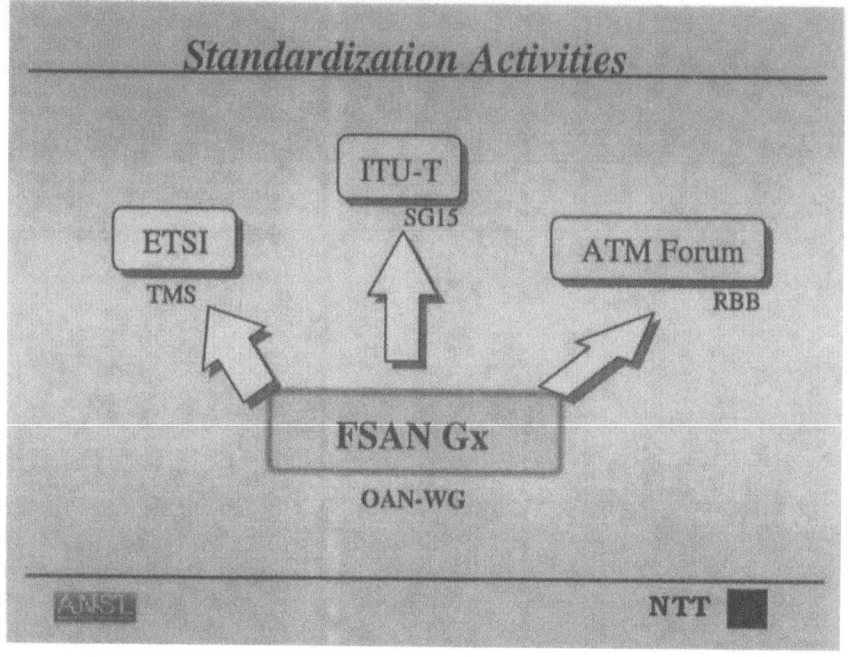

FSAN Conference

- **First Conference** June 1996 London

- **Second Conference** March 1997 Atlanta
  In conjunction with OAN Workshop

- **Third Conference** March 1998 Venice
  In conjunction with ISSLS '98

- **Fourth Conference** November 1998 Sydney
  In conjunction with OAN Workshop

http://www.labs.bt.com/profsoc/access/index.html

NTT

Resource Allocation Strategies for Multimedia Traffic in an ATM-based PON

Raffaele Bolla, Franco Davoli and Stefano Ricciardi
Department of Communications, Computer and Systems Science (DIST)
University of Genoa
Via Opera Pia 13, I-16145 Genova, Italy
e-mail: (lelus, franco, steric)@dist.unige.it

Abstract.
The provision of multimedia services to a large customer base is considered, over an ATM-based Passive Optical Network (PON). A distributed and hierarchical decision making architecture is defined, based on the physical system organization, where the Optical Access Units (ONU) and the Optical Line Termination (OLT) operate jointly, in order to make the best possible use of the transport capacity of the access network, while at the same time maintaining the Quality of Service (QoS) requirements of different users and service classes. In this framework, bandwidth allocation and admission control functions, in both directions of flow, are investigated, with the aim of defining management and control laws and algorithms that achieve the above goals. At the same time, the control architecture also reflects the multilayer hierarchy introduced by the presence of multiple teletraffic time scales, by essentially decoupling the problem of ensuring QoS at the cell-level, at least for connection-oriented traffic (cell loss probability, cell transfer delay). The models, the cost functions and the control strategies used in the parametric optimization are described, and numerical results are reported and discussed.

1. Introduction

The availability of a broadband integrated services transport network, as the one based on the Asynchronous Transfer Mode (ATM), is only one of the conditions needed to provide multimedia telecommunication services to business and residential customers. As a matter of fact, an access network that allows ubiquitous distribution of multimedia services to an enormous number of potential users is an equally important and essential requirement, whose deployment, involving huge investments, is currently a major interest of most operators. In this context, several options are being investigated [1-5]; among others, an international initiative stemming from operators and manufacturers has lead to the definition of a basic set of common requirements, presented in the Full Services Access Network (FSAN) Requirements Specification [6]. Among the main contributions, the reference architecture for supporting ATM over a Passive Optical Network (PON) is defined, its basic functional blocks are identified, and Transport, physical requirements, signaling and control and Operations, Administration and Maintenance (OAM) aspects are discussed. Key architectural elements in this structure are the Optical Line Termination (OLT) and the Optical Network Units (OLUs), where management and control of the downstream and upstream ATM cell flows is implemented.

In this context, we investigate in this paper bandwidth allocation and Call Admission Control (CAC) functions, in both directions of flow, in the presence of different ATM service classes, with the goal of defining management and control laws, aimed at maintaining desired grades of Quality of Service (QoS), at the cell level and above. More specifically, as regards traffic modeling, a mix of CBR, VBR and ABR/UBR service classes will be considered, with different source models, along with their statistical description at the cell-level. The latter will be used to derive the region in the space of connections of the connection-oriented classes, within which cell-level QoS is satisfied (Service Separation with Dynamic Partitions [7]). Within this region, the dynamics of connection-oriented traffic will be described in terms of a multidimensional Markov chain, whereas connectionless traffic (e.g., "short-lived" IP flows) will be modeled by means of "long-tail" distributions, with the generated cells entering a finite buffer with deterministic service. As regards the control architecture, a hierarchical structure will be defined, similarly to the one used in the context of hybrid TDM in [8], where each service class is served by an independent controller, parametrized by the bandwidth allocated to it, and a central agent, playing the role of a coordinator in the hierarchical control scheme, adjusts the bandwidth partitions, trying to minimize a global cost that reflects other QoS requirements (call blocking probability, connectionless data loss). The control actions will be considered both on the downstream and upstream flows. In the latter case, the coordinator will be supposed to reside in the OLT, whereas the ONUs will host local controllers, dedicated by user and service class, acting on the basis of local information only.

The paper is organized as follows. We outline our model and the cell-level requirements in the next section. The third section is dedicated to the admission control level, and to the definition of the cost function of the bandwidth allocation level, in the case of a "geographically" centralized control, exerted by the OLT, where each agent of the lower hierarchical level is devoted to a specific service class. In Section 4, we extend this model to the case of geographically decentralized controllers, each residing in an ONU, where the above mentioned two-level hierarchy is replicated, and a third level is introduced (at the OLT), to coordinate their actions, by allocating quasi-static bandwidth partitions to the ONUs, as needed. Section 5 reports some numerical results and Section 6 contains the conclusions.

2. Traffic modeling

We suppose the traffic in the network to be divided into $H+1$ classes. Of these, H are either CBR or bursty VBR (on-off) sources, characterized by statistical parameters like peak rate, average transmission rate and average burst length, as well as by QoS requirements, like cell loss probability and cell delay. We indicate with $B^{(h)}$ [cells], $P^{(h)}$ [bits/s], $M^{(h)}$ [bits/s] and $b^{(h)} = P^{(h)}/M^{(h)}$, the average burst length, the peak bit rate, the average bit rate and the burstiness, respectively, of a source of the h-th class, h=1,...,H (obviously, CBR sources are included in this description, with $b^{(h)}=1$). We let $\lambda^{(h)}$ and $1/\mu^{(h)}$ represent the average arrival rate and the average duration of connections of class h, respectively, and $\rho^{(h)} = \lambda^{(h)}/\mu^{(h)}$.

Moreover, we suppose to have an asynchronous packet flow, which represents *connectionless* traffic; this flow is supposed to originate from the superposition of a number of on-off sources, whose sojourn time Y (in slots) in the active state follows a Pareto distribution, i.e.,

$$\Pr\{Y = y\} = cy^{-(\alpha+1)}, \quad 1 < \alpha < 2, \quad y \geq 1 \tag{1}$$

where c is the normalization constant and α is a parameter. The Pareto distribution is well-known for its "heavy-tail" property, and has actually been used to model self-similar traffic [9-11]; more specifically, the aggregation of a large number of sources of the above mentioned type has been shown to give rise to self-similar traffic [12]. The packets (after segmentation into ATM cells) receive a variable rate service (ABR or even UBR). We model the queueing of cells generated in this way as a $Z/D/C_{asy}/Q^{(H+1)}$ system, where Z is the aggregated self-similar process, C_{asy} the capacity available for the connectionless traffic, and $Q^{(H+1)}$ [cells] is the dimension of the buffer dedicated to it. In the following, the upper bound on the overflow probability derived in [12] will be used.

At each ATM multiplexer, traffic class h, h=1,...,H, is assigned a separate buffer of length $Q^{(h)}$ [cells], whose output is statistically multiplexed on the outgoing link by a scheduler, which substantially divides the part of channel capacity assigned to connection-oriented traffic C_{co} [bits/s] into "virtual" partitions $C^{(h)}$ among the classes, whose sum amounts to C_{co}. The simplest way to maintain the partitions is by serving the buffer in a Weighted Round Robin fashion; other possibilities include the assignment of a slot (time to transmit a cell) to a cell of class h randomly, with probability $\Omega^{(h)} = C^{(h)}/C_{co}$ [13], or the use of a technique like Generalized Processor Sharing [14]. The overall queueing system at the cell level is depicted in Fig. 1.

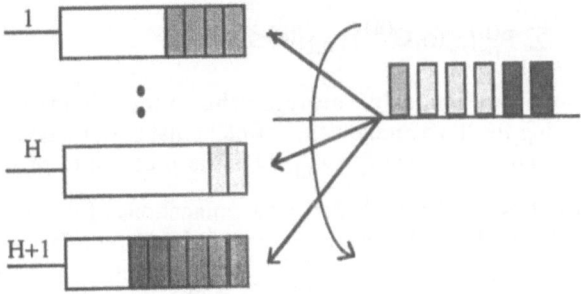

Fig. 1. Cell-level multiplexing under Service Separation.

Connection requests are also processed on a per-class basis. Given a model for the traffic sources of a class, the cell-level performance requirements for this traffic type (e.g., in terms of average cell loss and delayed cell rate) allow to define a region in call-space (which will be referred to as "Feasibility Region", or FR), where they are certainly satisfied. As we mentioned in the introduction, this region corresponds to the CAC method named "service separation with dynamic partitions" in [7, p. 147].

Clearly, the points on the boundary of the FR correspond to the maximum numbers of Virtual Circuit (VC) connections $\left[N_{max}^{(1)},...,N_{max}^{(H)} \right]$ that are compatible with the given cell-level QoS constraints. We can associate each $N_{max}^{(h)}$ with the minimum amount of bandwidth $C_{min}^{(h)}$ that is necessary to support that number of connections with the given QoS guarantees.

The computation of the FR has been the object of several studies and can be effected in different ways, either by analysis, given a model of the traffic sources, or by simulation. A more general view includes the characteristics and the optimization of the scheduling process [15], through the related concept of "Schedulable Region". Using any approach based on equivalent bandwidth (involving, in our Service Separation context, only homogeneous sources) yields a straightforward boundary of the FR. In any case, it is worth noting that, in the context of the methods to be considered in the next section, the FR itself will be just a tool to describe the CAC schemes. The specific technique to ensure QoS satisfaction at the cell-level might be changed (always within the framework of Service Separation), without affecting the access control general procedure.

However, to fix ideas, we refer here for the computation of the FR to the model we have used in [16] and in previous works (see [17] and the references therein), where maximum thresholds are set for the averaged values of the cell loss rate ($P_{loss}^{(h)}(n, C^{(h)})$) and of the delayed cell rate ($P_{delay}^{(h)}(n, C^{(h)})$), with n calls in the active state out of $N^{(h)}$ accepted calls and a bandwidth $C^{(h)}$ assigned to traffic class h ; more specifically

$$\sum_{n=1}^{N^{(h)}} P_{loss}^{(h)}(n, C^{(h)}) v_{n,N^{(h)}} \leq \epsilon^{(h)} \tag{2}$$

$$\sum_{n=1}^{N^{(h)}} P_{delay}^{(h)}(n, C^{(h)}) v_{n,N^{(h)}} \leq \delta^{(h)} \tag{3}$$

where $\epsilon^{(h)}$ is an upper limit on the average value of the cell loss rate and $\delta^{(h)}$ has the same meaning for the average value of cells that suffer a delay longer than a fixed upper bound. The quantity $v_{n,N^{(h)}}$ is the probability of having n active connections of class h, with $N^{(h)}$ accepted connections of the same class, which is given by a binomial distribution; an Interrupted Bernoulli Process (IBP) [18] is used to model the state of a call, from which $P_{loss}^{(h)}(n, C^{(h)})$ and $P_{delay}^{(h)}(n, C^{(h)})$ are derived. $N_{max}^{(h)}$ and $C_{min}^{(h)}$, h=1,..., H, can then be computed from (2) and (3).

As an alternative approach, any one based on equivalent bandwidth (e.g., [19]) could be used, yielding similar results (see [17]). A possible shape of the FR for two VBR traffic classes is shown in Fig. 2, where the dotted lines represent different boundaries of the region, corresponding to different values of the capacity C_{CO} actually assigned to the connection oriented traffic.

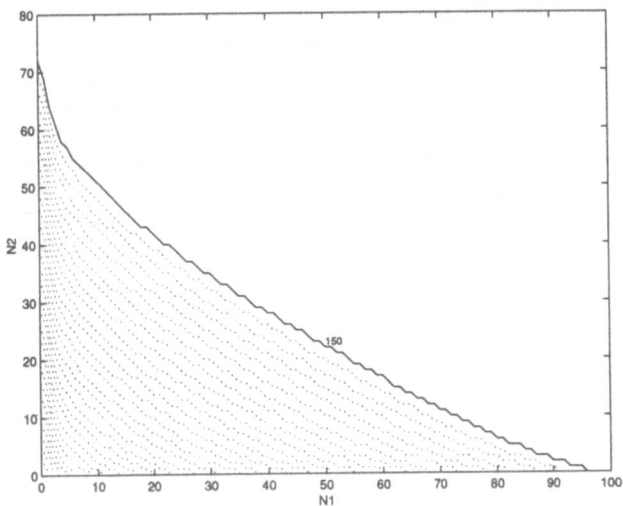

Fig. 2. Feasibility Regions for two connection-oriented traffic classes (boundaries parametrized by the assigned capacity C_{co}; the traffic characteristics are those given in Section 5).

We let

$$N_A(k) = col\left[N_A^{(h)}(k), \; h = 1,...,H\right] \qquad (4)$$

where $N_A^{(h)}(k)$ is the number of connections in progress at the generic instant (slot) k for class h. The vector in (4) represents the state of the system at instant k (the VC-profile in [7]).

3. Call Admission Control and capacity allocation - I: centralized case

At this point, we are ready to consider QoS performance measures and constraints of connection-oriented traffic at the call level, having essentially decoupled this problem from the lower level one for this traffic type (at the same time, we will take into account the presence of the connectionless data flow). We first do this under the assumption that the entire control architecture resides in the OLT, even though maintaining a hierarchical structure on a per-class basis. In particular, in doing so, we consider the servicing of ATM cells to be controlled only by the OLT, which assigns bandwidths to service classes, irrespectively of the ONU they originate from. In the model to be described in Section 4, we consider explicitly the assignment of bandwidth to the individual ONUs, which, in turn, take separate and independent bandwidth assignment, scheduling and CAC decisions on the traffic pertaining to them.

Moreover, in the present section there is no difference between the two directions of flow, the control is completely symmetric and we refer to a single direction.

Obviously, a possible way of performing the CAC operation for connection-oriented traffic is that of accepting an incoming call as long as the VC-profile is within the FR. This, however, does not allow to take into account specific call-level requirements. Again, there are several methods that can be used, and [7, ch. 4] contains an excellent overview. In line with our previous work in the ATM context [17], we have chosen to adopt admission control policies for our connection-oriented traffic that belong to the class of Complete Partitioning (CP) ones; such policies define a "rectangular" sub-region within the FR, by dividing the available capacity among the classes in a "static" way. In other words, given a point $N_{max}^* = col\left[N_{max}^{*,(h)}, h=1,...,H\right]$ on the boundary of a specific FR (as defined in the previous section), a new connection of class h at time k is accepted only if

$$N_A^{(h)}(k) + 1 \leq N_{max}^{*,(h)} \tag{5}$$

Even with this restriction, there are several optimization criteria that can be followed, in order to place the vertex N_{max}^* of the rectangular acceptance region [17]; in the present case, the problem is further complicated by the presence of the connectionless traffic, whose requirements must also be taken into account. So, the basic (and obviously related) questions are: i) how do we allocate the bandwidth resource between the connectionless packet flow and the connection-oriented service classes? ii) how do we find the "best" vertex N_{max}^{opt} on a certain FR boundary? and, finally, iii) can this procedure be made adaptive, in order to follow slow variations in the load proportions among the various classes?

Before considering the choice of N_{max}^{opt}, we may note that the connectionless traffic can always be allocated all the bandwidth unused by the connection-oriented ones, in a way analogous to movable boundary schemes in TDM networks [8]. We can do this through the knowledge of the VC-profile $N_A(k)$, by calculating (with any valid method, as mentioned in the previous section) the minimum bandwidth $C_{min}^{(h)}(k)$ that is necessary to ensure cell-level QoS to the $N_A^{(h)}(k)$ connections of class h in progress. By letting

$$C_{min}(k) = \sum_{h=1}^{H} C_{min}^{(h)}(k) \tag{6}$$

we can assign (through the scheduler) the connectionless traffic the *residual bandwidth*

$$R(k) = C - C_{min}(k) \tag{7}$$

where C is the total transfer capacity of the link; this assignment can last until either a new call is accepted or a connection terminates. It can be noted that, given the connection-oriented traffic characteristics and the bandwidth C_{co} globally assigned to it, the corresponding FR can be constructed off-line, and the residual bandwidth $R(k)$ can be determined for each of its points.

Even with the above described assignment, it is however clear that different "static" partitions (determined by the value of C_{co}) may be necessary to combine the requirements of the various classes. These will be determined by the optimization procedure that leads to the choice of N_{max}^{opt}. By first setting $C_{co}=C$ and then gradually decreasing (in discrete steps) the bandwidth globally allocated to connection-oriented traffic, we obtain the family of FRs, whose boundaries are depicted in Fig. 2 above. Now, for each corresponding boundary $S(C_{co})$, we can compute the point $N_{max}^*(C_{co}) \in S(C_{co})$ that minimizes a cost function involving the stationary call blocking probabilities $P_{block}^{(h)}\left(N_{max}^{(h)}\right)$; owing to the service separation assumption, these probabilities are given simply by the Erlang B formula, where $N_{max}^{(h)}$ is the number of servers. Among the various possibilities, we have chosen the following cost function (named Balanced Erlang Scheme (BES) in [17], as it tends to equalize blocking among the classes):

$$J_1(N_{max}) = \max_h \left\{ \vartheta^{(h)} P_{block}^{(h)}(N_{max}^{(h)}) \right\} \tag{8}$$

If, for each service class, we also want to take into account a constraint on the blocking probability, say

$$P_{block}^{(h)}(N^{(h)}) \le \Gamma^{(h)}, \qquad h=1,...,H \tag{9}$$

we can modify the cost function (8) as follows. Let $\overline{N}$ be the point whose coordinates satisfy relations (9) with equality; then we can consider

$$\bar{J}_1(N_{max}) = \begin{cases} \tau_1 \sum_{h=1}^{H} \left[P_{block}^{(h)}\left(N_{max}^{(h)}\right) - \Gamma^{(h)} \right]^2, & N_{max} < \overline{N} \\ J_1(N_{max}), & N_{max} \ge \overline{N} \end{cases} \tag{10}$$

where τ_1 is a constant (the larger τ_1, the higher the penalty for not matching the constraint).

Given $N_{max}^*(C_{co}) = \arg\min \bar{J}_1[N_{max}(C_{co})]$, we consider, for each value of C_{co}, the following cost function for the asynchronous, connectionless traffic:

$$J_2\left(N_{max}^*\right) = \sigma \overline{P}_{loss}^{H+1}\left(N_{max}^*\right) \equiv \sigma \sum_{s \in S_R(N_{max}^*)} P_{loss}(R_s)\pi(s) \tag{11}$$

where R_s represents the residual capacity corresponding to the value s of the VC-profile $N_A(k)$ (the system's state) and $S_R(N_{max}^*)$ is the rectangular region within the FR, whose vertex on the FR's boundary is N_{max}^*. With the given CAC rule, the state of the system $N_A(k)$ is a multidimensional Markov chain over $S_R(N_{max}^*)$, whose stationary distribution has been indicated by $\{\pi(s)\}$. On the other hand, as we have already mentioned, we use for the term $P_{loss}(R_s)$ the upper bound determined in [12] on the buffer overflow probability, given the state value s and, hence, the residual capacity R_s. More specifically, let $a_Y \equiv E\{Y\} = c\sum_1^\infty y^{-\alpha}$ be the mean of the Pareto distribution (1), P_{over} be the probability of buffer overflow and let λ indicate the intensity of the number of sources becoming active at a generic instant in the aggregated process [12], with $\lambda a_Y < R_s$. Since

$$P_{over} \leq \frac{c\lambda}{\alpha(\alpha-1)(R_s - a_Y\lambda)}\left(Q^{(H+1)}\right)^{-\alpha+1} \tag{12}$$

asymptotically with $Q^{(H+1)}$ [12], we take the term in the right hand side of (12) to represent $P_{loss}(R_s)$ in (11).

Essentially, in determining both P_{loss} and the stationary distribution $\{\pi(s)\}$, we use the same decoupling procedure as in [8] and [20]; $\{\pi(s)\}$ is readily computed from the transition probability of the multidimensional birth-death process describing the call dynamics [8].

Now, we may want to enforce a constraint on the maximum cell loss that is tolerable for the connectionless traffic, i.e.,

$$\bar{P}_{loss}^{H+1}(N_{max}^*) \leq \varepsilon^{(H+1)} \tag{13}$$

which, in turn, translates into the minimum bandwidth C_{asy} to be given the connectionless traffic

$$C_{asy} = \min_{N_{max}^*}\left[C - C_{co}\left(N_{max}^*\right)\right], \quad \text{s.t. } \bar{P}_{loss}^{H+1}(N_{max}^*) \leq \varepsilon^{(H+1)} \tag{14}$$

where $C_{co}(N_{max}^*)$ represents the bandwidth corresponding to the boundary $S(C_{co})$, which N_{max}^* belongs to. To take this constraint into account, we can modify (11) as

$$\tilde{J}_2\left(N_{max}^*\right) = \begin{cases} \sigma_1\bar{P}_{loss}^{H+1}\left(N_{max}^*\right), & N_{max}^* \in S(C_{co}), \ C - C_{co} \leq C_{asy} \\ \sigma_2\bar{P}_{loss}^{H+1}\left(N_{max}^*\right), & N_{max}^* \in S(C_{co}), \ C - C_{co} > C_{asy} \end{cases} \tag{15}$$

with $\sigma_2 \gg \sigma_1$.

A view of the optimal points N_{max}^* generated by the BES criterion, as well as of the capacity bounds corresponding to constraints (9) and (13) is given in Fig. 3 for our numerical example with H=2. The constraints values are $\Gamma^{(1)}=\Gamma^{(2)}=0.05$ and $\varepsilon^{(3)}=10^{-5}$.

We can now turn to the determination of the "globally" optimal point N_{max}^{opt} that minimizes the sum of the two cost functions introduced above, namely,

$$N_{max}^{opt} = \arg\min_{C_{co}}\left\{\tilde{J}_1\left[N_{max}^*(C_{co})\right] + \tilde{J}_2\left[N_{max}^*(C_{co})\right]\right\} \tag{16}$$

The optimal point and the corresponding rectangular region in our case are shown in Fig. 4.

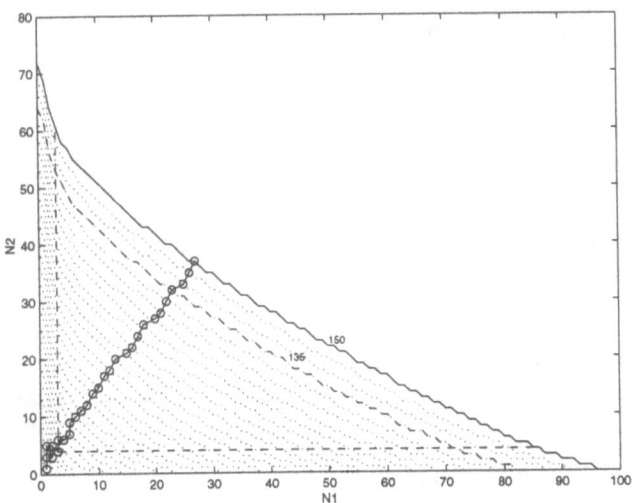

Fig. 3. Sequence of optimal points N^*_{max} for the BES criterion.

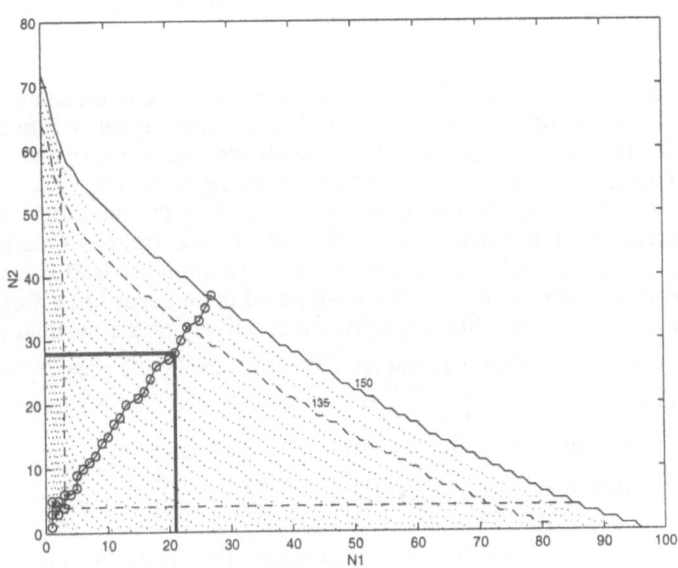

Fig. 4. Optimal region and constraint set.

It is worth noting that all calculations related to the FRs only depend on the cell-level parameters, i.e., on the characteristics of the sources, which determine their belonging to a specific class; for a given classification, all the FRs can be computed off-line. On the other hand, the calculations regarding the optimization do depend on the call level traffic intensity and on the connectionless traffic load, and they should be performed on-line in case of variations in these parameters. In this respect, however, it must be noted that variations in such traffic parameters may occur on a much slower time scale with respect even to connection request dynamics; this fact, along with possible computational simplifications, along the same lines of [8], allows to consider the control structure we have outlined to be embedded within a parameter adaptive scheme.

Before turning to consider the decentralized case, we may touch one further point, regarding the possible asymmetries in the traffic flows in the two directions. More specifically, we note that it is not difficult to extend the previous treatment to the case where the asynchronous flow in one direction is much larger than in the opposite one (for instance, it may well happen that the downstream asynchronous traffic intensity is much larger in the presence of WWW browsing applications). We will actually do so in the next Section. It is however worth noting that a second, perhaps more important, type of asymmetry may arise, according to the kind of services considered: this is the case of connection-oriented services with (possibly widely) different bandwidth requirements in the two directions (e.g., Video on Demand (VoD)). In the present paper we do not yet address this problem, which is currently under investigation.

4. Call Admission Control and capacity allocation - II: decentralized case

We turn now to briefly address the second model we have mentioned, namely, one where the control of bandwidth and CAC is explicitly taken into account at the ONUs. The OLT assigns the ONUs bandwidth partitions (which, however, may be changed adaptively), within which they apply control structures of the type described above. To this aim, with respect to the notation used in the previous section, we will index all capacities and buffers (for all traffic types) also by ONU (with the only exception of the asynchronous traffic flow in the downstream direction, whose cells are supposed to be queued in a single buffer, which is assigned all residual capacity, irrespectively of the destination ONU). Let N be the number of ONUs, and let $\overline{C}^{(i)}$, $i=1,...,N$, be the capacity assigned to the i-th ONU, such that $\sum_{i=1}^{N} \overline{C}^{(i)} = C$. For each ONU, a cost function can be defined, of the same type as that leading to (16), namely

$$\tilde{J}^{(i)}\left(C_{co}^{(i)}\right) = \tilde{J}_1\left[N_{max}^{*,(i)}\left(C_{co}^{(i)}\right)\right] + \tilde{J}_2\left[N_{max}^{*,(i)}\left(C_{co}^{(i)}\right)\right] \qquad i=1,...,N \qquad (17)$$

which will be used in the "local" optimization. The functions appearing in the right hand side of (17) correspond to (10) and (15), respectively, with the obvious modifications, and with $\overline{C}^{(i)}$ substituting C. As regards the asynchronous connectionless traffic, only the upstream buffer at the ONU is taken into account here.

The function $\tilde{J}$ to be minimized at the OLT in this case, with respect to $\overline{C}^{(i)}$, $i=1,...,N$, can be constructed as follows

$$\tilde{J}\left(\overline{C}^{(1)},...,\overline{C}^{(N)}\right) = \sum_{i=1}^{N} \tilde{J}^{(i)}\left(C_{co}^{(i)}\right) + \tilde{J}^{(2)}\left[N_{max}^*\left(C_{co}\right)\right] \qquad (18)$$

where the function $\tilde{J}_2$ is the analogous of (15), with $C_{co} = \sum_{i=1}^{N} C_{co}^{(i)}$ and C_{asy} computed accordingly (see (14)).

5. Numerical results

We complete the numerical results already presented above with some further evaluations of the performance indices under different traffic loads. The previous figures and the ones we present here refer to the data below. The model adopted is the centralized one.

$H=2$; $C=150$ Mbits/s

Connection-oriented traffic (VBR)

Cell level parameters (for the definition of the FR)   class 1   class 2

	class 1	class 2
Peak bandwidth	5 Mbits/s	2 Mbits/s
Average bandwidth	0.3 Mbits/s	1 Mbit/s
Delay constraint	2200 slots	2200 slots
P_{loss} upper bound	10^{-8}	10^{-5}
P_{delay} upper bound	10^{-5}	10^{-5}
Buffer length	10 cells	10 cells

Call level parameters (base values)class 1   class 2

	class 1	class 2
Traffic intensities	18.2 Erlangs	18.6 Erlangs
P_{block} upper bound	0.05	0.05
Weight $\vartheta^{(h)}$	1	1

Connectionless traffic

Intensity of the aggregated process λ	0.1
Pareto parameter α	1.4
P_{loss} upper bound	10^{-5}

With these figures for the traffic intensities, the bandwidth is partitioned, on the average, so that about 2/3 (100 Mbits/s) is assigned to the connection-oriented traffic and 1/3 (50 Mbits/s) to the connectionless one. We now present some results obtained by varying the traffic intensities, and then computing the ensuing optimal assignments and the corresponding blocking and buffer overflow

316

probabilities. The load values in the following figures correspond to the initial ones multiplied by the number on the x-axis.

We first consider the effect of increasing the connectionless traffic load. Fig. 5 shows the corresponding blocking probabilities for both connection-oriented classes (which tend to be equalized by the BES criterion); as should be expected, they increase, but, for the range of values considered, they still remain below the constraint value.

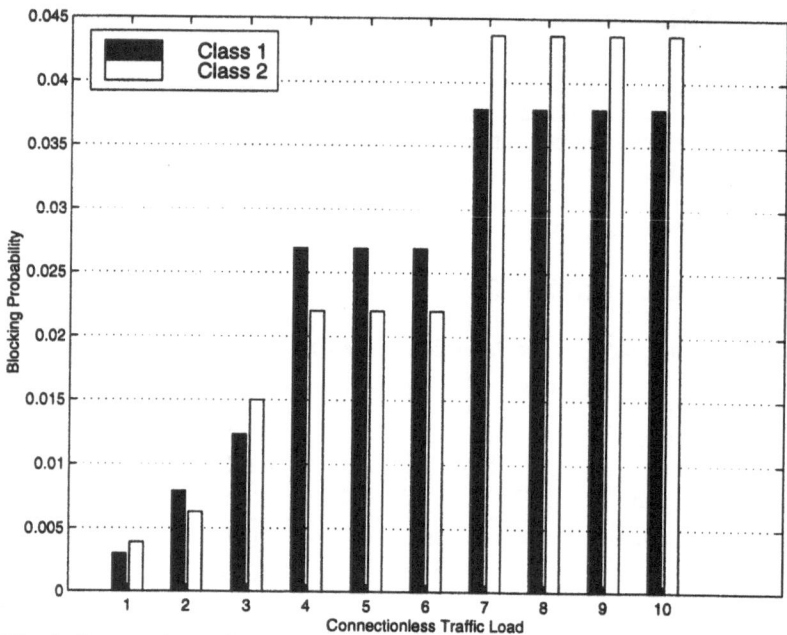

Fig. 5. Connection-oriented traffic blocking vs. connectionless traffic load.

The related behaviour of the buffer overflow probability is shown in Fig. 6.

The effect of increasing the load in one of the connection-oriented classes is shown in Figs. 7 and 8. Specifically, Fig. 7 reports the behaviour of the blocking probabilities of both connection-oriented classes versus the offered load of class 1, and Fig. 8 reports the corresponding increase in the connectionless traffic buffer overflow probability, due to the shift in the capacity assignments. It is worth noting that, in this case, the constraints can no longer be satisfied for the higher load values; the solution obtained in this case represents a sort of "best compromise", according to the criteria adopted.

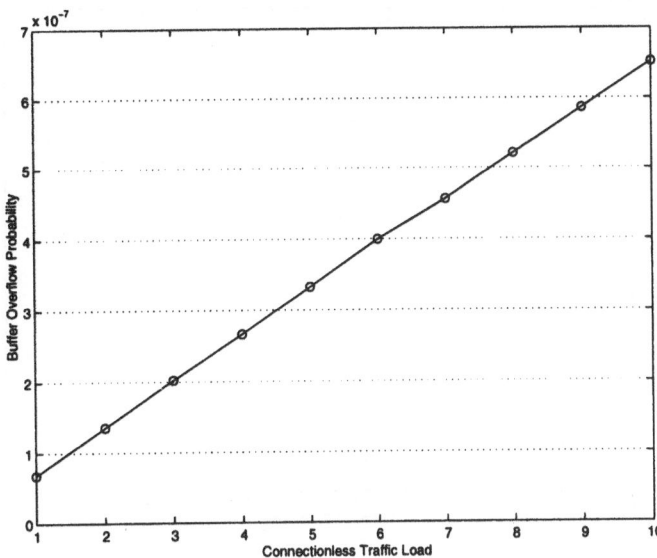

Fig. 6. Connectionless traffic buffer overflow probability vs. connectionless traffic load.

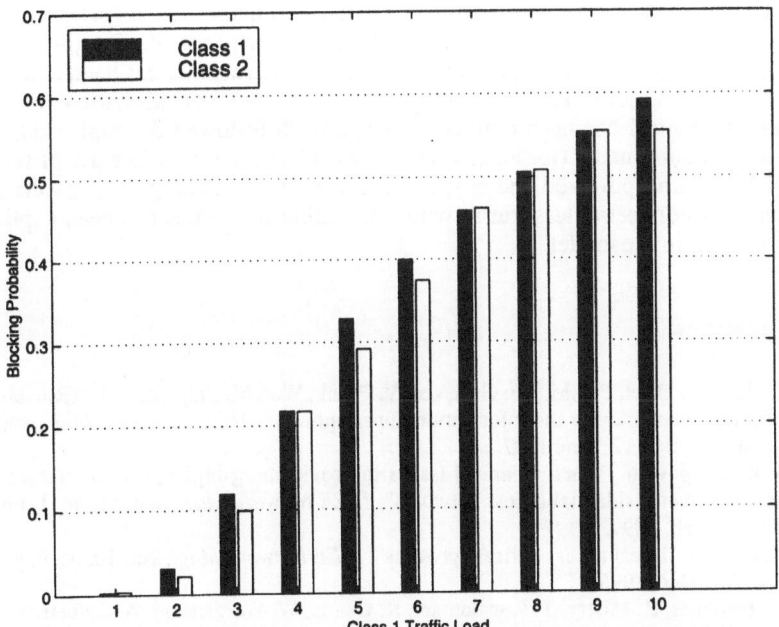

Fig. 7. Connection-oriented traffic blocking vs. class 1 traffic load.

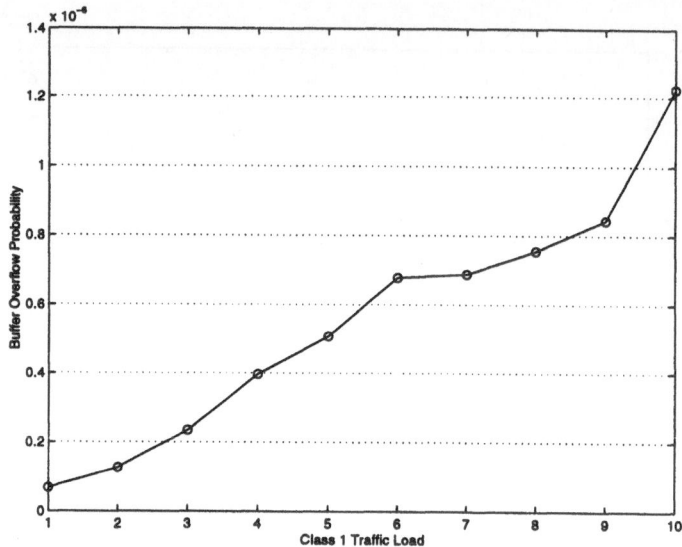

Fig. 8. Connectionless traffic buffer overflow probability vs. class 1 traffic load.

6. Conclusions

We have introduced and analyzed a control scheme suited for multimedia access multiplexers, operating in an ATM-PON environment. The presence of multiple traffic classes with possibly conflicting requirements has been taken into account, and a joint strategy for resource (bandwidth) allocation and call admission control has been defined. The approach followed decouples cell- and call-level requirements (for connection oriented traffic) and makes use of service separation with optimized and adaptive bandwidth partitioning. In particular, the presence of connectionless traffic with self-similar properties has been explicitly incorporated in the model.

References

1. C.-J.L. van Driel, P.A.M. van Grinsven, V. Pronk, W.A.M. Snijders, "The (R)evolution of access networks for the Information Superhighway", *IEEE Commun. Mag.*, vol. 35, no. 6, pp. 104-112, June 1997.
2. V.K. Bhagavath, "Open technical issues in provisioning high-speed interactive data services over residential access networks", *IEEE Network Mag.*, vol. 11, no. 1, pp. 10-12, Jan./Feb. 1997.
3. N.J. Frigo, "Local access optical networks", *IEEE Network Mag.*, vol. 10, no. 6, pp. 32-36, Nov./Dec. 1996.
4. D. Faulkner, R. Mistry, T. Rowbotham, K. Okada, W. Warzanskyj, A. Zylbersztejn, Y. Picaud, "The Full Services Access Networks initiative", *IEEE Commun. Mag.*, vol. 35, no. 4, pp. 58-68, April 1997.

5. I. Van de Voorde, G. Van der Plas, "Full Service Optical Access Networks: ATM transport on passive optical networks", *IEEE Commun. Mag.*, vol. 35, no. 4, pp. 70-75, April 1997.

6. Full Services Access Network Requirements Specification, J.A. Quayle, Ed., Issue 1, 1998 (available at http://www.labs.bt.com/profsoc/access/index.htm).

7. K.W. Ross, Multiservice Loss Models for Broadband Telecommunication Networks, Springer Verlag, London, UK, 1995.

8. R. Bolla, F. Davoli, "Control of multirate synchronous streams in hybrid TDM access networks", *IEEE/ACM Trans. on Networking*, vol. 5, no. 2, pp. 291-304, April 1997.

9. W.E. Leland, M.S. Taqqu, W. Willinger, D.V. Wilson, "On the self-similar nature of Ethernet traffic (extended version)", *IEEE/ACM Trans. Networking*, vol. 2, no. 1, pp. 1-15, Feb. 1994.

10. V. Paxson, S. Floyd, "Wide area traffic: the failure of Poisson modeling", *IEEE/ACM Trans. Networking*, vol. 3, no. 3, pp. 226-244, June 1995.

11. V. Paxson, "Empirically derived analytic models of wide-area TCP connections", *IEEE/ACM Trans. Networking*, vol. 2, pp. 316-336, Aug. 1994.

12. B. Tsybakov, N.D. Georganas, "Overflow probability in an ATM queue with self-similar input traffic, *Proc. IEEE Internat. Conf. Commun. (ICC'97)*, Montreal, Canada, 1997.

13. R. Bolla, F. Davoli, A. Lombardo, S. Palazzo, D. Panno, "Hierarchical dynamic control of multiple traffic classes in ATM networks", *Europ. Trans. Telecommun.*, vol. 5, no. 6, pp. 747-755, Nov.-Dec. 1994.

14. A.K. Parekh, R.G. Gallager, "A generalized processor sharing approach to flow control in integrated services networks: The single node case", *IEEE/ACM Trans. Networking*, vol. 1, pp. 344-357, 1993.

15. J.M. Hyman, A.A. Lazar, G. Pacifici, "Real time scheduling with quality of service constraints", *IEEE J. Select. Areas Commun.*, vol. 9, pp. 1052-1063, Sept. 1991.

16. R. Bolla, F. Davoli, M. Marchese, "A global control system for integrated admission control and routing in ATM networks", *Proc. IEEE Globecom '95*, Singapore, Nov. 1995, pp. 437-443.

17. R. Bolla, F. Davoli, M. Marchese, "Bandwidth allocation and admission control in ATM networks with service separation", *IEEE Commun. Mag.*, vol. 35, no. 5, pp. 130-137, May 1997.

18. J.P. Cosmas, G.H. Petit, R. Lehnert, C. Blondia, K. Kontovassilis, O. Casals, T. Theimer, "A review of voice, data and video traffic models for ATM", *Europ. Trans. Telecommun.*, vol. 5, no. 2, pp. 11-26, March-April 1994.

19. R. Guérin, H. Ahmadi, M. Naghshineh, "Equivalent capacity and its application to bandwidth allocation in high speed networks", *IEEE J. Select. Areas Commun.*, vol. 9, no. 7, pp. 968-981, Sept. 1991.

20. S. Ghani, M. Schwartz, "A decomposition approximation for the analysis of voice/data integration", *IEEE Trans. Commun.*, vol. 42, pp. 2441-2452, July 1994.

DSL Technologies (HDSL2, ADSL and VDSL), Improved Algorithms to Meet Wide Market Acceptance

L. Frenkel
Orckit Communication, Israel

Abstract

Digital Subscriber Line (DSL) technology is copper transmission technology that uses the existing copper wire infrastructure in order to provide high rate services. In this presentation, we shall overview the main characteristics of the DSL channel, describe the existing modulation techniques (HDSL, ADSL, VDSL), and overview the main considerations involved in the development of the new HDSL2 standard, which is currently under study at ANSI committee T1.

1. DSL System Components

Figure 1 describes the top level topology of the DSL access system.

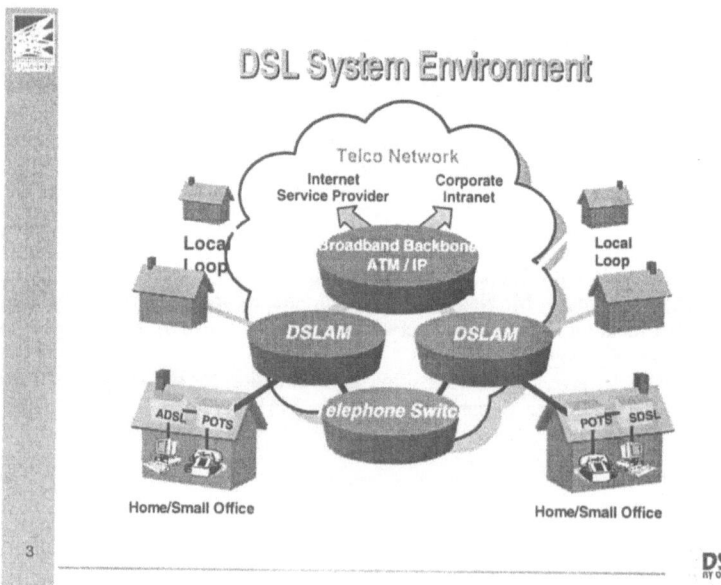

Figure 1

The signal of conventional telephone modems is confined to 3400 Hz (the voice bandwidth), and transported by the general switched telephone network (GSTN) to a similar modem (this connection is not shown in the slide). In contrary, the

DSL modem uses higher frequencies to provide a point to point connection to the central office, where a DSL access multiplexer (DSLAM) hands of the traffic to the service provider trough a backbone network [3].

2. DSL Channel

The DSL channel is limited by attenuation and crosstalk, both increase with frequency.

Figure 2 shows a frequency response of a subscriber line which consist of a single 3-km long section. Typical subscriber lines can consist of several sections, which are connected in manholes, for example. These sections may not have the same diameter or insulation. In addition, in order to provide plant flexibility, bridged taps can be used. These are segments of twisted pairs that are connected in shunt to the working twisted pairs, and increase the phase and amplitude distortion.

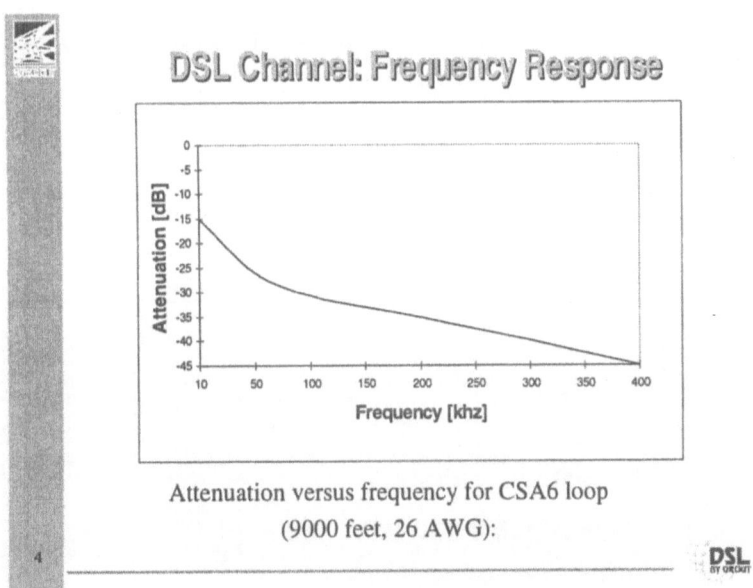

Figure 2

A typical copper cable bundle typically includes 25 or 50 twisted pairs. The transmitted signal (upper left side of figure 3) generates two kind of crosstalk: Interference to the to a receiver on the same side of the line is called near end crosstalk (NEXT), and interference to the to a receiver on the other side of the line is called far end crosstalk (FEXT).

The combined attenuation and crosstalk limits the usable bandwidth. This bandwidth is typically in the range of 100 kHz to 20 MHz depending on the length of the line.

A detailed mathematical analysis the DSL channel can be found in [4].

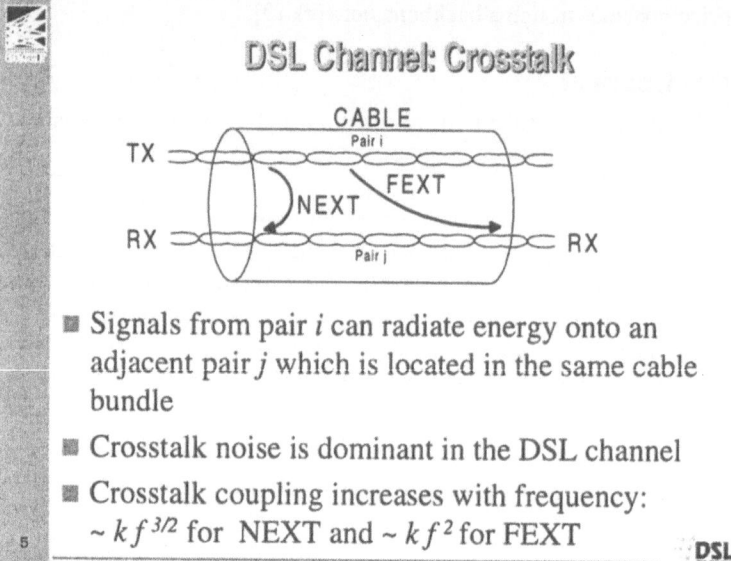

Figure 3

3. DSL Modulation Techniques

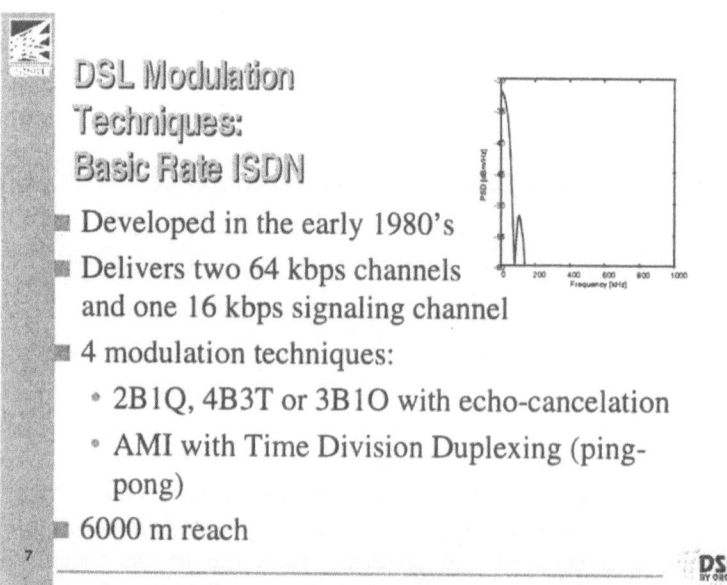

Figure 4

Basic rate ISDN was developed in the 1980's. This service was aimed to provide two channels of 64 kbps ("B" channels) and a single 16 kbps "D" channel, which was used for signaling. Including additional overhead, the bit rate was 160 kbps. In order to comply with the a requirement of a 6 km reach, a bandwidth efficiency of more than one bit per Hz was required. Consequently, simple line codes like 2B1Q and 4B3T where used, utilizing frequencies below 100 kHz (Figure 4).

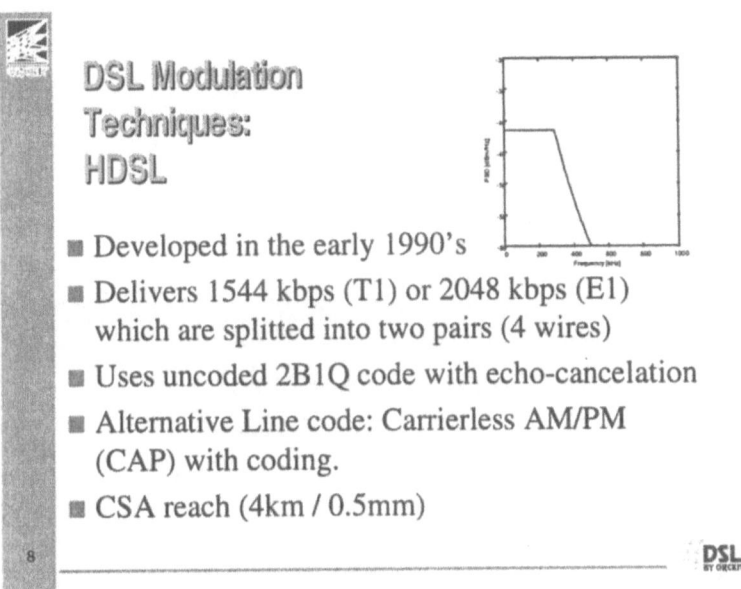

Figure 5

In the early 1990's, some vendors encouraged the use of 2B1Q at higher speeds as an alternative way to provision T1 and E1 services without repeaters [3]. The technique consisted of splitting the 1,544 kbps service into two pairs. This technique was referred as High bit rate DSL or HDSL. The HDSL was designed to reach 4km for 0.5mm wire, and 3km for 0.4mm wire.

Lower frequencies (below ~150kHz) are characterize by low attenuation and low crosstalk. ADSL (Asymmetric DSL) was designed to use these lower frequencies both for upstream and downstream transmission, and use higher frequencies (up to 1.1 MHz) only for downstream (see Figure 6). Since the downstream receiver was not impaired by ADSL NEXT at high frequencies, high bits rates (up to 8Mbits/sec) were possible on the downstream direction.

Two ADSL modulation techniques where suggested: single carrier CAP/QAM, and Discrete Multitone (DMT). Finally, DMT was selected by the standards. In this technique, the 1.1 MHz bandwidth was splitted to 256 tones, spaced 4.3 kHz apart. Each 250 micro-seconds, an IFFT is used to transform 256 complex QAM symbols (tones) to 512 real valued samples at a rate of 512x4.3125 = 2208 ksps. A cyclic prefix of 32 samples is appended, and the samples are converted to an analog signal. At the receiver, an IFFT recovers the 256 QAM symbols. The DMT method is extremely flexible, since any PSD mask can be used by distributing the power between the carriers, and the rate can be optimized by selecting optimal constellations for the different tones.

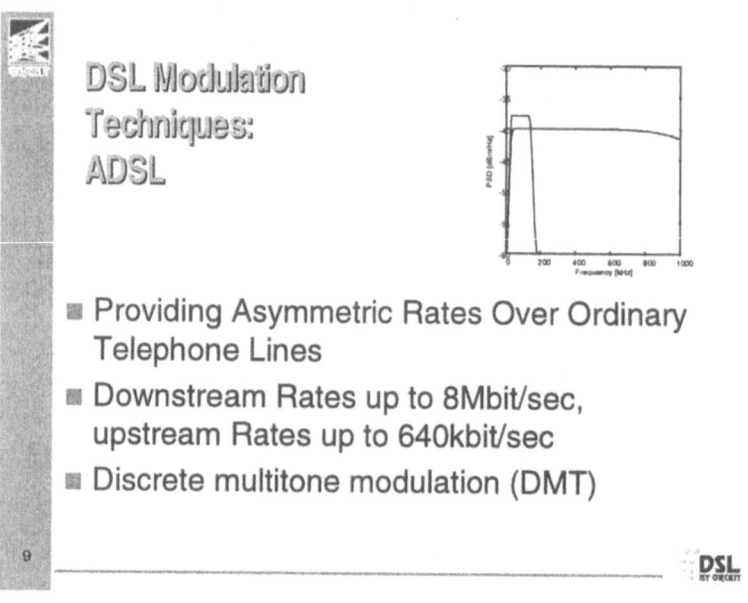

Figure 6

A passive POTS (plain old telephone service) splitter is typically used to separate the ADSL signal from the POTS band (Figure 7). New "Lite ADSL" standard is now under development, in which the modem will be plugged directly into an existing wall jack (Figure 8). This is a challenging requirement since telephone devices can introduce non-linearities and dynamic changes in the loop response. "Lite ADSL" will also is also designed to compromise rate for complexity, reaching a rate of 1500 kbps and utilizing a bandwidth of 550 kHz.

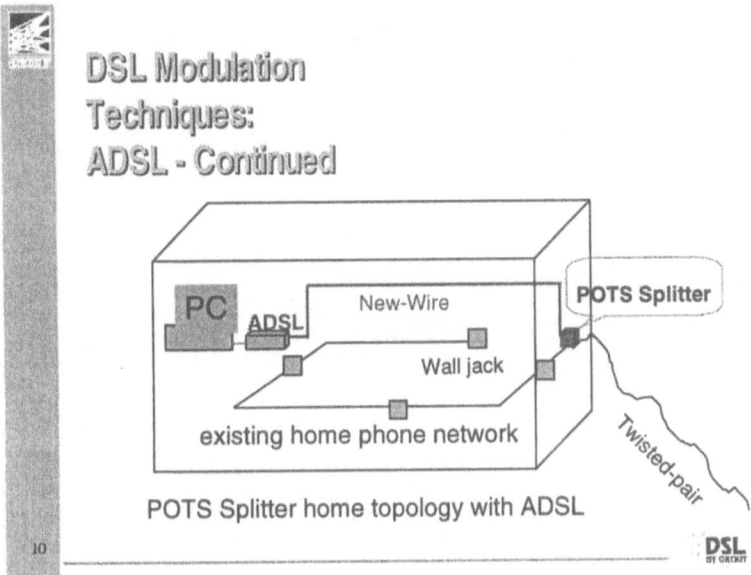

Figure 7

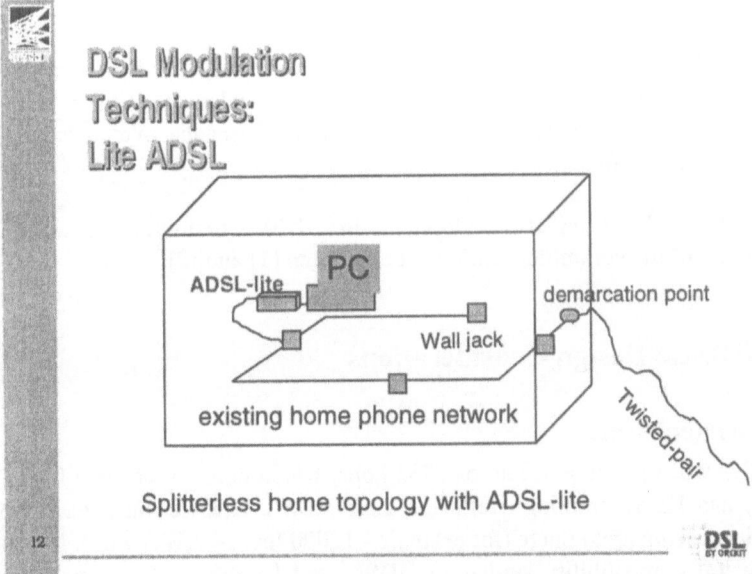

Figure 8

VDSL is under development in ETSI and ANSI. The idea is to use a fiber link to a street cabinet or to an office building, and then use a VDSL modem for the last lag (Fig. 9). VDSL modems will reach 300m-1500m with typical rates of 13, 26 and 52 Mbps. QAM modulation will be used, utilizing frequencies up to 18 MHz.

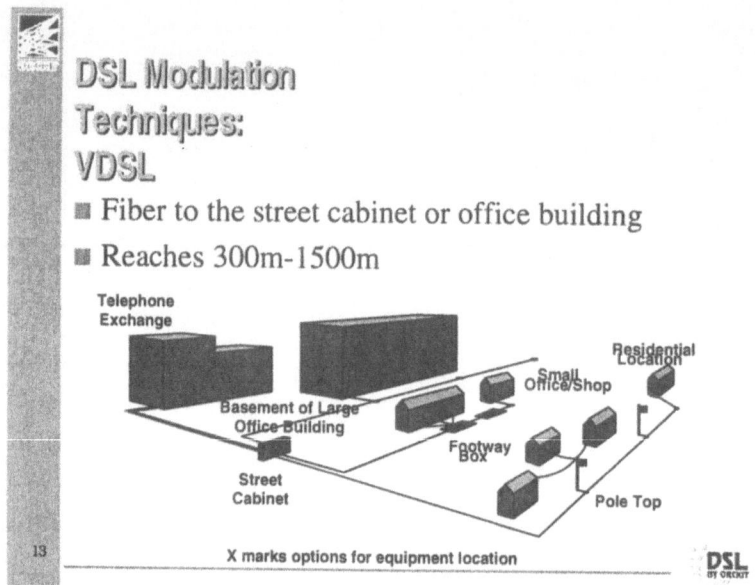

Figure 9

4. HDSL2

The HDSL2 modem can be considered as "the V.34 of DSL". Both systems (V.34 for general switched telephone networks (GSTN) and HDSL2 for DSL) exploit state of the art technology in order to push performance over a band limited channel to the limit.

HDSL2 has been under development in ANSI Committee T1 since 1995. Summary of the committee work can be found in [1] and [2].

5. HDSL2 Design Considerations

51. Requirements

HDSL2 will operate at a rate of 1552 kbps, which consists of the T1 rate (1544 kbps) and HDSL framing overhead. Like HDSL, HDSL2 must reach the CSA (carrier serving area) range (for example - 12000 feet 24 AWG wire). It must have a spectral compatibility similar to HDSL, and be robust to crosstalk from all existing services (basic rate DSL, HDSL, ADSL, VDSL). The latency must be lower than 500 micro-seconds, including framing, coding and all DSP functions. All of these requirements must be met, using a single pair instead of two pairs. That is, HDSL2 must multiply the rate by a factor of two (per link) comparing to HDSL, while keeping the same performance.

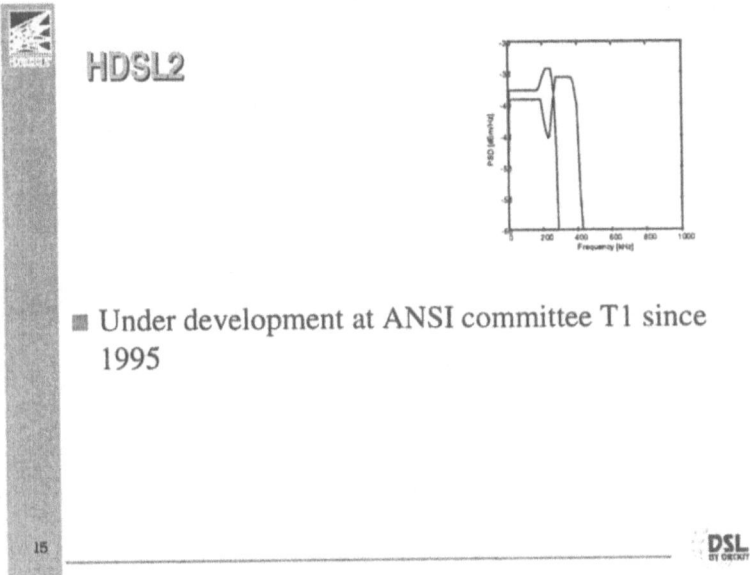

HDSL2

■ Under development at ANSI committee T1 since
1995

15

DSL

Figure 10

52. Spectral Shape

In order to meet the requirements, the HDSL2 bandwidth had to be increased with respect to HDSL. However, increasing the bandwidth alone, even with some power boost, could not gain the required performance. The reason is, that in the DSL channel, increasing power do not mitigate self NEXT, since the self crosstalk is increased proportionally. The PSD is also limited by the requirement for spectral compatibility, that is, DSL, HDSL and ADSL must not be impaired by HDSL2 transmission.

Both symmetric echo-canceled transmission and frequency division multiplexed transmission (FDM) approaches were considered by ANSI Committee T1. Symmetric transmission proved to have a self crosstalk limitation of 2-3 dB short of the requirements. In contrast, FDM was not limited by self-Next but by crosstalk from other services. It was also limited by crosstalk into other services, due to higher transmit frequencies involved. This made the FDM solution even less desirable.

The final PSD mask that was selected was named "OPTIS": Overlapped PAM transmission with Interlocked Spectra. This spectra adopts the ADSL concept of using low frequencies (up to ~250 kHz for HDSL2) both for upstream and downstream, with echo cancellation. The symbol rate for the downstream and upstream is similar, with higher excess bandwidth used for the downstream.

328

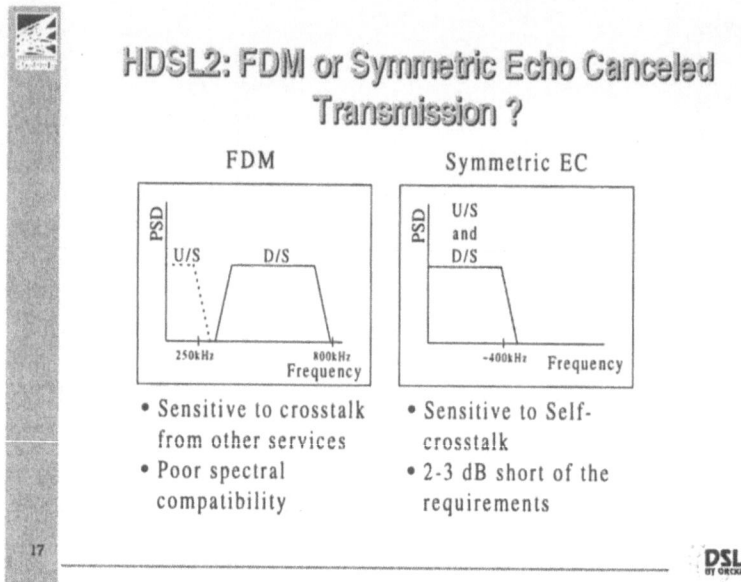

Figure 11

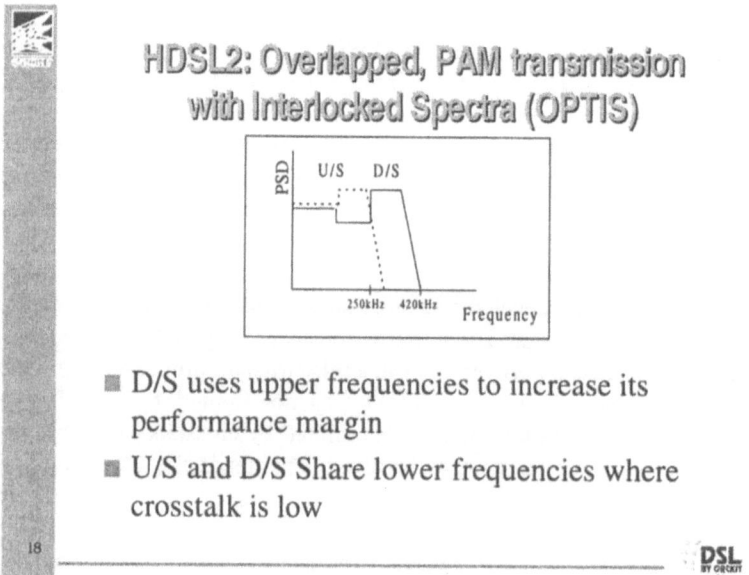

Figure 12

53. PAM or QAM ?

The SNR at the detector, which determines the performance of the modem, is given by :

$$SNR(f) = \sum_{n=-\infty}^{\infty} \frac{S(f + n \cdot f_{baud})|H(f + n \cdot f_{baud})|^2}{N(f + n \cdot f_{baud})}$$

SNR at the detector $= C \cdot \int_0^{f_{baud}/2} \log(1 + SNR(f))\, df$ [dB]

where $SNR(f)$ is the "folded" SNR, $S(f)$ is the transmitted signal PSD, $H(f)$ is the channel frequency response and $N(f)$ is the noise PSD [5]. An interesting characteristic of the DSL channel is that PAM modulation is advantageous over QAM/CAP when large excess bandwidth is used for the downstream. For PAM, the self-next free region of the downstream ((1) in Figure 13) is "folded" by sampling operation at the detector with a shift of equal to the symbol rate $f_{baud}=f_{PAM}$ into a low SNR region (2). In contrary, for QAM, the folding is with shift of $f_{baud}=f_{QAM}= f_{PAM}/2$, and region (1) is shifted to a high SNR region (3). Therefore, in this case, the final SNR at the detector which is computed at by the formula above is higher for PAM.

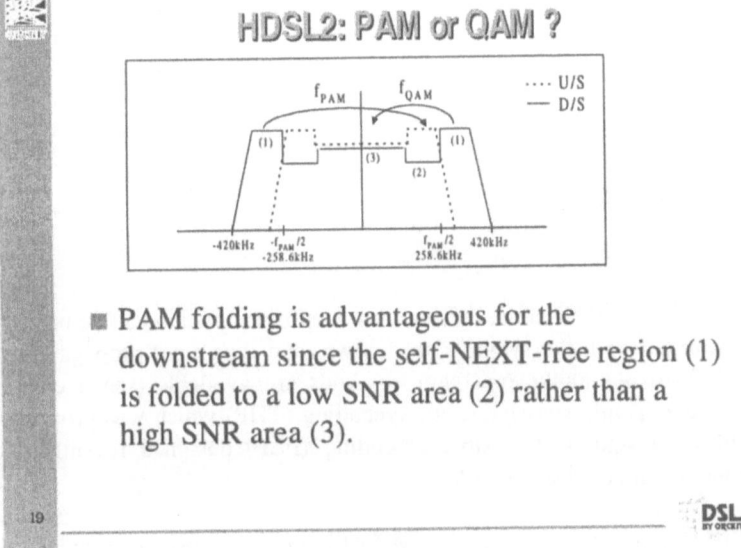

Figure 13

54. Coding

A code gain of at least 5 dB was required for HDSL2 in order to meet the requiremets. Near maximum likelihood (NML) detectors, which were used by some vendors for HDSL, could not reach this performance. With latency limitation for the code of about 250 us (or 400 bits), concatenated codes and turbo codes gave poor results. Finally, 1-D trellis code (TCM) with 512 states was found to have the best performance/latency tradeoff.

330

55. Precoding

A TCM decoder can not be used effectively with conventional FFE/DFE receiver. The reason for this is that the SNR at the slicer input is very low when operating with a TCM code, since higher constellation is used (16 PAM instead of 8PAM for HDSL2). This introduces symbol errors at the slicer output, which are then multiplied by the DFE feedback (see Figure 14).

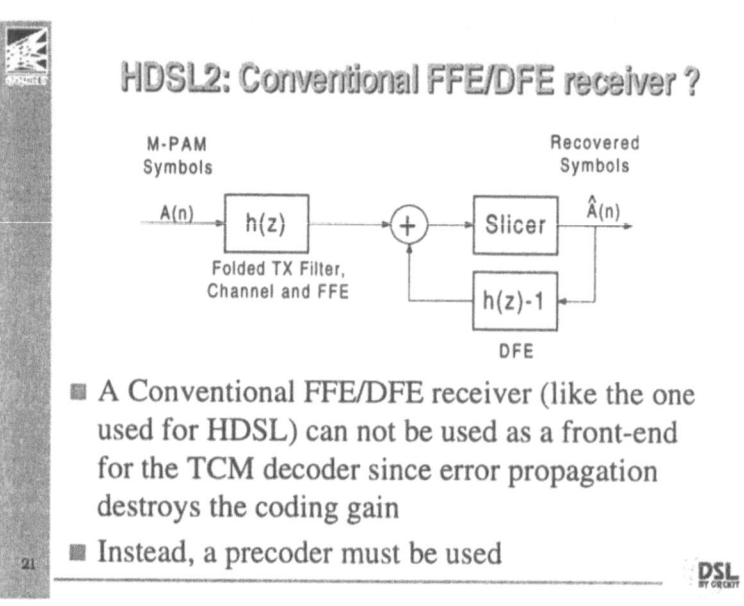

Figure 14

Instead of using a DFE at the receiver, the feedback function can be performed at the transmitter were the symbols are known, and thus error propagation can be avoided. Two alternative precoding methods are available (see a comparative study in [6]): Tomlinson-Harashima precoding (THP) which was proposed more than 20 years ago, and flexible precoding (FLP) designed recently for the telephone line modem standard V.34.

For THP (Figure 15), the incoming M-PAM symbols A(n) are summed with the feedback filter and a modulo function maps the result into the range [-M, +M] (For HDSL2: M=16 and A(n) = ±1, ±3, ... ±15). The idea behind the THP scheme can be understood if the modulo operation in the transmitter is replaced by a linear model in which an additional signal d(n) = 2*M*b(n) is added to the symbols A(n) (where b(n) is a series of integers). The extended symbol A(n)+d(n) is then going though the nearly transparent folded channel and reaches the slicer input. The modulo operation in the receiver (Figure 15) is then used to remove d(n) and recover the symbol A(n). Note that modulo operation at the transmitter is required in order to limit the transmitted power.

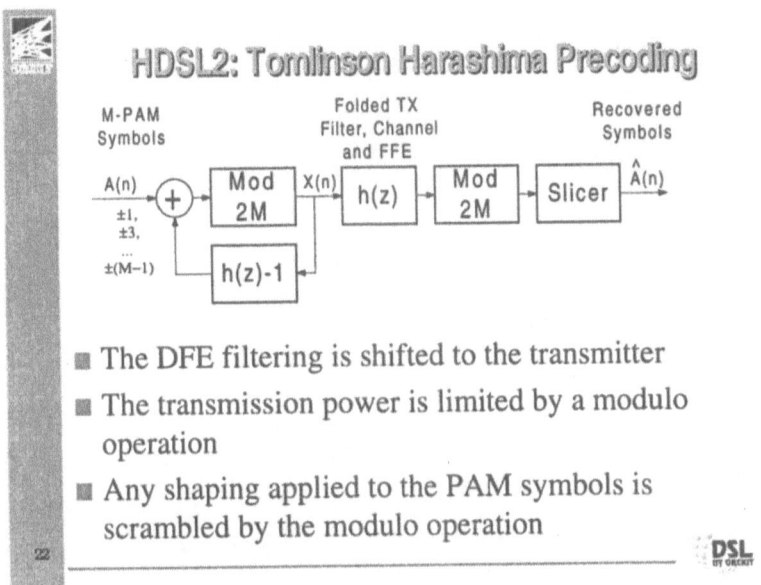

- The DFE filtering is shifted to the transmitter
- The transmission power is limited by a modulo operation
- Any shaping applied to the PAM symbols is scrambled by the modulo operation

Figure 15

It is well known that when the symbols A(n) are produced with a nearly Gaussian probability density distribution (pdf) instead of a uniform pdf, then a "shaping gain" can be achieved. The main disadvantage of THP is the scrambling of the incoming symbols A(n) by the modulo operation, which results in a uniform distribution of the output symbols X(n), and loss of any shaping gain. The FLP scheme was designed to preserve the pdf of the original symbols A(n) and thus allow the utilization of simple shaping schemes for the mapping of the information bits to the symbols A(n). Instead of using a modulo operation into the range [-M,M] in the forward path, FLP is using a modulo operation into the range [-1,1) in the feedback (**Figure 16**).

Again, the method can be understood by removing the modulo operation, and using a linear model in which a signal d(n) = 2*b(n) is added to the original symbols A(n), where b(n) is an integer. Since d(n) is on the same grid of A(n), a slicer (slicer1) can be used to recover the signal A(n)+d(n). Then a filter with the response of the inverse channel (1/h(z)) is used to recover X(n). But X(n) is was the result of adding a dither in the range of [-1,1) to the original symbols A(n), and therefore a slicer (slicer2) can be used to recover A(n).

332

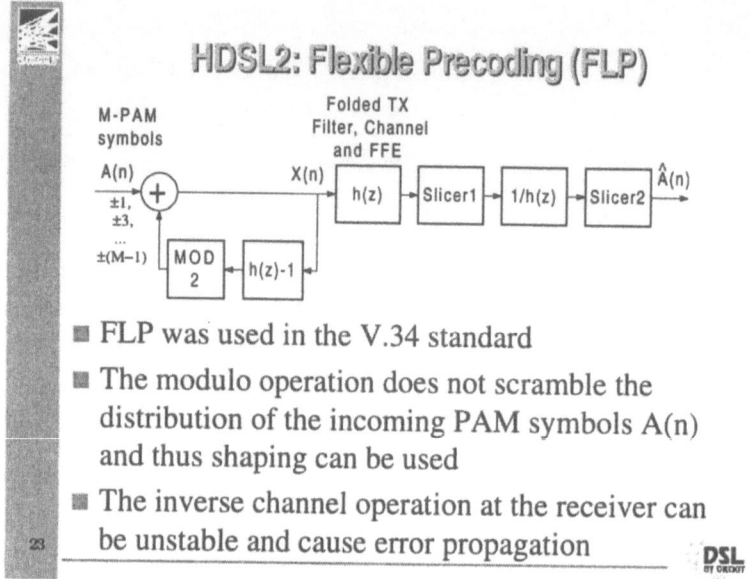

Figure 16

Although FLP was considered a great achievement of the V.34 standard, it was not found appropriate for the DSL channel. The main reason for this was that the DSL channel is extremely distorted, and includes a null at DC. Therefore, error propagation and instability at the receiver inverse-channel filter limited the performance. Finally, THP was chosen as the precoding scheme, and signal shaping was not used.

6. Conclusions

The DSL system utilizes higher frequencies in parallel to the POTS ("plain old telephone service") to transport high bit rates. Then a DSLAM is used to bypass the telephone switch. The DSL channel itself is limited by attenuation and crosstalk, both increase with frequency. In this presentation we have briefly described the DSL system and the DSL channel, and listed various DSL modulation techniques which are used to address different applications, rates and reach. Finally, we have described the main considerations which were involved in developing the HDSL2 standard, which used state of the art technologies to push performance over the DSL channel to the limit.

References

1. ETSI TM6 Madrid meeting Jan 1998, TD11 (Siemens & Adtran), "Next Generation HDSL"

2. G. A. Zimmerman, "HDSL2 Tutorial: Spectral Compatibility and Real-World Performance Advantages", presented at ETSI meeting, Lulea Sweden, June 23, 1998.
3. The DSL Sourcebook - 2nd Edition - http://www.paradine.com/sourcebook_offer/sb_html.html
4. J.J. Werner, "The HDSL Environment", IEEE Journal on Selected Areas in Comm., Vol 9, No 6, August 1991
5. J. Salz, "Optimum mean-square decision feedback equalization," Bell Syst. Tech, J., vol 52, no 8, p. 1341, Oct. 1973
6. R. F. Fischer, J. B. Huber, "Comparison of Precoding Schemes for Digital Subscriber Lines", IEEE Tran. on comm., vol. 45, no. 3, March 1997.

UMTS Access Network Architecture and Relevant Dimensioning Issues

F. Badini, E. Berruto, R. Menolascino, F. Piolini[1]

[1] CSELT - Centro Studi e Telecomunicazioni Torino (Italy)

Abstract

This paper describes the access network architecture proposed for third generation mobile systems (UMTS). The functions and the relevant network nodes are described along with some hints on the typical UMTS features as macrodiversity and soft handover. The second part of this paper presents some issues on the dimensioning and optimisation of an access network shared between fixed and mobile operators. This scenario seems to be promising especially in the fixed-mobile convergence perspective.

1. Introduction

Third generation mobile systems will provide customers with an enhanced list of services ranging from plain telephony to multimedia applications. Novel air interfaces, ensuring a higher efficiency with respect to the existing radio access schemes, have been designed, evaluated and proposed to the standardisation bodies (ETSI and ITU). At the moment, the UMTS proposal is grounded on the coexistence of two different air interfaces namely a Wide Band CDMA (W-CDMA) and a hybrid system that makes use of TDMA and CDMA capabilities (TDD-CDMA). The former access scheme is recommended for the paired bandwidth while the latter has been proposed for the unpaired bandwidth usage (asymmetrical services).

The services bit rate will increase if compared with the current systems capabilities. The typical bit rates expected in the UMTS environment are: 144 kbps in rural areas where users can move at high speed (highways); 384 kbps in urban and suburban areas and 2 Mbps in indoor environments or outdoor with limited users' mobility degree.

Major impacts are expected on the access network architecture to sustain the capabilities granted by the new air interfaces proposed in UMTS. The macrodiversity, for instance, is the capability of having multiple paths linking a single user's mobile terminal to several radio base stations to improve the overall radio link quality. This new function requires to install new functionality in the

Radio Network Subsystems (RNS) and to envisage new interfaces linking the Radio Network Controllers (RNC) together.

Furthermore, the convergence between fixed and mobile services might promote the interest in sharing some parts of the access network between fixed and mobile operators. This aspect has been the key issue being investigated in the ACTS Programme Concertation Activities where STORMS[1] and RAINBOW[2] have played an important role [1]. In this context, the impacts on the network planning procedures have been studied and led to the implementation of a simple software prototype tool which can assist operators in the planning and optimisation process of the access network.

2. UMTS Terrestrial Radio Access Network

The **UTRAN** (UMTS Terrestrial Radio Access Network) represents the part of a mobile network providing a radio connection to a mobile user. The separation between the Access Network (AN) and the Core Network (CN) aims at keeping a clear split between access and core network functionality. While in second-generation mobile systems these two functional categories are not sharply allocated into well-defined physical network nodes, this separation is strongly required in UMTS. The reason is mainly to guarantee an independent evolution of the access and core segments, as well as to allow a given AN to provide access to different CNs types.

The functional separation is modelled through the **Access Stratum** concept (Figure 1), consisting of a functional grouping including all the layers embedded in the UTRAN, part of the layers contained in the User Equipment (UE) and in a CN node. The Access Stratum boundary separates the layers **independent** from the access technique from those that are **dependent** on the particular access technique undertaken in the network. This border is located at the UE and in a CN node.

---

[1]   AC016 STORMS (Software Tools for the Optimisation of Resources in Mobile Networks) is in charge of investigating how the network planning activities will change passing from second to third generation mobile systems. It is also responsible for developing a set of software tools that might be used to assist the UMTS network planning process.

[2]   AC015 RAINBOW (RAdio INdependent Broadband On Wireless) will define an access network and core network architecture independent from the particular radio access scheme adopted by the operator. It also investigates about the transition, or migration path, from current to third generation mobile systems.

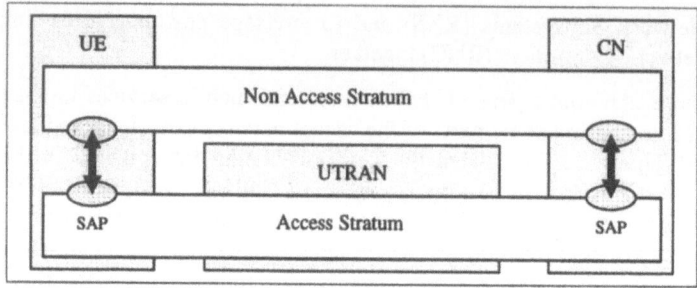

Figure 1: Access Stratum

Furthermore, the Access Stratum provides some services to the non-access stratum through Service Access Points (SAP). Among these services, there is the provision of signalling connections for control signalling between the UE and the CN, and the provision of radio access bearer for user data transfer. In particular the offered radio access bearer services are divided into two categories:

- **Restricted** radio access bearer services;
- **Unrestricted** radio access bearer services.

Unrestricted radio access bearer services are dedicated to data requiring bit sequence integrity, whereas **restricted** radio access bearer services are reserved to data that can be source coded/decoded within the Access Stratum. Different transport layers may support these bearer services in the Access Stratum.

The UTRAN architecture, outlined in ETSI/SMG2, consists of a set of Radio Network Subsystems (RNSs) connected to the CN through the *Iu* interface (Figure 2).

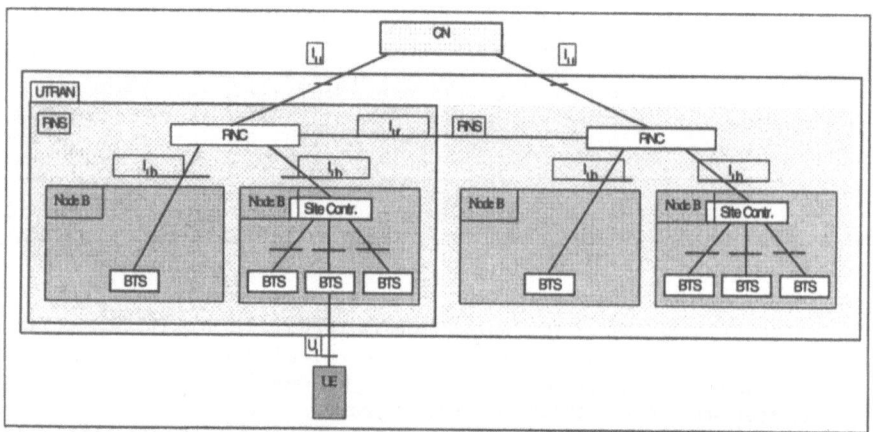

Figure 2: UMTS Terrestrial Radio Access Network

Each RNS is responsible for the radio resources management in its set of cells and offers the functionality to allocate and release the specific resources used to establish a connection between a UE and the UTRAN. Different RNSs can be interconnected together through the *Iur* interface. Both the *Iu* and *Iur* are logical interfaces. In particular the *Iur* interface can be conveyed over physical direct connection between RNSs or via any suitable transport network.

At the time being, ATM and AAL type 2 have been selected as the standard transport layer for **Soft Handover** data streams on the *Iur* and *Iub* interfaces. Other protocols, such as Frame Relay and AAL5, are under evaluation for data streams not using Soft Handover, while ATM and AAL type 5 is foreseen to support connections for signalling control.

Among the others, a RNS provides the functionality to support the **macrodiversity** technique (soft handover). This technique, introduced to improve the radio link quality, allows the UE to be attached to the UTRAN through more than one Base Transceiver Station (BTS) at the same time for the same connection. In this way, the same information flow (traffic and signalling data) is carried in parallel along a number of different paths between two common points, one located at the UE and the other inside the RNS. In these points, the received information is re-combined according to suitable algorithms and synchronisation requirements. The data stream transmitted to the UE is replicated onto the different paths. It is worthwhile mentioning that there may be several entities inside a RNS that combine the different information streams, depending on physical RNS configuration. Furthermore, although macrodiversity can be considered as an optional optimisation technique for TD-CDMA (Time Division Duplex - TDD mode), it is mandatory for W-CDMA (Frequency Division Duplex - FDD mode).

An important issue, deriving from the introduction of the macrodiversity technique, is the transport of uplink radio frames quality estimations and synchronisation data. In fact the macrodiversity selection/combining function uses BTS quality estimations of the uplink radio frames, while for the downlink there is the need for accurate time synchronisation between the soft handover branches. The quality estimations size of the uplink radio frames ranges from few bytes in case of simple selection algorithms, to n times ($n=4$ or 8) the size of a radio frame in case of more complicated combining algorithms. This fact can have a considerable impact on the traffic load of the link supporting the *Iur* and *Iub* interfaces.

When macrodiversity is applied, several RNSs could be involved in the same connection between a UE and the UTRAN. In this case the RNSs assume different roles depending upon the specific connection (see Figure 3).

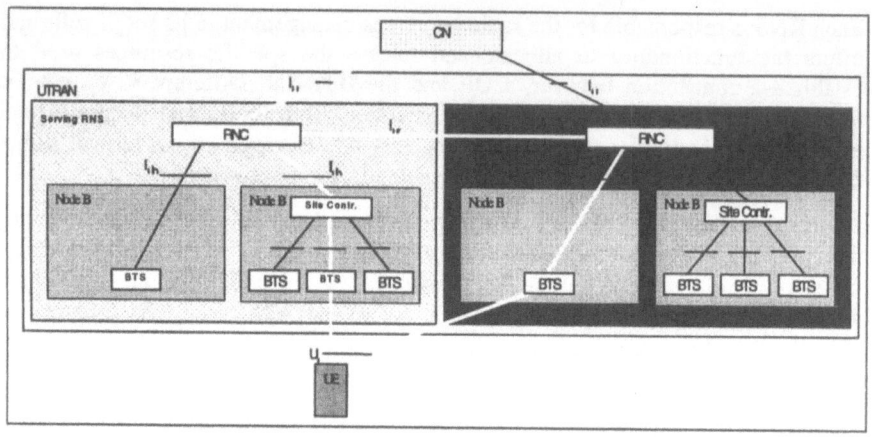

Figure 3: RNSs' roles during macrodiversity

The RNS that terminates the *Iu* interface for this connection is referred as **Serving RNS**; the other RNSs are indicated with the term **Drift RNS**.

The role of each RNS involved in the connection may change along the same connection, as a result of an inter RNS handover (Figure 3 and Figure 4). This procedure is called **Serving RNS Relocation** and is performed when the Serving RNS role needs to be taken over by another RNS. Note that a Serving RNS relocation always implies that the *Iu* interface connection point is moved to the new RNS.

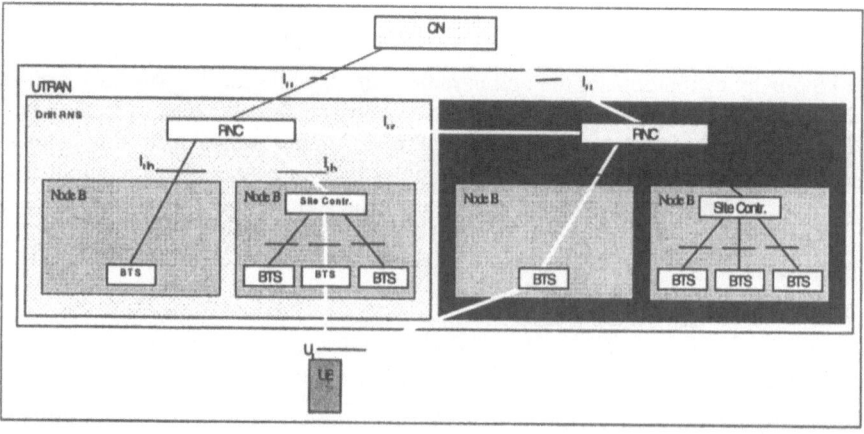

Figure 4: Serving RNS relocation

Although the Serving RNS relocation is triggered by the Access Network, it needs some support from the Core Network to be executed. Furthermore this procedure may follow an inter RNS handover, but the handover should be performed independently, i.e., the *Iur* interface supports the soft handover without necessarily performing a Serving RNS relocation.

The internal architecture of a RNS consists of a **Radio Network Controller** (RNC) and one or more abstract entities currently called **Node B**, connected to the RNC through the *Iub* interface.

A RNC is the equipment located inside the RNS. It is in charge of controlling the use and the integrity of the radio resources. For example, it is responsible for the handover decisions and execution and comprises a combining/splitting function to support macro diversity between different Nodes B.

The Node B is a logical node responsible for radio transmission/reception in one or more cell to/from the UE and terminates the *Iub* interface towards the RNC. Although the internal structure of Node B is unlikely to be included in the specifications, it can consist either of a single BTS or of a set of BTSs connected to a Site Controller. This additional equipment comprises the combining/splitting function to support macro diversity between cells belonging to the same Node B.

The described RNS architecture applies both to the FDD and to the TDD mode, with the exception of the internal node B architecture. However it is not mandatory that a RNS supports both modes. In fact FDD and TDD modes could have different application scenarios, since each mode has its own characteristics and can be optimal for a specific environment. For example FDD could be used for macro-cell/micro-cell coverage and TDD for pico-cell (indoor) coverage.

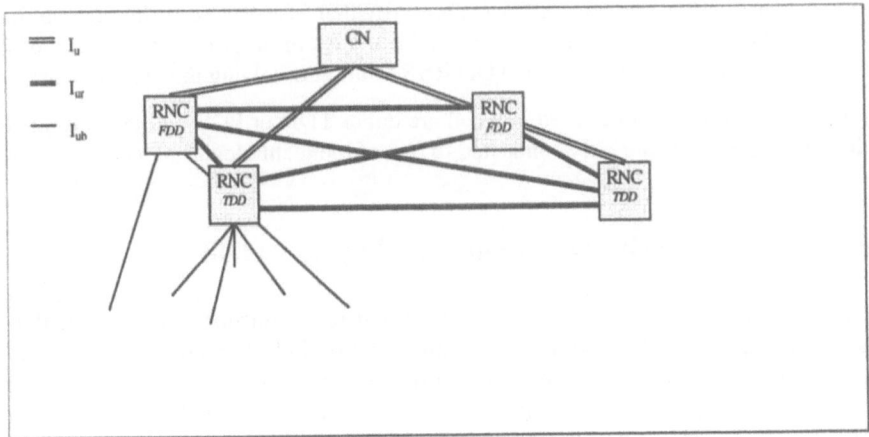

Figure 5: Scenario applying single mode RNSs

In order to allow different levels of flexibility in the network implementation, it is foreseen that both single mode and dual mode RNS could be developed. This means that it should be possible to have either single mode RNC, controlling Nodes B of a single radio technology (see Figure 5), or dual mode RNC, controlling Nodes B of both radio technologies (see Figure 6).

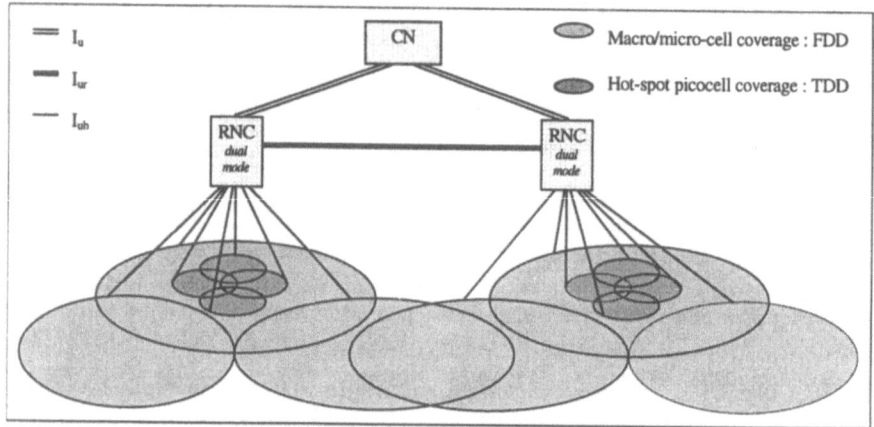

Figure 6: Scenario applying dual mode RNSs

In all cases, the *Iu* interface is independent from the radio mode; this means that the same functional division between the core network and the radio access network, as well as the same services, are provided irrespective of the modes. In addition the *Iur* interface supports both radio modes in order to allow inter mode handover between FDD RNS and TDD RNS without involving the core network.

Finally the radio resources used in a cell are either TDD or FDD and it is assumed that dual-mode UEs are only connected to one of the technologies at a time.

3. Access network dimensioning and optimisation

Once the access network has been described from a functional viewpoint, it is worth introducing some issues on the planing and optimisation aspects. Nowadays the convergence between fixed and mobile services has become a hot topic discussed in many venues. In this perspective, it is therefore likely to assume that also the access network will be involved in this convergence attempt: a possible scenario could be settled by fixed and mobile operators, or service providers, sharing the fixed access infrastructure.

From the technology viewpoint, Passive Optical Networks (PONs) might be used both for distributing typical fixed services (VOD, Teleshopping, Teleteaching, Internet, High Bit Rate data transfer, etc.) and for interconnecting some UMTS nodes B with RNC facility [1].

PONs can be profitably used in a mobile scenario since next generation systems (UMTS/MBS³) will provide customers with advanced multimedia services thus requiring high capacity and large bandwidth in the AN. The deployment cost is an issue to be thoroughly investigated as well. This paper proposes a scenario where a fixed and a (or several) mobile provider(s) distribute their services on the same access network thus sharing both the deployment and the management costs.

It's worth mentioning that the PON technology will be used only between nodes B and RNC and not in the CN (Core Network) where other techniques are expected (e.g., SDH).

3.1 Reference Scenario

The reference scenario is shown in Figure 7. An APON (ATM PON) can be used to link some UMTS nodes B deployed in a indoor/outdoor environment (either domestic or business) or in a typical public (outdoor with microcell layout) environment [2].

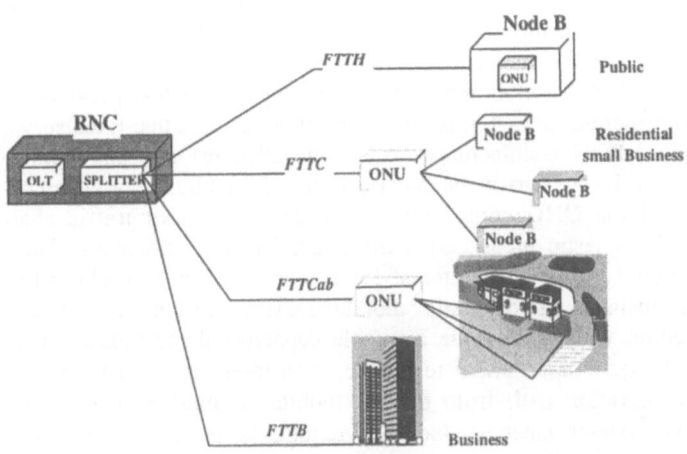

Figure 7: Extended FSAN scenario

The same APON infrastructure is used to distribute typical services of a fixed network operator or service provider, i.e., VOD, Internet, etc. using, for instance, a VDSL technique over a twisted pair physical carrier. This integrated network

---

³　MBS stands for Mobile Broadband Systems. MBS system grant services whose bit rate ranges between 20 and 155 Mbps. Such services usually rely upon a wireless ATM access.

arrangement reduces the overall operators' deployment costs and, at the same time, the management overhead can be shared among different operators or service providers.

The major architectural choice concerns the ONU allocation in the mobile part. In particular two alternatives have been considered and described in the next paragraphs.

3.1.1 Node B connected to the ONU

In this scenario several nodes B are connected to a single ONU as shown in Figure 8.

Figure 8: Short term implementation   **Figure 9: Long term implementation**

The physical carrier linking the ONU to the node B can be either a twisted pair or a coaxial cable using ADSL, HDSL or VDSL technique. This network arrangement seems to be attractive, especially in the short term, since a UMTS operator can deploy a mobile network exploiting, as much as possible, an existing fixed infrastructure. In fact, it is quite sensible to assume that fixed services, based on advanced PON architectures, will be installed prior to any third generation mobile network. The narrow band capabilities offered by the last drop between the node B and the ONU cope with the limited amount of traffic characterising Residential/Domestic areas and some small Business premises. Therefore this scenario could be of some interest for a service provider wishing to offer, for instance, business or domestic mobile services in an area where a fixed infrastructure, with some spare available capacity, already exists. This scenario seems to be quite appropriate to handle, with minor modifications and at a low cost, the migration path from current mobile communication systems to next generation systems since it does not require to perform expensive hardware changes in the node B equipment.

3.1.2 ONU within the node B

The ONU within the node B should be a piece of equipment appropriately designed to support the UMTS characteristics. It is assumed to be cheaper than the typical large ONU commonly used to serve fixed networks. The ONU in the node B arrangement suits the FTTH (Fibre To The Home) implementation: the optical physical transport is used up to the node B front end (no copper or twisted pair is used to interconnect the node B to the ONU) thus ensuring a wide bandwidth availability (Figure 9). The wide band availability allows operators to offer multimedia services on the mobile networks and therefore this arrangement really cope with the third generation systems specifications. It is worth noting that the ONU in the node B arrangement requires undertaking some extensive civil works to bring the fibre up to the node B site. However, the initial high investments will provide the node B with the flexibility necessary to support the fast growing traffic patterns expected in Public and Large Business areas.

3.2 Integrated Access Network Planning Issues

It is now necessary to illustrate the technological issues to be handled when sharing the access network between fixed and mobile services. This paper only deals with the transport issues and tries to give some hints on possible solutions.

3.2.1 Technological Constraints

The architectural choice made in this paper is derived by the work performed in the FSAN context. This scenario is compliant with the results achieved in the BAF project, developed under the relevant RACE Programme and, obviously, compliant as much as possible with ETSI/SMG2 specifications.

The transport network topology is built upon a star topology with OLT and Splitters allocated in the RNC node. Such architecture is grounded on the second alternative previously reported, i.e., one ONU per each node B (see Figure 9). The current technology allows the implementation of OLT devices handling up to 32 ONU nodes. However, future developments expected in the SuperPON context will allow the deployment of a larger number of ONUs (and hence nodes B) as expected in a typical UMTS micro and pico cellular scenario.

In this paper the APON capacity is assumed to be compliant with the SDH hierarchy (e.g., 155 Mbps STM-1 like). A possible planning strategy could suggest Operators to select an overall STM-4 capacity APON and reserve a STM-1 cluster to the UMTS services. The remaining capacity will be dedicated to handle fixed services.

3.3 Bandwidth Estimation

One of the major issue to handle when dealing with the planning of a fixed infrastructure for third generation mobile systems is the coexistence of a wide variety of services characterised by different bandwidth requirements and constraints. The equivalent bandwidth estimation has been proposed in the literature to assist the dimensioning process of ATM networks. With a few adaptations, such a methodology can be profitably used also in the mobile context.

The approach is based on the specification mentioned in [3, 4]. The assumption is that cell scale QoS will be provided by a rough dimensioning of the buffer, taking into account that small buffers are suitable to audio and video sources, while large

344

buffers fit data sources. An equivalent bandwidth formula can be used in both scenarios.

Any type of service has its own arriving and service rate. These two parameters allow estimating the traffic intensity in Erlang. Furthermore, each service requires an equivalent bandwidth W_i calculated as shown in the following flow chart.

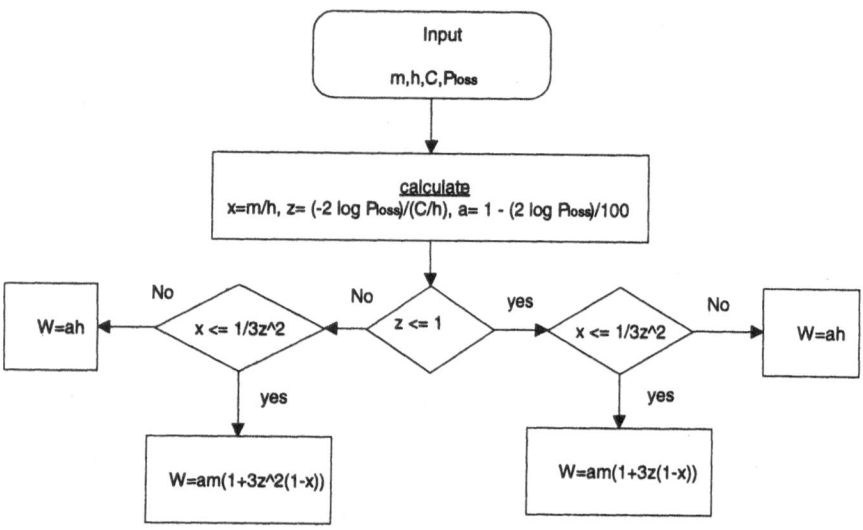

Figure 10: Translation process of QoS parameters into bandwidth estimation

Where:

- m is the average service bit rate;
- h is the service peak bit rate;
- C is the total link capacity (bandwidth);
- $Ploss$ is the service cell loss probability;
- W is the equivalent bandwidth.

Using these assumptions, it is possible to derive a figure representing the bandwidth requirement for each ONU.

3.4 Dimensioning Steps

The network-dimensioning task can be modelled into an optimisation problem and solved using a Tabu Search[4] heuristic approach [5, 6]. The initial scenario is shown in Figure 11 where some 40 node B sites have been deployed in a typical

---

[4]   The Tabu Search is a heuristic algorithm extensively used in network planning and optimisation software tools.

metropolitan environment (high traffic patterns). The position, type and dimensioning (number of channels) are supposed to be the output of a complex network planning process [7].

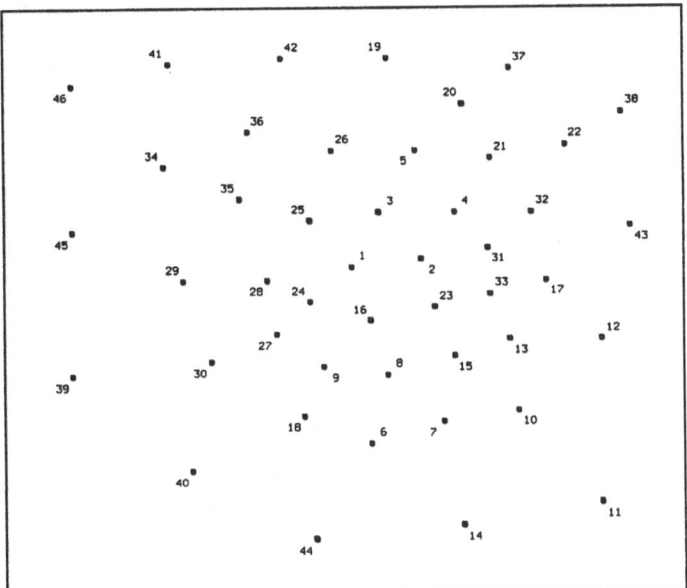

Figure 11: Initial scenario

Once the nodes B have been suitably characterised in terms of offered traffic and bandwidth requirements, it is necessary to determine the minimum number of PONs necessary to accomplish an appropriate interconnecting network and to assume the initial not optimised arrangement[5]. The PON capacity (155 Mbps in this example) is a constraint along with the nodes B geographical location. As a matter of fact, nodes B should be grouped into PONs in order to equally balance the global offered traffic and avoid useless (and expensive) long links or loops.

A rough estimate can be done by simply mapping the global offered traffic (in Erlang) into an equivalent global bandwidth (see section 3.3). Of course, such a mapping is achieved by looking at the service mix characteristics offered by each node B (voice, data, fax, etc.). The example proposed in this document suggests that two PONs should be adequate to appropriately manage the network arrangement shown in Figure 11. Therefore, the initial solution, used to feed the Tabu Search algorithm, is shown in Figure 12.

---

[5]   To start the optimisation process, the Tabu Search algorithm needs a not optimised initial solution.

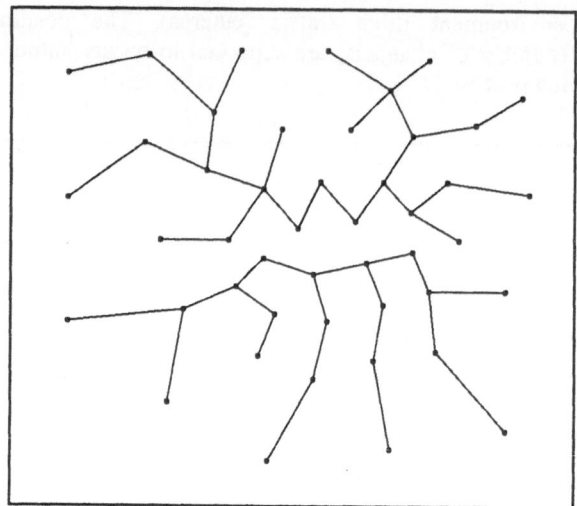

Figure 12: Initial solution

The PON arrangement, in the initial solution, has been achieved selecting a star topology. Civil works in a metropolitan scenario have been assumed to be ten times greater than fibre costs and this parameter has been used to tune the following optimisation algorithm based on the mentioned Tabu Search approach. Cost rates are some parameters duly included in the model that can be tailored and customised by the Operator according to his specific needs. .

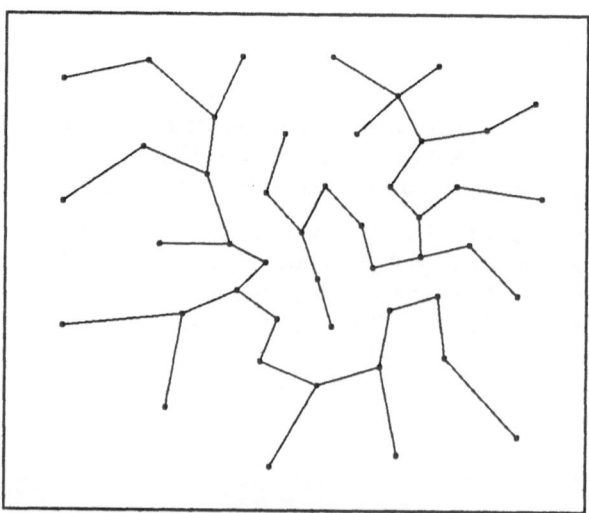

Figure 13: Final and Optimised Network Arrangement

It is worth mentioning that the network optimisation process is focused on three objectives (in order of priority):

1) to ensure that each PON groups a number of nodes B not exceeding the given capacity figure (155 Mbps in this example);

2) to balance the traffic quota associated to each single PON thus avoiding that, in the final configuration, there are highly loaded and scarcely loaded PONs;

3) to reduce, as much as possible the links length between nodes B.

Figure 13 shows the optimisation process result. The comparison with Figure 12 demonstrates that some major network re-arrangement were definitively necessary to meet the optimisation objectives described above.

The final mobile network-planning step to be performed is the RNCs optimum allocation. The candidate sites to host the RNC equipment should be selected in order to minimise the relative distance with all the other graph nodes. This task is again performed automatically by determining the graph median (see Figure 14).

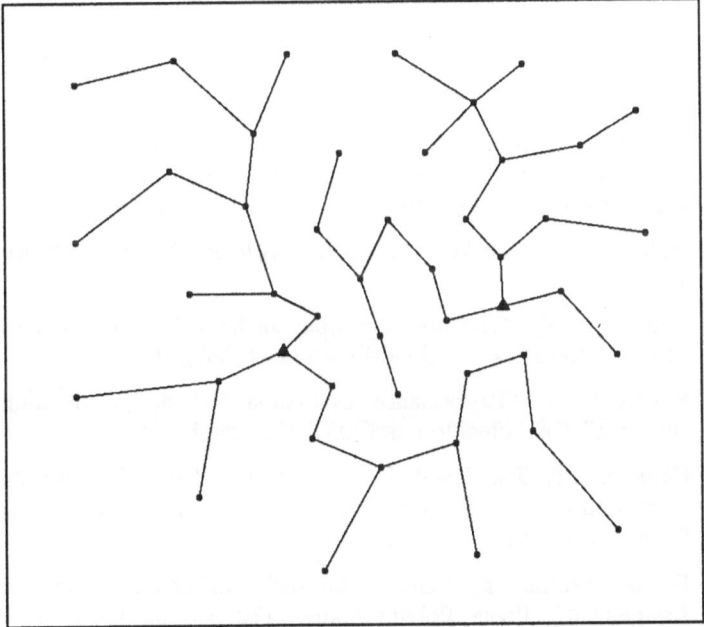

Figure 14: RNC allocation in the final configuration

This step concludes the network planning procedure of a UMTS access network based on a PON arrangement. Other issues, specific of the radio environment,

348

need to be addressed as, for instance, the radio frequency allocation but they do not impact on the access network dimensioning.

4. Conclusions

Third generation mobile systems will be grounded on a novel access network configuration where radio access dependent layers will be clearly separated from radio access independent layers. This feature allows defining a generic network architecture that could be used to support multiple radio access schemes solutions as currently proposed for the UMTS.

This paper also describes some new network properties as, for instance macrodiversity and soft handover and the functionality necessary to support such features.

The second part of the paper is dedicated to dimensioning and optimisation issues arising when an access network structure is shared among different providers, both fixed and mobile. In this perspective, a software tool has been described able to assist network planners in the dimensioning process.

5. References

[1]   F. Badini, E. Berruto, R. Menolascino, M. Mittrich P. Vetter. Chain BAM Final Report. " Sharing the transport network between fixed and mobile systems". May 1998

[2]   FSAN Full Service Access Networks, Atlanta, Georgia (US), March 2-5 1997.

[3]   Lindberger K.: "The use of simple methods for integrated call scale streams", COST 242 Technical Document (065), 1992.

[4]   Robert J. W. "Performance evaluation and design of multiservice networks". CEC Final Report COST 224, October 1991.

[5]   Emile Aarts, Jan Karel Lenstra. "Local Search in Combinatorial Optimisation". Wiley Interscience Series in Discrete Mathematics and Optimisation, 1997.

[6]   D. de Werra, Y. Gay. "Chromatic Scheduling and Frequency Assignment". École Polythechnique Fédérale de Lausanne, ORWP, 89/06, 1991.

[7]   CEC Deliverable A016/NTU/DS/L/078/a1. "Final report on resource dimensioning", 31 August 1998.

Wireless Multimedia Assessment with Traffic, Access and Transmission Protocols in Actual Environment at Millimeter Waves

O.Andrisano (*), D.Dardari (**), G. Mazzini (***)
(*) DEIS-CSITE, (**) CNIT
University of Bologna, V.le Risorgimento 2, 40136 Bologna, Italy.
E-mail: {oandrisano,ddardari}@deis.unibo.it
(***) DI, University of Ferrara, via Saragat 1, 44100 Ferrara, Italy.
E-mail: g.mazzini@ieee.org

Abstract. A complete study by means of a semi-analytical procedure is proposed to evaluate the performance of a wireless star polling network in actual environment at 60 GHz. In order to underline this point, the paper takes into consideration a real traffic, measured on an Ethernet LAN; a model for traffic routing by properly selecting intra and inter wireless communications; a medium access control realized by means of polling structure with a central hub on the ceil of the room; a transmission system based on OFDM to counteract the selective fading effects; a propagation channel characterized with a ray tracing model. All these topics are jointly considered through the paper and the performance is reported as a function of the service availability, based on the access time for real-time traffic, and the average access time. The strong impact of the actual environment through the service sensitivity to error probability is underlined

1 Introduction

In the next few years, increasing attention to Wireless Local Area Networks (WLAN) is expected due to the benefits, primarily mobility and installation flexibility with respect to fixed LAN, that WLAN can offer to the end user [1] [2] [3]. More generally, the demand of wide-band multimedia services arising from wireless users has stimulated particular interest in the design of high quality, high speed (beyond 10 Mbit/s, up to 155 Mbit/s e.g. ATM [4]) wireless networks able to provide access to services offered by fixed networks.

The use of millimeter waves (e.g. 60 GHz) is attractive due to the large bandwidth available and the large wall attenuation which leads to low co-channel interference between WLANs and easy cell planning [5] [6]. Recently, some studies and proposals for high-speed transmissions have been oriented to the use of multi-carrier modulation, such as Coded Orthogonal Frequency Division Multiplexing (COFDM), in order to counteract the propagation frequency selectivity and exploit the inherent diversity of a wide-band channel [3, 7, 8].

In this paper we give a semi-analytical procedure in order to characterize the performance of a WLAN with a star topology in a realistic environment as the traffic and propagation are concerned. The investigation is carried out in five topics in which the paper is structured, with a top-down approach: traffic modeling, traffic routing, medium access control, channel model and transmission scheme. Finally, some results and conclusions are presented by mainly selecting the service availability probability as performance index.

2 Traffic Generation

The classical model used in the literature for traffic generation is based on the Poisson distribution [9]; this means that the source is modeled through a two state Markov chain (one for activity and one for inactivity) in which the time spent in each state has an exponential probability density function (p.d.f.). This model is in good agreement with classic teletraffic behavior and has been used for many years in order to design circuit-switched based systems.

Recently, some measurements taken on Ethernet [10] have shown that the actual traffic nature is not in agreement with the Poisson model. The main consideration is due to the memory introduced by the query and response iterations on packet networks, where the client server architecture is widely used. This correlation is not taken into account by the Poisson model that requires memoryless time. Measures show that both the single source and its aggregation could be modeled as self-similar one. This behavior can be taken as reference for multimedia services.

Some different self-similarity definitions can be given; let us refer to the asymptotic one [10], here revised. Let τ be the elementary observation interval; let $x_i = x(i\tau)$ be the i-th stationary process sample such that $x_i \in \{0, 1\}$; let $x_i^{(m)}$ be the i-th sample of the aggregate process defined as $x_i^{(m)} = 1/m \sum_{k=1}^{m} x_{mi+k}$, where $x_i^{(1)} = x_i$; let $C^{(m)}(k)$ be the auto covariance function of the aggregated process defined as $C^{(m)}(k) = E[x_i^{(m)} x_{i+k}^{(m)}] - E^2[x_i^{(m)}]$, that does not depend on i by virtue of the stationarity of the process.

We assume that a process is asymptotically second-order self-similar if an m' exists such that for all $m > m'$, for k large enough and for $\beta \in]0, 1[$:

1) $\frac{C^{(m)}(k)}{C^{(m)}(0)} \approx \frac{C^{(m')}(k)}{C^{(m')}(0)} \approx Ak^{-\beta}$,

2) $\frac{C^{(m)}(0)}{C^{(m')}(0)} \approx \left(\frac{m}{m'}\right)^{-\beta}$,

where A is a constant that does not depend on k and m.

The traffic sources are characterized by two parameters: the activity index, $\lambda = E[x_i]$, and the self-similarity index, $H = 1 - \beta/2$, usually called Hurst parameter.

In this paper we will use an hour of Ethernet observation as single station traffic source with given λ and H parameters. This observation has been

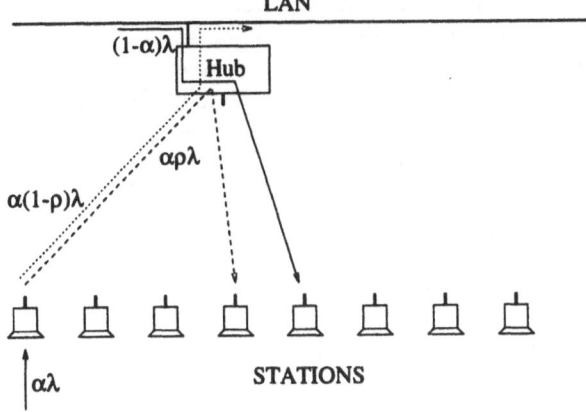

Fig. 1. Traffic route parameters

done on an Ethernet 10base2 cable in our department, separated from other networks by means of a bridge.

The measure on Ethernet is used to extract the couple (T_i, B_i) for each packet, where T_i is the starting time of the packet transmission and B_i the byte length. Let us observe that the measures are made with low traffic load, in order to avoid collision and retransmission from CSMA-CD Ethernet protocol. We checked the self-similarity property on the collected samples with estimated parameters $\lambda = 1.4\ 10^{-2}$ and $H = 0.86$.

In order to modulate the traffic load, without loss in the self-similarity behavior (maintain the Hurst parameter), we introduce a time scale factor, SF.

3 Traffic Routing

Given the traffic source, the problem is to build a model for the packet routing. By referring to figure 1, let us consider the presence of N_s stations and model the Hub as a set of N_s equivalent stations. The equivalent $2N_s$ stations have a total average generation $N_s \lambda / SF$.

Let us define $\alpha =$Prob[*packet born on a WLAN station*], that is the probability that a generated packet is relative to a station; $1 - \alpha$ is the probability that the packet is born outside the WLAN. This parameter gives a model of the inter LAN communications.

Let us define $\rho =$Prob[*packet born on a WLAN station is addressed to a WLAN station*], that is the probability that a WLAN generated packet is addressed to a WLAN station; so $1 - \rho$ is the probability that the packet goes outside the WLAN. This parameter gives a model of the intra LAN communications.

Let us assume an uniformly random address scheme, due to lack of measures and models for routing decisions, S the source and D the destination entities, respectively, so that the routing is described by the couple (S, D). S and D span in the range $0 \ldots N_s$, where 0 identifies a generic station outside the WLAN and involving the inter-LAN communications; moreover, the range $1 \ldots N_s$ identifies a WLAN station. A possible algorithm for selecting the (S, D) couple for each packet considered is the following. Let fr be a real random number in the range $[0, 1]$, $Fr(A)$ a function giving a natural random value in the range $0 \ldots N_s$ that is outside the set A. Then

$$S = \begin{cases} Fr(\{0\}) \text{ if } fr < \alpha \\ 0 \quad\quad\quad \text{otherwise} \end{cases} \quad D = \begin{cases} Fr(\{0\}) & \text{if } S = 0 \\ Fr(\{0, S\}) \text{ if } fr < \rho \text{ and } S \neq 0 \\ 0 \quad\quad\quad \text{otherwise} \end{cases} \quad (1)$$

4 Medium Access Control Protocol

A centralized controlled access scheme is here considered due to its ability to efficiently deal with both asynchronous and synchronous traffic. This protocol has received some attention in recent literature [11, 12, 13, 14, 15].

It is based on the token passing strategy, where the channel is utilized only by the station which holds the token. Furthermore, a low user mobility is considered in order to have a stationary channel during packet and token transmissions.

For multimedia traffic the access time, T_{acc}, represents a very significant parameter that has to be checked. The access time is defined as the time required to transmit with success a packet from the source S to the destination station D. When D and S are two WLAN stations, the access time takes into account both the up-link (source WLAN terminal to Hub) and down-link (Hub to destination WLAN terminal) with related queueing times.

As the channel introduces errors due to different transmission impairment, to obtain error free transmissions, a stop-and-wait ARQ strategy is introduced. The time to correctly deliver information is analytically computed by using the bit error probability, P [16]: $T = T_G * TX$, where T_G is the time to send a packet or token of length X bits, $T_G = (X + TA)/C + 2\tau$, TA is the length of ACK or token, here assumed to be of equal length, C the link bit rate, τ the actual propagation delay, TX the total number of transmissions given by $TX = 1/(1 - P)^X$.

In order to avoid large access time on each single link, a threshold on transmission attempts number, STX, is adopted. In particular, if the estimate of TX overcomes this threshold, no further packet transmission attempts are performed until a 'good' channel condition is met. In this case the hub takes the control and tries the next station. In the case of real-time traffic blocked packets could be discarded as they becomes not significant. On the contrary, in the investigated case of data traffic, packets remain in queue.

An interesting index, considered in this paper, is the service availability defined as

$$\text{service availability} = \text{Prob}[T_{acc} > T_{th}] \qquad (2)$$

where T_{th} is a threshold dependent on the particular service.

5 Channel Model

Millimeter wave bands (e.g. 60 GHz) are under investigation for high speed wireless systems design due to the large bandwidth available for multimedia applications [6].

We consider the band around 60 GHz for which the electromagnetic field is practically confined in each environment due to the high wall attenuation. This makes the system more resistant to external interference, even if a larger number of transmitters is required to cover wide scenarios. A typical indoor environment has been considered whose propagation analysis results can be found in [17, 18]. A database has been derived by means of Ray-tracing modeling in order to evaluate the transmission system performance. In particular, a large number of configurations have been tested by moving the receiver in a grid of 1700 points with 0.25 meter step. The transmitter (HUB) was put in the middle of the room at ceiling height. For each configuration both narrowband and wide-band characteristics are available for path-loss and channel frequency dispersion estimates.

It has been checked that major link difficulties are found when transmitter and receiver are not in visibility (NLOS condition), for example, when obstacles are encountered. In these cases the propagation channel experiences both a large additional path loss (over 20 dB) and frequency selectivity (delay spread up to 20 ns in this scenario) [18].

In order to counteract propagation effects special attention has to be payed to transmission system definition. We considered sectored antennas at the receiver end to improve the link budget and reduce frequency selectivity. It represents a trade-off between antenna flexibility and complexity. The selection strategy adopted is based on the measure of the power received from each sector. An ideal sectored radiation pattern is considered for the terminal antenna, whereas an omni-directional antenna is placed at the Hub.

6 Transmission System

As far as the transmission scheme is concerned, COFDM received particular attention recently for high speed indoor transmissions [3, 7]. The OFDM modulation scheme is able to deal with large channel selectivity without complex equalization technique.

The functional scheme considered is reported in figure 2 which is based on what proposed in [7] for this application at $C = 155$ Mbit/s. It uses $N = 32$

TRANSMITTER

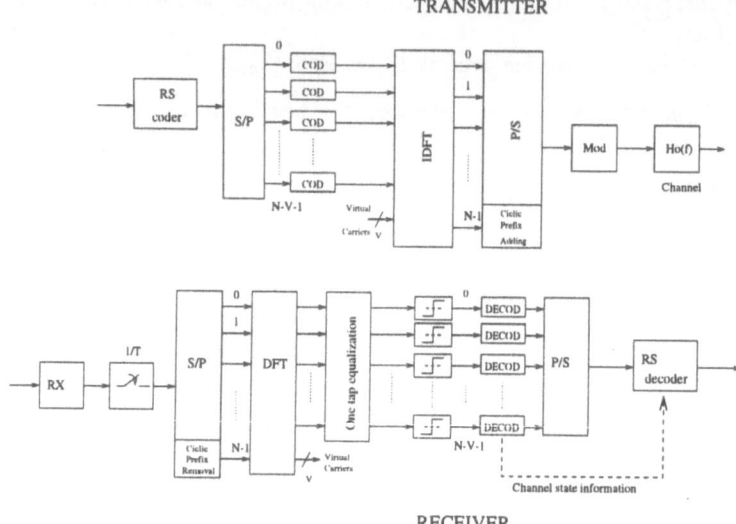

Fig. 2. Coded-OFDM system considered.

sub-carriers where $V = 6$ of them are reserved as virtual to avoid aliasing effects due to filtering. At each sub-channel a 4-QAM signaling is adopted. Sub-carriers are frequency spaced by $\Delta f = 1/NT$ where T is the symbol time [21], and a cyclic prefix $D = 8$ is added after the IDFT block to avoid inter-symbol interference due to channel distortion. This value leads to a guard time T_g approaching the maximum delay of the propagation (about 50 ns).

A two level coding technique is considered based on a RS(104,61) block outer code and a Parity check (8,7) inner code for error detection. The inner code is applied in each sub-channel with the purpose, at the receiver, to detect errors and classify 'bad' from 'good' sub-channels (channel state information). This information can also be utilized to realize the ARQ scheme. The RS outer code utilizes the channel state information and it is able to make erasure detection decoding with a good system performance improvement [19, 20, 7]. The decoding technique is based on hard-decision. The total coding rate is equal to $R_c = 0.51$ leading to an useful word length $F = 427$ bits.

The performance of this scheme can be evaluated by following the method-ology introduced in [7] which gives, by means of a low complexity iterative algorithm, the word error probability, P_w, after decoding by starting from the knowledge of the uncoded bit error probabilities $\{p_n\}$ at the output of each sub-channel. For each point under study in the room the channel trans-fer function $P(f)$ is known from the propagation database. The uncoded bit error probability related to the n-th sub-channel is given by (assuming ideal frequency and time synchronisms, perfect channel parameter estimation)

$$p_n = \frac{1}{2}\text{erfc}\sqrt{\frac{E_b}{N_0}|H_n|^2\frac{N}{N+D}R_c}, \qquad (3)$$

where E_b/N_0 is the signal-to-noise ratio and $H_n = H_0(n\Delta f)$ are the DFT samples of the discretized low-pass equivalent channel normalized impulse response; transfer function $H_0(f)$ is the low-pass equivalent version of $P(f)$ normalized to the mean received power, already taken into account in the ratio E_b/N_0, in the considered point.

By means of ray-tracing simulation and the semi-analytical approach, a one hour of propagation delays and bit error probability couples, (τ_i, P_i), has been evaluated. Users are considered on random basis placed in the grid. During the simulation, they move very slowly (2.5 cm/s) compared with the polling cycle time in order to have a quasi-stationary scenario. The direction of movement is chosen randomly in $[0, 2\pi]$.

For each user configuration, delay and channel transfer function are estimated from the propagation database. The word error probability is evaluated as previously stated. The actual packet and token length transmitted is bounded to a multiple of the single word length F. The packet error probability is then obtained directly from the word error probability.

7 Numerical Results and Conclusions

A simulator has been developed to merge all the described effects. In order to have random use of traffic and channel configurations, their usage has been randomized in sample selection. In particular the randomization has been done for each possible $N_s + 1$ source (N WLAN stations and one wired external) and the channel for each possible N_s links, respectively.

Common system parameters assumed are: $C = 155\text{Mbs}$, $TA = F = 427\text{bits}$, $N_s = 10$, $STX = 5$. Two scenarios has been investigated: the former by varying the scale factor (i.e. varying the traffic load) $SF = 1, 0.01$ by fixing the routing probability to $\alpha = \rho = 0.5$; the latter by varying the routing with the couples $(\alpha, \rho) = (0.2, 0.2), (0.2, 0.8), (0.8, 0.2), (0.8, 0.8)$ and with the original scale factor $SF = 1$.

For each scenario the service availability is reported in figure 3 and 4 as primary performance index as a function of the target time threshold, T_{th}, required by the particular service. Curves related to a free error channel ($P = 0$) are also reported for comparison and to show the strong impact of the channel on service availability. Let us observe that by changing the traffic load (figure 3), by means of the scale factor, no particular difference in the behavior can be noted. This effect is due to the low total load with respect to the high channel capacity and to the necessity of bounding the transmitted packet length at multiple of F. In fact, the packet bounding is due to the use of redundancy (coding, training sequence, synchronization), necessary to have a reliable link. This makes the amount of bits in the token and ACK

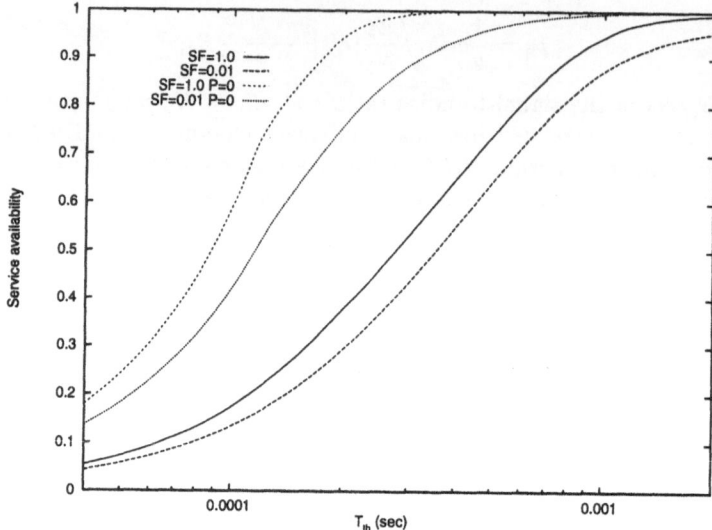

Fig. 3. Service availability (former scenario). $\alpha = \rho = 0.5$.

SF	T_{acc}	tot_pkt	$queued\_pkt$	$\%ARQ$
1.0	$1.26 \ 10^{-3}$	232241	1645	0.209
0.01	$5.63 \ 10^{-2}$	23212213	156977	0.185

$SF \ (P=0)$	T_{acc}	tot_pkt	$queued\_pkt$	$\%ARQ$
1.0	$9.58 \ 10^{-5}$	232241	0	0
0.01	$1.59 \ 10^{-4}$	23212213	0	0

Table 1. Former scenario table (actual and error free channel).

transmissions number comparable with that of useful packets: therefore, it introduces a limitation in the traffic impact on system performance that is to be expected for each indoor wireless access. This investigation is far from the classical studies on wired networks where no particular limitation on the packet format is required. More effort has to be done to study high speed modulation schemes able to deal with variable and shorter packet lengths.

For each scenario considered, a table containing the average access time, T_{acc}, the total number of packets, tot_pkt, involved in the simulation, the number, $queued\_pkt$, of packets left in the queues at the end of the simulation and the fraction of blocked ARQ with the fixed threshold STX, $\%ARQ$, are reported. The high value of $queued\_pkt$ is due to the presence of several WLAN terminal configurations where the link is not available in force

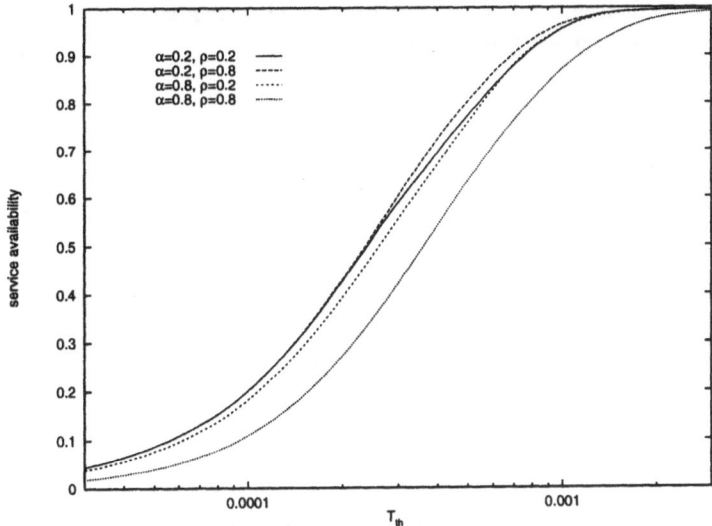

Fig. 4. Service availability (latter scenario).

α	ρ	T_{acc}	tot_pkt	$queued\_pkt$	$\%ARQ$
0.2	0.2	$1.14\ 10^{-3}$	235798	1607	0.205
0.2	0.8	$1.16\ 10^{-3}$	235712	1752	0.217
0.8	0.2	$1.16\ 10^{-3}$	214535	1504	0.216
0.8	0.8	$1.74\ 10^{-3}$	214323	1999	0.209

Table 2. Latter scenario table.

of the propagation at millimeter waves. The same effect is also noted on $\%ARQ$, with more emphasis due to the continuous token polling attempts in all channel conditions, independently of waiting packet number. A more useful strategy could be to perform a selective polling with rate adaptive to the channel condition.

As can be noted in table 2, there is not a great sensibility of the mean access time, T_{acc}, on routing parameters α and ρ. The service availability is a more suitable index by means of which the routing effects can be evaluated (see figure 4).

An analysis of the integration of indoor channel and modulation with coding has shown a step-like behavior as regards the bit error probability. This makes the performance weakly sensible on the ARQ threshold STX.

Finally, even if more work has to be done, this paper gives a first step for an integrated study that takes into account different topics, ranging from propagation to traffic.

This work has been founded by the University of Bologna (Progetto Reti Locali ad Onde Millimetriche).

References

1. K.Pahlavan, T.H.Probert, M.E.Chase "Trends in Local Wireless Networks", IEEE Comm. Mag. , vol.33, no.3, march 1995, pp.88-95
2. R.O.LaMarie, A.Krishna, P.Bhagwat, J.Panian, "Wireless LANs and Mobile Networking: Standard and Future Directions", IEEE Comm. Mag. , vol.34, no.8, august 1996, pp.86-94
3. A.Falsafi, K.Pahlavan, G.Yang, "Transmission Techniques for Radio LAN's - A Comparative Performance Evaluation Using Ray Tracing", IEEE Journal of Sel. Areas in Comm, vol.14, no.3, april 1996, pp. 477-491
4. E.Ayanoglu, K.Y.Eng, M.J.Karol "Wireless ATM: Limits, Challenges, and Proposals", IEEE Personal Communications, vol.3, no.4, august 1996, pp.18-34
5. P.Marinier, G.Y.Delisle, L.Talbi, "A Coverage Prediction Technique for Indoor Wireless Millimeter Waves Systems", Wireles Pers. Comm., vol.3, no.3, 1996, pp.257-271
6. T. Ihara, T. Manabe, M. Fujita, T. Matsui and Y. Sugimoto, "Research Activities on Millimeter-wave Indoor Wireless Communications Systems at CRL", ICUPC'95, Nov. 1995, Tokyo.
7. D.Dardari, V.Tralli, "Performance and Design Criteria for High-Speed Indoor Wireless COFDM systems at 60 GHz", IEEE GLOBECOM 1997, Phoenix.
8. O. Andrisano, M.Chiani, D.Dardari, A.Zanella, " Service Availability of Broadband Wireless Networks for Indoor Multimedia at Millimeter Waves", submitted to ISSSE'98, Pisa, Italy.
9. L.Kleinrock, "Queueing Systems", John Wiley & Sons, 1975.
10. W.E. Leland, M.S.Taqqu, W.Willinger and D.V.Wilson, "On the Self-Similar Nature of Ethernet Traffic (Extended Version)", IEEE/ACM Transactions on Networking, Vol.2, No.1, Feb. 1994.
11. R.L.Davies, R.M.Watson, A.Munro and M.H.Barton, "Ad-Hoc Wireless Networking: Contention Free Multiple Access Using Token Passing", in proc. of VTC '95, Chicago, Illinois, USA, pp. 361-365.
12. Z.Zhang and A.S.Acampora, "Performance of a Modified Polling Strategy for Broadband Wireless LANs in a Harsh Fading Environment", in proc. of GLOBECOM 91, Phoenix, Arizona, USA, pp. 1141-1144.
13. A.Hoffmann, R.J.Haines and A.H.Aghvami, "Performance Analysis of a Token Based MAC Protocol with Asymmetric Polling Strategy ('TOPO') for Indoor Radio Local Area Networks under Channel Outage Conditions", in proc. of ICC'94, New Orleans, Louisiana, USA, pp. 1306-1310.
14. A. Giovanardi, G. Mazzini, "Time Division Token Based WLAN's: a Proposal and Analysis", in proc. of ISSSTA'98, Sept 1998, Sun City, South Africa.
15. O. Andrisano, D. Dardari, G. Mazzini, "An Integrated Approach for the Design of Wide-band Wireless LAN", in proc. of ICT'98, Jun 1998, Greece.
16. M.Schwartz, "Telecommunication Networks: Protocols, Modeling and Analysis", Addison-Wesley, 1987.

17. D.Dardari, L.Minelli, V.Tralli and O.Andrisano "Wideband Indoor Communication Channel at 60 Ghz", PIMRC'96, Oct. 1996, Taipei, Taiwan.

18. D.Dardari, L.Minelli, V.Tralli and O.Andrisano "Fast Ray-Tracing Characterization of Indoor Propagation Channels at 60 GHz", Proc. VTC'97, May 1997, Phoenix, AZ

19. S.Lin, D.J.Costello, "Error Control Coding: Fundamentals and Applications", Prentice-Hall

20. A.M.Saleh, L.J.Cimini, "Indoor radio Communications Using Time-Division Multiple Access with Cyclical Slow Frequency Hopping and Coding", IEEE Journal of Sel. Areas in Comm, vol.7, no.1, january 1989, pp. 59-70

21. S.B.Weinstein, P.M.Ebert, "Data Transmission by Frequency Division Multiplexing Using the Discrete Fourier Transform", IEEE Trans. on Comm., vol.19, no.5, october 1971, pp.73-83

Error Propagation Stop Principle for CD3-OFDM : a Comparison with Perfect Coherent Demodulation

Ludovic GRANDJEAN, PIERRE COMBELLES AND DAMIEN CASTELAIN
CCETT-CNET/DMR/DDH 4, rue du Clos-Courtel
35512 Cesson-Sévigné FRANCE
email : ludovic.grandjean@cnet.francetelecom.fr

Abstract

This paper presents an improvement on the Coded Decision Directed Demodulation (CD3) which is applicable for some Orthogonal Frequency Division Multiplex (OFDM) signals. The CD3 channel estimation is based on a data-oriented method which may be affected by error propagation. This effect can be reduced by the addition of a correction block (*eps*: Error Propagation Stop) in the CD3 receiver chain. The performance of the *eps*-CD3 has been evaluated by computer simulations in the presence of Additive White Gaussian Noise (AWGN), Doppler effect (mobile environnement) and impulsive noise. A comparison with perfect coherent demodulation shows a good robustness of *eps*-CD3 against strong disturbances in fixed or mobile conditions.

1 Introduction

Since the last decade, some multicarrier signal modulation systems, using C-OFDM, have been developed for television and radio digital transmission in the European projects DAB and dTTb. The latter have ended at the norms : DAB for digital audio radiodiffusion, DVB-T for terrestrial digital television.

The DAB and DVB-T norms respectively use differential and coherent [with reference carriers insertion (or pilots)] demodulations. However, another multicarrier demodulation method has been proposed by the RAI research centre, called Coded Decision Directed Demodulation (CD3). This RAI receiver authorises a richest signal in pilots. The goal of this paper is to propose and analyse an improvement of this data-aided demodulation process, by reducing the error propagation phenomena.

Section 2 describes the general principles of OFDM. Part 3 presents the basis equations of OFDM coherent demodulation with data exploitation. Section 4 reports the OFDM-CD3 receiver and the feedback loop operation. Part 5 underlines the drawback of the closed loop : the propagated errors. Section 6 presents one solution at this issue : the Error Propagation Stop (*eps*) principle. Finally, part 7 describes the measurements with *eps* process and its results improvement.

2 COFDM presentation

The OFDM is a classic multicarrier modulation scheme, where the overlapping of the orthogonal subcarriers allows a high spectrum efficiency. In addition, the inherent large time duration of the corresponding modulation symbols enhances the capability of the system to deal with large echoes. Therefore, one of the advantages of the OFDM[3] is to be effective in a multipath environment with the presence of several propagation paths between the transmitter and the receiver. This characteristic yields both frequency fadings in received spectrum and ISI which can degrade the received performance in giving high BER .

In combining the OFDM scheme with channel convolutional coding and frequency interleaving, fadings could be fought. When a carrier is fallen in a channel notch, the frequency interleaving is scattering the bits errors at the input of the Viterbi decoder which appear like some random isolated errors. Then the decoder can correct most of them. Moreover, the input soft metrics in Viterbi should take account of the reliability information reflecting the channel response characteristics : a lower weight is given to the bits falling in channel fadings.

OFDM systems are made robust against echoes by inserting a guard interval before each symbol, longer than the lengthiest path delay. The guard interval content is the symbol end repeat in this interval. So if a suitable synchronisation of the receiver is carried out, the information from one given symbol is isolated. ISI is eliminated and orthogonality is then keeping. Within each symbol, after FFT, equalisation may be performed on each carrier independently.

3 OFDM coherent demodulation with data exploitation (CD3) basis equations

The CD3 (Coded Directed Decision Demodulation)[6] coherent demodulation uses the channel estimation. The channel response is obtained after received data exploitation. As insertion reference carriers are useless with CD3 [compared with coherent demodulation by insertion reference carriers (Cf. DVB-T Norm[2])], the debit gain is therefore from 5 up to 15 %. The format constellation is QAM . The complex equaliser output is equal to :

$$\underline{z}(n,k) \quad = \frac{\underline{y}(n,k)}{\hat{\underline{H}}(n,k)} \quad \approx \quad x(n,k)e^{j\phi(n,k)} + \underline{\nu}(n,k) \tag{1}$$

But in the hypothesis where the channel is almost stationary between n and (n-1) time moments[1], the equaliser output can be obtained by :

$$\underline{z}(n,k) \quad = \frac{\underline{y}(n,k)}{\hat{H}(n-1,k)} \quad \approx \quad \underline{x}(n,k) + \underline{\nu}(n,k) \tag{2}$$

If the transmitted OFDM symbol is known with very few errors in (n-1) by the receiver (second hypothesis), the channel estimation becomes :

$$\hat{\underline{H}}(n-1,k) = \frac{\underline{y}(n-1,k)}{\hat{\underline{x}}(n-1,k)} = \ H(n-1,k) + \underline{\epsilon}(n-1,k) \approx \ \hat{\underline{H}}(n,k) \tag{3}$$

where $\underline{y}$ is the received signal, $\hat{\underline{x}}$ the estimate of the transmitted signal and $\underline{\epsilon}$ the component of gaussian noise.

It is important to note that the noise $\underline{\epsilon}(n-1,k)$ is reduced by smooth operation (frequential filter) at the output of channel estimation. The transmitted signal $\hat{\underline{x}}(n-1,k)$ is estimated from the equalised signal on previous OFDM symbol in (n-1), thanks to the CD3 closed loop (Cf. figure 1). This estimation of $\hat{\underline{x}}$ is made reliable by decoding then coding. To summerise, Demodulated and decoded $\hat{\underline{x}}(n-1,k)$ is used in order to estimate the response $\hat{\underline{H}}(n-1,k)$. This response is then applied to demodulate $\hat{\underline{x}}(n,k)$. The idea of reinjecting the transmitted symbol for estimating the channel (on prior symbol) needs the use of a closed loop in the CD3 receiver.

4 OFDM-CD3 receiver and closed loop operation

4.1 OFDM-CD3 receiver

The reception part treats received datas after crossing the hertzian channel. The upper part of the CD3 reception chain articulates around the inverse functions of the emission part :

---

[1]n is the time domain index (OFDM row symbol) and k the frequential index (column carrier).

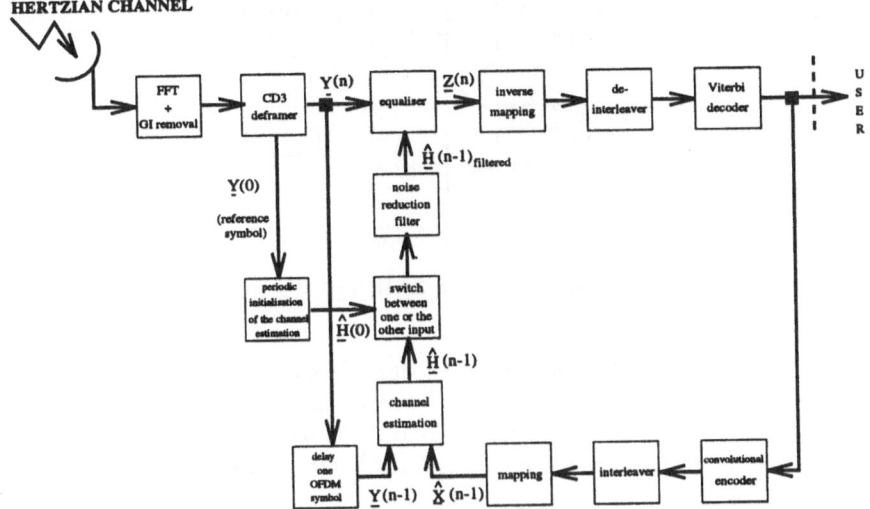

Figure 1: CD3 receiver.

4.2 Loop operation

The CD3 demodulation loop operation process is :

1. <u>step 0</u> :

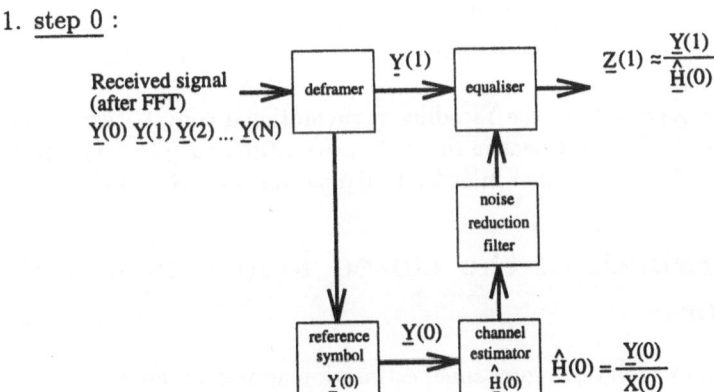

Figure 2: CD3 loop step 0.

loop start (reference symbol $\underline{X}(0)$[2] already known) (Cf. equations 2 et 3 given $\underline{Y}$ and $\underline{H}$) ;

2. <u>step 1</u> : decoding of the equalised signal $\underline{Z}(1)$ and then regeneration of the transmitted signal $\underline{\hat{X}}(1)$.

---

[2]X, Y, Z are some frequential vectors of one symbol OFDM.

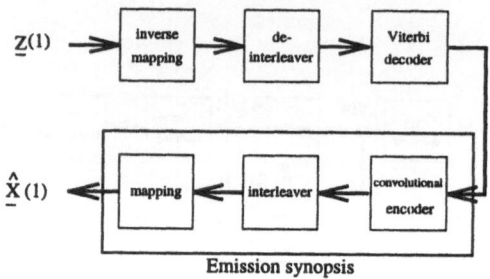

Figure 3: CD3 loop step 1.

The Viterbi decoder enables the exiting of some reliable data in the system working limit (BER $= 2.10^{-4}$ needed at the output of the Viterbi decoder). Because of the closed loop, CD3 demodulation has also another running limit. For a BER $\geq 10^{-3}$, the loop takes down ;

3. step 2 : looping by calculated channel estimation using the previous symbol.

   The stability of the CD3 chain is ensured by channel estimation initialisation for each frame (reference symbol periodicity). The following basis equations are used (Cf. equations 2 et 3, figure 1) :

$$\underline{Z}(n) = \frac{\underline{Y}(n)}{\hat{\underline{H}}(n-1)} \quad \text{and} \quad \hat{\underline{H}}(n-1) = \frac{\underline{Y}(n)}{\hat{\underline{X}}(n-1)} \tag{4}$$

where $\hat{\underline{X}}(n-1)$ is the recoding (convolutional coder) -remapping of $\underline{Z}(n-1)$. We must bear in mind that the estimated $\hat{\underline{H}}(n-1)$ is filtered before the equalisation in order to reduce the additive noise.

5 Drawback of the closed loop : propagated errors

Some errors or variations in channel estimation appear because of :

- Doppler effect : channel varies in time (strong variation from one OFDM symbol to another one) which represents a normal variation ;

- poor decisions in the Viterbi decoder that then reinjects a false information (pathologic variation).

If an erroneous decision is taken by the Viterbi decoder, these effects are more serious than a simple error because this error is reintroduced and affects following decisions. This principle is called error propagation effect.

More precisely, the decoding error will be found in the regenerated transmitted signal $\hat{\underline{x}}_e(n-1,k)^3$.

Therefore, the channel estimation will be erroneous :

$$\hat{\underline{H}}_e(n-1,k) \quad = \frac{\underline{y}(n-1,k)}{\hat{\underline{x}}_e(n-1,k)} \tag{5}$$

At the next symbol, the equaliser output will inject the error again, reduced eventually by the frequential filter, at the Viterbi decoder input :

$$\underline{z}_e(n,k) \quad = \frac{\underline{y}(n,k)}{\hat{\underline{H}}_e(n-1,k)} \tag{6}$$

Even if the interleaver can avoid an avalanche effect by spreading out the false bits, the Viterbi decoder could sometimes not correct these errors. The errors reinjection must be checked and stopped.

If the Viterbi decoder does not succeed for correcting and if no stop process is applied, the errors propagation never shut down until a new initialisation symbol. So we have looked for attenuating this phenomenon by the *eps* (Error Propagation Stop) process.

6 Error Propagation Stop (*eps*) principle

If an error occurs, it is possible to change the defected channel estimation by a more correct one (the estimation found on prior symbol for instance). Artificially, the loop instability is limited. The critical point of this study is to determine the error detection criterion.

These errors are determined by checking that the variation in phase or in complex amplitude between two consecutive symbols does not exceed a particular threshold.

6.1 *eps* principle diagram

The two inputs of the correcting process are : the estimator output before filtering at the moment n $\hat{\underline{H}}(n,k)$ and the frequential real filter output at (n-1) $\hat{\underline{H}}^f(n-1,k)$. This configuration of input gives the best results. No operation of the *eps* block is performed on the channel estimation gained from the first symbol (reference symbol) which initialises the process at the beginning of each frame.

A test determines if $\hat{\underline{H}}(n,k)$ is probably false or not. If the test value is inferior to an optimal threshold, $\hat{\underline{H}}(n,k)$ is keeping. But on contrary, the estimation $\hat{\underline{H}}(n,k)$ is replacing by the estimation determined on the prior symbol $\hat{\underline{H}}^f(n-1,k)$.

The two possible complex difference simulated tests used are :

---

[3] e symbolising the fact that the carrier estimation is false.

- no Doppler effect taking account :

$$S_a(n,k) \; = \; \|\hat{\underline{H}}(n,k) - \hat{\underline{H}}(n-1,k)\|^2 \tag{7}$$

- with implicit Doppler effect taking account :

$$S_{ad}(n,k) = \; \|\hat{\underline{H}}(n-1,k) - \frac{\hat{H}(n,k) + \hat{H}(n-2,k)}{2}\|^2 \tag{8}$$

The change is made in the controlled switch block by the error detector (Cf. figure 4). The simulations show that the *eps* process prevents the system from taking down when some errors occur : the OFDM-CD3 is running back on the status where it was before the noisy carrier(s) reception.

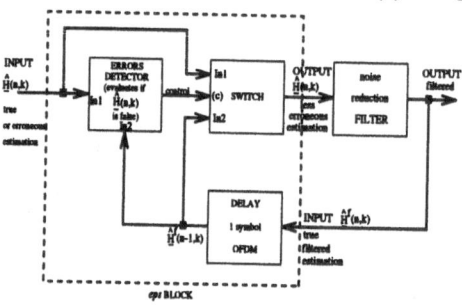

Figure 4: Diagram of the error propagation stop block.

The insertion of the *eps* block in the CD3 chain (Cf. figure 1) is :

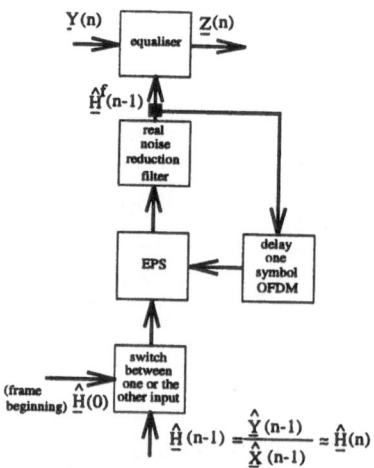

Figure 5: Insertion of *eps* block in the CD3 chain.

When there is some Doppler effect, a natural deviation exists between $\hat{\underline{H}}(n,k)$ and $\hat{\underline{H}}(n-1,k)$. A simulated test should predict this variation (between two k carriers of two consecutive symbols) due to Doppler and correct this effect in isolating the low phase variations (due to Doppler) from risky variations (due to errors).

In the case where the channel is constituted by a preponderant path, it is possible to correct the errors due to the Doppler effect : the channel estimation change is achieving with the Doppler phase displacement between symbols.

6.2 Doppler effect determination in the case where one path is preponderant

Receiver mobility gives Doppler effect and is taken account by transmission channel model.

x(t) is the emitted signal in the channel. For mobile reception, the multipaths channel show some fadings and has a Rayleigh statistic. We suppose that this impulsive response h(t) has a maximal delay inferior to the guard interval Δ. y(t) is the received signal after crossing the channel [2]. The L paths Rayleigh channel modelisation gives :

$$x(t) \quad \overset{\mathcal{C}}{\longmapsto} \quad y(t) = [x * h](t) \tag{9}$$

$$h(t) = \alpha \cdot \sum_{i=1}^{L} \rho_i \cdot e^{-j.2.\pi.(\theta_i + f_{di}t)} \cdot \delta(t - \tau_i) \quad \text{with} : \quad \alpha = \frac{1}{\sqrt{\sum_{i=1}^{L} \rho_i^2}} \tag{10}$$

where :

1. ρ_i expresses the amplitude modification due to the mean channel fading on the path i ;

2. θ_i is the phase modification in regard with propagation time dispersion of multipaths ;

3. f_{di} is the Doppler effect because of receiver mobility and is worth :

$$f_{di} = f_o \cdot \frac{v.cos\gamma_i}{c} \tag{11}$$

with :

- f_0 is the carrier frequency [central frequency of the RF signal] ;
- v_i is the receiver velocity ;
- γ_i is the incidence angle.

4. τ_i gives the relative delay to the i path.

If we neglect the gaussian additive noise, we obtain :

$$y(t) = \alpha . \sum_{i=1}^{L} \rho_i . e^{-j.2.\pi.(\theta_i + f_{d_i}t)} . x(t - \tau_i) \qquad (12)$$

The Doppler effect expresses the receiver velocity. In applying the Fourier Transform $[h(t) \xrightarrow{T\mathcal{F}} H(\nu, t)]$ and in supposing that the Doppler dephasing is almost constant during one time symbol , the channel frequential response is determined, or time-variant transfer function[4] :

$$H(\nu, t) = \alpha . \sum_{i=1}^{L} \rho_i . e^{-j.2.\pi.(\theta_i + f_{d_i}t + \nu\tau_i)} \qquad (13)$$

In the majority of practical cases, one path can be preponderant on the others. If this path has a Doppler frequency, it is going to interfere with the signal and give a proportional dephasing to this frequency and a ICI (Inter Carriers Interference) noise. The extreme case where only one delay exists is considered in the following. From the equation 13, the Doppler effect can be corrected by intervening on the channel estimation. In this case, the channel response is modified by a fixed Doppler frequency given a phase deviation $\varphi_d = 2.\pi.f_d.T_S$ (where $T_S = t_S + \Delta$)[4], this deviation being estimated. When some errors occur, the used estimation $\hat{H}(n-1, k)$ should take account of the Doppler phase displacement between the n and the (n-1) estimations. So the Doppler effect is limited by anticipation :

$$\underline{\hat{H}}_c(n-1, k) = \underline{\hat{H}}(n-1, k)_{\text{corrected}} = \underline{\hat{H}}(n-1, k).e^{-j\varphi_d} \qquad (14)$$

This phase displacement is calculated between two consecutive OFDM symbols. In the general case of Rayleigh channel[5], the frequency Doppler estimation in order to modify eventually each channel response carrier is impossible. Each delay could have a different Doppler. It is no use to expect of determining a constant dephasing carrier by carrier and then modifying the estimated value. In this point of view, the Doppler effect seems like noise. The channel turns with an irregular step carrier per carrier and may not be adjusted. Besides, if an error is detected, the erroneous estimation is replaced by the one of the prior symbol ; the estimation is then less erroneous but the Doppler effect is accentuated.

---

[4]t_S : useful symbol duration, Δ : guard interval duration, T_S : total OFDM symbol duration ;

[5]Multipaths channel with a great number of independent paths. P1 channel of DVB-T norm[2] is used for the simulations.

7   Simulation results with *eps* process

7.1   64QAM case

Simulations are made for a constellation 64QAM and a rate code of 2/3. The desired BER are almost 2.10^{-4} after decoding. For a constant Signal to Noise Ratio (SNR) level (the required level without correcting block *eps*), the BER and the threshold are evaluated for each Doppler frequency. The threshold is optimal when the BER is minimum.

The correction must also consider the eventual channel estimation modification on the prior symbol. For stationnary channel, the test used is the quantity 7 and in order to compensate the slow phase variations (or Doppler effect), the amount is 8.

In adjusting the SNR, for a required BER of 2.10^{-4}, three different cases were simulated :

- without *eps* ;

- with *eps* : the threshold is optimised for each Doppler frequency ;

- with *eps* and impulsive noise[6] : one noisy symbol in two when there is no Doppler effect. These symbols are considered as lost by the system. Besides, with the Doppler effect, a more favourable case is taken with one noisy symbol per frame (every 68 symbols). This symbol is also not scheduling into the BER.
   The conventional model for impulsive noise characterisation follows a Poisson law[1] for the stream arriving impulses. A time domain impulsion on baseband versions of the transmitted signal $\underline{x}(n, k)$ is spread over all the FFT output carriers $\underline{y}(n, k)$ (OFDM case). Therefore, when the pulse energy is very high, the carriers may be modified up until they achieve a complete and erroneous OFDM symbol.

Figure 6 compares the SNR performance of *eps*-CD3 for the three previous cases and for different Doppler shifts :

---

[6] Impulsive noise comes for instance from high intensity equipments (train, bus...) and is an additive disturbance.

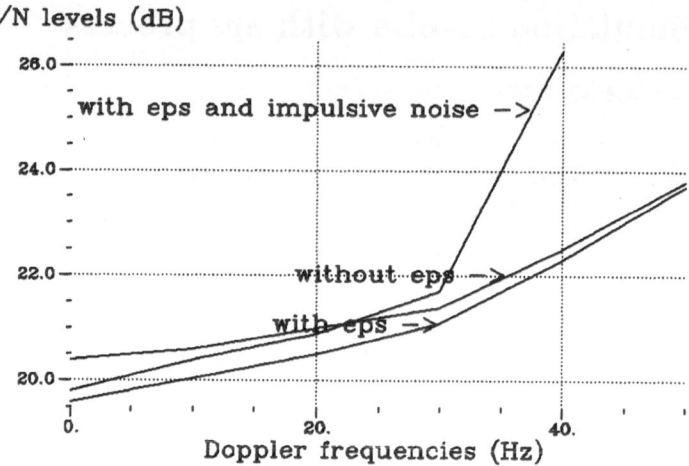

Figure 6: Influence of *eps* on performance of CD3 system [Case : 64QAM constellation, 2/3 rate, Rayleigh channel, bandwidth filter of $\Delta = 1/32t_S$, 2K FFT].

Without any Doppler effect, the obtained SNR is 19.6 dB. Therefore the gain is 0.8 dB in comparison with the case without *eps*. The optimal threshold is 0.001 for the equation 7. The errors are arranging in binary packets after decoding and there is no isolated error. The observed rate carriers correction is 96.8 %. In the same conditions, DVBT norm (ideal coherent demodulation) SNR is 19.3 dB. After *eps* correction, CD3 demodulation is inferior of 0.3 dB of ideal coherent demodulation for stationnary Rayleigh channel (BER $\approx 2.10^{-4}$, MAQ64 constellation, code rate 2/3, filter bandwidth optimised for $\tau_{max} = \frac{t_S}{32}$).

In supposing that the <u>channel is stationary</u>, the symbol per symbol channel estimation is almost inhibited and thus the closed loop. In fact, the estimation is copied from the estimation furnished by the reference symbol ($\approx 97\%$). Without taking account of Doppler effect, *eps* limits the CD3 estimation efficiency. On the other hand, it must be recalled that the CD3 demodulation is designed to follow Doppler effect. Because of the channel stationarity (without any Doppler effect and phase noise), CD3 has no use because the symbol reference permits the operation.

If we now introduce Doppler effect in the simulation, we observe that CD3 chain unhooks at around +/−60 Hz. *eps* does not reduce this limit which seems intrinsic to the system. But results show an amelioration up to 0.5 dB ($f_d = +/−10$Hz) for a correction of 8.8 % of the estimations (or *eps* use rate). The amelioration is higher for the low Doppler frequencies. The correcting block reduces the required SNR for obtaining some 2.10^{-4} BER :

DF	0	10	20	30	40	50
SNR (dB)	19.6	20.1	20.5	21.1	22.3	23.7
MCMP	96.8	8.8	4.1	3.5	3.2	<1

Table 1: *Eps* influence on CD3 system performances [Case : 64QAM constellation, 2/3 rate, Rayleigh channel with a filter with a bandwidth optimised for $\Delta = t_S/32$, 2K FFT] ; DF : Doppler frequencies (+/−**XHz**) ; MCMP : Modified carriers mean percentage.

Indeed, if the channel response estimation is affected by the Doppler effect, displacement by a detected error[7] is limited, per principle, only to the very erroneous carriers whatever the used algorithm.

7.2 16QAM case

For a 3/4 code rate, the optimal threshold is increasing up to 0.015 for a complex detection without any Doppler. The SNR is 17.3 dB. The gain is 0.5 dB in comparison with the result without correcting block : difference with ideal demodulation becomes 0.6 dB in the 16QAM case.

The 64QAM threshold is smaller than in 16QAM : 0.001 against 0.015. For a QAM constellation, if an error has happened on the binary stream, the inverse mapper could make the mistake of choosing a complex position instead of the good one. From a correct point to a false one, the mistake will be mathematically either an angular difference or an amplitude difference[5]. The angular or the amplitude difference is less important between two narrow points in 64QAM than in 16QAM.

7.3 QPSK case

The environment simulations is a QPSK constellation and a 1/2 code rate. The channel has four paths whose delays are inferior to the guard interval ($\Delta = t_S/4$). These measurements allow one comparison with DVB-T chain [no ideal channel estimation from reference carriers] :

	DVB-T	CD3 without *eps*	CD3 with *eps*
SNR (dB)	7.5	8.0	7.9

Table 2: Comparison between DVB-T and CD3 systems for a Doppler frequency of +/−100 Hz (Case : QPSK, 1/2 code rate, multipath channel (4 paths) with AWGN, filter with a bandwidth of $3t_S/16$).

The *eps* amelioration is 0.1 dB with 5.9 % mean estimation copy for this QPSK case.

---

[7] Detection of some impulsive noise on the channel estimation.

7.4 Doppler phase displacement estimation

Simulations are made for a 64QAM constellation and a 2/3 rate code. The simulated channels are either gaussian with a delayed path of $0,95\,\Delta$ (where $\Delta = 1/4t_S$) or of rician[8]. The filter bandwidth is equal to the guard interval. When there is some Doppler effect and in cases where one preponderant path exists (Rice or Gauss channels with one delayed path) (Cf. part 6.2[9]), the results are increasing of 0.1 dB in comparison with the simple change. The Doppler frequency is $+/-50$ Hz :

Channels	SNR (dB)		
	without *eps*	simple change *eps*	*eps* with change and Doppler correction
Gauss	21,8	21,3	21,2
Rice	19,0	18,9	18,8

Table 3: CD3 system efficiency ($f_d = +/-50Hz$) in taking account the Doppler dephasage (Case : 64QAM constellation, 2/3 code rate, Gauss with 1 delay or Rice channels , filter with a bandwidth optimised for $\Delta = t_S/4$, 2K FFT).

8 Conclusion

The CD3 demodulation advantages are :

- the Bit rate is increased from 5 to 15 % versus DVB-T;

- the simulation results are comparable to the ideal coherent demodulation : an average difference of 0.7 dB without *eps* process whatever the channel and 1.0 dB for the specific Rayleigh channel (arithmetic mean for the 1/2, 2/3 et 3/4 code rates) ;

CD3 system principle of coherent demodulation with data exploitation is a closed loop operation. In a strong noisy environment, this loop facilitates the errors propagation (or reinjection) after decoding symbol per symbol : this is the main drawback of the CD3 chain, that limits the system operation. It could explain the observed difference between the two coherent demodulations.

In adding an Error Propagation Stop (*eps*) block, the number of channel estimation errors is reduced in CD3 chain :

---

[8]F1 channel with 21 paths of DVB-T norm[2]. The directed path is amplified by comparison with the others paths : the Rice factor K (power ratio between the directed path and the reflected paths) is 10 dB ;

[9]The possibly erroneous carrier is replaced by the prior symbol carrier, but modified by the estimated Doppler dephasage between the OFDM symbols (n-1) and n.

- case without Doppler effect :

  The correction is relatively important on stationary channel with or without impulsive noise (one lost symbol over two but not accounted for the BER). Results are close to the ideal demodulation with reference carriers insertion (amelioration in 64QAM by comparison with CD3 without *eps* of up to 0.8 dB in a Rayleigh channel) ;

- case with Doppler effect :

  A gain of 0.5 dB maximum ($f_d = +/- 10Hz$) appears from results by contrast with no correcting *eps* block (Rayleigh channel, 64QAM constellation, 2/3 rate code).

  For channels with at least one preponderant path (gaussian or Rice) and if a correction *eps* is applied with Doppler dephasage, the amelioration is 0.1 dB in 64QAM in regard with the simple change *eps*.

But in a mobile environnement, the impulsive noise (one lost symbol per frame OFDM) exclusively limits the CD3 chain operation for the Doppler frequencies inferior to $+/- 40Hz$ (Rayleigh channel, 64QAM constellation, code rate 2/3).

References

[1] A. D. SPAULDING AND D. MIDDLETON. **Optimum Reception in an Impulsive Interference Environnement-Part I : Coherent Detection.** *IEEE Transactions on Communications*, Vol. 25(No.9):pp. 910–923, September 1977.

[2] European Telecommunication Standard. **Digital Broadcasting Systems for Television, Sound and Data Services ; Framing structure, channel coding and modulation for digital terrestrial television,** March 1997. ETS 300 744.

[3] M. ALARD AND R. LASSALLE. **Principles of Modulation and Channel Coding for Digital Broadcasting for Mobile Receivers.** *EBU Review - UHF satellite sound broadcasting*, (Technical No.224):pp. 47–69, August 1987.

[4] J.G. PROAKIS. **Digital Communications.** MC GRAW-HILL BOOK COMPANY, NEW YORK, 1989.

[5] W. WEBB AND L. HANZO. **Modern Quadrature Amplitude Modulation Principles and Applications for Fixed and Wireless Communications.** PENTECH PRESS, LONDON, 1994.

[6] V. MIGNONE AND A. MORELLO. **CD3-OFDM : A Novel Demodulation Scheme for Fixed and Mobile Receivers.** *IEEE Transactions on Communications*, VOL. 44(No.9):PP. 1144–1151, SEPTEMBER 1996.

Performance of a RS-coded QPSK cable modem for TDMA communication over the return channel of HFC networks

Mark Van Bladel and Marc Moeneclaey
Communication Engineering Lab, University of Ghent, Belgium

Abstract

This paper deals with communication over the return path channel of a cable TV (CATV) network designed as a HFC network. First, the main impairments are identified to be ingress noise and impulse noise. A TDMA receiver scheme suited for this application is then derived. It is shown that the performance of such a TDMA system substantially improves by using equalization to combat peaks in the ingress spectrum. With respect to impulse noise, error probability calculations are presented to determine the level of error correction that is required in order to obtain the target bit error rate of 10^{-8}.

1. Introduction

Telecom operators experience more and more competition these days from cable TV operators when it comes to offering new telecommunication services. One can hereby think of fast internet, home shopping, banking, video on demand, telephony services etc... In order to allow for this type of applications the network configuration should evolve from a coaxial to a hybrid fiber coax (HFC) structure. In the HFC architecture optical node units (ONU) are connected to the headend by means of single mode optical fiber, while the part between the ONU and a group of subscribers is over coaxial cable. This latter part of the structure is implemented as a tree and branch network. The downstream signals are transmitted over a trunk line, which is a cascade of a number of trunk amplifiers. These amplifiers compensate for the frequency dependent cable losses. At various points on the line, a bridger amplifier provides so called feeder lines for the distribution of the signals to the subscribers. Each feeder line contains a small number of line extenders that are connected by 75 Ohm coaxial cable. Between the line extenders are located taps, that draw off a portion of the feeder signal for the subscriber. This network structure is now being modified to allow for bidirectional services. The return path must also be equipped with amplifiers. In general unity gain is desired from the first return amplifier as seen coming from the subscriber to the ONU. Usually one adopts the design convention that a maximum of 7 return path amplifiers in cascade is tolerated, which means at most 3 trunk amplifiers, 1 bridger amplifier and 3 distribution amplifiers. The scheme described here is shown in Figure 1.

It is the CATV operator's choice which frequency band to use for the return path for the interactive services as well as the number of channels and the channel bandwidth. Some operators only want to use the spectrum between 5 and 35 MHz, whereas others go from 5 to 65 MHz. Anyway, in this paper results are presented for channels of 1 and 2 MHz bandwidth and with several values for the central frequency.

Two very important impairments are encountered in HFC networks, namely ingress noise and impulse noise [1]. Ingress noise is slowly-varying noise, originating from nonlinearities in the CATV system, from interference generated by a device at the subscriber location or from signals radiatively coupled to the return path from outside the network. Impulse noise can be caused by a whole range of phenomena : it can be generated by naturally occurring sources (like lightning) as well as by man-made sources (engine ignitions, power switching,...). Another (but less critical) impairment is the non-ideal channel transfer function which is different for every user. Figure 2 shows a typical amplitude transfer function (deviation from the mean) over the complete band between 5 and 65 MHz.

Various multiple access techniques can be considered to establish reliable two-way communications over HFC networks. Hereafter, we shall consider a TDMA structure with QPSK modulation. In a TDMA system each channel is shared by many users, to whom are assigned time slots for sending their information packets. Also the DVB and the DAVIC proposal for a return path channel standard are based on a TDMA solution.

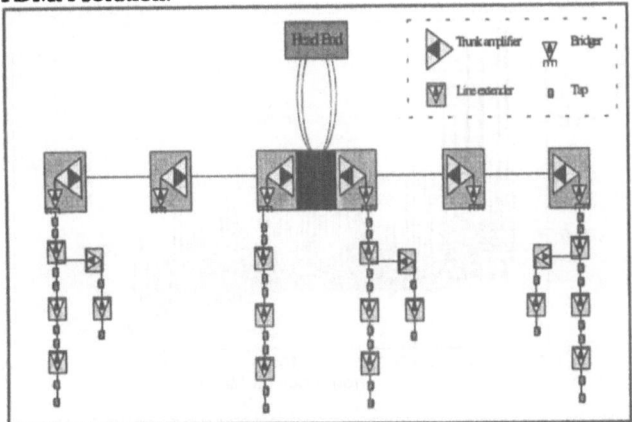

Figure 1 : typical HFC network structure

The organisation of the rest of the paper is as follows. Section 2 describes the influence of the ingress noise, derives the receiver structure and shows the benefit equalization can bring. In section 3 the impulse noise is dealt with. The required error correcting capability is investigated as a function of the input signal power. Finally, some conclusions are drawn in section 4.

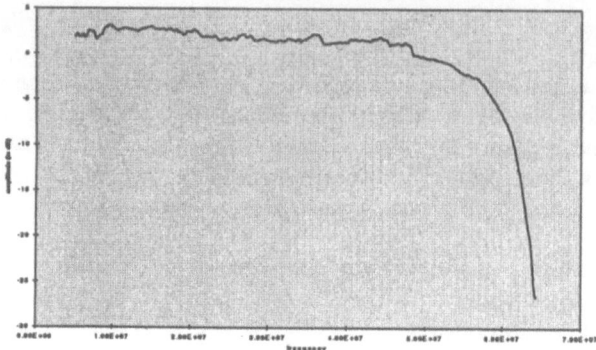

Figure 2 : amplitude transfer function

2. Effect of ingress noise

In this section we shall discuss the benefits and drawbacks of some TDMA receiver schemes. In the simulations in this section the channel transfer function is normalized, such that at the input of the receiver filter the signal power spectral density is -100 dBm/Hz. The ingress noise is modeled as coloured Gaussian noise with a power spectrum equal to the measured spectrum shown in Fig. 3. Note that this is a worst case noise spectrum with respect to the shape : all the peaks observed during the 24 hour measurement are present. The amplitude levels can be considered as typical.

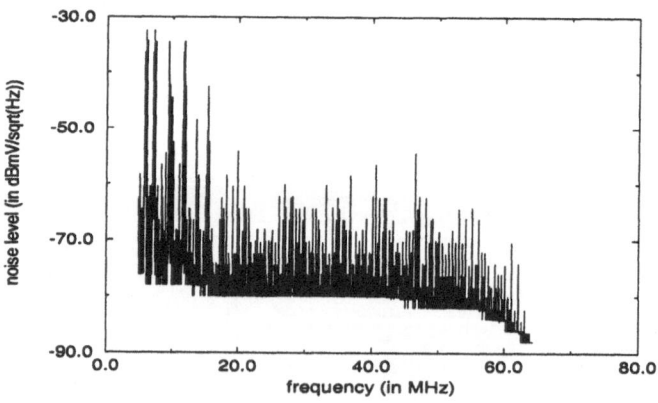

Figure 3 : ingress noise spectrum

Consider now the receiver structure depicted in Fig. 4. The transmit filter P(f) is a square root raised cosine filter. The receiver filter is matched to both the channel $H_{ch}(f)$, the ingress noise spectrum $S_n(f)$ and the transmit pulse. This structure yields the theoretical optimal ratio of useful signal power P_s to noise power P_n at the input of the decision device (matched filter bound) [2]. Table 1 shows this optimal P_s/P_n for some channels, as well as the ratio of P_s to the intersymbol interference power P_{isi}. The calculations were done with a roll-off factor of 0.3. In the fourth column the total ratio $P_s/(P_{isi}+P_n)$ is given.

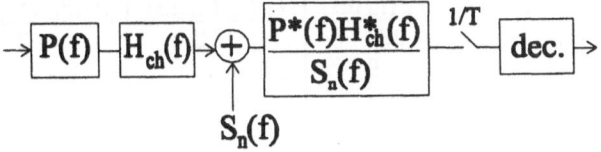

Figure 4

Approximating the ISI by a Gaussian process, the transmission of uncoded QPSK with a BER < 10^{-3} would require that $P_s/(P_{isi}+P_n)$ > 9.8 dB (ignoring implementation loss). It follows that such transmission cannot be achieved for the channels located below 15 MHz; for the channels located at higher frequencies, such transmission is possible but the implementation margin is rather small. The values of $P_s/(P_{isi}+P_n)$ in Table 1 are limited by the ISI.

channel (in MHz)	P_s/P_n (in dB)	P_s/P_{isi} (in dB)	$P_s/(P_{isi}+P_n)$ (in dB)
9 - 11	22.3	0.8	0.8
13 - 14	26.7	7.1	7.0
21 - 23	29.1	11.3	11.3
47 - 48	29.9	12.4	12.3
50 - 52	30.5	11.3	11.3

Table 1

Given the relatively flat shape of the channel transfer function, it follows that the ISI is mainly caused by the receiver filter. For this reason as well as because of the complexity of the receiver filter transfer function in Fig. 4, we now consider the receiver structure in Fig. 5. The receiver filter is a fixed filter matched to the transmitted pulse.

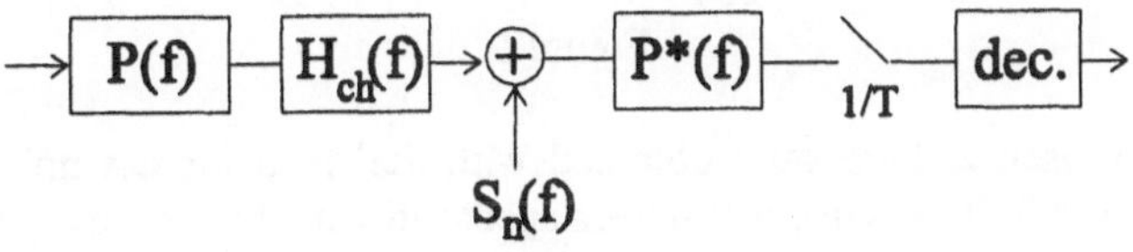

→ $P(f)$ — $H_{ch}(f)$ →⊕— $P^*(f)$ — $1/T$ — dec. →

$S_n(f)$

Figure 5

Results are given in Table 2. The relatively flat channel characteristic in Fig. 2 now gives rise to high P_s/P_{isi} ratios in all cases. As the receiver filter is not matched to the ingress noise spectrum (nor the channel transfer function), the resulting values of P_s/P_n are smaller than the matched filter bound. In frequency bands not too heavily affected by ingress, large values for P_s/P_n are obtained though. Communication over channels in the frequency band below 15 MHz still remains problematic in presence of a heavily peaked ingress spectrum like that of Fig. 3.

channel (in MHz)	P_s/P_n (in dB)	P_s/P_{isi} (in dB)	$P_s/(P_{isi}+P_n)$ (in dB)
9 - 11	2.5	32.8	2.5
13 - 14	14.9	40.3	14.8
21 - 23	27.6	47.9	27.6
47 - 48	27.5	39.8	27.2
50 - 52	29.7	42.8	29.5

Table 2

Attempting to obtain higher P_s/P_n ratios, we bring into the scheme a fractionally spaced MMSE decision feedback equalizer as shown in Fig. 6. Sampling is performed at twice the symbol rate (2/T). Table 3 shows the resulting performance for various lengths of the feedforward and feedback filters. In the table FF and FB denote the number of taps in the feedforward and feedback filter, respectively. For the channels located below 15 MHz, we obtain a considerable increase of $P_s/(P_{isi}+P_n)$ as compared to Table 2, even for feedforward and feedback filters as short as 10 taps each. Comparing the figures for the channels 9-11 MHz and 13-14 MHz with those in Table 2, we observe that the equalizer strongly increases the ratio P_s/P_n. We have verified that this is accomplished by a feedforward filter transfer function that exhibits dips at the position of the peaks of the ingress noise spectrum; the feedback filter reduces the postcursor ISI introduced by the feedforward filter. For the channels with a large $P_s/(P_{isi}+P_n)$ in Table 2, the introduction of the equalizer yields only a small improvement. With exception of the 9-11 MHz channel, the values of $P_s/(P_n+P_{isi})$ in Table 3 are for all channels within a few dB from the matched filter bounds, shown in column 2 of Table 1.

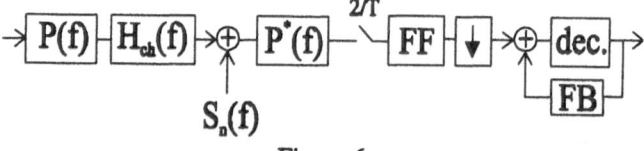

Figure 6

The results presented here were obtained with the fixed ingress noise spectrum shown in Figure 3. In practice the ingress spectrum will change slowly in time and so the adaptive equalizer must be capable of tracking these variations. We investigated a time-domain equalizer and a frequency-domain equalizer, both with least-mean square (LMS) equalizer tap updating. Although frequency-domain equalization is somewhat more complex than time-domain equalization, our simulations showed no substantial gain in convergence speed for the frequency-domain equalizer as compared to the time-domain equalizer.

channel (in MHz)	equalizer (FF/FB)	P_s/P_n (dB)	P_s/P_{isi} (dB)	$P_s/(P_{isi}+P_n)$ (dB)
9 - 11	10/10	14.2	21.7	13.5
	50/100	16.2	24.9	15.7
	∞/∞	16.9	∞	16.9
13 - 14	10/10	22.9	41.7	22.8
	50/100	24.4	49.7	24.4
	∞/∞	25.1	∞	25.1
21 - 23	10/10	28.0	53.1	28.0
	50/100	28.5	62.2	28.5
	∞/∞	28.8	∞	28.8
47 - 48	10/10	28.7	54.1	28.7
	50/100	29.0	61.1	29.0
	∞/∞	29.7	∞	29.7
50 - 52	10/10	29.8	59.2	29.8
	50/100	30.1	70.8	30.1
	∞/∞	30.3	∞	30.3

Table 3

3. Effect of impulse noise

Impulse noise measurements were carried out on a test network of cable operator Integan in Antwerp. From these data some statistical information was derived. The impulses exceeding the threshold level of 6 mV, are divided into a number of classes of pulse width, amplitude and interarrival time. From the measurements it was found the complex envelope of the pulses can be approximated by a sawtooth shape for the shorter pulses and a rectangular shape for the longer pulses.

Each pulse of the impulse noise gives rise to a burst of QPSK symbol errors. The effect of impulsive noise can be reduced by using error correcting coding with interleaving.

As the frequency band below 15 MHz is most strongly affected by impulsive noise, we consider in this section the 2 MHz channel between 9 and 11 MHz (for which also the ingress noise is largest). We investigate the performance of the receiver with decision feedback equalization (Figure 6), which is affected by the sum of ingress noise and impulsive noise. We assume the updating of the equalizer coefficients is not disturbed by the short pulses of the impulsive noise, so that the equalizer transfer function is the same as in the absence of impulse noise. Finally, we consider two ingress noise spectra, i.e. the pessimistic spectrum from Figure 3 and a less pessimistic spectrum that is obtained by reducing by 5 dB the spectrum from Figure 3.

First, we determine the receiver performance in the absence of coding. Then we investigate the effect of applying a (N,53) Reed-Solomon (RS) code with 53 information bytes (i.e. one ATM cell) and N-53 check bytes; such a code can correct at most (N-53)/2 byte errors [2].

The receiver performance in absence of coding is obtained in the following way.

(a) We determine the average number of QPSK symbol errors caused by a single noise pulse with a given pulse duration and amplitude, in the presence of ingress noise. This is repeated for all occurring combinations of pulse duration and amplitude.

(b) The contribution of transmitted QPSK symbols, affected by both ingress noise and impulse noise, to the QPSK symbol error rate is computed from the result in (a), taking into account the relative frequency of the combinations of pulse duration and amplitude.

(c) From the measured statistics of pulse duration and interarrival time, we determine the fraction of transmitted QPSK symbols that is not affected by impulse noise. From this fraction, the contribution of these QPSK symbols to the QPSK symbol error rate is computed.

(d) Combining the results from (b) and (c), the average QPSK symbol error rate is derived. Assuming an infinite interleaving depth, the resulting byte error probability P_{byte} is obtained (1 byte = 4 QPSK symbols).

The uncoded byte error probability as a function of the power spectral density (psd) of the received QPSK signal is shown in Figure 7, for the two considered ingress noise spectra, assuming a decision feedback equalization (dfe) with 10 taps in the forward filter and 5 taps in the feedback filter. In order to indicate the considerable performance advantage that results from the dfe, we have also included the byte error probability when no dfe is used (i.e. the receiver from Figure 5). Also included are the byte error probability curves in the absence of impulse noise; comparing these curves with those for ingress noise plus impulse noise reveals that there is a treshold value for the signal psd, above which the performance of the receiver is completely determined by the impulse noise. Under the treshold the ingress noise is the main impairment.

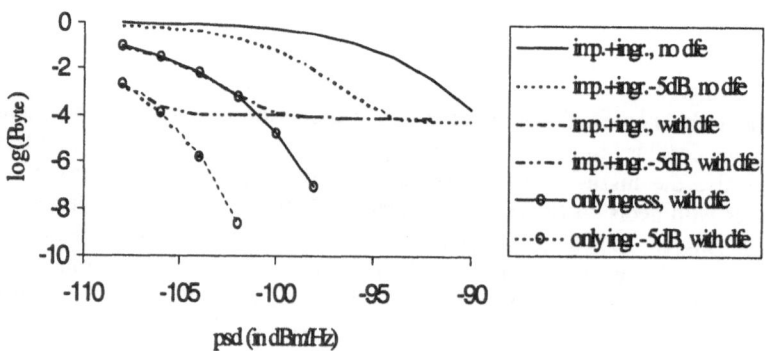

Figure 7 : byte error probabilities before coding

The uncoded byte error probability is then used to calculate the byte and bit error probabilities after Reed-Solomon decoding [2]. Results are presented in the following way : we determine the shortest (N,53) Reed-Solomon code offering a BER after decoding smaller than 10^{-9} for some values for the signal power spectral density at the input of the receiver. Results are shown in Table 4, again employing the equalizer with 10 taps FF and 5 taps FB.

psd	ingress spectrum		ingress spectrum – 5 dB	
(dBm/Hz)	N	bit error prob.	N	bit error prob.
-94	57	$3.91\ 10^{-10}$	57	$4.14\ 10^{-10}$
-96	57	$5.00\ 10^{-10}$	57	$5.09\ 10^{-10}$
-98	57	$6.92\ 10^{-10}$	57	$7.16\ 10^{-10}$
-100	59	$3.53\ 10^{-12}$	57	$9.70\ 10^{-10}$
-102	61	$4.52\ 10^{-12}$	59	$2.70\ 10^{-12}$
-104	69	$6.28\ 10^{-11}$	59	$3.41\ 10^{-12}$
-106	85	$2.47\ 10^{-10}$	59	$5.80\ 10^{-11}$

Table 4 (FF 10 taps, FB 5 taps)

From the table we see that if the input signal power does not exceed the treshold value, the required error correcting capability for the two considered ingress spectra is the same. As the performance is determined by the impulse noise, the 5 dB difference in ingress noise levels is not of any influence. Below the treshold, when the performance is determined by the ingress noise, the required length N of the RS codeword rapidly increases, and the two considered ingress spectra need substantially different codeword lengths.

In Table 5 the byte error probability before coding, the number N of bytes per RS codeword and the bit error probability after decoding are given for various equalizer lengths. The input signal psd is kept constant at –100 dBm/Hz. The calculations are carried out with the ingress spectrum of Figure 3, reduced by 5 dB.

FF / FB	P_{byte} (before coding)	N	P_{bit} (after decoding)
5 / 5	$1.00\ 10^{-4}$	57	$7.75\ 10^{-10}$
50 / 5	$1.19\ 10^{-4}$	59	$3.04\ 10^{-12}$
5 / 10	$9.77\ 10^{-5}$	57	$7.17\ 10^{-10}$
10 / 10	$1.08\ 10^{-4}$	57	$9.59\ 10^{-10}$
25 / 10	$1.35\ 10^{-4}$	59	$5.16\ 10^{-12}$
50 / 10	$1.36\ 10^{-4}$	59	$5.30\ 10^{-12}$
10 / 50	$1.08\ 10^{-4}$	57	$9.59\ 10^{-10}$
50 / 50	$1.55\ 10^{-4}$	59	$8.86\ 10^{-12}$

Table 5

From Table 5 we observe that longer equalizers do not necessarily give rise to a smaller error probability. This is because the longer equalizers tend to smear out the pulses of the impulse noise, so that each pulse affects more QPSK symbols.

4. Conclusions

In this paper a TDMA receiver structure suitable for communication over the return channel of a CATV network is investigated. We indicated the benefit from decision feedback equalization in combating the effect of the peaks in the ingress spectrum.

With respect to impulse noise a method to evaluate the required error coding was presented. Again it was shown that equalization substantially improves the system performance for channels in a frequency range where ingress is large. However, the equalizer length has to be chosen carefully, as an equalizer with too long an impulse response can amplify somewhat the detrimental effect of the impulse noise. In most cases, (59,53) or (61,53) Reed-Solomon codes will be sufficient to obtain a bit error rate of less than 10^{-8}.

Acknowledgment

This paper gives an overview of the work performed at the University of Ghent within the framework of the European ACTS Interact project. We would like to express our gratitude to the partners in this project, namely Alcatel SESA (E), Barco (B), Bosch (D), CCETT (F), EBU (B), Integan (B), ITC (GB), Nozema (NL), Retevision (E), SAT (F), TBS (F), TDF (F) and UPM (E).

References

[1] "CATV Return Path Characterization for Reliable Communications", C. Eldering, N. Himayet and F. Gardner, IEEE Communications Magazine, August 1995, pp.62-69

[2] "Digital Communications", J.G. Proakis, McGraw-Hill, New York, 1989

How self-similar processes persistence determines the queue length distribution moments

E. Costamagna[1], G. Iacovoni[1,2], M. Isopi[3]

[1] Department of Electronic Engineering, Pavia University, Pavia, Italy

[2] CoRiTel Research Center on Telecommunications, Rome, Italy

[3] Department of Mathematics, La Sapienza University, Rome, Italy

e-mail: giovanni@comel1.unipv.it

Abstract

In this paper we compare the queue performances of three different traces with gaussian marginals generated using the discrete-time Fractional Gaussian Noise algorithm [1]. We have introduced some additional features in one out of the three traces generation process to have less dependence on very large time scale (as it is the case in video traffic).

We highlight the issues at stake when we look at the way the correlation structure influences the moments of the queue length.

1. Introduction and definitions

It has been recently shown [4] that real packet traffic properties can be parsimoniously captured by a modelling approach based on fractional gaussian noise (FGN). This is true in particular for data and video traffic.

Let $X = \{X(i), i \geq 1\}$ be a stationary process and let X^m be the aggregated process obtained by averaging the original series over non-overlapping blocks of size m. X is *exactly self-similar* if:

$$X = m^{(1-H)} X^m \qquad \forall m \geq 1 \qquad 1/2 \leq H \leq 1 \qquad (1)$$

where the equality is in the sense of finite-dimensional distributions. If $H = \frac{1}{2}$ the process has short range dependence (SRD), whereas $H > \frac{1}{2}$ characterizes a process with long range dependence (LRD).

FGN process is exactly self-similar, and it is completely specified by the triplet (μ, σ, H), where μ and σ are the first and second order moments of the process marginal (which is gaussian), while H is the Hurst parameter that captures the degree of the process dependence.

In our modelling context X is the number of bytes generated in the time unit i.

In the following we use three different traces with gaussian maginals to drive our simulations. All of them are 171 000 points long. Since the Bellcore people made an intra-frame coding of Star Wars available on the Net [2], this is quickly becoming a benchmark length for performance simulations.

All the three traces exhibit long range dependence of various sorts. They all have gaussian marginals with the same mean and variance.

In the following we will look at sources with various correlation structures, meant to mimic data and video traffic and see how they influence the queue length distribution.

2. Traces description

The traces were generated by implementing the usual discretization of the so called *harmonizable* representation of fractional brownian motion (FBM), i.e. discrete-time Fractional Gaussian Noise (dFGN) [3].

Here we skip all technical details and just say that the harmonizable representation of FBM consists of writing it as a weighted average of its past values, where the weights are given by a "memory kernel" which decays to zero as you look further away into the past. (We refer the reader to [1] for a description of the procedure and a comparative discussion of algorithms used to generate FGN).

We generated three traces by making use of the dFGN algorithm. Their statistical properties are shown in the following table:

	μ (bytes/time unit)	σ (bytes/time unit)	H
TRACE 1	27791	6254	0.5
TRACE 2	27791	6254	0.6
TRACE 3	27791	6254	0.83

Trace2 and Trace3 are both FGN. Trace1 is in some sense akin to Trace3. It was also generated with the same input parameters of Trace3, except that we put a much smaller cutoff on the memory kernel in the discretization of the harmonizable representation. This resulted in an effective H = 0.5.

Looking at the Variance-Time plot (fig. 1) for the three traces one can see clearly what happens. Trace2 and Trace3 are both a remarkably good approximation of a self-similar process for all the time scales that appear in the simulation. Their Variance-Time plots are very well fitted by just one straight line with a slope of (2H-2). Trace1 is different. It looks self-similar with H = 0.83 up to a certain time scale, but blocks on longer time scales are effectively Markovian. Its Variance-Time plot is well fitted by two straight lines, the second one corresponding to an effective H = 0.5.

From a phenomenological point of view one observes the same difference between data traffic and video traffic [4]. While the former looks self-similar at all time-scales in the same way (the Variance-time plot can be approximated by a straight line), the latter shows a low-lag correlation structure which results in a Variance-time plot with two different slopes corresponding to small and large aggregation levels. The video behaviour can therefore be qualitatively captured by Trace1.

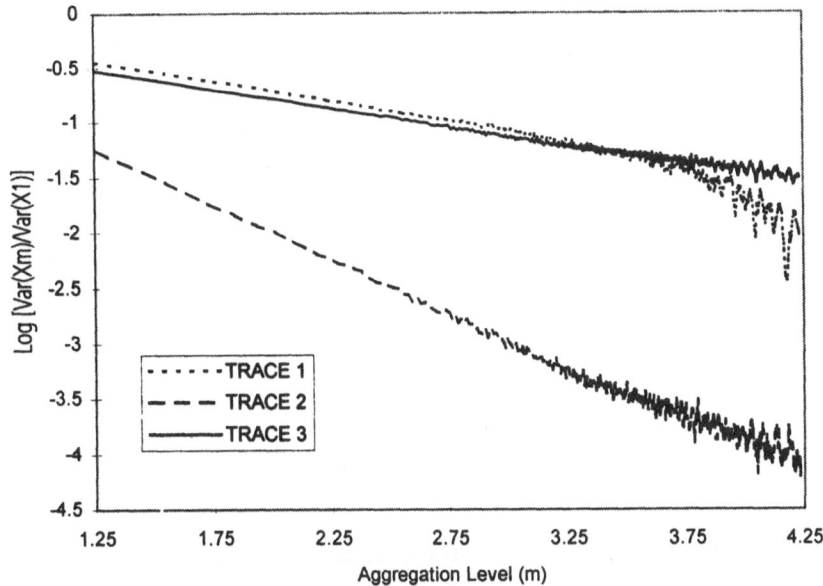

Fig. 1 Variance-time plot for the three traces

In the course of this paper we will concentrate on contrasting what happens to Trace1 and Trace3. Trace2 is mostly provided for comparison purposes.

On a heuristic level it is clear that persistence is the feature of LRD processes which most affects queueing performances.

For the traces considered here we use the following crude measure of persistence. We consider the graph of the trace and fix a threshold corresponding to a given load for the queue. Choosing load = 1 results in a threshold value = μ (see fig.2). We then look at each period the trace spends over this threshold. We compute the area of the part of the graph above the threshold for each of these periods. This is a random quantity whose (sample) mean depends only on the marginals of the trace and is therefore the same for all traces. Not so for its higher moments which also depend on the correlation structure of the trace. We take here the second moment as a simple indicator for persistence; high levels of emission tend to last longer, but there are less of them.

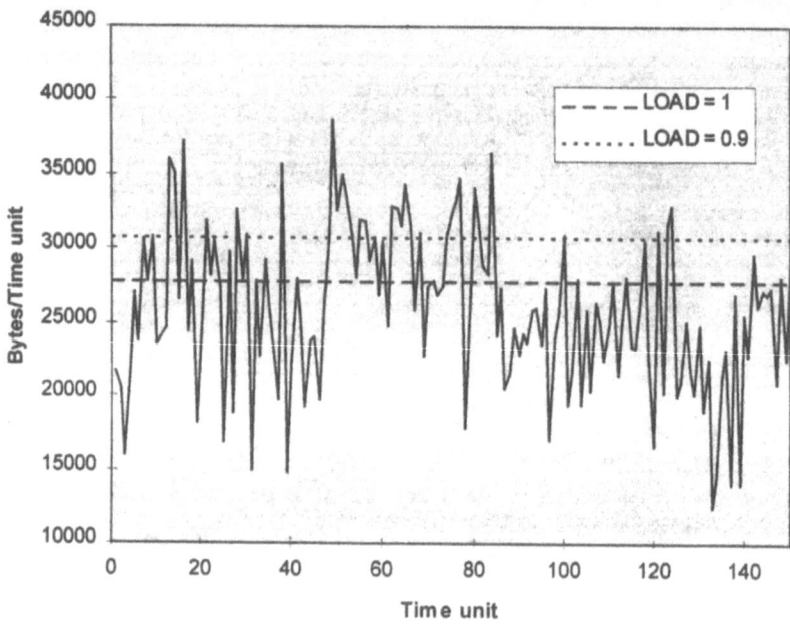

Fig. 2 Sample plot of Trace3 with respect to load =1 and load = 0.9

3. Simulation description and results

We consider an ATM queue of infinite waiting room with each of the three traces as the input process. To this aim a flow with a bit rate given by each trace is packetized into ATM cells introducing AAL5 overhead (one AAL5 PDU made of 8 ATM cells). The cells are then evenly spaced over a time unit period; this, however, is not critical since it is well known that the high frequencies in the power spectrum are negligible in terms of queueing behaviour [5]. The number of generated ATM cells for each trace is approximately 10^8, with a mean of 595 cells/time unit.

Cells are then processed into the queue with deterministic service time, FIFO mode. The outgoing data frequency is varied so as to consider a load ranging from 0.1 to 0.95, with step size of 0.05.

For each simulation we measured the following quantities:

1) The queue length distribution (probability density function and survival function)
2) The first 10 centered moments of the queue length distribution
3) The cells delay distribution (probability density function and survival function)

All the results are available at http://www.coritel.it. Here we show:
- fig. 3: mean queue length plot of the three traces versus load (ranging from 0.6 to 0.95)
- fig. 4: standard deviation queue length plot of the three traces versus load (ranging from 0.6 to 0.95)
- fig. 5: plot of the first 10 centered moments of the queue length distribution for the three traces with load = 0.6
- fig. 6: plot of the first 10 centered moments of the queue length distribution for the three traces with load = 0.9
Lower loads are not considered here since they do not pose severe problems in terms of buffer occupancy; these cases are generally very well handled by markovian models.

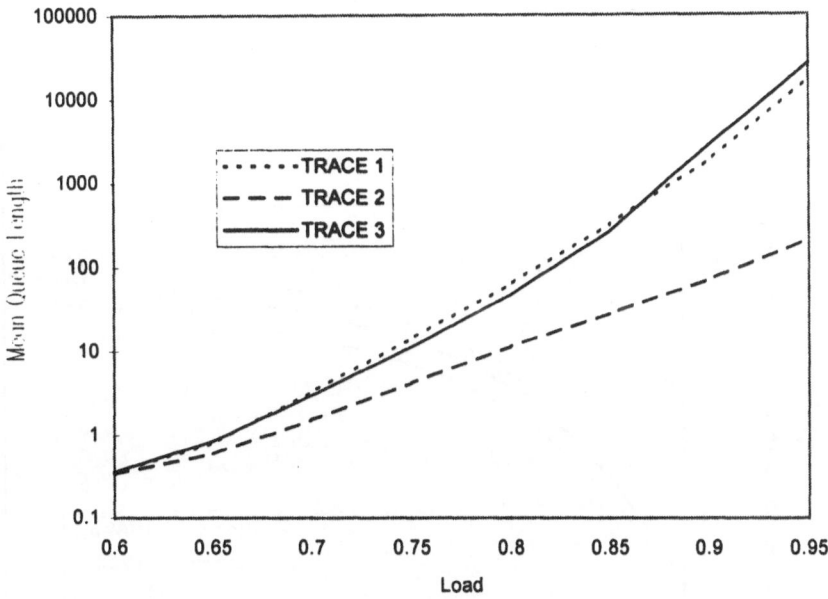

Fig. 3 . Mean queue length versus load

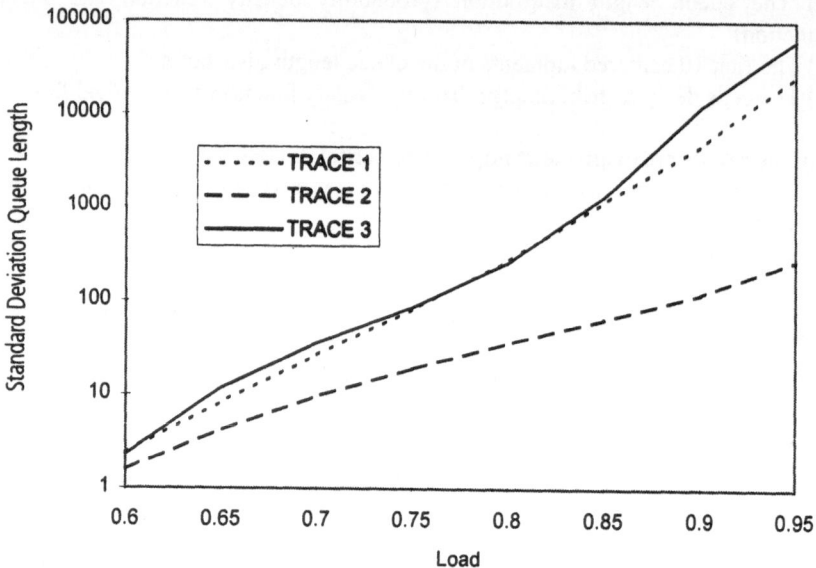

Fig. 4 Standard deviation queue length versus load

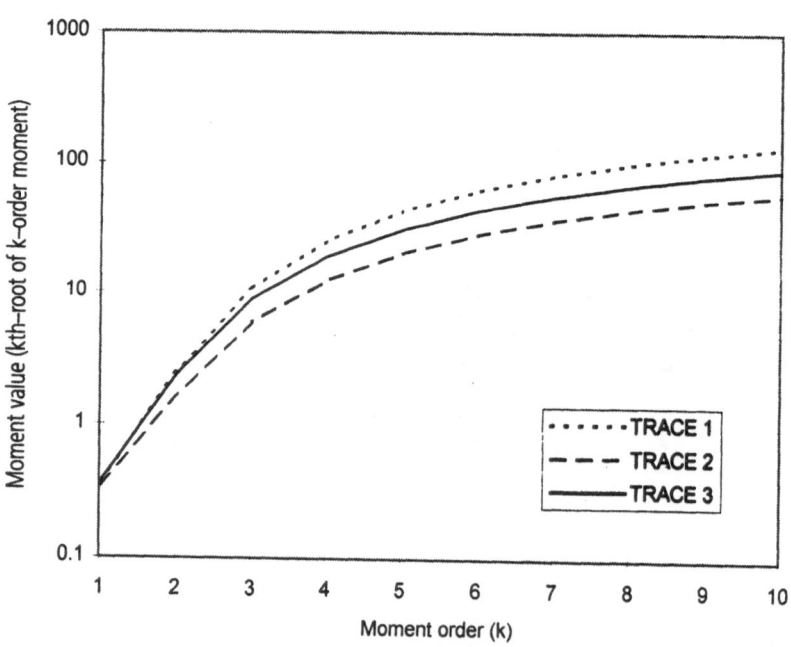

Fig. 5 K^{th}-root of k-order moment for LOAD = 0.6

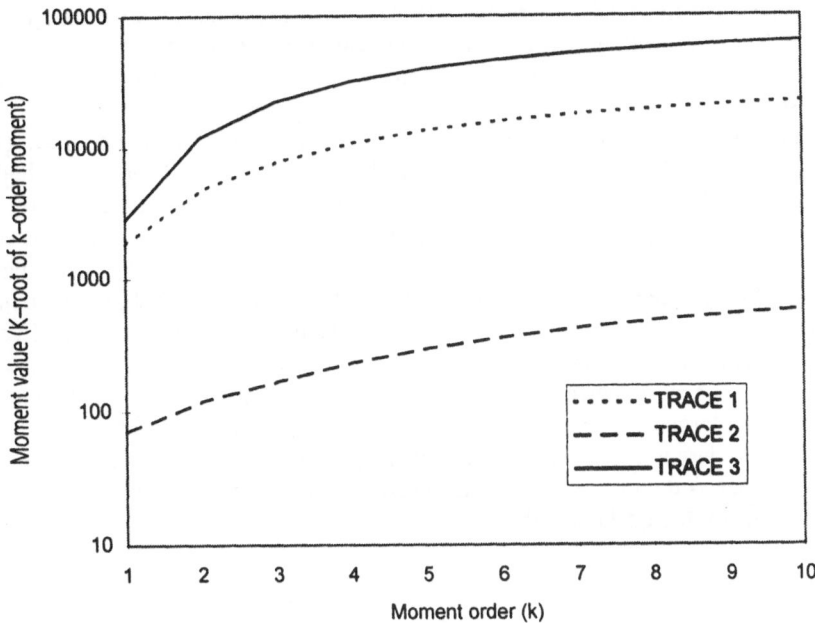

Fig. 6 Kth-root of k-order moment for LOAD = 0.9

4. Discussion

The asymptotic behaviour of the stationary distribution for a G/D/1 queue under self-similar input is given by the so called Norros bound which states that the tail behaviour of the queue length Q is at best Weibullian:

$$P(Q > x) \sim e^{-\gamma x^\beta}, \tag{2}$$

with γ and β constants depending on H; $\beta < 1$[6]
This bound was proved to be asymptotically tight by Duffield and O'Connell. [7]. Knowing the tail behaviour one can get the high moments of the distribution of the queue length with good approximation, but knowledge about the low moments (mean and variance in particular) cannot be inferred from tail behaviour.

Now we illustrate our simulation results and then comment on them.

Except when the load is very low, Trace2 gives moments of lower magnitude already starting with the mean. This is apparent in figures 3 and 4.

This is not surprising, as it is widely believed that the Weibullian picture (2) is a reasonable description, even far from the asymptotic region [4].

Comparing Trace1 and Trace3 is more interesting. They give raise to mean (fig. 3) and variance (fig. 4) which are basically the same for all loads, but higher moments differ more and more as the load is increased (compare fig.5 with fig. 6). As noted previously, the most apparent feature of LRD processes is persistence and we introduced before a simple way to capture this feature that in our opinion most closely agrees with eyeballing evidence. To this aim, fig. 7 shows the plot of the standard deviation of the trace area above the thresholds; each threshold value corresponds to a load value ranging from 0.6 to 0.95. Trace1 and Trace3 have a very similar persistence at various loads with Trace1 actually exhibiting a somewhat larger one for all loads. At low to medium loads (up to 0.6, not shown here) this larger persistence is enough to give higher moments for the queue length, but as the load increases the situation is reversed and Trace3 gives bigger moments for the queue length.

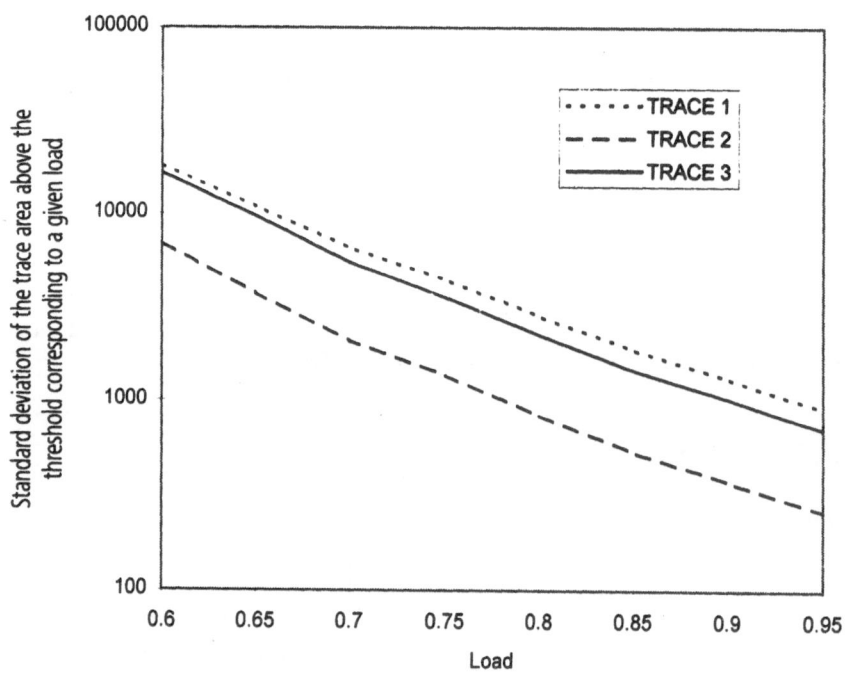

Fig. 7 "Persistence measure" plot

Thus, the idea under the persistence measure of fig. 7 does not completely capture the behaviuor of the tail of the queue length distribution, which is to say that a truly LRD process gives a heavier queueing behaviour.

In order to comment on this phenomenological observation we refer the reader to a paper by Lazar et al.[8]. There they introduce a finer notion of stability for queues by looking at the *conditional mean* of the input process within a certain time interval, called a scene. Even if context unaware definitions of scene are to a certain degree subjective, it is enough for our purpose to think that a scene change occurs when there is an abrupt change in the emission level. A discussion on this issue can be found in [8] and references therein.

According to [8] a queue is called *strictly stable* if the conditional mean load in each scene is less then 1. Otherwise a stable queue is said to be *weakly stable*. They demonstrate that under strong stability conditions LRD in the input process can be effectively neglected for performance evaluation.

By looking at the traces one sees that the examples we are considering here are only weakly stable for loads which are larger than 0.6. As in [8], we also conclude that LRD comes into play only in these cases.

Furthermore we observe that mean and variance of the queue length are influenced almost only by what happens in the intermediate time scale, which corresponds to the leftmost part of fig.1, with $H = 0.83$. Higher moments, on the other hand, depend strongly on very long time scales: see rightmost part of fig. 1, Trace1 with $H=0.5$, Trace3 with H still at 0.83.

As Trace1 was synthesized to reproduce qualitative features of the correlation structure in video traffic and Trace3 that in data traffic, we conclude that first and second order descriptors for the queue length in the two cases tend to be the same, while higher order descriptors (which are very relevant for QoS issues) may differ in a marked way.

Acknowledgements

We wish to thank to Alessandra Brunetti and Valerio Manca for their help with simulations.

References

[1] F. H.-M. Chen, J. Mellor, P. Mars, "Comparisons of simulation algorithms for self-similar traffic models", in *Proc. 13th IEE UK Teletraffic Symposium*, Glasgow, U.K., 1996

[2] M. W. Garret and W. Willinger, "Analysis, modelling and generation of self-similar vbr traffic", in *Proc. ACM Sigcomm '94*, London, U.K., 1994, pp. 269-280

[3] B. B. Mandelbrot and J. W. Van Ness, "Fractional Brownian motions, fractional noises and applications", *SIAM Review*, vol. 10, pp. 422-437, 1998

[4] A. Erramilli, O. Narayan, W. Willinger, "Experimental queueing analysis with long range dependent packet traffic", *IEEE/ACM Trans. Networking*, vol. 4, n. 2, pp. 209-223, 1996

[5] S. Li and J. D. Pruneski, "The linearity of low frequency traffic flow: an intrinsic property in queueing system", *IEEE/ACM Trans. Networking*, vol. 5, n. 3, pp. 429-443, 1997

[6] I. Norros. "A storage model with self-similar input", *Queueing Syst.*, vol. 16, pp. 387-396, 1994

[7] N. G. Duffield and O' Connell, "Large deviations and overflow probabilities for the general single-server queue, with applications", in *Math, Proc. Cambridge Philos. Soc.*, 1995, pp. 363-375

[8] P. R. Jelenkovic, A. A. Lazar, N. Semret, "The effect of multiple time scales and subexponentiality in MPEG video streams on queueing behaviour, *IEEE J. Select. Areas Commun.*, vol. 15, n. 6, pp. 1052-1071, 1997

Part 4

Multimode Multimedia Terminals

Multimode multimedia terminals: issues and trends

Giuseppe Coppola
Philips Research Monza, Italy
E-mail: coppola@monza.research.philips.com

Abstract

The vision of providing to consumer users simultaneous access to a variety of digital audio/video and "native" data services through a unique flexible, reconfigurable, interoperable device, namely a multimode multimedia terminal, is pulling simultaneous progresses in all the relevant technical and standardisation fields.

Indeed, many research and development activities are currently carried out around the world focusing on a large set of SW/HW advanced enabling technologies, while standardisation groups and consortia are specifying architecture and interfaces of such a multimedia common platform.

Such a "digital age" scenario isn't obviously expected to happen overnight, even because on top of technology and standardisation issues, ease-of-use and cost-effectiveness represent important barriers to break in order to enable mass consumer market.

This short note reports an overview of the key issues and trends for various aspects of the multimedia terminals, as an introduction to specific studies and results that will be presented during the Multimode Multimedia terminal session at the 10th Tyrrhenian International workshop on Digital Communication.

1.   Introduction

The implementation of end-to-end interactive and multimedia service chain relies upon the evolution of a collection of advanced multimedia technologies as joint audio/video/data content creation, retrieval and rendering, multimedia streams transport, access and delivery, end-user applications, RF/baseband signal processing.

Take-off of mass multimedia services will strongly depend from the manufacturer ability to develop terminal platforms including all these advanced technologies while satisfying ease-of-use and cost-effectiveness user requirements as well. Consequently, many research and standardisation activities are currently carried out all over the world.

This short note reports an overview of the key issues and expected trends concerning various aspects of the multimedia terminals, as an introduction to specific studies and detailed results that will be presented during the Multimode Multimedia terminal session at the 10th Tyrrhenian International workshop in Digital Communication.

2. Multimedia systems scenario

In the last couple of years some infrastructures and relevant consumer products based on digital technologies for video/audio compression and transmission have become available, nevertheless penetration of such terminals within the consumer market have shown to be very low and surely growing slower than what was expected in the beginning.

One of the main reasons is probably due to intrinsic characteristics of the typical residential user, which is moved to substitute his already owned boxes and update his own in-house infrastructures only if he perceives some real added value in doing that, that is only if there are clear advantages with respect to the previous status.

Ordinary users, though surely attracted by multimedia interactive services as increasing user demand for communication (mobile phones/E-mail) and personal fruition of available information (Web Browsing) are showing, would be really concerned by other aspects as ease-of-use, dimensions and obviously costs. Moreover, once experienced these new possibilities, they would like to have access at these services anytime and anywhere (at home and away) requiring the same quality and claiming the re-use of the same terminal equipment.

This full convergence is very difficult to happen because of extremely heterogeneous requirements as the services to be offered, the configuration of entertainment networks in the user's home, the compatibility among many interfaces and protocols, the interoperability with various access networks, the mobility characteristics of the terminal (nomadic, mobile, fixed) to name a few, furthermore coupled to the need for cost and dimension optimisation.

However, though envisaging co-existence of several platforms, maximisation of interoperability and interconnectivity aspects represent issues to address, in turn requiring evolution of programmable SW/HW architectures and adopting of common standardised specifications.

These agreements are needed not only to define physical interfaces and protocols, but also for the identification of system independent application interfaces for the delivering of multimedia applications. In this respect high level interfaces as MHEG-5/6 or Java, to name a few, have been already proposed and adopted in some specific consortia. Currently both special working groups (as DVB-MHP) or major CE companies (HAVi) are discussing requirements and selection of such a common API, which should provide interface between multimedia and interactive applications and, via an interactive run-time engine, actual HW/SW resources available.

3. Multimedia terminal architecture

As stated, due to heterogeneity of the scenarios it's rather difficult to identify a terminal architecture which can be suitable for all various requirements, because for instance terminals could be fixed or mobile, networked or not, based on PC or TV kind of services. Anyway, some basic issues or trends are common as the tendency towards a SW configurable architecture and a clear distinction between network dependent and independent functions by making use of clearly distinguished SW and HW layers.

An example of this modular architecture approach has been already identified within consortia as DVB, where a terminal model is assumed to be constituted by a network interface unit (NIU), which mainly performs channel coding/decoding and Medium access control protocols especially developed for each medium, and the Set Top Unit (STU) which performs all the baseband source decoding function and the processing of end-to-end interaction channel content ones, clearly independent from the access network.

Although their clear distinction, those basic modules have been so far mainly thought to be integrated within the same box, where at most another NIU module could be integrated to connect to different access network.

Alternative and to some extent longer term vision is base on the concept of having 2 separate boxes, one (the NT gateway) placed somewhere in the house (preferably hidden), and the other (the STU) placed close to the user terminal display, connected by a proper (standard) interface.

Such an approach relies on the evolution of in-house networks, when many terminal devices in the house would be able to share the same gateway to broadband access network. This approach should not add complexity once the standard for in-house interconnections would be more predictable and diffused than the number of possible technologies to connect to the external world. In this context wired high-speed serial interface as IEEE1394, able to deliver up to 400Mbit/s or USB, though with more limited bandwidth (maximum of 12Mbit/s) are already available. As far as in-house networking and interconnection aspects are concerned, however evolution towards a complete wireless solution (through infrared or radio frequency links) or at least a mixed wireless/wired scenario with wireless used for some specific terminals where mobility advantages can counterbalance data rate limitations, is generally envisaged. A major issue here concerns the availability and allocation of radio spectrum, and the consequent system and physical layer specifications.

Whether integrated or not, multimedia platform units surely require proper HW solutions both at a chip and at a board level.

As far as network interface unit is concerned due to the wide bandwidth requested, so far the only viable solution is represented by the presence of specific ICs for physical interfacing and channel coding. Strong pushing to the re-usability of similar channel coding and physical interfaces here from the semiconductor companies is understandable, though this is pushing sometime the attempt to re-use of some already developed functions for different media, which probably works, but it's far from the optimum media exploitation.

Other network dependent functions (MAC protocol, data link) should be anyway performed by programmable devices.

The core of each of these devices is therefore expected to manage control and configuration of all its network and services interfaces, while possibly performing at the same time, sophisticated signal processing (image, graphics, audio) and data processing algorithms (bitstream manipulation, data extraction).

This processor should therefore combine the processing power of a general purpose DSP, with the simultaneous handling of various tasks and applications typical of a powerful microcontroller. Some of this specific engines, often referred to as media processors, though obviously different in terms of architecture has been already designed by major semiconductors companies (TI, Philips Semiconductors).

These represent some first and very important steps, towards next generation of these devices, that would be possibly improved and allows evolution toward a system-on-chip (SOC), once interactions between processor HW architecture and the full SW architecture stack (drivers, OS, middleware, applications, baseband signal processing) would be better understood in actual application contexts.

The availability of such flexible HW devices, possibly integrated in the future on a single chip and as much as possible SW programmable/reconfigurable, is surely a necessary condition to meet the requirements of future multimedia terminal.

Obviously, key characteristics as flexibility, reconfigurability, expandability are required also at a board level, thus implying strong constraints on combined HW and SW architecture design.

The embedding of all SW functions onto a single core controller, requires also very efficient real time operating systems (RTOSs) able to perform both time critical and not-critical tasks with high efficiency. In this respect typical RTOSs (Qnx, VxWorks, pSOS) could be threatened by MS Windows CE evolution, especially as far as PC-based terminal (palmtop) are concerned, and even more by Sun's Personal Java virtual machine concepts, which could be preferable in the entertainment domain.

Indeed, in the end-to-end interactive service provision, many enabling SW technologies as robust/secure software downloading, easy-to-use interfaces and interaction tools (speech recognition), domain-specific & intelligent search engines, agents, and filters are needed as well.

Contents protections from unwanted reception (conditional access, cryptography) and from unwanted duplication and copying (watermarking, fingerprinting) is another area which has to be somewhat embedded within the multimedia terminal.

Another issue that can have implication on both HW/SW is the need for user/services data storage within the multimedia terminal. The need for this feature is debated and sometime neglected, due to the cost constraints on terminals and the assumption that future broadband access networks would be able to provide enough capacity.

However, also in this case, so far various storage technologies have been exploited within specific context with minimum or null interoperability or reusability. A proper and balanced combining of the basic storage technologies (HDD, magnetic tape, optical disk) would probably be beneficial for the terminal potentialities, though if externally connected.

4. Conclusions

The implementation of an easy-to-use, low-cost, open, flexible, reconfigurable, interoperable multimedia platform, strongly depends from the specific development of a large set of advanced enabling HW/SW technologies joint to agreement and standardisation of common APIs and SW/HW interfaces.

A brief overview of the current scenario and research/standardisation activities, accompanied by some (personal) thoughts/comments, have been reported in order to introduce relevant and more detailed results which will be showed during the Multimode Multimedia terminal session at the 10th Tyrrhenian International Workshop on digital communication.

References

1. ACTS Interactive Digital Multimedia Services Domain,
   http://www.infowin.org/ACTS/ANALYSYS/CONCERTATION/MULTIMEDIA.html

2. D.Lappe, "Evolution towards a global mobile multimedia communication system", Proc. of IEEE ComSoc '96 Workshop on "Wireless Multimedia Communication"

3. DAVIC 1.3 specifications, Parts 1 through 14,
   http://www.davic.organization/DOWN1.htm

4. EN 301 192, December 1997, DVB specification for data broadcasting (DVB-Data), ETSI Secretariat

5. ETSI 300 468 Ed.2: Digital Broadcasting systems for television, sound and data

services, Specification for Service Information (SI) in Digital Video Broadcasting (DVB) systems, 1997.

6. J-P.Evain, "The multimedia Home Platform - an overview", EBU Technical review, n.275, Spring 1998

7. M.Luise, S.Pupolin, "Broadband Wireless Communications: Part 4: Broadband Multimedia terminals, systems and services", Springer, 1998

Specification of digital interactive TV systems

Albert J. Stienstra
Philips Business Electronics, Eindhoven, The Netherlands

Abstract

Standardization is essential for successful development of a large market for digital TV systems and services. Due to the novelty of the technologies involved and the initially very small market having fast growth, the consensus in industry is that standardization via formal procedures in the existing standards bodies takes too long. Promising opportunities in digital audio, video and multimedia systems and applications have made it possible to organize specification consortia in a phase, preceding formal standardization. In these consortia many industry players from different parts of the value chain participate, with their best engineers and managers as delegates. Three clearly visible examples of these consortia are Digital Video Broadcasting in Europe, the Advanced Television Systems Committee in the USA and the Digital Audio- Visual Council with a global approach. In this paper a short description is given of the work by DVB. Ongoing and near future work on the Multimedia Home Platform and Home Networks is described in a little more detail, with its relation to the work in DAVIC and ATSC. Finally some conclusions are presented.

1. Introduction

The paper gives an overview of the specification work by DVB, DAVIC and ATSC in digital interactive TV systems. The word specification is used here to differentiate the work done by the consortia from that by formal standardization bodies such as IEC, ISO or ITU.

As distinct from the computer industry, in the telecommunication and broadcasting areas standardization is recognized as key for the development of a large consumer market. However, the formal standards bodies that used to do this have long-established voting procedures that take a very long time. Digital technology now makes greater strides than can be followed by processes with such a large time constant.

The consortia use consensus, voting only when it cannot be avoided. Because of this, a consortium specification takes much less time than an official standard does. As a result of this and boosted by the fact that top-level expertise is scarce, industry is more inclined to send its experts to the consortia, causing the time needed for a specification to be reduced even more when compared to a formal standard. The role of the standards bodies is developing more and more in the direction of maintenance of the standards, developed from the Publicly Available Specifications (PAS) made by the consortia.

Within a period of ten years from now a large segment of TV broadcasts will be digital, having replaced the present analog TV services. Digital transmission will increase channel capacity by almost an order of magnitude. Inclusion of data in the digital audio and video formats will be very easy. Channel capacity and data transmission will open up many opportunities for interactive TV services.

The Digital Video Broadcasting project (DVB), the Digital Audio- Visual Council (DAVIC) and the Advanced Television Systems Committee (ATSC) are the main consortia involved in the specification of digital interactive TV. All three have based their work on the MPEG-2 set of standards [1] for digital television.

Section 2 describes DVB as one of these specification consortia, its way of working and the status of the specifications. Ongoing work on multimedia home platform and home networks is dealt with in some more detail. Section 3 describes aspects of the relation between DVB and other consortia. Section 4 provides some conclusions.

2. DVB

2.1 Organization and technical work

DVB started end 1993, as a sequel to the Eureka 95 HDTV project. DVB's members are organized in four constituencies: broadcasters, network operators, equipment manufacturers and regulators. The members have signed a memorandum of understanding, addressing among other things the intellectual property rights and the commitment to use the DVB specifications. The technical work is driven by commercial requirements, to achieve an optimum fit to the developing market. The specifications are published as DVB Blue Books, which are sent on to ETSI and CENELEC for transposition into European standards. In 1995 the European Commission published a directive to the member states [2], stating that digital TV services to viewers in Europe shall use transmission systems standardized by a recognized European standardization body, such as ETSI and CENELEC. These mutually supporting circumstances have led to rapid development of an extensive set of standards. Although the original focus of DVB was on digital TV (standard definition or SDTV) in Europe, the attention is now on all main regions of the world, with solutions for SDTV as well as HDTV, at 25 and 30 Hz frame rates.

At this time DVB has published the following specifications:
- Digital transmission systems for satellite [3], cable[4], terrestrial UHF [5], microwave [6], [7] and satellite master antenna TV systems (SMATV) [8]
- Guidelines how to use the MPEG-2 standard [9]
- Service Information [10], Teletext Insertion [11]and Subtitling [12]
- Data Broadcasting [13]
- Equipment interfaces for receivers [14] and professional equipment [15] (e.g. satellite up-links, cable head-ends)
- For conditional access (CA): Common scrambling [16], SimulCrypt (multi-provider CA) [17] and a Common Interface [18]

- Transmission systems for interaction channels on telephone (including cordless) [19, 20], cable [21], GSM [22] and microwave systems [23]
- Network independent protocols for interaction control and command [24]
- Contribution and primary distribution on SDH [25] and PDH [26]
- Single frequency networks [27]

Clearly, this set of specifications deals with the major part of a standard platform for digital interactive TV services. To complete this platform for interactive services, DVB is now working on the application programming interface (API) for interactive content and on home networks, in the context of a Multimedia Home Platform.

2.2 Multimedia Home Platform

Figure 1 shows the general software architecture of a receiver decoder, with the location of the broadcast API in relation to the other components. With a standard high-level API as indicated, implementation of the other software elements may be manufacturer-specific. The resident application may consist of an electronic program guide, Word-Wide-Web access or similar. The run-time software consists of the communication protocol stack, presentation engines, virtual machine etc.

DVB defines the Multimedia Home Platform (MHP) as the set of physical layer and software interfaces to the user terminal and its peripherals on the home network, with a generic broadcast API as a key enabler of interactive applications. Both the broadcast API and the home network play an important part in the convergence with other data services, like those on the Internet. Therefore protocols are considered that are being used on the Internet, such as IP for data transport and networking, HTML as a presentation format and Java as a software platform, with a virtual machine and special Java APIs for broadcast.

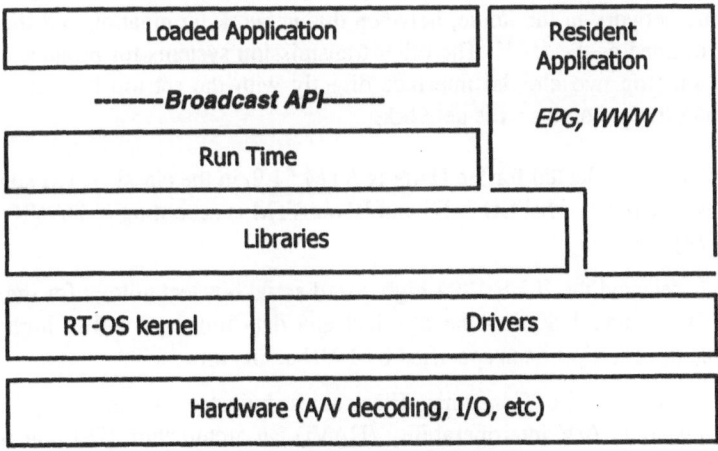

Figure 1. Decoder software structure with API.

IP is already specified as the network protocol in the DVB interaction protocol stack [24]. This protocol can also be encapsulated in the MPEG-2 Transport Stream according to the DVB-Data specification [13], however, it serves no purpose to use IP for carriage of the MPEG-2 packets on broadcast satellite, cable or terrestrial channels.

HTML has been developed for asynchronous applications on the World-Wide-Web. It needs extensions to handle A/V streams and synchronous events, for effective use on a DVB broadcast channel. Also, the footprint of full HTML is rather too large for the more low-end interactive TV applications and associated equipment. Even if memory cost will rapidly go down, software footprint remains an issue, especially when download is contemplated via a broadcast channel, where bandwidth is a scarce resource. For digital TV, a specific HTML flavor "Broadcast HTML" is required.

DVB has decided to define a specific DVB-Java platform, optimized for interactive digital TV applications and equipment. Full Java as used in PCs requires large memories and high performance processors. DVB will specify a Java Virtual Machine with associated libraries and APIs. A plugin interface will be defined for compatibility with existing systems, such as the MHEG-5 API specified by the UK-Digital Terrestrial Group. Unfortunately it has been impossible to accept the UK-DTG specification as a basic profile in DVB.

2.3 In-home digital network (IHDN)
DVB has adopted the home network architecture developed by DAVIC. A distinction is therefore made between the home access network (HAN) and the local area network in the home (HLN). Both subnets connect to the DVB receiver, see fig.1 «Elements of the IHDN».
Figure 2 shows the broadband telecommunication access network (bottom cloud) interfacing with the DVB domain via a Home Access Network Termination (HAN NT). The network in the home, between this network termination and the set top box, is defined as the HAN. The other transmission systems for broadcasting and interaction (top two clouds) interface directly with the set top box, without the intervention of a home access network.

The technology selected for the HAN is ATM 51.2, in the classic ATM star topology. Carriage of MPEG-2 TS and IP on ATM is according to DAVIC 1.3, Part 7 [28].

DVB has selected the IEEE 1394 high-speed serial bus technology for the IHDN-HLN. This standard defines the physical and data link layers. The higher layer protocols have not yet been specified by DVB at this time.

Major players in the Consumer Electronics industry are in the process of specifying home A/V interoperability (HAVi), an architecture [29] that provides interoperability between devices from different vendors connected to a network. It features distributed system control in a peer-to-peer architecture, that is able to handle legacy devices and networks. HAVi presents an API to applications on the home network, similar to what the DVB MHP offers to applications on the broadcast network. Clearly, HAVi and the MHP protocols will have to live harmoniously together in consumer electronics equipment.

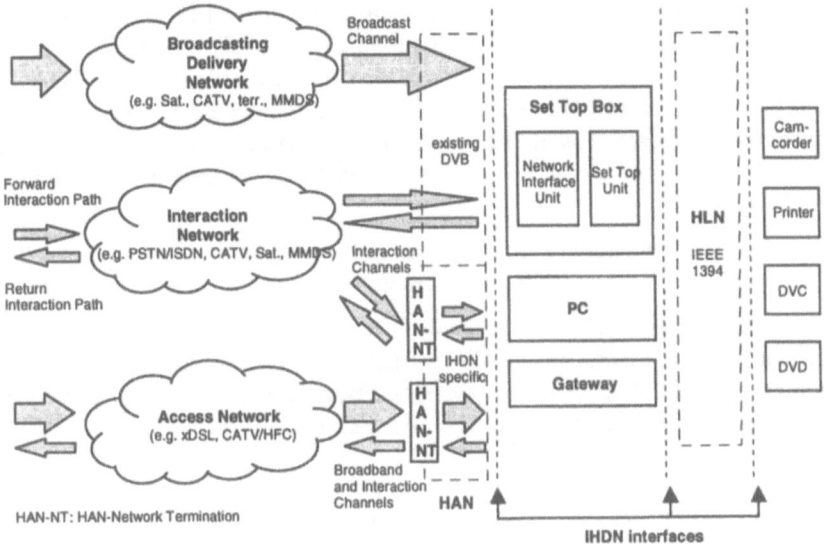

Figure 2. Elements of the IHDN.

3. Relation between DVB and other consortia

3.1 ATSC

The Advanced Television Systems Committee (ATSC) started its work in the early nineties. What DVB is doing in Europe, the ATSC is doing for North America. A significant difference between the two approaches is that the ATSC is dedicated to terrestrial UHF broadcasting systems (almost) exclusively for HDTV, while DVB addresses all broadcast media; satellite, cable, terrestrial UHF and microwave, in a coherent system approach. Although specified by DVB as an option, in Europe broadcasters have no interest in HDTV. Here the best opportunities are identified in a multiplex of standard definition TV and data services per transport stream. Recently however, Australia has selected DVB-T for transmission of HDTV services.

ATSC has finalized the specifications for digital TV transmission on UHF and defined a number of video display formats to be used by TV services carried by the MPEG-2 transport stream. The audio format selected is not from the MPEG-2 standard; it is a proprietary system from Dolby, known as AC3. Although ATSC embarked on the definition of Service Information long after DVB, unfortunately a different set of metadata describing the TV services was decided upon.

In the definition of an API for interactive services, ATSC and DVB are on the same level. As in DVB, ATSC is going in the direction of a Java platform with virtual machine, libraries and Java broadcast APIs. Because of a deadlock DVB was unable to select a suitable basic resident content decoder, such as the open standard MHEG-5, for use in terrestrial interactive TV services. Now an opportunity is created for ATSC to do this, e.g. by the development of a specific broadcast flavor of HTML. Elements from MHEG-5 could be used for this, however to simply select MHEG-5 would be very difficult in the USA due to the overwhelming interest in the developing market based on Internet and Internet-related protocols.

Terrestrial TV services are expected to start in the USA in the 4th quarter of 1998, the same time frame when DVB-T services will start in Europe.

In North America the cable issues are dealt with by the Open Cable consortium, while the satellite market in the US is fragmented by several operators, using different incompatible transmission systems. Efforts are being made to coordinate between ATSC and Open Cable to come to a harmonized API.

3.2 DAVIC

The Digital Audio- Visual Council (DAVIC) started in 1994, about a year after DVB, aimed at video on demand services to consumers via ATM networks. DAVIC's goal is to define a set of interfaces enabling mass deployment of interoperable applications and equipment, on a global scale. Like DVB, DAVIC is a consortium of more than 200 member companies from different parts of the value chain. However, the representation of telecommunication and computer companies in DAVIC is larger than in DVB, which is favorable for the definition of high-level software interfaces.

It was soon realized by DAVIC, that the first mass deployment of interactive services with high quality video will be through digital broadcast media and not via ATM networks. Therefore the DVB transmission systems for satellite and cable were adopted as access networks. A special set of tools for interactive broadcast services has been defined. Some of these tools, notably the mid-layer protocols for interaction control, were adopted by DVB. This was caused by the rapid growth of Internet use by consumers, making the broadcasters in DVB realize that interactive TV services might constitute an interesting opportunity. DAVIC also specified a content decoder, Java virtual machine and APIs. Some of these are of crucial importance for interactive TV.

DAVIC has recently embarked on the 1.5 series of its specification, working on system interfaces employing local storage and/or using IP. The major applications on these systems are categorized under the headings «TV Anytime» and «TV Anywhere». Tools developed for these systems will most likely also be of use for DVB.

4. Conclusions

Consortia now do the standardization of digital interactive TV systems and applications, with the formal standards bodies in the second echelon for maintenance.

DVB has specified most of the interfaces needed for interactive digital TV, with maximum interoperability between satellite, cable and terrestrial transmission systems.

In North America the broadcast market is more fragmented than in Europe, with different systems being used for digital terrestrial-, satellite- and cable TV.

Although ATSC is only focused on digital terrestrial HDTV, it now has the opportunity to specify an API for basic interactive TV, because of a deadlock occurring in DVB.

DAVIC through its focus on interactivity has been leading in developing tools for interactive services, which have been and are being adopted by DVB and ATSC.

Acknowledgements

The author thanks colleagues in DVB, DAVIC and ATSC for the stimulating period in the previous couple of years, when the foundation of digital interactive TV systems was laid.

Figure 2 of this paper has been derived from work in the IHDN group of the DVB Technical Module.

References

1. ISO/IEC DIS 13818-1, -2 and -4; prepared by JTC1/29 of ISO/IEC, 1 Rue de Varembé, PO Box 56, CH-1211 Geneva 20, Switzerland
2. Directive 95/47/EC; 23.11.95 EN Official Journal of the European Communities No L 281
3. European Telecommunication Standard ETS 300 421, December 1994; Digital broadcasting for television, sound and data services; Framing structure, channel coding and modulation for 11/12 GHz satellite services (DVB-S); ETSI Secretariat, 06921 Sophia Antipolis Cedex, France
4. ETS 300 429, December 1994; DVB Framing structure, channel coding and modulation for cable systems (DVB-C); ETSI Secretariat
5. ETS 300 744, March 1997; DVB Framing structure, channel coding and modulation for digital terrestrial television (DVB-T); ETSI Secretariat
6. ETS 300 748, October 1996 DVB Framing structure, channel coding and modulation for multipoint video distribution systems above 10 GHz (DVB-MS), ETSI Secretariat
7. ETS 300 749, April 1997, Framing structure, channel coding and modulation for multipoint video distribution systems below 10 GHz (DVB-MC)
8. ETS 300 473, May 1995; DVB Satellite Master Antenna Television (DVB-SMATV) distribution systems; ETSI Secretariat
9. ETR 154, September 1997, 3rd ed., DVB implementation guidelines for the use of

MPEG-2 systems, video and audio in satellite, cable and terrestrial broadcasting applications (DVB-MPEG), ETSI Secretariat

10. ETS 300 468 , October 1995; Specification for Service Information in DVB Systems (DVB-SI); ETSI Secretariat

11. ETS 300 472, May 1995; Specification for conveying ITU-R System B Teletext in DVB bitstreams (DVB-TXT); ETSI Secretariat

12. EN 300 743, September 1997, DVB subtitling systems (DVB-SUB), ETSI Secretariat

13. EN 301 192, December 1997, DVB specification for data broadcasting (DVB-Data), ETSI Secretariat

14. prEN 50201 (DVB-IRD), to be published by CENELEC

15. EN 50083-9, ed.2 June 1998, Interfaces for CATV/SMATV headends and similar professional equipment for DVB MPEG-2 Transport Streams (DVB-PI), CENELEC

16. ETR 289, October 1996, Support for use of scrambling and conditional access (DVB-CS), ETSI Secretariat

17. TS 101 197, June 1997, DVB SimulCrypt, Part 1; Head-end architecture and synchronization (DVB-SIM), ETSI Secretariat

18. EN 50221, February 1997, Common Interface specification for conditional access and other DVB decoder applications, CENELEC

19. ETS 300 801, August 1997, DVB Interaction channel through PSTN/ISDN (DVB-RCT), ETSI Secretariat

20. EN 301 193, August 1998, DVB Interaction channel through DECT (DVB-RCDECT), ETSI Secretariat

21. ETS 300 800, August 1998, DVB Interaction channel for CATV distribution systems (DVB-RC), ETSI Secretariat

22. draft EN 301 195, June 1998, DVB Interaction channel through GSM (DVB-GSM), ETSI Secretariat

23. draft EN 301 199, August 1998, DVB Interaction channel for local multipoint distribution systems (DVB-RCL), ETSI Secretariat

24. ETS 300 802, November 1997, Network independent protocols for DVB interactive services (DVB-NIP), ETSI Secretariat

25. ETS 300 813, December 1997, DVB Interfaces to PDH networks (DVB-PDH), ETSI Secretariat

26. ETS 300 814, March 1998, DVB Interfaces to SDH networks, (DVB-SDH), ETSI Secretariat

27. TS 101 191, April 1997, Megaframe for single frequency network synchronization (DVB-SFN), ETSI Secretariat

28. DAVIC 1.3 specifications, Parts 1 through 14, http://www.davic.organization/DOWN1.htm

29. The HAVi architecture, Sony, Philips, Hitachi, Sharp, Matsushita, Thomson, Toshiba, Grundig, http://www.sv.philips.com/news/press/

MHEG-5 common application interface for broadcast and interactive services

Pietro Marchisio
CSELT, Torino, Italy

Abstract

As the world is becoming more connected, there is an ever increasing demand for new multimedia services. Their emergence is being favored by deregulation and converging technologies. In this scenario, standards are essential for delivering multimedia applications across heterogeneous platforms. The ISO MHEG-5 standard specifies an encoding format for multimedia applications independently of service paradigms, system architectures, and network protocols. It targets both the broadcast and the interactive domains, and is effective also on low-cost terminals, like set-top-boxes. The paper gives an overview of MHEG-5, describes the implementation of a run-time presentation engine, and illustrates the use of the standard in the emerging domain of interactive television.

1. Introduction

Standards and well-defined interfaces are essential for building and exploiting complex computing and communications systems. They simplify the design and construction of individual hardware and software components and help ensure satisfactory interworking. The interoperability requirement becomes even more important in the multimedia arena, where broadcast and interactive technologies converge almost daily, especially in the emerging scenario of interactive television.

With "multimedia standard" we indicate a declarative format to represent the "composition level" of an application, which presentation to users generally involves a series of scenes. A scene is a composition of aural and visual ingredients, with specific properties and behavior, in both dimensions of space and time, including capability to manage user interaction and other external events. Due to the declarative approach, authors are no longer burdened with the writing and testing of procedural code.

The ISO MHEG-5 standard [Ref 1] specifies a declarative format to distribute multimedia applications across platforms of different types and brands, including terminals with limited resources, like set-top boxes adopted by interactive TV. In this context, upcoming applications are expected to offer scenes showing a blend of broadcast video and other information, retrieved for example on the Web at user's convenience.

The paper provides the rationale of MHEG-5, illustrates the main features of the standard, and describes the implementation of a player of applications. The player is compliant with the DAVIC (Digital Audio Visual Council) specifications, and constitutes a common interface towards broadcast and interactive delivery channels. Finally, an example of MHEG-5 application in the interactive TV domain is also provided.

2. The MHEG-5 Standard

MHEG encompasses the family of standards issued by ISO/IEC JTC1/WG12/SC29, coding of audio, picture, multimedia and hypermedia information. MHEG-5 provides "Support for base-level interactive applications". See [Ref 7] for the complete list of MHEG standards.

2.1 Requirements

A standard at the "composition level" should be able to represent multimedia applications in final-form, which means no additional processing is needed to restructure the exchanged information for presentation.

A generic multimedia application consists of sets of self-contained objects based on synchronization and spatial-temporal relationships of multiple media formats, structural composition, event-action associations, navigation and user interaction. Controlling the playback of time-dependent contents, like streams of multiplexed audiovisual data, requires specific support. These streams demand VCR control functions (play, pause, and so on), as well as capability to manage events generated during presentation. For example, the playback of a stream might synchronize the rendering of text subtitles and graphics animations.

Moreover, since the standard aims to guarantee interoperability across heterogeneous platforms, also the encoding syntax should be specified.

2.2 Object-oriented approach

MHEG-5 allows to shape page-oriented applications that consist of a set of scene objects. At most one scene is active at one time and navigation within an application is performed in terms of transitions between scenes. Inter-application navigation is also possible, being an application able to "launch" another application. Scenes provide support for spatially and temporally coordinated presentation of audiovisual contents comprising elementary graphics, bitmaps, texts and audiovisual streams. Interaction is performed via graphic controls such as buttons, sliders, text entry boxes and hypertext selections.

Every scene, as well as an entire application, is a self-contained entity that represents its local behavior by means of links that are event-action associations. Events can be generated: by the user, by the expiration of timers, by the playback of streams as well as by other conditions internal to the execution process.

Since MHEG-5 adopted the object-oriented approach, all features are specified in terms of classes of objects. The standard specifies each class in terms of three kinds of properties:

- *attributes* that make up an object's structure
- *events* that originate from an object, and
- *actions* that target an object to accomplish a specific behavior or to set or get an attribute's value

2.3 Major Classes of objects

There are two major superclasses from which other classes inherit properties: the Group and the Ingredient. These two are, in their turn, subclasses of a common Root that specifies an object identification mechanism and an object availability status.

The Group class handles the grouping of ingredient objects as a unique entity of access and interchange. A Group can be specialized into the Application and the Scene classes. A Group cannot include another Group.

An MHEG-5 application consists of one Application and a set of Scene objects. The Application contains objects common to all the Scenes and is in charge of activating the first Scene of the presentation. A Scene has an associated presentation space where the rendition of its ingredients is performed. At most one Scene is active at any one time. Navigation in an application is done in terms of transitions between Scenes.

The Ingredient class provides the common behavior for all objects that can be included in an Application or a Scene. There are two major kinds of ingredients: Presentable and Link. Presentable ingredients support the rendition of bitmap, line-art, text, stream, audio, video, run-time graphics, and so on. The list also includes some specific user interaction classes: entry-field, slider, hotspot, and so on.

A Link is used to represent behavior as an event-action association. Different kinds of events can be tested by a link: the availability of a content to be presented, a user interaction, a 'time-code' crossed during playback of a stream, and so on. When the event occurs, the link fires, and the associated sequence of actions is executed. More than 100 actions are specified by the standard: running a presentation, changing the speed of a stream, its direction or its current time-code position (i.e. seek to a given point in the presentation), modifying the scene's layout, managing user interactions, controlling link activation, and so on.

A class not directly involved in presentation tasks, but essential for calling an external piece of procedural code, is the Program class. It allows, for example, the activation of a Java script with capability of passing parameters.

There are also a number of classes that can be considered optional for a basic implementation of the engine. The Variable class stores values to be tested by links. The TokenGroup, TemplateGroup and List are helpful in contexts where navigation of a "logical token" is required, to highlight for example the "focus" among a list of selectable items. This simplifies the representation of selection menus, fill-in-forms, and so on.

2.4 Encoding

Two encoding formalisms are supported, and referred to as "basic" notation and "alternative" notation. The basic notation relies on a syntax expressed in ASN.1/DER (Abstract Syntax Notation One / Distinguished Encoding Rules). It produces a compact encoding which is not, however, human readable. The alternative notation is human readable, and therefore enables authors to create applications without the need of an authoring tool.

2.5 Distributing Applications

MHEG-5 aims to distribute applications in a client/server architecture. Normally, these applications reside on the server and only the portions that are required at a given point in the presentation are downloaded to the client. Figure 1 shows the MHEG-5 client-server model.

It is responsibility of the client to have a compliant player (also referred to as Run-Time Engine - RTE) that interprets the application parts, performs their presentation to the user, and handles the local interaction with the user. Whereas applications are machine independent, the client's software depends on the actual platform, and should be optimized in terms of both speed and reduced footprints for the specific hardware it runs on. This is particularly important in the set-top-box environment, where a strict memory management is required to ensure that the engine does not allocate more memory than it was pre-configured.

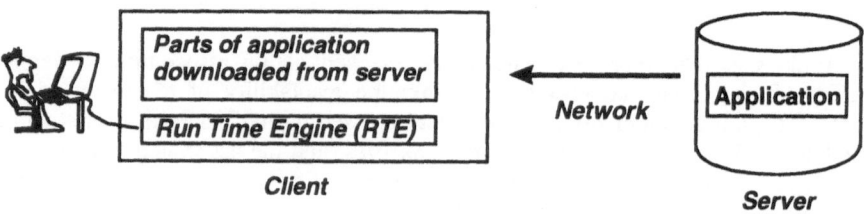

Fig. 1 MHEG-5 Client-Server model

The use is not limited to "storage and retrieval" services. In the broadcast environment, for example, the set of channels transmitted on a broadcast network can be considered as a virtual server where the download mechanism relies on a carousel (cyclic rebroadcast) of all portions of an application.

2.6 MHEG-6: the MHEG-5 API

Most of "retrieve and navigation" applications can entirely be encoded in MHEG-5, since their scope is confined within representation and presentation of information. There are, however, profiles which also require extended data processing (arithmetic and logic operators, flow control, and so on) and communication functions with the external environment. These features demand a procedural code extension to the declarative model.

To provide a "complete solution" for application representation, also the definition of the procedural code syntax and its execution model need to be covered by a standard. This is the scope of MHEG-6 [Ref 2], which extends the declarative approach of MHEG-5 with a Java virtual machine, and provides an API for internal and external communication.

Compliance with MHEG-6 is based on:

- a Java Virtual Machine, and
- an MHEG-5 API to allow Java code to call upon MHEG-5 objects.

3. Relationships with the DSM-CC, DAVIC, and the DVB

To achieve full interoperability among heterogeneous systems, there are important relationships between MHEG-5 and other standards and specifications to be considered.

3.1 DSM-CC

The DSM-CC (Digital Storage Media Command and Control) [Ref 4] specifies a set of protocols providing the following functions and operations in a distributed system:

- Network session and resource control
- Configuration of a client
- Downloading to a client
- VCR-like control of the MPEG stream
- Generic interactive application services
- Generic broadcast application services

The generic reference model provided by MHEG-5 can be enhanced by the use of the DSM-CC to access application objects, part of them, and media data.

DSM-CC Data Carousel. Particularly relevant for ITV applications is the Data

Carousel scenario of the DSM-CC Download Protocol. It is built on the DSM-CC download protocol where data and control messages are periodically re-transmitted following a pre-defined periodicity. The download server sends periodic download control messages which allow a client application to discover the data modules being transmitted and determine which, if any, of these modules must be acquired. The client typically retrieves a subset of the modules described in the control messages. The data modules are transmitted in blocks by means of download data messages

3.2 DAVIC

DAVIC [Ref 3] is an international industry consortium whose purpose is to promote the emerging digital audio-visual applications and services, first for broadcast and interactive use. Its activities result in internationally agreed specifications of open interfaces and protocols. As DAVIC aims to maximize interoperability on the basis of available specification, MHEG-5 and MHEG-6 were selected as the standard set-top-unit high-level interface for multimedia applications. DAVIC also adopted the DSM-CC for client server interworking.

3.3 DVB

The DVB provides a complete solution for digital television and data broadcasting across the range of delivery media, where audio and video signals are encoded in MPEG-2. The specification includes an open Service Information system, known as DVB-SI [Ref 5], which provides the elements necessary for the development of a basic Electronic Program Guide (EPG) to support navigation amongst the new digital television services.

DVB MHP (Media Home Platform). This working group pursues a solution that covers the whole set of technologies necessary to implement digital interactive multimedia in the home, including protocols, common API languages, interfaces and recommendations. A DVB API is being defined as a set of high-level functions, data structures and protocols that represent a standard interface for platform independent application software.

4. MHEG-5 player

This chapter addresses the implementation of the MHEG-5 player, which provides a common API towards broadcast and interactive channel protocols. The player was developed as the client of ARMIDA [Ref 6], a client-server system for interactively retrieving multimedia applications from remote databases via broadband networks. ARMIDA was developed as a prototypical implementation of the specifications issued by DAVIC 1.0 for systems supporting interactive multimedia services and applications. It also receives, decodes and displays TV programs conforming to the DVB standard.

The player consists of a "pure" RTE that implements semantics of MHEG-5 and issues I/O and data access requests to other specialized components that are optimized for the specific run-time platform. Figure 2 depicts the software architecture of the player, which runs on a commercial PC equipped with Windows95/NT.

4.1 Run-Time Engine

The RTE is based on two main tasks. The first, preparatory to the presentation, performs accessing and decoding of MHEG-5 objects in their internal form. The second, which is the actual presentation, consists of an event-loop, where events raise requests to the Presentation Layer and actions with an effect internal to the engine.

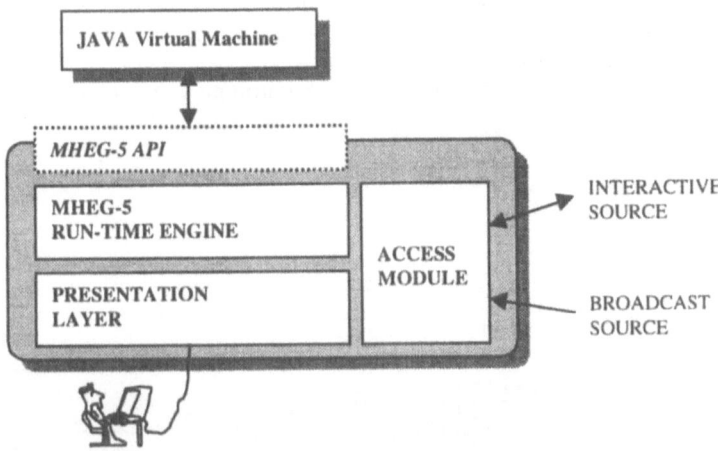

Fig. 2 MHEG-5 player architecture

Retrieving, decoding and activation of objects. At activation time, the RTE receives the identifier of the first Application object to launch. The RTE retrieves this object via the Access Module, and then decodes it. This process also implies decoding all the Ingredients contained in it. At this point the Application is activated and is expected to perform a "transition to" the first Scene object that, in its turn, is retrieved, decoded and activated. As soon as exactly one Application and one Scene object is active, the "context" is ready to start the event-loop that coordinates the presentation.

Generation and processing of events: the event-loop. There are two main categories of events: synchronous and asynchronous. Asynchronous events are generated externally to the RTE. This means by the user, by the availability of a content, by a timer or by the presentation of a stream (e.g. when a predefined

416

"time-code" position is crossed). Since the execution model is single-thread, these events need to be buffered into a queue. Synchronous events are, on the other hand, generated internally to the RTE as the direct result of the execution of certain actions, and are processed immediately. Essentially, the event loop performs the following steps (see Figure 3):

1. the first asynchronous event in the queue is moved to the LinkProcessor; if the queue is empty, an asynchronous event is awaited;
2. the LinkProcessor searches for a Link which condition matches with the event. If no Link is found, go back to 1;
3. the ActionExecutor processes the LinkEffect of the Link, that means executing a sequence of actions;
4. as a direct result of an action being executed, a synchronous event may occur (e.g. a "run" action generates the "is running" event on the target object). This event shall be dealt with immediately: all its effects (actions) are completely processed before executing the next pending action of the current Link. To accomplish this, go to step 2;
5. when all the actions pertaining to the asynchronous event are processed, go to step 1

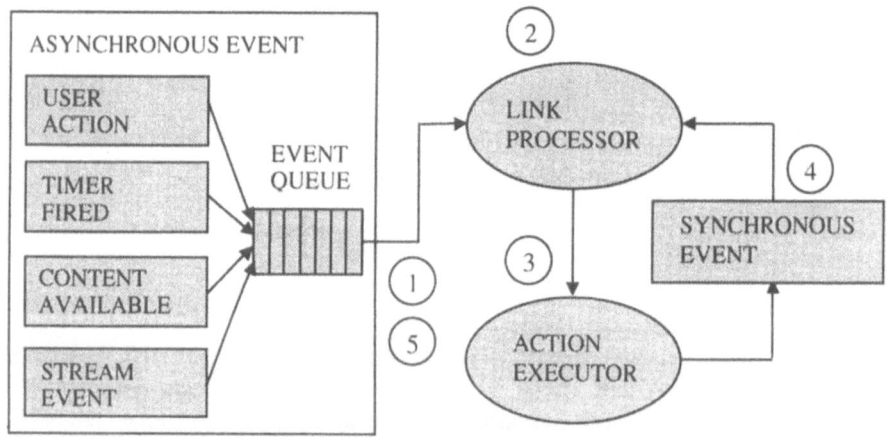

Fig. 3 Managing events within the RTE

The two basic tasks of retrieving-decoding and event processing are interleaved. In fact, the actions that launch a new Application/Scene object change the current "context" of execution and, consequently, the event-loop is reinitialized with a new set of active links.

4.2 Presentation Layer

The Presentation Layer (PL) manages windowing resources, deals with low-level events and performs decoding and rendering of contents from different media to the user. This functionality is available to the RTE via an Object-Oriented API that encapsulates all I/O platform specific aspects. Essentially, the basic MHEG-5 Presentable classes have their counterpart at the level of this API that makes provision for: initialization/termination, data access and decoding, setting specific attributes (e.g. text font, color, and so on), performing spatial and temporal controls. There exists also an informative flow from the PL back to the RTE, implemented in terms of callback functions that notify user interaction and stream events.

Data access requests are turned to the Access Module that downloads information into a buffer under PL control to feed the proper content decoder. Thanks to the integration of decoders available off-the-shelf, the player allows to render text, bitmaps and audiovisual streams encoded in different formats. The LeadTools™ library (from Lead Technologies) allows the decoding of several types of bitmaps, including Gif (fixed and animated), Jpeg, Bmp, and so on. ActiveMovie™ (from Microsoft) manages the presentation of streams. It supports both audiovisual multiplexed formats (MPEG, AVI, QuickTime) and audio formats (Wave, Midi, Au, and so on).

4.3 Access Module

The Access Module is used by both the RTE to get objects and the PL to access content data. It provides a consistent API for accessing information from different sources, either based on the interactive or on the broadcast paradigms. The selection of a particular delivery strategy is out of the scope of MHEG-5, and hence remains an implementation issue. The following access modes are supported.

Interactive access via DSM-CC. The adoption of the DSM-CC allows the straightforward implementation of the following strategies:

- bulk download, for bitmaps, texts and MHEG-5 objects
- streamed download, for audio and audio-visual streams

The protocol offers full interactive control of the playback of streams in terms of VCR-like commands ("play", "pause", and so on). These commands are exchanged between client and server as effect of user interactions. Higher quality MPEG-2 streams can be delivered to the client via ATM network.

Internet access. Object and content access requests can also be issued on the Web via HTTP. However, this widespread protocol still suffers from limitations when high-quality video clips are progressively downloaded, since the underlying networks may not yet provide adequate quality-of-service (QoS).

Selecting TV Programs. The basic use of the broadcast channel is achieved by allowing to select and display digital TV programs referred to by MHEG-5 Stream objects. When a Stream is activated, the Access Module requires a DVB channel selection component to select the addressed channel, with the final effect of displaying the MPEG-2 stream in the presentation space of the active Scene.

Broadcast Carousel. A more advanced use of the broadcast channel is based on the DSM-CC data carousel, which makes possible a certain degree of interactivity. All information transferred is cyclically rebroadcast in a carousel, and is available to the RTE on request.

5. Interactive TV Applications

Several MHEG-5 applications can be delivered to interactive TV users. Among them are Electronic Program Guides (EPGs), enhanced teletext, and applications related to specific television programs. The latter allows, for example, the user to get additional information on sports event or select a view from multiple video streams via the MHEG-5 application.

Let us consider the EPG, which is an application offered by the broadcast service provider to give information about the available services and coming events. An EPG can be encoded as an MHEG-5 application that is broadcast using a DSM-CC object carousel, and also made available on the Web to increase the number of potential users. Figure 4 shows the look-and-feel of an MHEG-5 scene belonging to an EPG application.

Fig. 4 Look-and-feel of an MHEG-5 scene

This scene provides the preview of a TV program in terms of a video clip, textual information, bitmaps and other animated contents. A set of buttons and hotspots can be used to navigate towards related scenes that contain links to video streams as the means for selecting the television service. The scaling facilities of the MHEG-5 video class can also be used to present the video in a small window on the screen, while presenting other information to the user.

6. Conclusions

MHEG-5 reached the status of international standard in 1996, and is presently undergoing a "technical corrigendum" process, which is based on the results of a set of implementations. A list containing relevant projects and other related activities is maintained by the MHEG-5 Users Group (MUG) [Ref 8], an unofficial forum aimed to disseminate information and enable discussion on the standard.

There are now sufficient MHEG-5 implementations in existence to conclude this "prototype phase" as completed. It allowed to validate the soundness of the specification and the capability to operate in distributed environments, especially those compliant with DAVIC. Recently, MHEG-5 entered a new phase where prototypes are being turned into products to fit the tight requirements of commercial platforms. Most prominent one is the domain of interactive television, where some players are planning the launch of pilot services based on MHEG-5. They recognized that the adoption of such a standard presents a set of advantages, which include a reduced per-unit cost of the run-time engine, a more validated product based on extensive product testing, and a greater return on investment.

The aspect of conformance testing of MHEG-5 is also receiving attention. A specific Working Group, known as MHEG-7, has been set up with the mandate of providing a specification of "interoperability and conformance testing" of MHEG-5 engines and applications. The objective is twofold: providing developers of tools and applications with clear unambiguous guidelines, and promoting the widespread development and use of MHEG-5 applications and tools. First outcomes are scheduled for 1998.

References

1. ISO/IEC International Standard 13522-5: Information Technology - Coding of multimedia and hypermedia information - Part 5: Support for base-level interactive applications, 1996.
2. ISO/IEC International Standard 13522-6: Information Technology - Coding of multimedia and hypermedia information - Part 6: Support for enhanced interactive applications, 1998.
3. DAVIC 1.0 Specifications, Revision 5.0, December 1995.
4. ISO/IEC International Standard 13818-6: Information Technology - Generic coding of moving pictures and associated audio information, Part 6: Digital Storage Media Command and Control, 1996.
5. ETSI 300 468 Ed.2: Digital Broadcasting systems for television, sound and data services,

Specification for Service Information (SI) in Digital Video Broadcasting (DVB) systems, 1997.

6. S. Dal Lago, G. Franceschini, P. Marchisio, M. Mesturino, E. Polese, G. Venuti, "ARMIDA™: Multimedia Applications across ATM-based Networks accessed via Internet Navigation", *Multimedia Tools and Applications*, Vol. 5, No. 2, September 1997.

7. M. Echiffre, C. Marchisio, P. Marchisio, P. Panicciari, S. Del Rossi, "MHEG-5 - Aims, Concepts, and Implementation Issues", *IEEE Multimedia*, Vol. 5, No. 1, January-March 1998.

8. MHEG-5 Users Group Home Page: http://www.fokus.gmd.de/ovma/mug.

New concepts for Multimode Multimedia Terminal Architectures

Petri Mähönen
VTT, Technical Research Center of Finland,
Wireless Internet Laboratory, P.O. Box 1100, FIN-90571 Oulu, Finland

Abstract

The advanced multimedia terminals of future are an interesting area of technology, which may be heading towards a market explosion similar to that of the mobile phone market. This perspective on the research and market trends in the field focuses on Last Mile broadband (wireless) terminals. However, I am not maintaining that there will be a single terminal type dominating the market in the future, but rather a large blend of diverse terminal platforms and standards will be co-existing. The most promising approach to coping with the increasing market diversity can be found in enhancing the adaptivity of the software in the terminals.

1. Introduction

This paper summarises some of the trends of the work done on terminal R&D at VTT Electronics and elsewhere. In recent years, there has been an explosion of activity and interest in the applications of multimedia systems, especially in the area of digital video and television. There has also been an increasing rush to move from old analog services to digital ones and to create new "value-added" services to boost revenues. Designing easy-to-use and cost-effective multimedia terminals, such as set-top-boxes, for end-users has been one of the most significant trends in this development. It is important to note that although the technology itself for powerful multimedia set-top-boxes has been around for rather a long time, the end-user price is still too a high allow a breakthrough in the market. Thus, multimedia R&D should be able to meet the demands of low-price and high performance at the same time.

Wireless telecommunications networks have been deployed rapidly in recent years all over the world. In Europe the Global System for Mobile Communications (GSM) is entering into the phase 2+, which will include packet switched data services. Around 2002-2005 we can expect to see the third generation mobile systems (UMTS) that will offer mobile data services with user bitrates up to 2

Mbps. These advances in the wireless telecommunications are also pushing the technology and user-demand towards more efficient wireless multimedia terminal [1].

One of the key factors in the success of multimedia and digital video has been the rapidly increasing processing capacity of microprocessors and the emergence of a diversity of standards for audio, video and overall-systems architecture (e.g. MPEG2, DVB and DAVIC). Standards are needed for providing a common framework allowing the interoperation of different systems [2,3].

In addition to telecommunication systems also digital broadband networks, including fibre-based networks and wirelesses broadband extensions, are being deployed around the world, thus providing a huge market potential and infrastructure support for terminals and multimedia services. Government deregulation processes, especially the EU telecommunication deregularization in 1998, is likely to foster competition and to provide more investment money for the infrastructure building in the field. Finally, the progress in ASIC implementations and increasing computing power have enabled new applications and terminal technologies faster than ever thought possible.

2. Multimedia Terminal Systems

2.1 Classification of Terminals

The terminal equipment is, of course, the crucial part of the communication system. Without the access terminal, the network would be useless. The design of the terminal equipment for high-speed networks has become more and more difficult task. Difficult technological and architectural problems are encountered not only with mobile wireless terminals with multipurpose set-top-boxes. Indeed, it would be misleading to focus only on set-top-boxes [4] or any other single terminal type. In fact, the term multimedia terminal will be used in a very generic sense in this text, denoting several different kinds of equipment. There are many various ways of categorising terminals. Generally speaking following criteria can be used to classify different terminal technologies:

- Wireless vs. wired terminals; terrestrial vs. satellite systems
- Networked vs. non-networked systems, e.g. interactivity
- External type: Set-top-box / computer / embedded / PDA / mobile
- Functionality: multimedia "player", PDA, computerised terminal etc.

Some terminals are just limited hardware platforms offering the user only a restricted selection out of a large number of possibilities. On the other hand, one may expect to see the appearance of a class of "super-terminals", incorporating an extensive set of technologies. One of the most essential requirements on the future terminals is the capacity of supporting a wide variety of devices and network standards. The main problem area in the terminal R&D is the *need to support different standards, services and networks; and same time quickly answer to changes in the markets.*

2.2 Diversity of Standards

The standardisation process is essential for securing not only the coherence of the R&D work worldwide but also for the compatibility between different vendors and service provides. Unfortunately, the standardisation is impeded by two inherent weaknesses at the moment:

- *Standardisation time and need of consensus*: A common argument used against official standardisation bodies maintains, somewhat rightfully, that the time needed for reaching an official and final standard is so long that the standard will be obsolete once completed. Several standardisation organisations, like ISO and ITU, are reacting to this criticism by means of re-engineering their standardisation processes. However, it should be borne in mind that usually the main reason for the slowness is found in the need of reaching a reasonable agreement within technical committees to finalise the standard text. The future fast track approaches, which will give a possibility for single or multiple partners to officially standardise their industry *de facto* standards through ISO, are indeed extremely tempting. However, the reason for which the formal standardisation bodies still have a competitive edge is their high level of expertise and sophistication of process, usually yielding an outstandingly high quality of work, compared, e.g., to some "industry group standards". In addition, multinational standards do, after all, cover a remarkably extensive market area.

- *The excessive complexity of standardisation market:* The number of standards relevant to multimedia terminals is becoming overwhelming. There is no clear way out of this problem. The diversity and the number of standards produced for multimedia use is in some sense excessive, while, on the other hand, it is also clearly showing the complexity of the overall research area. However, this complexity, in which a single device design might need an input with 50 different standards, will inevitably lead to growing demands on the staff of companies and laboratories, and often force companies to use specialised sub-contractors and/or to form industrial partnerships.

One of the key standards for digital multimedia has been the development of the MPEG2. In MPEG2 several areas have been left *intentionally* unspecified. These unspecified details can be found in the areas of conditional access, transport layer features (specifications of the number of PIDs required to be supported) and stream filtering requirements. At the transport stream level, for instance, there are various differences between service providers, concerning the number of PIDs, private section filtering, (dec)scrambling, graphics, MAC-layer etc. The decision to leave these areas open made the standardisation process considerably easier. Also DAVIC and DVB standards are very important in the development of multimedia systems.

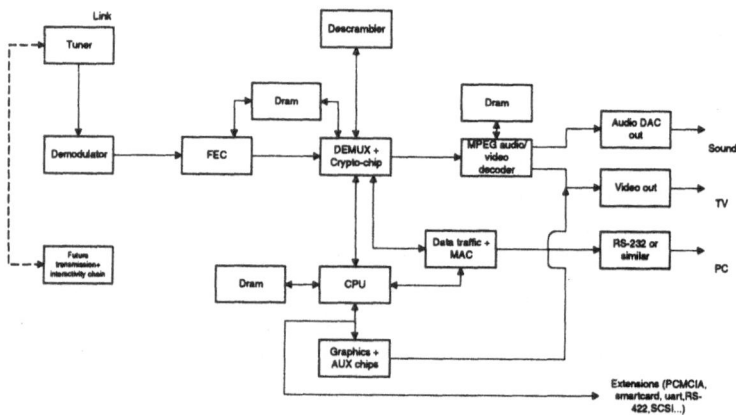

Fig. 1 Typical functions of a standard set-top-box for multimedia services.

2.3 Implementation Feasibility and Problems

At present, multimedia terminal manufacturers and researches are facing problems with non-finished standardisation, open implementation issues and different types of terminal platforms. For example, set-top-box manufacturers should somehow be able to separate different design properties for each service provider from the fixed-hardware and provide these aspects as soft-configurations, using a common operating platform. In essence, a successful multimedia terminal should be able to

• Support multiple standards and service providers without any significant extra cost. In ideal conditions, set-top-boxes should be backward compatible with analog broadcast services, while supporting digital broadcast and interactivity.

• A multimedia terminal should be versatile and flexible, supporting data, text, audio, video and graphics based services. Only in the case of limited "mobile or nomadic" systems customers will be prepared to compromise to some extent.

• The subjective quality of the service must be very high, while the average expectations of reliability and ease of use are setting high demands for manufactures.

• To guarantee the appeal to the customer, a proven upgrade path should be readily available, as the standards and services are evolving all the time.

A typical multimedia terminal consists of a large number of diverse basic building blocks, including video (de)coder, audio (de)coder, MPEG-2 demultiplexer, channel (de)modulator, forward error correction capacity, graphics, hardware, firmware and different peripherals. At the moment, hard disk is not a common component in these terminals as it is still too expensive. Note that although the present day terminal technology basically needs decoders only, future terminals with interactive features will also *be calling for coders to provide interactive services at customer premises.* Figure 1 shows a typical block diagram of a present day multimedia terminal.

In case of mobile communication terminals situation is more difficult. Those products should be small, save power, but must be economical – and provide multimedia services. In VTT we have begun to study in several projects slowly moving hand-held multimedia terminals that can *receive fast multimedia data and transmit limited bitrate multimedia streams.* The main research challenges with multimedia networks, and indeed also with terminals, can be listed as following [5]: heterogeneity of applications, transport media and terminal performance, flexible mobility/nomadicity, privacy issues, need of subjective quality, low latency time requirements and traffic efficiency. It is also important to note that many of present multimedia standards, such as MPEG-2, are not well suited or designed for wireless multimedia use.

Not only the diversity requirements, but also the increasingly competitive market is making the engineering and R&D tasks in this field extremely demanding. The needs for shortening the product cycle and introducing competitive value-added services quickly are likely to make the development process even more demanding. At present, the average product cycle time is no more than about 12-16 months in certain communication business areas. For the project management, this may pose serious problems, as rapid development is likely to favour small and quick task forces, while the complexity of products calls for an extensive expertise. This demand is leading to a situation in which companies are increasingly focusing on their core competence and outsourcing for the necessary extra R&D and engineering tasks. This approach implies modular design.

The heterogeneity in networks also means that a set of proven standardised methods does not necessarily work well in all situations. For example, the MPEG video was designed for relatively reliable transport channels – it simply discards corrupted data and tries to hide it away – thus not being the best match for wireless applications [3,5-7].

3. Building blocks for multimedia terminals

3.1 Software terminal systems

The feasibility of realising highly different terminal architectures in practice depends on our ability to generate an implementation with versatile adaptivity and configuration possibilities. In practice, this means that we have to develop systems, in which only the functions that are absolutely necessary are carried out by means of specialised hardware (which should be also very cheap to allow exchangeable modularity). The rest of the system will be realised with software running in fast processor(s). The goal of providing a terminal technology serving multiple standards and applications can be referred to as "software terminal technology". I have chosen in this paper to use term *software terminal* as the paradigm for our research at VTT. In essence, this could also denote software radio. However, usually researchers in our field associate software radios with adaptive antennae, baseband and with multi-mode mobile communication terminals (phones). Software terminal is a more generic term, just stating that we are trying to implement highly configurable terminal equipment. The diverse functions are performed and/or also modified and enabled with software (e.g. the configuration of FPGA can be seen as the software process where a certain interchangeable binary mask is downloaded into the system).

The requirement definition process for multimode multimedia terminal must be very clear. First, the common system requirements will be split to *functional* (memory, processor...) and *non-functional* (cost, power...) parts. Further, the functional domain will include *system independent* and *system dependent* parts. In case of multimedia terminals system independent area should be maximised. It should include upper-level protocols, but also as many software radio and downloadable codec issues as possible. A Considerable amount of work has already been carried out to specify the interfaces and performance characters necessary for implementing this sort of future terminals. The rapid advancement in memory and CPU manufacturing has made it possible to introduce fast and economical processors and large memories into the terminals.

Fig. 2 The "super-chip", a single chip cable modem demodulator, is the an example of increasing silicon capacity (photo: courtesy of the Stanford Telecom, U.S.A.) The chip has been used also for wireless applications.

Fortunately, several companies have already started designing and producing high-performance chips and chip-sets for terminal manufactures. In future, the terminal development, at least in the case of set-top-boxes, might involve a combination of acquiring the appropriate chip-sets for specialised tasks, while performing most of the general tasks by means of fast CPUs. Let us now summarise:

• **New CPU choices:** In a software radio/terminal system the CPU is a very crucial part of the system. At the moment, most of the CPUs are older low-performance processors in all systems; some set-top-boxes are still using M680x0s, handheld units are using ARMs etc. While it is still hard to say which architecture will dominate the field, there are several interesting contenders for the title: StrongARM, Intel x86, Java chips (Sun), "ColdFire" (Motorola) and some RISC solutions.

• **Software tools:** Software support is very important for the compatibility and design of the application layer. There are several competing "Open Architectures", e.g. PowerTV and OpenTV (Sun-Thomson Alliance). The aim of these open "standards" is to define a common reference platform for vendors. However, the situation seems to be rather hopeless at the moment - it is even worse than the diversity afflicting the Unix operating system several years ago. Also the "openness" of these 'standards' are sometime dubious at the best. One of the research areas we have been focusing on is defining a common basic architecture for configuration and system downloading. This would support the possibility of using several diverse "multimedia operating systems" in a single platform. In near future, Java technology along with other virtual machine implementations will provide extremely interesting research areas in terminal technology. VTT will be pursuing these goals in its future research. Already we are using Java-technology to provide hardware independent implementations of the architecture and simulations.

The terminal operating system concept is a very powerful paradigm. According to our basic research ideology there will be a dedicated or commercial RTOS (like VxWorks, pSOS) in the system, governed by a scheduler kind of terminal OS, capable of loading different applications, protocols, TV operating systems etc. onto the system. Hence, only terminal operating system, access channel and media access layer functions will be needed as a minimum "firmware" configuration. Moreover, we are also looking at the possibilities to dynamically configure most of the MAC-layer functions.

As regards nomadic and mobile multimedia terminals, our need to minimise the power consumption will also affect the software design. Instead of the - rather naïve - approach of working with power-saving electronics only, I would strongly recommend optimising the software, as well, so as to minimise the overall terminal power consumption. A good example of this kind of work is provided by the research done on the optimisation of energy consumption with error control algorithms.

• **Modem chips:** In case of modulation and "tuners" there are already several reliable ASIC manufacturers around. Stanford Telecom being an example: the company is providing single chip solutions for modulators/demodulators and spread-spectrum systems. For usual terminal manufacturers wishing to concentrate on terminal functionality and software, the possibility of buying quickly baseband and RF technology at relatively low expense is a very useful option. This trend is not that important for, say, handheld terminal manufacturers, because of their highly stringent demands. However, in our laboratory, we have been very pleased to be able to use third party products in implementing most of the baseband and RF technology for terminals. For example, the chip in Figure 2 is capable of a full demodulation for QPSK and 16/64/256 QAM, while also including a FEC block for Reed-Solomon and Viterbi error decoding.

• **MPEG-2 processing etc:** There are lots of different solutions for encoding, decoding, stream selection etc available. One example of an overall chipset provider is LSI Logic. This is an area where, thanks to competition, prices have been decreasing and the availability of products increasing very quickly.

3.2 Towards new designs

The above discussion has been a highly generic, focusing mostly on high-end set-top-box type appliances. However, as we have noted earlier, there are a huge number of different kinds of terminal equipment available. For many networks, set-top-boxes or similar home access points may be important, while significant R&D and business growth will be also provided by various nomadic and mobile terminals. In this area, the strict requirements on power consumption and mobility are likely to make R&D problems very demanding.

The area of wireless multimedia is one of the main interests of our research work at VTT. In particular, VTT Electronics is focusing on software terminals and networking technology. The main justification for this kind of terminal research is the vision that fixed and wireless networks will merge and coexist (sometimes called as Negroponte's switch). Terminal equipment might be able to support both narrowband and broadband systems at the same time, i.e. the system operates as a narrowband terminal when in wireless mode, but at home it might configure itself to be an ethered broadband system. Various wireless PDA systems, such as Nokia Communicator 9000, are only the first step towards such lightweight systems. Furthermore, a two-way video communication is most likely to be augmented to near-future products. As mentioned the next generation mobile communication standards, such as UMTS, are already aiming at transfer rates of 2 Mbps.

Even the small handheld type of terminal may serve as a fully functional multimedia device, if it is connected to e.g. virtual reality glasses. In the software terminal research, one of the goals is to develop small portable equipment (roughly the size of a large PDA) with a connector for VR glasses, while the applications include full DVB/DAVIC compatible services with the mobile phone and web browser system. Although this kind of equipment is clearly at the extreme limit of technology and it will not by any means be cost-effective or energy-efficient at the moment, this vision will enable lots of interesting research. In Figure 3, we show the futurological CyPhone terminal, which is designed in the consortium lead by VTT and the University of Oulu. The Cyphone is UMTS based communication device, which is providing augmented reality features to user and also includes a fast up-link for image transmission. The whole basic architecture and idea is designed as a component model in virtual reality (see later text).

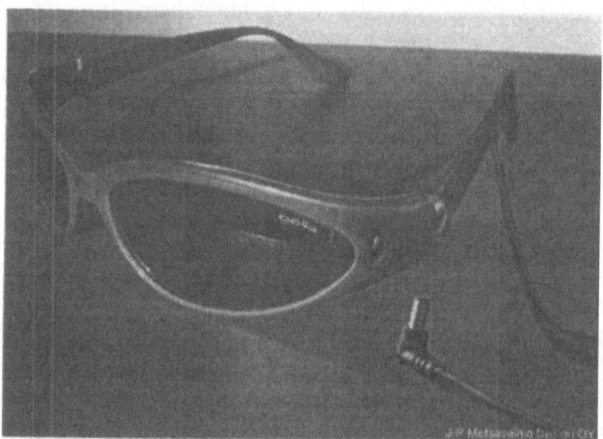

Fig. 3 The vision of the future CyPhone multimedia terminal, which is using fast UMTS-link to provide augmented reality information to user. It is also capable to send slow speed image information upstream (Photo by courtesy of prof. Petri Pulli).

Network computing is a hot topic thanks to Java/Jini and the advances in HW/SW technology. We regard network computing as being closely related to the field of software terminals. However, lots of questions still remain open; like the performance of Java, its suitability for lower-level communication programming etc. While it seems impossible for network computing to become a serious 'killer' challenge for home computing or standalone set-top-boxes, it will certainly offer an interesting contribution to the development of technology in this field. In addition, through Java kind of technologies, network computing might lead to some formal or de facto HW and SW standards. In a broadest sense, Jini technology and BlueTooth consortium are pushing the technology towards ubiquitous network environment, where every electronic device is working as a limited terminal platform.

The EU IV Framework CABSINET-project has been one of the projects, where we have been doing a terminal development. In this project, we are designing and implementing a two-layer LMDS-network for broadband communications and interactive TV at 40 GHz. The terminal equipment being largely of the "software enabled" type. Due to the fast schedule and unfinished specifications in project, there was a need to start part of the terminal implementation work as soon as possible. Hence we used our software terminal approach in real project. Thus even before there was a final hardware specification and platform, we had some of the software implemented. In fact, an ordinary PC can work as a multimedia terminal as well as a TV set (although with poorer functionality). All the major functions of the terminal are carried out by software, e.g. MAC-layer is realised through software that can be downloaded. Forward Error Correction block is also accomplished with software, in fact FEC-SW is running in the same processor as MAC, and it can be seen as a task in the "terminal operating system". This R&D project has evolved in to a new PCI-backbone based configurable set-top-box architecture designed in our laboratory.

Virtual development process: One of the main interests in the terminal development is the possibility of making fully functional digital virtual reality models of products. VTT is currently running a research project on developing VR-based haptic modelling tools for the R&D-work of electronic products. In virtual modelling, most of the design can be carried out with digital means; the electronics design is combined with a fully haptic VR model of the product, thus enabling the developers to test and market the product even before the final physical prototype model has been completed. I feel that this sort of R&D process is becoming increasingly common in future, as it will considerably shorten the design time of the terminal product. A fictional design for a pen-shaped and sized futuristic mobile phone by VTT is shown in Figure 4. The digital model of this VTT product is fully functional in the sense that it is connected to a real GSM phone network and the VR-user can make real phone calls using this virtual model. It is also an excellent example how the same downloadable software codec, in this case GSM-voice, can work in different environments (in this case it is same for GSM-phone, pen-shaped phone and CyPhone as well as our Internet PCs).

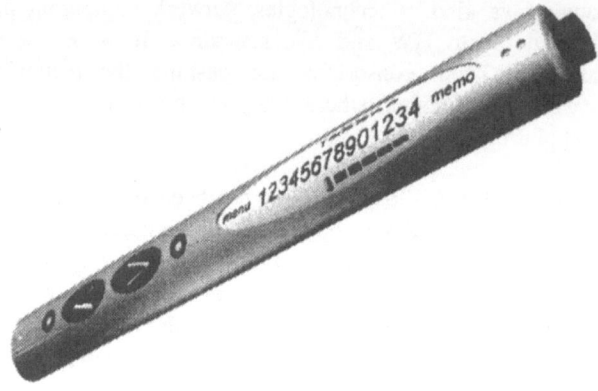

Fig. 4 The virtual model of a future pen-shaped phone terminal. The virtual and haptic model features all the functionality of the final product. Note that the basic system independent implementations, such as voice-codec can be same in pen-phone, CyPhone and set-top-box alike (Photo courtesy of VTT Electronics, VIRPI project / Prof. P. Pulli).

The main lesson in the engineering towards the software terminal development is the critical balance between the configurability and efficiency. Hence, in the organisational level one have to start to collect different level *digital libraries* for terminal R&D and manufacturing:

(a) *Software algorithm libraries* include our generic implementations of upper-layer protocols, MAC, error-coders etc.
(b) *Channel coding libraries*: different algorithmic libraries.
(c) *Digital VHDL models* for design toward possible hardware optimised ASIC realisations.
(d) *Design libraries* of generic terminal architectures.
(e) *Virtual reality component libraries*: the highest level approach to collect most of the models under the common virtual reality environment. As discussed this will enable the functional testing of the product even before it is manufactured.

The complexity of the telecommunications and multimedia technologies are trends that imply need to use same approaches as in computing; modularity and communication of modules through the bus architecture. The development in electronics and microprocessor technology particular means that traditional mobile communications efficiency and performance optimisation will be less important in future. This will make it more possible to build software terminal architectures. The modularity we have described avoids many problems of standardisation and need to define strictly terminal applications beforehand. The overall platform (like the terminal operating system) must be still standardised. However, after platform is agreed, it is possible to download different standard, or even proprietary, modules. Our feeling is that the modularity is desirable by end-user because it makes terminals more flexible, and indeed they would be evolving towards computer technology like platform. However, the end-user approval is not the only reason towards the modular terminal technology. The complexity of the telecommunications and multimedia standards and technology is reaching the critical state, where there are so many (competing) vertical technologies (networks/applications/services/protocols) that none is able to understand the overall global system.

4. Conclusions

This article presents some considerations on the current research on wireless broadband terminal systems, focusing on the technical, economic and design-related issues in the field. The main focus is on the problem of diversity of the market, standards and service providers. The envisioned solution is to provide substantial support for diverse research programs focusing on the field of software radio/terminal, aimed at developing highly configurable terminals.

We argue that the virtual reality based digital simulations of terminal equipment will lead to better products with less-cost in shorter time. The design of terminals should be as software-component based as possible, to guarantee good maximal configurability and reuse of development work. We claim that the modular platform approach could deliver us almost limitless new possibilities and real convergence of telecommunications and computer technologies.

Acknowledgements

The work has been funded by VTT (Technical Research Center of Finland). I would like to thank all my colleagues at VTT, for their invaluable feedback and collaboration in various projects. I especially wish to acknowledge Antony Jamin from the CABSINET project. Thanks are also due the EU CABSINET project members, within which some of the research has been carried out. I am also grateful to Professor P. Pulli for providing the information and picture material on the VIRPI project. The Wireless Internet Laboratory is operated as the research program by VTT Electronics and collaborating companies & institutes.

References

1. G. Brasche and B. Walke, 'Concepts, Services and Protocols of the New GSM 2+ General Packet Radio Service', IEEE Comm. Magazine, vol. 35, no. 8, Aug 1997, pp. 94

2. Electronics & Communications Engineering Journal, vol. 9, No 1, February 1997, IEE Special Issue on Digital Video Broadcasting

3. L. Chariglione, 'MPEG: A Technological Basis for Multimedia Applications', IEEE Multimedia, vol. 2, Spring 1995, pp. 85

4. J. McGarvey, "Competition Heats Up Early Digital Set-Top Market", Interactive Week, 16/1/1995

5. P. Haskell and D.G. Messerschmitt, 'In favour of an enhanced network interface for multimedia services', IEEE Multimedia Magazine 1996

6. G.M. Parulkar and J.S. Turner, 'Towards a framework for high-speed communication in a heterogeneous networking environment', IEEE Network, March 1990, pp. 19

7. N. Morinaga, M. Nakagawa, R. Kohno, 'New Concepts and Technologies for Achieving Highly Reliable and High-Capacity Multimedia Wireless Communication Systems', IEEE Comm. Magazine, vol. 35, No 1, pp. 34

Software Radio Technology: a Practical Example

C.D. Taylor

Telecom MODUS Ltd, Cleeve Road, Leatherhead, Surrey, UK, KT22 7SA.
Email ctaylor@t-modus.co.uk

Abstract

The work of the MMITS Forum and European ACTS programme has broadened interest in software radio technology. As faster DSP engines develop along with bus architectures that support rapid access to memory, software radio techniques are becoming a practical reality for the multimode multimedia terminal. This paper gives an overview of the practical difficulties in building such a terminal and looks to some solutions available in the immediate future.

The paper begins by identifying the key drivers towards software radio and the goals of the ACTS FIRST project. It considers candidate operating systems for a multimode software radio terminal and the possibility of using the Wireless Application Protocol as a software download mechanism. The paper goes on to describe the design of a software radio terminal demonstrator and associated single-channel basestation, that support algorithms for the DCS 1800 and UMTS Mode 1 (TD-CDMA) air interfaces, showing how algorithmic toolbox techniques may improve the functionality of multimode terminals.

Much of the work reported in the paper has been conducted in the collaborative ACTS FIRST (Flexible Integrated Radio Systems Technology) project, which is performing research into applying software radio techniques to intelligent multimode terminals.

1. Introduction

The 3rd generation of mobile communications proposes to deliver integrated computation and communication in the form of intelligent multimedia communicators with considerable computing power, memory and networking facilities, serving business and personal users on the move [1]. Some elements of this system, such as palm-top personal computers or personal mobile radio voice and data communicators are already commercially available, but integration is required to providing near-ubiquitous radio services at a low cost [2, 3]. This paper addresses the design of the radio-access side of such a mobile multimedia terminal, with emphasis on the physical layer, and the means of providing a simple user interface and download mechanism.

The additional air interfaces required for such ubiquitous coverage may be excessively expensive if radio chipsets are stacked together in the terminal, so what comes after the tri-band chipset (e.g. GSM + DCS1800 + UMTS), with its 3 LNAs, multiple SAW filters, power amplifiers and duplexers? Software radio (radio parameters controlled by software in a flexible DSP) can support the functionality of a multimedia terminal by providing access to variable-bandwidths over different air-interfaces [4, 5, 6, 7]. The collaborative European Flexible Integrated Radio Systems Technology (FIRST) project has been researching suitable software radio techniques foe a mobile multimedia terminal with the following aims:

- To show it is feasible and cost-effective to design and deploy Intelligent Multimode Terminals (IMTs), with the ability to deliver multimedia services to mobile users
- To show adaptation between 2nd and 3rd generation standards
- By embracing the software radio concept to provide flexible radio access in DECT, DCS1800 and UMTS frequency bands.
- Demonstrators use advanced adaptive RF front-ends, flexible SHARC DSP baseband, and TCP/IP multimedia data/voice ports.
- Research into advanced RF, baseband, terminal and coding techniques
- To generate practical proposals for wideband linear RF architectures
- Baseband - VLSI technology, processing architectures and integration
- Algorithmic research into channel coding and CDMA
- Standards participation for UMTS software radio

The GSM-based DCS 1800 and UMTS TD-CDMA air-interfaces were chosen to demonstrate the ability of a terminal implemented in DSP to migrate from 2nd to 3rd Generation standards. To fulfil the above objectives, the approach taken by this project has been to break down the functionality of the terminal into a set of algorithmic toolboxes, as shown in Figure 1. This toolbox definition of the architecture of a complete multimode terminal to the level of interfaces and functional blocks was the basis for the technology innovation and terminal manufacture activities within the project. The advantages of re-configurable architecture compared to wideband transceiver have been taken as a requirement that is fundamental to the project. At the system design phase, which began in 1996, major assumptions were necessarily made about the future of the emerging UMTS air-interface standard. An overhead in DSP hardware was allocated to allow for future inter-standard adaptation work. To take advantage of this proposed technology, new concepts in air interface sensing, selection and download are required.

The software radio demonstrator built by the project will be explained in more detail later in this paper, but first let us examine the type of operating system and user application interface that may be suitable for a UMTS multimedia terminal.

3. Potential OS for a multimedia mobile terminal

There are several new candidate operating systems that may be suitable for a 3rd generation multimedia terminal, of which two are suggested here. Microsoft Windows CE is an operating system (OS) platform for a broad range of communications and mobile-computing devices capable of sharing information with Windows-based PCs, and connecting to the Internet. The OS is a 32-bit, multitasking, multithreaded system with an interface based on Windows 95. It has been designed to be power-efficient (essential in an intelligent mobile terminal), compact and portable to enable the development of wireless communication devices such as digital pagers and cellular smart phones and purpose-built Internet access devices.

An alternative to the Windows CE is the EPOC32 OS developed by Psion [8, 9, 10, 11]. EPOC 32 is also optimised for low-power, portable machines including smart phones and handheld computers. It is a 32-bit C++ multitasking OS for mobile ROM-based computing platforms and includes information management and communications applications. Support is built into the platform for rich text, wide characters and colour graphics. It also allows embedded graphics and voice recording with a combination of pen and keyboard input. The modular design separates the operating system and middleware from the user interface, enabling designers to produce devices with their own user interface and applications. Java's strong Internet support will make it easy to develop Internet centric applications for EPOC 32. In addition, support for Java applets will also be provided in the EPOC 32 mobile web browser [12].

4. Microbrowser technology as a potential software download mechanism

Work has already begun on methods for downloading software into wireless terminals using a protocol based on the standard WWW communication protocol. This is known as the Wireless Application Protocol (WAP). A microbrowser in the wireless terminal co-ordinates the user interface, in a method analogous to a standard web browser[13]. WAP is designed to span multiple air interfaces [14] – GSM, IS-95, IS136 and 3rd Generation schemes, using as few resources as possible on the handheld device. It compensates for this by enriching the functionality of the network, using microbrowser technology based on standard Internet protocols. Its architecture allows for inclusion of scripting, graphics and animation. The specifications extend current mobile networking technologies (such as digital data networking standards) and Internet technologies (such as XML, URLs, scripting, and various content formats). WAP is able to exploit the opportunities that the Internet combined with mobile communication present. Most of the technology developed for the Internet has been designed for desktop and larger computers and medium to high bandwidth data networks. However, mass-market, hand-held wireless devices generally present a more constrained computing environment because they have:

- less powerful CPUs and/or reduced instruction sets
- less memory (both ROM and RAM)
- restricted power consumption
- smaller display size
- different input devices (e.g. non-QWERTY keypad)

Similarly, wireless data networks present a more constrained communication environment compared to wired networks due to fundamental limitations of power, spectrum availability, and mobility. In order for WAP to be effective, it is necessary that solutions are:

- **interoperable** - terminals from different manufacturers communicate with services in the mobile network
- **scaleable** - mobile network operators are able to scale services to customer needs
- **efficient** - provides quality of service suited to the behaviour and characteristics of the mobile network
- **secure** - enables services to be extended over potentially unprotected mobile networks while still preserving the integrity of user data; it must protect the devices and services from security problems such as denial of service

Figure 2 shows the WAP programming model for a wireless terminal connected via a gateway to a server. This route can be used to download software into the User Equipment (UE) to change application software or the type of radio access of the Mobile Equipment (ME), after user authentication using the USIM. In fact, any of the algorithmic toolbox parameters of Figure 1 can be changed in this manner.

5. Video coding for a 2nd/3rd generation mobile terminal

Very Low Bitrate image coding has the aim of developing video codecs for low to medium resolution with very low bitrate video at around 5 - 20 kbit per second. It is important to maintain a high robustness in the codec technique for use over a mobile radio channel. The voice channels of 2nd generation mobile systems (such as the Pan-European GSM, IS-54 and IS-95, and Japanese systems) range from 6.7 k bit to 13.6 k bit per second (kbps). These types of radio channels are all subject to Rayleigh fading errors. As delays of more than 0.3 seconds are unacceptable for real-time interactive video, the frame rate has to be 10 frames per second (fps) or higher. To produce a system that works reliably over mobile channels, robust video coding techniques may be employed such as multiple modulation schemes and packet dropping to reduce delay when the channel quality is poor. Latest techniques employ Fractal Coding, Discrete Cosine Transform (DCT) based coding, Vector Quantiser (VQ) based coding, Quad Tree (QT) and Parametric Quad Tree (PQT) based coding.

6. VLSI for a software-radio based multimedia terminal

There is currently a fundamental problem employing DSP engines to perform tasks that have previously been performed by custom ASICs – their integration density and thus their physical size. These devices are too large at present to implement a mobile terminal exclusively in DSP. Additionally, it is important to quantify the power dissipation versus the processing power for the terminal as a long-term power reduction strategy [15].

In order to address these questions, the FIRST project has conducted a study of the semiconductor processes likely to become available during the early lifetime of a multi-standard software radio product. A process technology roadmap was constructed recommending in the medium term the use of BiCMOS, and in the longer term the use of SiGe, for the creation of all components of a software radio with the possible exception of the PA. Complete VLSI integration of a fully flexible radio architecture on a single process is likely to become possible within 7 years.

To produce better integration, the research recommended novel DSP core paradigms such as SIMD and VLIW processor architectures, fixed block accelerator units and the integration of FPGAs with processor cores. This approach requires the development of new methodologies for optimisation of signal processing architectures to specific and generic algorithms. Such components are now becoming available. For example, Harris has had a software-definable downconverter available for over a year that allows the designer to define and apply multiple demodulation standards in a single device.

7. The FIRST software radio demonstrator

7.1 Targets for the demonstrator

- Build 4 software radios (2 mobile multimedia terminals and 2 single-channel base stations)
- Implementation of two modes : DCS 1800 and UMTS TD-CDMA
- To reuse a maximum number of modules (protocol, hardware, synchronisation module, etc.)
- End-to-end communication (voice and Internet)
- Exact implementation of the physical layer and simplification of other protocol layers
- To demonstrate new services (downloadability, air-interface selection and air-interface sensing)

7.2 RF design

The FIRST demonstrator employs 3 RF subsystems – the transmitter, receiver and synthesiser, and a separate interface module containing an FPGA (see Figure 3). The integrated RF module covers the DECT, DCS1800 and UMTS frequency bands, including the satellite (S-UMTS) band [16, 17, 18, 19]. An idealised RF architecture proposed by the project is shown in Figure 4, suitable for both narrowband and broadband air interfaces. The demonstrator channel bandwidth is limited to the 1.6MHz requirement of TD-CDMA, although a superior design would accommodate a wider channel bandwidth (target >5MHz for a multi-carrier air interface including W-CDMA). Such an extension to W-CDMA would necessarily require increased linearity (a suitable target would be 60dBc on two-tones). Careful design of low-cost high rejection analog filters was required in the

FIRST demonstrator, to avoid flexibility-reducing compromises in the upconversion and downconversion schemes.

Transmitter

To cope with multiple air-interface standards spanning a very wide frequency range, highly linear transmitter architectures are required. To this end, FIRST has performed comparative studies of Cartesian loop, adaptive predistortion, envelope elimination and restoration and feedforward transmitter linearisation techniques [20]. In practise, the FIRST demonstrator employs a transmitter composed of multiple amplifier devices combined by Wilkinson splitters (these were the best option available at the time of design - see Figure 5) and direct quadrature upconversion to RF. The project has also completed a feasibility study of an integrated bipolar wideband PA/upconversion mixer, available from the project website.

Receiver

Rather than employing a direct downconversion architecture that would have seriously limited the IF bandwidth, a dual-IF downconversion technique was chosen for the receiver design. This employed RF filters whose parameters were carefully chosen to maintain maximum flexibility in the frequencies accommodated by the Rx chain - especially important at the band edges in the cut-off part of the RF filters where their characteristics are rapidly changing. The Rx part of the receive chain achieved values for IP3 and Noise Figure of -16.5 dBm and 8 dB as mean values over the whole RF band. The IP3 value was less than the objective (-18.8 dBm) from the original design simulations. Careful adjustment of the Rx gain was required on all four Rx chains (O1 and O2) to equalise the gain obtained in the main part of the band, even if the gain was a little low at the upper frequencies. The receiver has proved reliable and flexible, but like the other RF components has not been optimised for size.

Synthesizer

In order to implement the multiple standards required, the demonstrator required a custom-designed fast-switching synthesiser to supply LO1 and LO2 for transmit and receive paths [21]. The frequency band covered is 1710 to 2766 MHz, with a step of 1 kHz. The synthesizer is programmed through a serial bus and has a switching time of < 25 ms to obtain $\Delta F_0/F_0$ accuracy lower than 10^{-6} whatever the initial and final frequencies (F_0 = desired LO frequency). Continuous improvements to the synthesizer have reduced harmonics and spurii and improved the phase noise.

7.3 Baseband design

The baseband subsystem employs flexible DSP modules using the Analog Devices SHARC processor, and TCP/IP multimedia data/voice ports. This subsystem has been reported in detail elsewhere [22] and will not be repeated here.

7.4 Demonstrator results to date

At the time of writing (mid-1998), the FIRST project has completed integration of the DCS1800 mode in the demonstrator and integration of the TD-CDMA mode is imminent. Problems encountered have been of the practical engineering type (cabling, earthing, FPGA integration, etc.) rather than conceptual. When complete, the demonstrator is likely to be the first of its kind to implement 2nd and 3rd generation air interfaces entirely in flexible DSP.

8. Future software design for software radio applications

Where possible, FIRST has planned to re-use the maximum number of software modules from DCS1800 in its implementation of TD-CDMA. The software integration methodology for porting C code onto the SHARC platforms for both DCS1800 and TD-CDMA modes is shown in Figure 6. A more advanced methodology is to use Software Object Components (highly configurable software objects), for which there are currently two competing standards:

- CORBA (Common Object Resource Broker Architecture)
- Microsoft COM (Component Object Model) and OLE (Object Linking and Embedding)

These approaches are more versatile than C++ objects, which are linked into a program at compilation time, allowing objects to be scripted together to perform the desired functions. Other research work [23, 24] has also recommended use of this approach for software radio development.

9. Conclusions

The demand for greater integration density and multiple air-interface standards is driving design towards the software radio concept. "Software radio" has become a collection of techniques, some of which are likely to be implemented in future multimedia terminals, and the ETSI decision for dual UMTS air-interface standards is likely to push forward software radio research. Although it simple in concept to bundle these techniques together, a practical software radio is deceptively difficult to implement due to the many conflicting technology-related compromises.

However, the FIRST project demonstrator has shown that it is possible to implement 2nd and 3rd Generation air interfaces in a software radio architecture on today's technology. Many companies are now researching software radio and it is recommended to them that object-oriented software design should be used for future software radio design. Also, further research is required into wideband RF architectures capable of both CDMA and TDMA, and into operating systems suitable for future multimedia mobile terminals. The Wireless Application Protocol may be a suitable microbrowser candidate for a software download mechanism for software radio mobile terminals, based on the commonly-accepted WWW application protocol.

Acknowledgements

The work described in this paper has been partly carried out under the ACTS programme of the European Commission, whose financial support is gratefully acknowledged. The author wishes to thank the partners within the FIRST project, whose work is represented – Bosch, CSEM, Dassault Electronique, ERA Technology Ltd, Motorola ECID, Motorola SPS, Orange, Thomson Communications, the University of Bristol and the University of Southampton. The author was manager of the FIRST project from 1995 – 1998.

References

1. Global Multimedia Mobility (GMM) report, ETSI, August 1996, ISBN 2-7437-1024-1.
2. C. Taylor, "Enabling technology for the software radio – 3G mobile applications", ACTS Mobile Communications Summit, Granada, Spain, November 1996.
3. C. Taylor, "Technologies for the flexible mobile terminal", ACTS Goes Asia Workshop, Singapore, September 1997.
4. Mitola, J., IEEE Communications Magazine - Special issue on Software Radio, 1995.
5. Proceedings of the European Commission Workshop on Software Radio, Brussels, May 1997.
6. C. Taylor, "Software Radio - New Techniques and Opportunities in Radio System Design", UMTS Conference, Aalborg, Denmark, March 1998.
7. C. Taylor, "Using software radio in 3rd generation communication systems", Proceedings of the ACTS Mobile Communications Summit, Aalborg, Denmark, October 1997.
8. Psion web site, "Psion licenses Java for its EPOC 32 platform", 13th March 1997.
9. Psion web site, "eMisis InfoCom to provide wireless content to devices based on Psion Software's EPOC 32 platform", September 1998.
10. Mobilis: Mobile Watch, "Mobile Java from Psion", April 1997.
11. Computer Shopper, "Psion Debuts Series 5 Organiser", September 1997.
12. Symbian WWW home page, "EPOC technology", September 1998.
13. Wireless Application Protocol Forum Ltd, "Wireless Application Protocol Architecture Specification", 30th April 1998.
14. C. Parrish, "Looking at the Wireless Application Protocol", Land Mobile magazine, June 1998, p30.
15. N. Drew, "IC technologies and architectures to support the implementation of software defined radio terminals", ACTS Mobile Summit, Rhodes, Greece, June 1998.
16. B. Friedrichs, Karimi, R., "Flexible multi-standard terminals for 2nd and 3rd generation mobile systems", ACTS Mobile Communications Summit, November 1996.

17. R. Karimi, B. Friedrichs, "Wideband digital receivers for multi-standard software radios", ACTS Mobile Communications Summit, Aalborg, Denmark, October 1997.
18. R. Karimi, "Software-definable implementation of a TDMA/CDMA transceiver", ACTS Mobile Communications Summit, Rhodes, Greece, June 1998.
19. H. Erben, P. Crichton, "Software radio architecture for UMTS", ACTS Mobile Communications Summit, Aalborg, Denmark, October 1997.
20. P. Kenington, D. Bennett, R. Wilkinson, "RF transceiver architectures for s/w radio design", ACTS Mobile Communications Summit, Aalborg, Denmark, October 1997.
21. W. Rebernak, D. Peris "A high resolution synthesizer for software radio", ACTS Mobile Communications Summit, Aalborg, Denmark, October 1997.
22. C. Taylor, "A DSP system for a Software Radio testbed", Proceedings of the ACTS Mobile Communications Summit, Rhodes, Greece, 7 – 10th June 1998.
23. A. Drewer "Software engineering considerations for software radios", ACTS Mobile Communications Summit, Aalborg, Denmark, October 1997.
24. A. Drewer, "The path to software radio products", ACTS Mobile Communications Summit, Rhodes, Greece, June 1998.

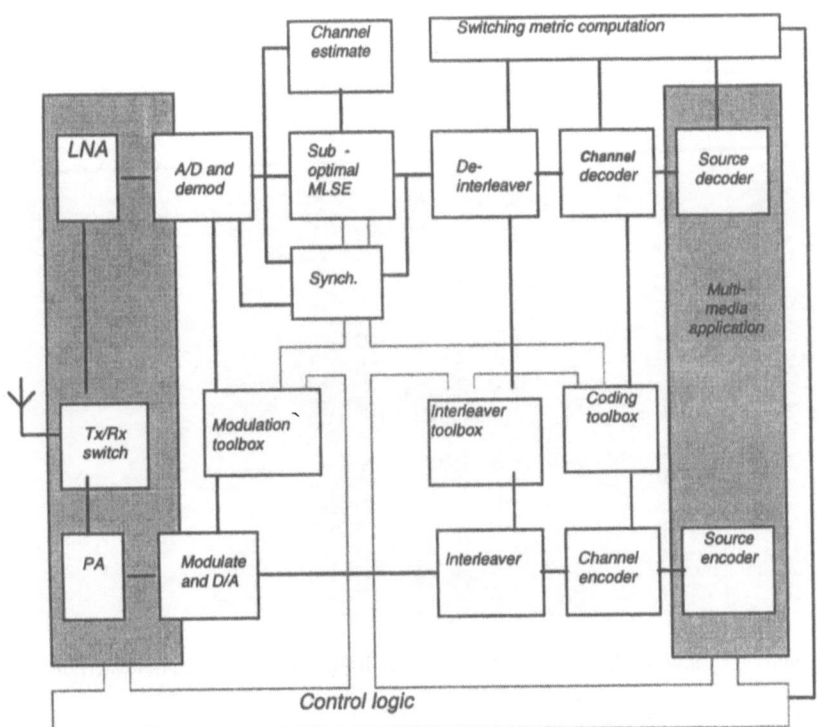

Figure 1 – Algorithmic toolbox partitioning of terminal

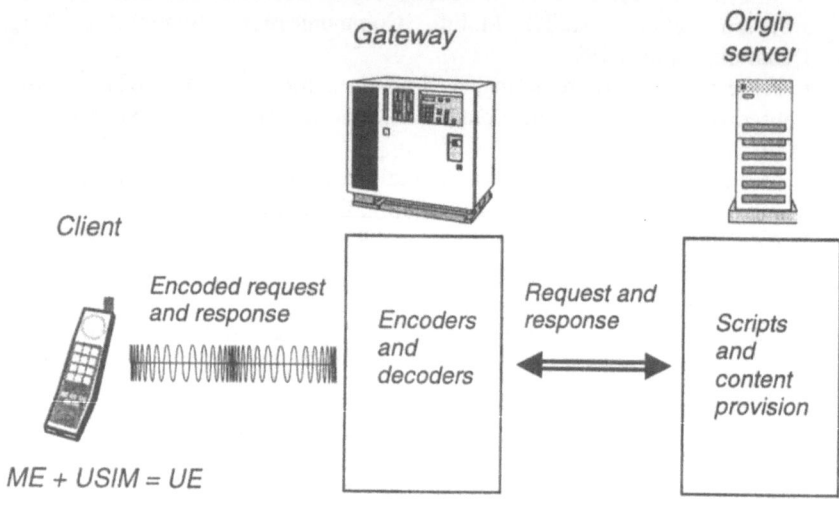

Figure 2 – Wireless Application Protocol (WAP) programming model

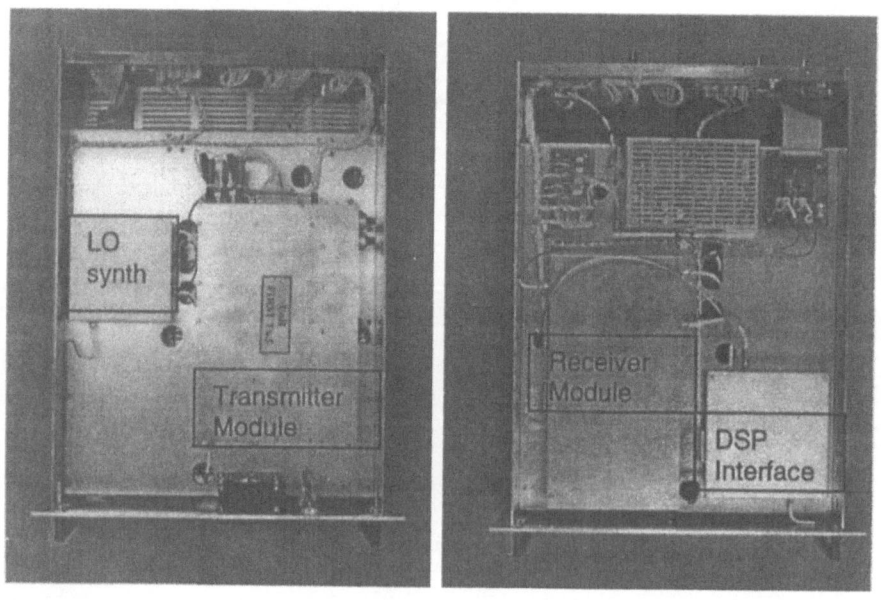

Figure 3 – Views of the RF transceiver module from above (left) and below (right)

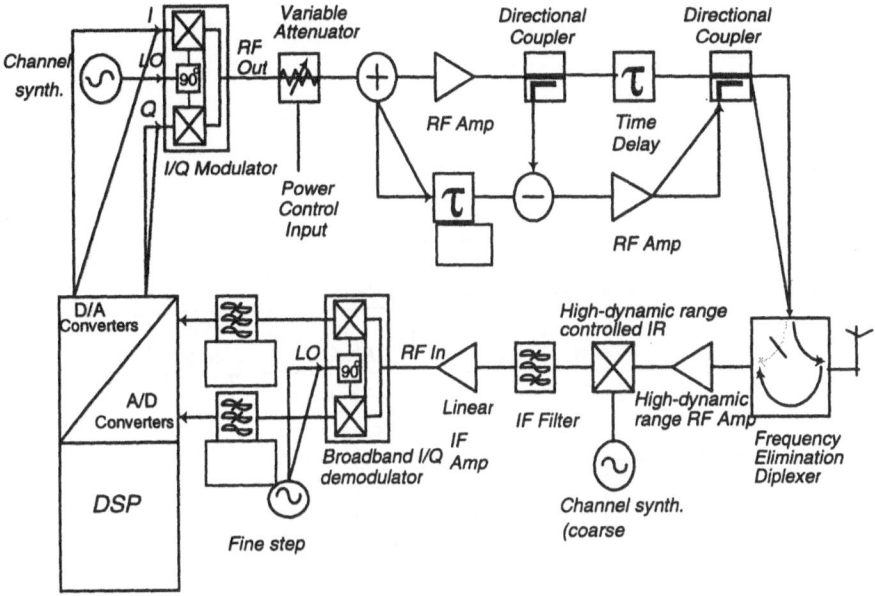

Figure 4 – Idealised RF architecture

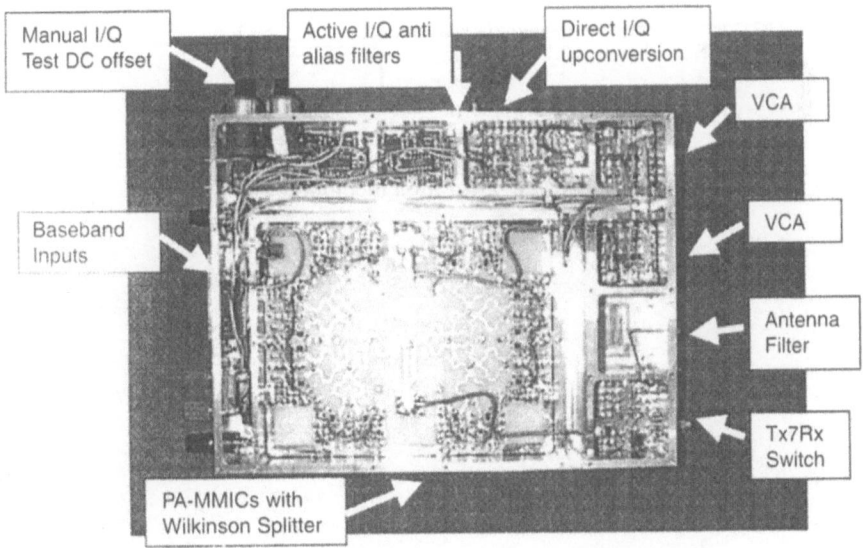

Figure 5 – Transmitter module and Tx/Rx switch

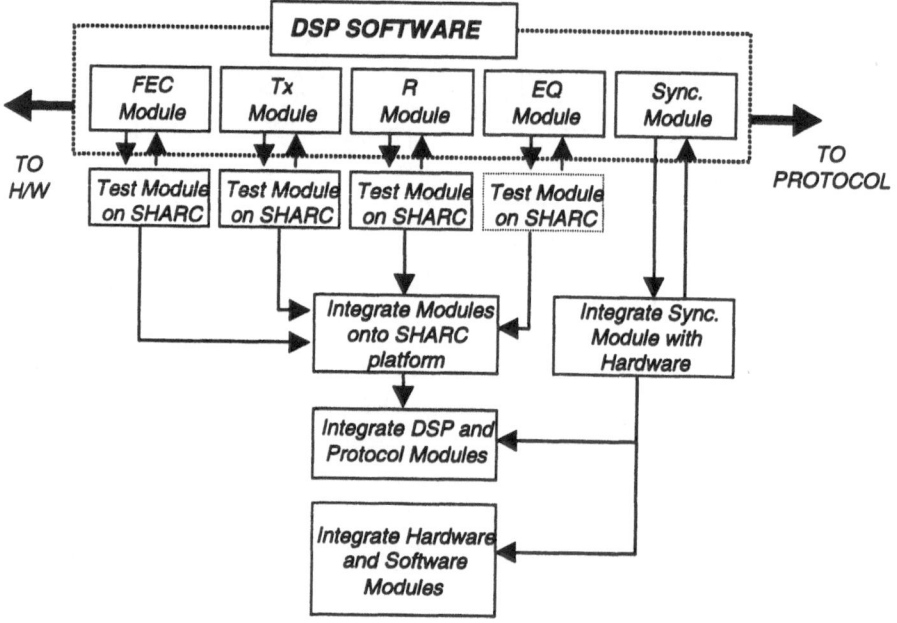

Figure 6 – Software integration methodology

A Flexible Software Architecture for Multimedia Home Platforms

S.Antoniazzi
Customer Equipment and Networks Lab.
Advanced Research Dept., Italtel spa, Italy

Abstract

Market perspectives for interactive multimedia services targeting home users have recently seen a dramatic and somewhat controversial evolution. This new scenario has opened an opportunity window for a different generation of home platforms to deploy such services. In this context, innovative user interfaces and design flexibility will play a key role, and such features are mainly provided by software architectures. The present paper aims at providing a feedback on the Italtel R&D experience so far in the interactive multimedia terminal field, including an overview of the adopted software approach as well as plans for future developments.

1. Introduction

In the past few years, the market perspectives for new interactive multimedia services targeting home users have seen a dramatic and somewhat controversial evolution. The main factors determining such an evolution can be summarized as follows: the growing success of the Internet as a new mass-media, in spite of its narrowband limitations, largely unexpected by most observers in the past; the lack of a clear strategy to provide end-users with broadband access infrastructures at acceptable costs; the difficulty in establishing a core of commonly agreed standards and technologies, with a limited attention paid for the application layer in particular; the reluctance of content providers to deploy high-quality multimedia titles on the envisaged networks.

In the context of such a market scenario, a significant effort is constantly devoted by Italtel to advanced research projects concerned with both terminal platforms and access network solutions. The present paper aims at providing a feedback on the Italtel R&D experience so far in the interactive multimedia terminal field, including an overview of the adopted software architecture approach and a discussion of the most critical issues which have been already identified.

It should be remarked that the presented approach and results reflect the work performed by Italtel in the context of the following European projects: MUSIST (ACTS AC010), OCEANS (ACTS AC323) and VRSHOP (ESPRIT 25024).

2. Multimedia home platforms

Personal computers have gradually grown in popularity even for residential users, with a very recent peak due to the Internet explosion as a new mass-media. Anyway, there are still some critical factors with PCs that represent a limit to their further expansion in the next years. In spite of any smart and aggressive marketing policy, PCs are still too expensive, complex, unreliable, huge and noisy devices to achieve the status of home appliances. Such situation could possibly get better in the future, but there are specific technical, historical and (overall) strategic reasons which drive PCs towards their Wintel-driven and business-oriented evolution path.

This scenario has opened an opportunity window for a new and different generation of home platforms to deploy multimedia services. Concerning Europe, the current industry standard is represented by DVB set-top-boxes, providing digital access to broadcast TV channels. The DVB specification is now evolving towards the inclusion of a new standard known as MHP (Multimedia Home Platform). While the main DVB focus is on broadcasting, it is quite evident that the MHP technologies under consideration (Java, HTML, etc.) are the same as those ones required by fully-interactive services.

Interactive services for home users are likely to play a strategic role in perspective. Based on easy and cheap access to the current Internet as a Trojan Horse, interactive multimedia platforms could penetrate millions of households with a potential mass-market which is theoretically larger than the home PC market. However, in order to be successful, the design of the envisaged new platforms has to take into account a number of heterogeneous factors.

First, the manufacturing cost must be kept significantly lower than the cost of a typical home PC. In fact, while PCs are usually purchased by end-users, set-top-boxes could be also distributed at a nominal price as part of specific commercial service packages. Then, on the hardware side, the physical device has to be reduced in size, made less noisy and fully reliable. The average rate of PC crashes appears as ridiculous if compared to home appliances such as TV sets, hi-fi systems and videogame consoles.

But innovative user interfaces and flexibility will play a key role, and such features are mainly provided by software design as discussed in the next sections.

3. User interface innovation for Internet access

User interfaces should be reconsidered from scratch to be tailored to residential customers. The users addressed are characterized by limited time and budget, and simply look for attractive features and services to be provided by cheap appliances which are quickly installed, easily used and with zero-administration effort. In such a context, the user interface of interactive terminals is felt as a key factor for the successful and widespread deployment of both current and future interactive services.

A base assumption is made that end-users neither have expertise in using conventional computers, nor are usually interested to learn to use them. In other terms, our vision is a "Grandma's interface to the information society". Therefore, a rethinking of the interaction metaphors designed for the traditional business-oriented Information Technology world is strongly required. Such a rethinking should be driven by the experience with completely different approaches, such as the TV, game consoles and multimedia kiosks.

In addition to their importance for customers, it is also becoming more and more evident that the user interface probably represents the most critical and complex task to be performed by advanced terminals. Hence, its impact on both software and hardware design is very high and it is expected to grow further in the future. Finally, in a longer term perspective, user interfaces will not be concerned just with remote control devices and visual/aural presentation techniques. In fact, the ultimate user interface of an interactive application will be potentially a very sophisticated software system, which could be able to exploit some kind of user model and carry on smart dialogues.

The first Italtel step in designing new user interface software tailored to set-top-box requirements has been a custom Web browser, called Nautilus [2], developed in the MUSIST project. Nautilus is a full-feature browser and provides a special user interface enabling interaction by a remote control device which is similar to those adopted for TVs and VCRs. Page scrolling and focus management is automatic and managed by a proprietary graph-based algorithm.

Nautilus is updated to the last W3C standards (HTML 4.0 [11], Document Object Model [10], SMIL [12]) and also supports dynamic HTML by an embedded ECMAScript [9] engine. The entire system has been developed in the Java language [7]. An EPG (Electronic Program Guide) application has been also developed on top of Nautilus, including a direct interface to DVB functions and DVB-SI (Service Information) tables. In our approach, TV program data are extracted by scripts through a custom API and HTML pages are automatically and dynamically generated.

Recent improvements to the base HTML approach include more precise control on look and feel (cascading style sheets), client-side dynamic capabilities (scripting language, document object model) and synchronized multimedia presentations (SMIL). A Java-based SMIL player has been already developed and integrated by Italtel in the Nautilus system. SMIL could provide a better framework than HTML for infotainment applications, but further investigation and enhancements are required, since SMIL currently does not provide a general-purpose user interface solution. In particular, there is no dynamic control mechanism so that the format can be used just for largely pre-defined multimedia presentations.

HTML and SMIL support is a strategic feature for the short/medium term market and DVB-MHP compliance in particular. Anyway, while simple and suitable enough for traditional hypertextual navigation applications (HTML was originally designed for exchanging scientific documents) the digital book metaphor is not actually very natural for the consumer mass-market, mainly composed of users accustomed with the TV/game look and feel.

As a general conclusion, front-end technologies in widespread use on the Internet still provide users with the feeling of standing in front of a computer screen. Hence, new user interface paradigms need to be investigated and further research is in progress to design even more innovative set-top-box front-ends. Virtual reality and software agents are the driving forces behind the OCEANS and VRSHOP projects. The former addresses core technologies and will be shortly discussed in the next section, while the latter is more concerned with electronic commerce applications and authoring.

4. The future: virtual reality and software agents

The main aim of the OCEANS project is to investigate a new computing and communication paradigm based on the 'agent' concept, which could provide a valuable contribution to push interactive online services towards a real mass-market. As well-known, there is no single definition for the agent concept. In fact, we feel that looking for such a definition probably does not make sense at all. Actually, several forms of interpretation of the agent paradigm do exist and will continue to exist in the future, each one tailored for some specific context or application domain.

The project has just started in March 1998. As a preliminary study, the following main agent domains have been identified:

- virtual reality agents: in this view, agents are seen as virtual models, obtained by computer graphics and aural techniques, of real-world (or just invented) entities such as virtual humans, animals and other kinds of active objects; a general trend towards a virtual reality approach is evident both on the Internet (with technologies such as VRML and Quicktime VR), in the research community [5,13] and within standardization bodies (ISO VRML97 [8], ISO MPEG-4);

- intelligent agents: research in this field is concerned with the attempt of emulating some capabilities of the human mind and its style of reasoning; this approach reflects the original concept of agents, developed during the last three decades in the Artificial Intelligence area; the approach is founded on the idea that some intelligent behavior can be obtained by the cooperation of a number of simpler components (agents) with very limited intelligence or no intelligence at all;

- distributed agents: the main objective is to exploit the processing power of computer networks by distributing complex tasks in different remote network nodes; the components of such systems (agents) are dedicated to specific subtasks and are expected to cooperate in order to achieve some overall common goal; this approach includes the possibility of agents to be dynamically relocated across network nodes; this specific features is known as mobile agents.

It should be noted that, while agents are always intended as interacting autonomous processes, the emphasis of the distributed agents domain is on loosely coupled systems (LANs, WANs); on the other hand, the previous two agent categories (virtual reality and intelligent agents) are applicable even without any kind of network distribution.

Such three domains reflect the most significant trends in the agent arena. Actually, some synergies and convergences between these domains can be observed. For instance, the Distributed Artificial Intelligence (DAI) research field is the result of combining concepts from intelligent and distributed agents. On the other hand, each domain requires its own tailored set of methodologies and tools, and requirements from different domains sometimes can be also conflicting.

The OCEANS project aims at integrating concepts and technologies from the above three agent domains and it is currently in the specification phase. The concept of an 'agent space' has been adopted, composed of the selected domains. Such a space and a (non-exhaustive) survey of relevant technologies are depicted in Fig.1. The project also intends to monitor the activities of agent-related standardization consortia, such as FIPA (Foundation for Intelligent Physical Agents [6]). Software prototypes of a multimedia home platform based on virtual reality and software agents will be demonstrated next year.

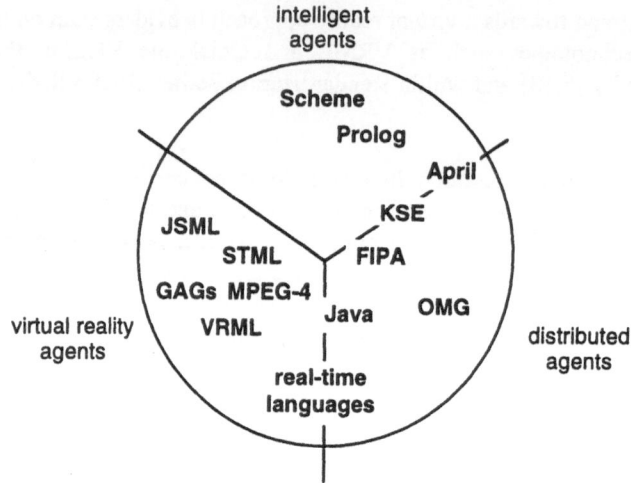

Figure 1 - Agent space, domains and relevant technologies.

5. A flexible software design

PCs became popular mainly because they can be easily adapted to different tasks and applications. Likewise, set-top-boxes which are strictly designed for a limited set of tasks (like current DVB products) will not have a bright future in an Information and Communication Technology world which evolves faster and faster. Due to the need for supporting both current standards (Internet, DVB-MHP) and future evolution (for instance 3D graphics and speech synthesis, some key technologies of intelligent agents and VR-based user interfaces) a very flexible software architecture has to be designed.

The overall architecture for the envisaged multimedia home platform is depicted in Figure 2.

The first layer on top of the set-top-box hardware is represented by an embedded real-time operating system (RTOS) and the required custom software drivers for devices such as graphics, audio/video decoders, machine-generated audio, remote control and network interface. A hybrid approach has been taken for the next layers. Based on our experience so far, we feel that a flexible platform should support both a traditional native execution environment based on C/C++ and a virtual machine environment based on Java.

The main benefit of Java support is platform independence of software, not just for easy porting but directly at final distribution level. The same software application, encoded in the Java virtual instruction set and using only a well-identified API (e.g. Personal Java for set-top-boxes), can be downloaded and executed on different terminals. The Java virtual machine also offers a safer execution environment w.r.t. C/C++, protecting terminals from software viruses and other undesired external attacks.

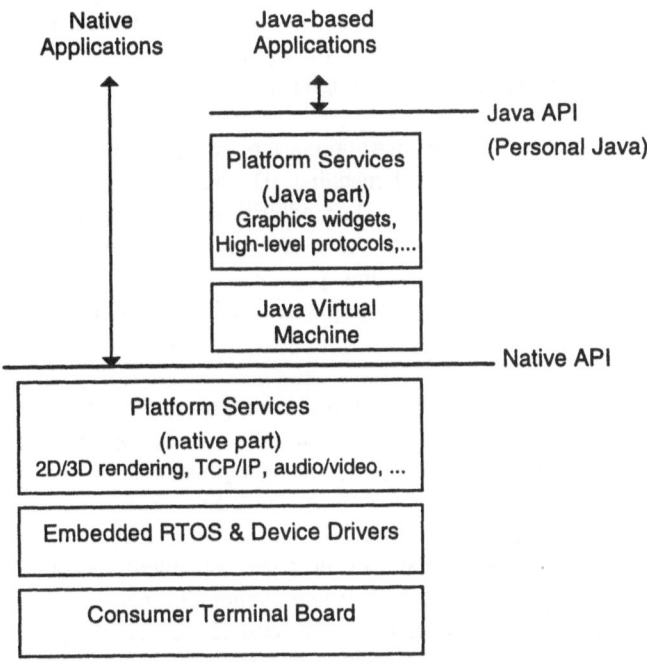

Figure 2 - Hybrid platform architecture.

On the other hand, some Java shortcomings should be reported as well. Java applications are currently slower and consume more memory than C/C++ software. The reduced speed is mainly due to the dynamic dispatch of subroutine calls (method invocation, in Java jargon) and the execution of an automatic memory management task (garbage collector) which runs concurrently to the main application. Just-in-time (JIT) compilers are usually adopted on PCs and workstations to increase Java execution speed, but they are not very suitable for embedded systems, since the JIT process implies a preliminary translation from the virtual instruction set to the native one. This process takes additional overhead time and also increases the final memory size used for instructions, typically by a twofold or threefold factor.

Anyway, JIT compilers have no impact on the main factors: dynamic dispatch and garbage collection. They simply reduce the overhead of instruction interpretation. On embedded system, the best approach to avoid such overhead will be to use the so-called 'Java chips', which directly execute the Java virtual instruction set in hardware. If complemented by some improved garbage collection technology, such a solution could provide performance more similar to current native applications. However, such advanced solutions are not feasible yet. In any case, there are speed-critical tasks, such as 3D low-level rendering, MPEG decoding and speech synthesis, which will remain out of the Java domain for a long time. Such services need to be provided by optimized native libraries, possibly interfaced to the Java environment through specific APIs.

Therefore, platform services are partitioned into two layers: the native layer and the libraries based on the virtual machine. The partitioning tries to avoid any duplication of functionality. Only real speed-critical functions are implemented in the native modules, delegating to Java code as many functions as possible, according to the target speed and memory constraints. In the same way, most end-user applications are intended to be implemented in Java (like the Nautilus Web browser).

Anyway, the presented architecture allows to support native applications as well. This opportunity is particularly valuable to port existing applications without a rewriting in Java language. Of course, native applications are most suitable for terminal-resident purposes, since their downloading is not recommended for security reasons and requires multiple versions of the distributable software, one for each different terminal platform in terms of microprocessors, operating systems, libraries and so on. Integration and cooperation between native and Java applications is also a final objective of the discussed software design.

6. Conclusions

We are currently in the last development phases leading to a first demonstration prototype of the multimedia home platform presented in this paper. We are confident that such software architecture is flexible enough to be exploited in multiple contexts, including Internet access, DVB-MHP compliance as well as very advanced future technologies.

References

1. Antoniazzi S., "A Virtual Machine Approach for Delivering 2D/3D Applications to Networked Terminals for Residential Users", International Workshop on Synthetic-Natural Hybrid Coding and Three-Dimensional Imaging, Rhodes, Greece, September 1997.
2. Antoniazzi S., Marmolin H., Schapeler G., Weickert B., "The MUSIST Browser and Navigation Concept", Proc. of European Conference on Multimedia Applications, Services and Techniques (ECMAST '98), Berlin, May 1998.
3. Antoniazzi S., Schapeler G., "An Open Software Architecture For Multimedia

Consumer Terminals", Lecture Notes in Computer Science, Multimedia Applications, Services and Techniques - ECMAST '97, Vol.1242, pp.621-634, Springer-Verlag, 1997.

4. Antoniazzi S., Schapeler G., "Interoperability for multimedia home platforms", Proc. of Interworking 98 Conference, Toronto, 1998.

5. Costa M., Feijo B., "An Architecture for Concurrent Reactive Agents in Real-Time Animation", Proc. of IX SIBGRAPI, October 1996.

6. FIPA97 Specification, November 1997, http://drogo.cselt.it/fipa.

7. Gosling J., Joy B., Steele G., "The Java™ Language Specification", Addison Wesley, ISBN 0-201-63451-1, August 1996.

8. ISO/IEC IS 14772-1: 1997, "The Virtual Reality Modeling Language", December 1997.

9. Standard ECMA-262, "ECMAScript: A general purpose, cross-platform programming language", June 1997, http://www.ecma.ch

10. World Wide Web Consortium, "Document Object Model (DOM) Level 1 Specification, Version 1.0 - W3C Proposed Recommendation", August 1998, http://www.w3.org.

11. World Wide Web Consortium, "HTML 4.0 Reference Specification - W3C Recommendation", December 1997, http://www.w3.org.

12. World Wide Web Consortium, "Synchronized Multimedia Integration Language (SMIL) 1.0 Specification - W3C Recommendation", June 1998, http://www.w3.org.

13. Zhukov S., Iones A., "Real-time Control of Intelligent 3D Agent Behavior in Synthetic Environment", Proc. of Eurographics-UK'97, Norwich, UK, 1997.

PHS Mobile Communication System for Simultaneous Voice and Image Transmission

Masaki YAMASHINA[1], Kento MIYAOKU[1], Yohichiro OHIRA[1],
Tomoyuki KANEKIYO[2]

[1] NTT Human Interface Laboratories, Hikarinook Yokosuka, Kanagawa, Japan

[2] NTT Multimedia Business Department, Chiyodaku, Tokyo, Japan

Abstract

This paper proposes, Mobile VISION, a PHS mobile communication system for simultaneous voice and image transmission. A trial system is developed to realize mobile multimedia communication. It consists of mobile terminals connected to a PHS 32 kbit/s digital bearer network and center equipment connected to ISDN. Experiments confirm that the trial system can simultaneously transmit voice, with the same quality as PSTN telephones, and motion pictures with the frame rate of about one per second, image size of 160 x 120 pixels, and compression rate of 4%. This system was used by the security department of the Organizing Committee for the XVIII Olympic Winter Games for monitoring remote sites.

1. Introduction

Personal Handy Phone System (PHS)[1] [2] has been providing a digital bearer service with the transmission speed of 32 kbit/s since April 1997 in Japan. This service is already being utilized for mobile computing such as remote access to FTP servers and mail servers. PHS has more than three times the transmission speed of PDC[3] or GSM[4], and is useful for mobile multimedia communication services. To realize the concept of "Mobile Multimedia Communication" in a practical system, a PHS mobile communication system for simultaneous voice and image transmission, named Mobile VISION, has been developed. This system can transmit voice, and motion pictures with low frame rate or a still fine picture simultaneously between terminals via PHS and the integrated services digital network (ISDN). In this system, to achieve a mobile terminal that can be put into practical use, a new Application Specific IC [ASIC] that can realize media multiplexing[5] and error free transmission over PHS, has been developed. A flow control scheme to continuously transmit encoded still images while minimizing the transmission time delay and channel wastage, was implemented in this system.

The Mobile VISION terminal can act as a wireless video phone which makes face to face communication possible in the mobile environment. This system can transmit voice and images from any remote site covered by PHS to any center site connected to ISDN. Mobile VISION was put into practical use for remote site monitoring by the security department of the Organizing Committee for the XVIII Olympic Winter Games, Nagano 1998 (NAOC). Conventional remote monitoring systems only provide image transmission from fixed points.   The newly developed system can transmit images from any point because the mobile terminal is easy to carry and the control center personnel can direct the person at the remote site by voice even during image transmission.

This paper first explains the PHS digital bearer service used by Mobile VISION. Functions needed to construct Mobile VISION are then studied from the viewpoint of system requirements. Next, system architecture, the technical features, and experimental results offered by the trial system are described.   Finally, the application of Mobile VISION to remote monitoring and user evaluation results are described.

2. System Concept

2.1 Network Configuration

PHS uses 32 kbit/s adaptive differential pulse code modulation (ADPCM) for speech coding and uses ISDN as the backbone network [2]. By bypassing the ADPCM CODEC in the PHS terminal and cell station (CS), PHS provides a 32 kbit/s digital bearer. The configuration of this network system is shown in Figure 1.   Since the basic transmission speed of ISDN is 64 kbit/s, CS adjusts the bit rate between 32-64 kbit/s using ITU standard I. 460.   For digital bearer service, the network doesn't perform any error control, so that the network provides the capability with constant bit rate transmission speed and constant time delay to the terminal. This is useful characteristic for mobile multimedia transmission and makes it possible to transmit voice without automatic repeat request (ARQ) control and to transmit data with ARQ control using just one digital bearer link.

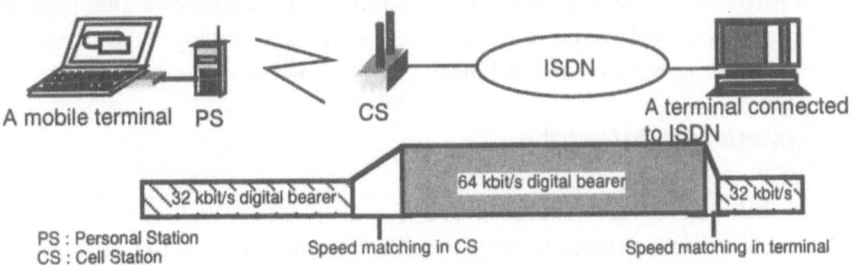

Figure 1 32 kbit Digital Bearer Data Transmission via PHS and ISDN

2.2 Voice Coding

Because PHS offers just 32 kbit/s, voice services in Mobile VISION are compressed using CS-ACELP [6] to the coded bit rate of 8 kbit/s.

2.3 Image Coding

There are two ways of coding motion pictures. One encodes a sequence of still pictures by using DCT and Huffman coding [7]. This method is commonly called "motion JPEG". The other combines intra frame coding (DCT and Huffman coding) and inter frame coding using motion compensation such as H.261 and H.263. The image coding method for Mobile VISION has to satisfy several requirements. First, the coding must not be computationally expensive since low power consumption is essential. Second, clear images must be provided at the coding speed of around 20 kbit/s, the capacity available for image transmission. Third, the transmission of either moving pictures or high resolution still pictures must be freely selectable at any time if the system is to be applied to remote monitoring. From these requirements, continuous still picture coding using DCT and Huffman coding for moving pictures was selected for Mobile VISION. This method can encode / decode continuous images by a low power CPU and encode a fine picture with low frame rate; the image transmission rate is within 20 kbit/s.

2.4 Media Multiplexing Control over wireless Channel

H. 223 Annex A, B, C has been standardized by ITU for media multiplexing in wireless channels. Annex A uses 8 byte flags that include CRC and Annex B uses 20 byte flags that include FEC to identify image data and voice data in a packet. Annex C converts the payload into data that resists errors. Because this standard was designed to be applied to digital and analog networks, it adopted a flag synchronization mechanism to identify media type and realize transmission control. On the other hand, Mobile VISION was designed to be applied to only digital networks such as PHS and ISDN. To fully utilize the transmission speed of 32 kbit/s, a new media synchronization mechanism was employed that locks the frame timing assigned to individual media ; flags are not used at either end [5].

3. System Architecture

3.1 Mobile Terminal Architecture

The mobile terminal must be easy to carry, have long battery life, and easy to operate in a mobile environment. Considering the progress in portable PC size, weight, power consumption, and user interfaces, it seems effective to construct a mobile multimedia terminal on a portable PC. Thus, the terminal in Mobile VISION is a PC that offers a stylus interface and two Type II PC card slots.

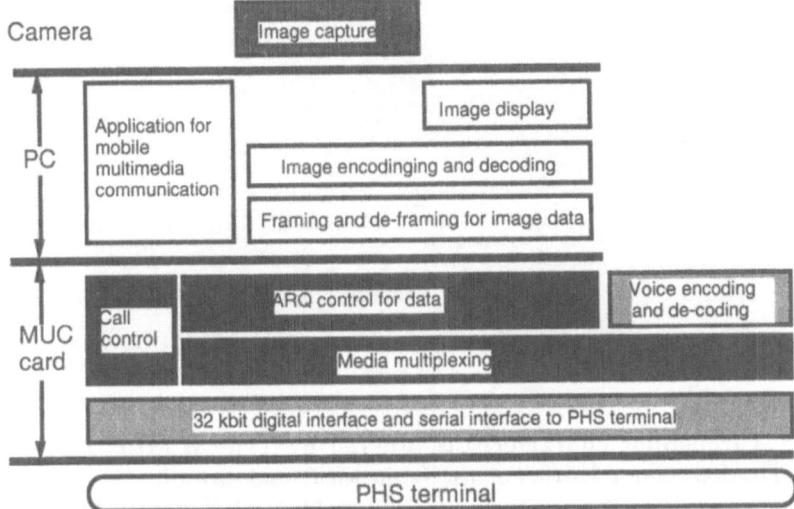

Figure 2 Protocol Stack on Mobile Terminal

Figure 2 shows the protocol stack of the mobile terminal. As shown in this figure, the CS-ACELP voice CODEC, interfaced to the PHS terminal through a 32 kbit/s digital interface, a media multiplexing function, and ARQ control function are implemented on a Type II PC card for multimedia transmission control over PHS (MUC card). Relatively simple PCs having low CPU power can be used because CS-ACELP encoding/decoding is performed by the CODEC embedded in the MUC card.

The image encoding / decoding process is implemented on the PC. This process needs about 20MIPS to encode / decode images continuously when the transmission speed is 20 kbit/s. A camera connected through the second Type II PC card captures the image. This configuration reduces the load on the PC's CPU, and thus reduces the power consumption of the terminal.

3.2 Type II MUC Card

To ensure that the MUC card conformed to the Type II PC card standard, a new ASIC was developed. This ASIC consists of a 32bit RISC CPU core, memory controller, an interface to the PHS terminal with 32 kbit/s digital I/O, a PC card interface and an interface to the CS-ACELP voice CODEC as shown in Figure 3. This ASIC allows the MUC card to be constructed with just a few chips such as ASIC, RAM, ROM, CS-ACELP CODEC LSI and an LSI including A/D, D/A and an amplifier for voice communication. The MUC card's power consumption is about 700 mW which is within the allowed power consumption of the Type II PC card standard.

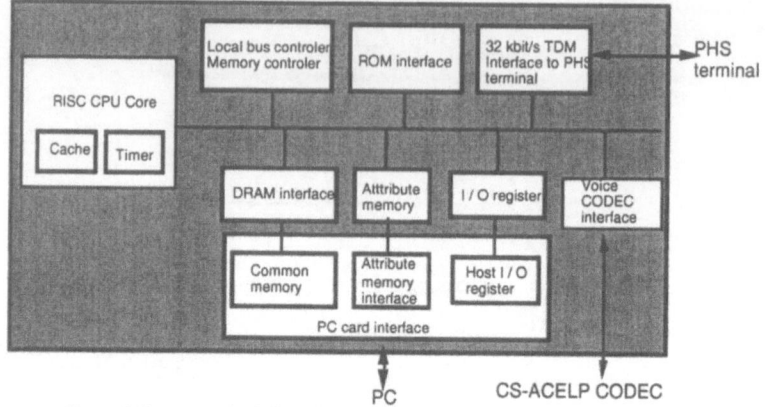

Figure 3 Structure of ASIC for Multimedia Communication Control

3.3 Throughput characteristics offered by Type II MUC card

The ARQ control function offered by the MUC card transmits the data from PC to the center or another terminal error free over PHS. ARQ is applied to the image data whose transmission speed can be 32 kbit/s if no voice signal is active. The ARQ control mechanism returns the identity of several corrupted frames in one control frame to the sender to decrease the transmission delay [8] [9].

Figure 4 shows the measured throughput characteristics offered by the MUC card. These characteristics were measured at the terminals. This test used a transmission frame with length of 20 bytes of which 4 bytes were used for the header and CRC. The results confirmed the parallel operation of multiplexing and ARQ processing offered by the MUC card. In addition, it was confirmed that a data throughput of around 18 kbit/s can be obtained in an application without any transmission protocol under the frame error rate (FER) of 1×10^{-1} when speech is multiplexed at 8 kbit/s.

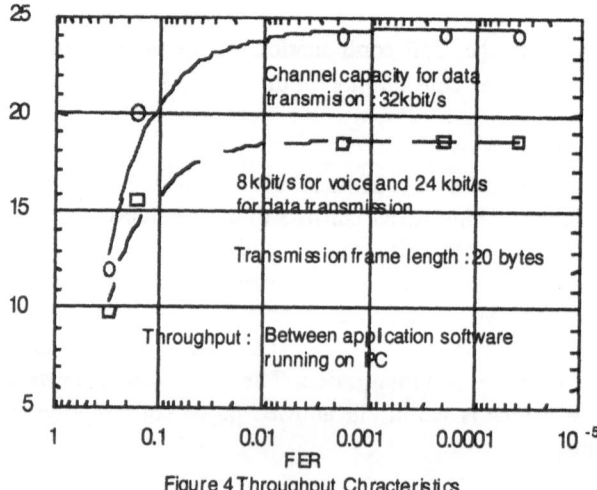

Figure 4 Throughput Chracteristics

4. Image Transmission Control over Wireless Channel

4.1 Minimizing image transmission time delay

In this system, still image sequences encoded by the PC are sent and received error free by means of ARQ. Because transmission delay would excessive if decoding did not start until all data had been received, each image frame is divided into several areas. After the data of an area is encoded, it is sent to the channel, and during this process the encoding of the next area commences. At the receiver side, area data decoding overlaps data reception. As shown in Figure 5, this approach strongly reduces the total transmission delay. If the image size is 160 x 120 pixels, there are five areas, and the PC uses a 75 MHz Pentium processor for image encoding/decoding, the time delay is reduced from 600 msec to 120msec.

4.1 Flow Control

A flow control mechanism is required which adjusts the encoding speed of continuous images to the transmission speed in order to minimize the channel idle time, T4 (Figure 5 describes the parameters involved) and the time spent by the encoded data waiting in the buffer of the transmitter's MUC card. In the flow control of Mobile VISION, the transmitter waits for a trigger signal from the receiver before capturing and processing the next frame as shown in Figure 6. The receiver issues the trigger upon receiving the encoded data for area m, where m is determined from the parameters of the network, and transmitting and receiving terminals as follows.

Tx is the time taken for the receiver to get the first area of the next frame from the time at which the trigger signal was sent to the transmitter.

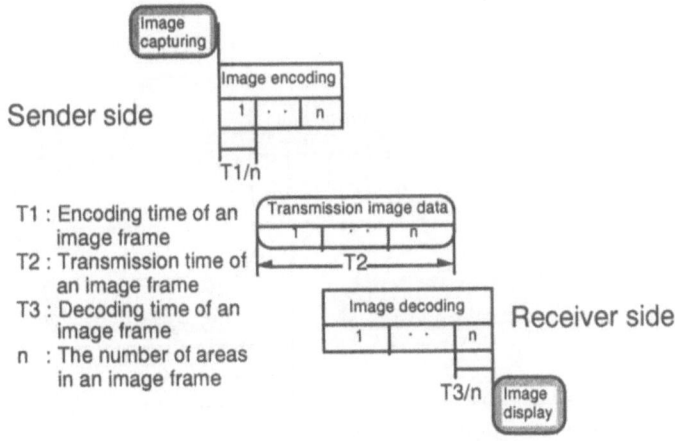

Figure 5 Encode and Decode Timing

$$TX = 2Td + T1 / n + TL$$

Here, T1 is the time to encode an image frame, Td is network delay time, n is the number of areas, and TL is the time it takes the trigger signal to pass across the network.

If the period (m / n) x T2, which is the time taken to receive the data from areas (m) to area (n) in an image frame, is equal or less than TX, the transmitter's MUC card buffer would hold no data and channel idle time (T4) would be suppressed to almost zero.

The image transmission time delay, TZ, is given by the following equation, when the data buffering time in the MUC card is ignored.

$$TZ = (T1 + T3) / n + T2 + T4$$

Here, T2 is the time to transmit an image frame and T3 is decoding time of an image frame.

5. System Performance

To evaluate the performance of Mobile VISION, the following data were collected while the Mobile VISION terminals were in use.

(1)Session establishment time

This time was measured from the time when an operator called the terminal to the time when the voice and image connection was established. When the PHS commercial network is used, the session establishment time is about 12 sec. Establishing a connection between the PHS terminal and the network took about 9 sec. Establishing the voice and image connection took about 300 msec when the FER was less than 1 x 10-3. Establishing continuous image transmission took about 2 sec.

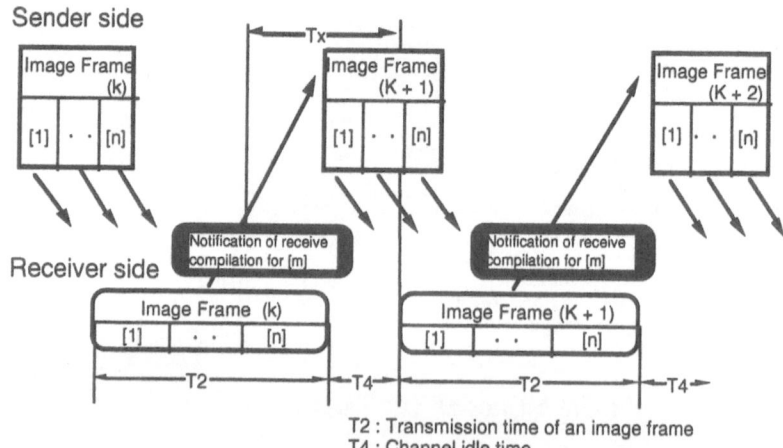

T2 : Transmission time of an image frame
T4 : Channel idle time

Figure 6 Flow Control for Continuous image transmission

(2)Image Transmission Frame rate and time delay

Image transmission characteristics of the experimental system were measured. In this test, the CS was connected to an ISDN network simulator and communicated with PHS terminals at the transmission speed of 32 kbit/s. Transmission errors were added to simulate FER levels from 1 x 10-4 to 5 x 10-1.

Figure 7 shows the relationship between image frame rate, image compression ratios, and FER. Image frame rate was measured as 1 / (T2 + T4) as shown in Figure 6. In these experiments, an image frame was divided into five areas. The sender was triggered to capture and to encode a new image by the notification of fourth area compilation in the previous frame. These results confirm that when FER is lower than 1 x 10-1 and the compression rate does not exceed 4 %, Mobile VISION can transmit moving pictures at the frame rate of more than 1 frame / sec, together with voice. In these conditions, if a 75 MHz Pentium is used, the image transmission time delay is about 1.1 sec.

(3)Voice delay

The voice delay between two mobile terminals is 90 msec if the PHS commercial network is used. This delay is equally due to CS-ACELP CODEC operation, network delay [2], and the multiplexing / de-multiplexing process of the MUC card.

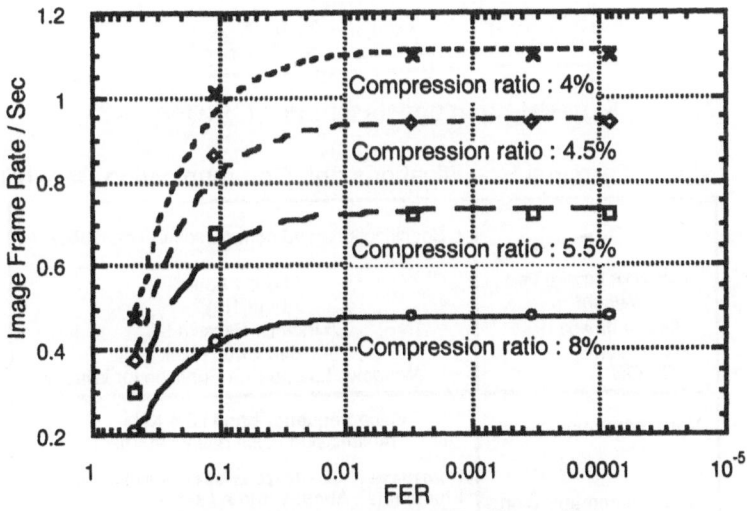

Figure 7 Transmission Image Frame Rate

6. The Trial System

6.1 The mobile terminal

The trial terminal consisted of a PC (75 MHz Pentium processor) , an MUC card, a camera with PC card interface, and PHS voice terminal with digital interface as shown in Figure 8. Table 1 summarizes the terminal's specifications and its multimedia communication capability.

Figure 8 Components for the mobile terminal

Table 1 Terminal Specifications and Communication Capability

	Specifications and communication capability
Continuous Running Time Weight Display device Resolution PC's OS	About 1 hour About 1Kg 7.5 Inch DFPassive LC 640 x 480 pixel Windows 95 and Pen Services for Windows
Voice communiccation	Voice frequency band : 3.4 KHz Transmission voice delay : 90 msec
Image communication	When Images of 160 x 120 pixels are transmitted continuously Frame rate : About 1 frame / sec Transmission image delay : About 1 sec When Images of 320 x 240 pixels are transmitted continuously Frame rate : About 0.25 frame / sec Transmission image delay : About 4 sec Available to change image size during communication
Communication mode	8 kbit/s for voice and 24 kbit/s for image transmission or 32 kbit / s for image transmission

6.2 Center Equipment

The basic center equipment consists of a Windows NT server PC with PCI Bus slots, four ISDN boards which can communicate with MUC cards (ISDN MUC board), and an audio mixer and a switch for voice mixing as shown in Figure 9. The center equipment can communicate with four mobile terminals simultaneously via ISDN and PHS. Images from the four remote sites are displayed simultaneously on the center's display. The center can rebroadcast the images so a mobile terminal can display the images from the three other remote sites sequentially.

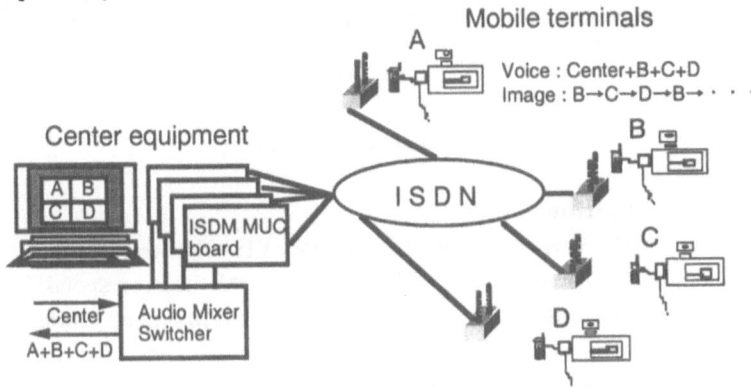

Figure 9 Configuration of Trial System

6.3. Trial use at the Nagano Winter Olympic Games

The security staff of NAOC used 40 mobile terminals over a period of 2 weeks. The PCs were placed in carrying cases and the camera was attached to the case as shown in Figure 10.

Images could be captured while watching the PC display and communicating with the center or other terminals by voice. The terminals were disbursed among eleven stadiums and three sets of center equipment were set in the headquarters of the security department. Mobile VISION was used to observe the degree of crowding, and to confirm the state of security. In some cases, mobile terminals were used to fill in the blind spots of the fixed security cameras.

Figure 10 Use by NAOC Staff

6.4 User evaluation

After the trial, user opinions of Mobile VISION were collected by a questionnaire and an interview. 80 % of the 40 users responded to the questionnaire. Evaluation results and useful opinions to improve Mobile VISION, are as follows.

(1)Usefulness of Mobile VISION for remote monitoring

45% of the users thought that Mobile VISION was "very useful or useful". 27.5% of the users thought that it was "useful in some situations". 7.5% of the users thought that it was "not necessary". From these results, it is possible to say that the simultaneous voice and image transmission function offered by Mobile VISION is useful for the mobile monitoring of remote sites. The judgement of "useful in some situations" means "multimedia monitoring would be necessary only when voice communication was insufficient". Several users complained about not being able to acquire a PHS connection. Obviously some sort of priority connection scheme is needed.

(2)User opinions about the mobile terminal

The biggest complaint was the difficulty of setting up the terminal. This is one drawback of using a general level OS. Other requests were to improve the terminal's LCD display, to enlarge the displayed image size, and to decrease the image transmission time delay when large size images were transmitted continuously.

We note that this was the first time for the staff to use mobile monitoring equipment and some resistance is to be expected.

7. Conclusion

Mobile VISION, a PHS mobile communication system for simultaneous voice and image transmission was introduced. The system requirements, hardware structure, the system implementation, the structure of MUC card, the Type II PC card for voice and image transmission over PHS, and the image transmission control mechanism over wireless channel were described. A trial system was developed that consisted of mobile terminals connected to a PHS 32 kbit/s digital bearer network and center equipment connected to ISDN. Field tests confirmed that the trial system could simultaneously transmit voice (PSTN quality) and motion pictures having the frame rate of about one per second (160 x 120 pixel images with compression rate of 4%). This system was put into practical use by the security department of NAOC and was evaluated as being useful for the mobile monitoring of remote sites.

Now, two kinds of field trial using Mobile VISION are being conducted. In the first trial, Mobile VISION is being used to coordinate remote construction work sites and headquarters using simultaneous voice and image communication. Mobile VISION is also being used for home medical care. When a nurse visits a patient's home, the nurse sends the patient's images to a doctor in a hospital and receives instructions from the doctor.

In future, it is planed to implement various functions into Mobile VISION such as a multimedia communication function using the 64 kbit/s digital bearer service of PHS and secure communication.

Acknowledgments

The authors would like to thank Mr. Nishino and Dr. Sakurai of NTT Human Interface Labs for their support and advice. The authors would also like to thank Mr. Tanda of NTT Personal for his cooperation.

References

1. Personal Handy phone System, RCR STD-28, Version 2, Association of radio Industries and Businesses, Feb. 1996
2. T.Tanaka et al : PHS Multimedia Transmission System, NTT Review, Vol. 9, No. 3, 1997
3. Personal Digital Cellular telecommunication System, RCR STD-27, Reserch & Development Center for Radio Systems , Apr. 1991
4. M. Rahnema : Overview of the GSM System and Protocol Architecuture, IEEE Communication Magazine, Vol. 31, No. 4, Apr, 1993
5. .M. Yamashina et al : PHS Mobile Multimedia Terminal System, NTT Review, Vol. 1, No. 3, 1997

6. N. Kitawaki : An-8kbit/s Speech Coding Method (CS-ACELP) Standardized by ITU, NTT Review, Vol. 8, No. 4, 1996

7. ISO10918-1 T.81 INFORMATION TECHONOLOGY-DIGITAL COMPRESSION AND CODING OF CONTINUOUS=TONE STILL IMAGES

8. D. Bertsekas et al : DATA NETWORKS, Prentice-Hall, 1987

9. M. Yamashina et al : A TCP Packet Transmission Control Method over Wireless Chanel, Proceeding of IEICE of Japan 1997.

Software Open Multimedia Interactive Terminal: an open terminal architecture to support multimedia high-quality services.

Roberto Becchini
SIA-Società Italiana Avionica SpA, Torino, Italy
URL: http://www.sia-av.it/
e-mail: becchini@sia-av.it

Gianluca De Petris
CSELT-Centro Studi E Laboratori Telecomunicazioni, Torino, Italy
URL: http://www.cselt.it/
e-mail: Gianluca.Depetris@cselt.it

Abstract

The existing complex and very rich scenario of multimedia services requires a new kind of terminal allowing people to enjoy them. The broad set of multimedia services that can be offered to potential users requires a terminal to be flexible, adaptable and integrated with the emerging technologies. The purpose of this paper is to describe the results achieved until now by the ACTS European project SOMMIT (Software Open MultiMedia Interactive Terminal), that aims at defining and validating a software terminal architecture that supports applications for the realization of multimedia services.

To validate the defined architecture a small set of multimedia applications has been developed (video on demand, terminal management, tourist guide).

1.   Introduction

In the last years, the quick development of digital technologies in the field of multimedia communications allowed the design and the implementation of a number of services and applications through which people can communicate by means of a very rich set of media. The kind of services that can be offered to the end-user by the telecommunication market ranges from the video on demand to the chat with other people inside a virtual 3D world. Consolidated technologies like MPEG-2, the availability of equipment and boards supporting broadband transmission over ATM, the effort devoted to ensure a suitable quality of service (QoS) with the Internet Protocol (IP) and the emerging of new telecommunication/IT technologies such as MPEG-4, play an important role in the design, development, distribution and use of the new services. But also the technology of distributed objects gives its contribution to such a development (the Object Management Group's CORBA architecture is an example).

The paper summarizes the work performed under an ongoing European ACTS project. The project, named SOMMIT (Software Open Multi-Media Interactive Terminal, project no. AC033), is being carried on by a partnership of European companies: CSELT (I), CNET (F), Finsiel (I), IBM (D), Philips (NL), SIA (I), Sony Objective Composer (B). The overall goal of the project is to define an open interactive multimedia terminal architecture that ensures the independence of the applications from the platform, the interoperability with the other components of the audio-visual contents delivery system (e.g. video servers), the portability of the applications across various platforms (PC and set-top boxes). The terminal architecture is designed to allow a full software implementation. These objectives are achieved by adopting, in the design of the architecture, standards and protocols defined by international consortia (e.g. ISO/IEC and DAVIC). Also a synergy with existing new technologies as, for example, OMG CORBA, has been adopted as a necessary way to achieve the maximum compatibility with different solutions existing in the market. The architecture definition has been immediately followed by a development of those modules of the architecture that have been considered strategic to build a small set of demonstrative applications. Thus, the defined and implemented architecture has been validated by means of suitable applications running on a demo platform. The software of the platform integrates, among others things, the developed modules.

This document describes the SOMMIT architecture requirement for an open interactive multimedia terminal, the adopted technical approach and, finally, the multimedia applications developed in order to validate the whole architecture. Particular attention will be given in this paper to the approach followed in the definition of the architecture to solve the problems that arise when different kinds of network access exist. The results reported in the paper have been obtained in close collaboration with several Colleagues of all Companies involved in the Project.

2. Technical approach

2.1 Architecture requirements

A set of functionality that a general multimedia terminal should provide to applications has been identified. This set of functionality has been partitioned in the following way:

- Decoding and control of time-based contents
- Local and remote file system access
- Remote management of the terminals
- Graphic capabilities
- Network connectivity
- Operating system related functionality

The need of integrating a terminal, with the above functionality, in a real communication and service provisioning system is the base of the definition of the requirements for a totally flexible and open terminal. With this view in mind a set of key objectives of the architecture has been identified:

- **Network and delivery media independence:** the architecture should not limit the communication capabilities of the multimedia terminal to specific protocols or network services (broadband or narrowband), nor should it limit the delivery of content via other media devices to specific hardware implementation.
- **Extensibility:** the architecture should be structured as a set of building blocks, each defining currently understood capabilities and features of an interactive multimedia terminal, without restricting itself to only supporting these currently known functions.
- **Scalability:** the architecture should allow sub-setting of the building blocks required in a multimedia terminal, based on the specific hardware and software platform needs and capabilities.
- **Application portability:** the architecture should allow applications to be executed on a SOMMIT compliant multimedia terminal, without requiring the application to be redesigned for the specific terminal. A SOMMIT compliant multimedia terminal is a platform on which the SOMMIT library has been ported.
- **Open interfaces and incorporation of standards:** to integrate the terminal in a real working environment the architecture should not reinvent interfaces and definitions needed as part of the architecture, if industry standards (such as defined in ISO, ITU, DAVIC, OMG and other standard bodies) exists that meet the needs of the architecture.

2.2 Terminal architecture

The terminal architecture is specified by means of a set of interfaces that export a number of functionality's. The architecture describes, in software terms, the building blocks required in an interactive multimedia terminal. These building blocks are intended to deliver functions required to support the following categories of applications:

- Unidirectional services (e.g. digital TV broadcasting)
- Asymmetrical services (e.g. video on demand, home shopping)
- Symmetrical services (e.g. computer supported cooperative work)

The building blocks (or modules) that compose the SOMMIT architecture are sets of objects, each exposing a specific interface.. The object oriented approach facilitates the achievement of the architectural objective of modularity, extensibility and scalability. To reach the greatest level of interoperability, the software architecture incorporates OMG compliant interfaces and technologies. The structure of the SOMMIT architecture is depicted in Figure 1.

The application at the top and the operating system at the bottom are not part of the SOMMIT architecture itself.

On top of the native OS API the set of modules of the SOMMIT architecture are shown. Each module offers to the application and to the other modules of the architecture itself a well defined set of functionality through the SOMMIT API.

Below the allocation of functionality to the individual modules is summarized:

- **Subsystem SOMMIT Kernel:** It provides the basic functionality of a generic operating system to the applications and to the other subsystems. SOMMIT Kernel isolates applications and the other subsystems from the underlying platform specific operating system and hardware peculiarity.

- **Subsystem SOMMIT Support:** It provides functionality normally not provided by a generic operating system to support easy application development.

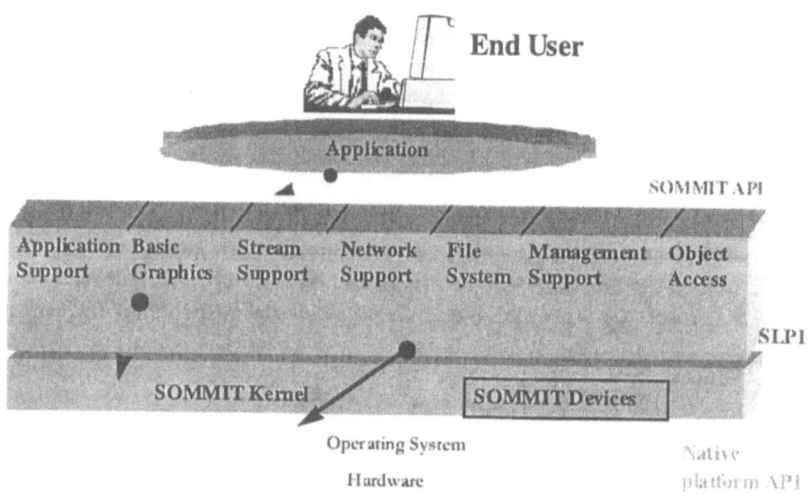

Figure 1: The overall structure of the SOMMIT architecture

- **Subsystem SOMMIT Graphics:** It provides basic graphic functionality.
- **Subsystem SOMMIT Stream:** It provides functionality to manage and control multimedia real-time data streams (audio and video streams).
- **Subsystem SOMMIT Network:** It provides functionality that are needed to the terminal to communicate with data servers and other external entities.
- **Subsystem SOMMIT Files:** It allows the applications and other subsystems a transparent access to files, independently where they resides (locally on the terminal or remotely on a server) and independently of the operating system that manages it.
- **Subsystem SOMMIT Management:** This subsystem allows suitable applications executed on the server side of the service provisioning system to manage and control remotely the terminal. This is done using the SNMP protocol.

- **Subsystem SOMMIT Objects:** It provides the applications and others subsystems functionality to access to relevant objects inside the terminal.

All these subsystems present the same interfaces to each other as they present to the applications. The interfaces are based on object-oriented technology and are described using the Interface Definition Language (IDL) defined by OMG.

Wherever it was possible, in the definition of the SOMMIT modules, standardized IDL definitions have been used. The objects belonging to the SOMMIT Subsystems communicate each other and with the outside world through the functionality offered by an Object Request Broker CORBA 2.0 compliant.

In order to stress the terminal capability to cope whit different kind of wideband access (e.g. ATM, DVB, etc.), protocols, and networks, in the following chapters a more detailed description of networking related subsystems is given.

2.3   Details on the module SOMMIT Network

This subsystem provides the functionality required to let the terminal communicate with servers and other external entities. The following areas are covered: **connection management, session management, isochronous traffic management**. One important feature of this module is that it allows to determine the quantity and quality of network resources needed between the Terminal and the Server system for a specific multimedia application. The interfaces that the subsystem defines allow a terminal to communicate with the outside world in a DAVIC compliant way. SOMMIT Network supports both connection oriented and connectionless communication services. For a set of common communication network types this services are provided in a transparent fashion. This means that the network and the protocol details are hidden to the application layer. An application will be able to request a communication service without the knowledge of how this service is provided over the network. The available communication services are:

- Connection and session establishment (over switched and non-switched networks, LAN, MAN, WAN)
- Network Resource configuration and management (session management)
- Media data download and generic file transfer support
- MPEG2 Transport Stream support

To achieve DAVIC compliance the MPEG-2 DSMCC protocols and service specifications has been adopted. An abstract object oriented communication layer is provided, which hides the protocol and network specific details and facilitates the use of both DSMCC and non-DSMCC compliant communication services.

SOMMIT Network support includes the logical entities: Terminal Configuration, Session and Call Management, Resource Management, Download Management.

In Figure 2 the SOMMIT Network System model is depicted. The SOMMIT Network subsystem is divided in 3 main components: the DSMCC User-To-Network/User-To-User APIs (ISO/IEC 13818-6), the Resource Manager and the Service Access Point.

SOMMIT Network, as part of the client software runtime, mediates between the client application and the remote service interface, providing a simplified interface to the applications. All service requests are directed to a Service Gateway (either local or remote) which acts as an entry point into a domain of services, by providing a Directory of Services. A service is a logical entity in the system that provides functions and interfaces to support one or more applications. To establish a session (and connections, as resources of the session) between a client and a server the set of DSMCC U-N core interfaces must be used. These interfaces use a set of messages designed to be carried on top of various communication protocols.

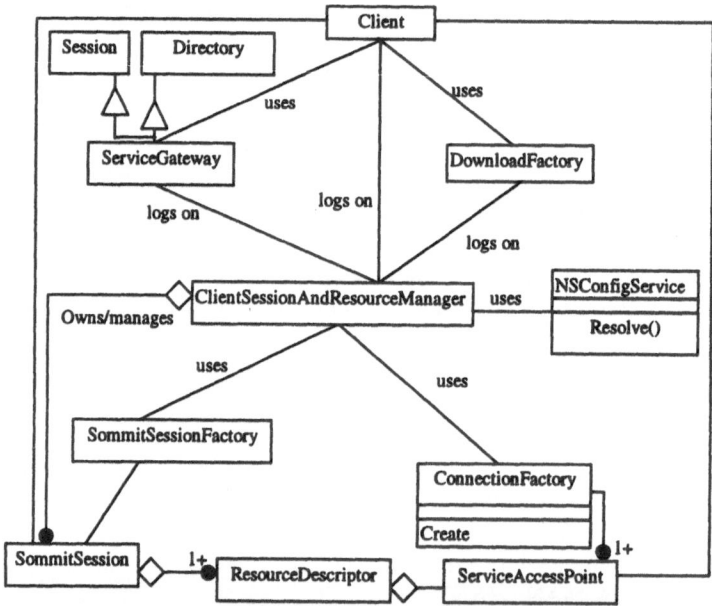

Figure 2: The SOMMIT Network system model

One of the scope of the SOMMIT Network subsystem is to encapsulate key network programming interface and therefore making programming at transport layer unaware of the actual transport layer used by the terminal. To achieve this goal the SOMMIT Network subsystem allows connection establishment and data transfer through a Service Access Point defined as an object-oriented component that can be specialized to provide support to the actual network programming interface in the terminal. SOMMIT defines a class hierarchy to support a number of transport layer network interfaces. The class hierarchy is shown in Figure 3. In order to communicate over different network with different protocols, SOMMIT Network operates an abstraction of the network address formats and types. The network addressing issues is addressed within SOMMMIT through the use of resource descriptors. Different kinds of resource descriptor are defined within

SOMMIT: IPDescriptor, AtmDescriptor, MpegProgramDescriptor, PhysicalChannelDescriptor and AtmSvcDescriptor. Each of these descriptors specify a particular addressing system.

Following communication hardware, protocols and layers supported by the SOMMIT Network subsystem are listed:

- Signaling over control channels (D-channel using Q931 for ISDN, ATM signal channel using Q2931)
- AAL5 over ATM
- TCP/UDP over IP
- BSD Sockets APIs (WinSock2.0.2 included)
- Secure Socket Layer
- UNO RPC
- DSM-CC

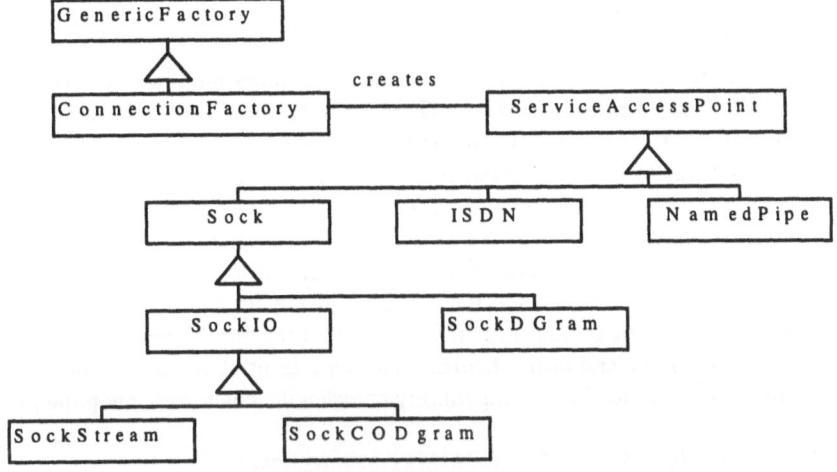

Figure 3: Class hierarchy of Service Access Point

2.4 Details on the module SOMMIT Streams

The SOMMIT Streams subsystem provides the functionality for processing digital and analog data streams. Processing data stream include: acquisition, demultiplexing, decoding and rendering operations. For the purpose of this subsystem, a stream is a continuous flow of data. SOMMIT Streams subsystem defines a generic framework composed of processing nodes and a structure for moving data from a processing node to another. Nodes offers their specific interfaces that depend on the processing performed (e.g. decoding, source of data, etc.) but is independent of the type of data moved (e.g. MPEG streams or AVI streams). Processing of data is performed along a processing chain which is composed of several objects each connected to the other via the communication mechanism. Stream are processed along the chain. As an example, to process an MPEG-2 Transport Stream, a network source object that originates the data is connected to a demultiplexer that extracts video data and passes it to a MPEG-2 decoder object. The decoder outputs the decoded content to a display object that interfaces to the actual hardware display.

A chain is in general composed of a source object, some processing objects and a sink object. A source object is the origin of the stream and is connected to a processing object, where the actual processing of stream data occurs. It is possible to pipeline several processing objects. The resulting stream will then be delivered to a sink object that consumes the data.

Another description of the SOMMIT Streams subsystem is to consider it as a processing graph that transforms one or more input streams in output streams.

While, for a matter of convenience, some specific source objects are defined, they actually are the interfaces to objects belonging to other subsystems. A network source object is a bridge to the interface belonging to the SOMMIT Network subsystem, that is the actual source of data. Sink objects interface corresponding objects that consumes data, in SOMMIT Files or SOMMIT Graphics subsystems.

Data transfer between SOMMIT Streams is modeled through the concept of port. A port is the endpoint of a directed connection. The source of data is modeled by an output port and the consumer of data is modeled by an input port. Processing objects may therefore be unaware of communication characteristic that are all handled by Port objects. A Port has a type associated to it. The type should be considered a tag of the port and not a feature of the stream. In fact there is not semantic checking of the correspondence between the stream delivered and port tag by the communication mechanism. The type is used when a connection is established in order to check compatibility between the connection endpoints.

2.5   Details on the module SOMMIT Management

The SOMMIT Management subsystem consists of three main parts:

- Gateway, which enables a SNMP manager to perform SNMP functions (get, get-next, set) on CORBA managed objects
- Managed Objects; SOMMIT Management provides the virtual specifications for scalar and indexed Managed Object classes
- Notifications environment

SOMMIT Management provides a library of functions that enable the SOMMIT subsystems, owners of Managed Objects, to request the sending of notifications to the manager and a library of functions to perform sending.

2.6   Details on the module SOMMIT Files

The SOMMIT Files subsystem provides the objects required for file system services, directory management, files input/output. All the interfaces provide an O.S. independent interface for File System services. It also offers ways of access to both local (to the terminal) and remote (on a server) files. Remote access is supported by the adoption of the DSMCC User-To-User core interfaces. These interfaces allow terminal to access to remote Domains of Services through a Service Gateway that acts as a primary access point to the services inside one or more remote servers. The SOMMIT Files subsystem supports also a means to perform buffered and formatted input/output to a rich set of objects: local files, remote files, byte arrays on the memory. The set of possibilities is extensible by adding new interfaces for new objects in the framework provided by the SOMMIT Files subsystem.

3. Test applications

In order to validate the architectural choices made by the project, some demo applications have been developed. These applications have been selected from the wide range of existing multimedia applications considering different peculiarity of each kind of application (asymmetrical services, unidirectional services, management services).

In the following paragraphs are briefly described the selected multimedia applications:

3.1 The VOD Angkor application

This application is basically a Video On Demand type of application dedicated to tourism. It is an asymmetrical, MHEG-5 compliant application. This service offers the user the opportunity of discovering the world and its riches: civilizations, natural sites, archeological sites, and so on. For the SOMMIT project a tour of the Angkor archeological site in Cambodia has been developed.

3.2 VCR Editor application

This application is basically a Video On Demand type of application. It gives the end-user the ordinary Video Cassette Recorder functionality, as well as some new possibilities. The idea is to give the user the possibility to edit a movie (creation of new clips).

3.3 Terminal management

This demo application allows the remote management of the SOMMIT Set Top Unit (Set Top Box or PC). The SOMMIT architecture provides the framework described in DAVIC 1.0 Specification Part 7 to allow remote system (named Network Management Station) to query and modify some STU aspect, as listed in Annex A STU MIB of above specification. Only the Level 1 manager (using the access network) is required to be demonstrated for the SOMMIT project.

3.4 Electronic Program Guide

It is an example of a DAVIC 1.1 compliant application supporting a unidirectional service: an EPG allowing the user to choose a program through multiple offer and/or to see information about the offered programs. This application allows the end user to browse among the offer of multiple multilingual broadcast TV programs, see information about the offered programs, select the desired language and watch the program.

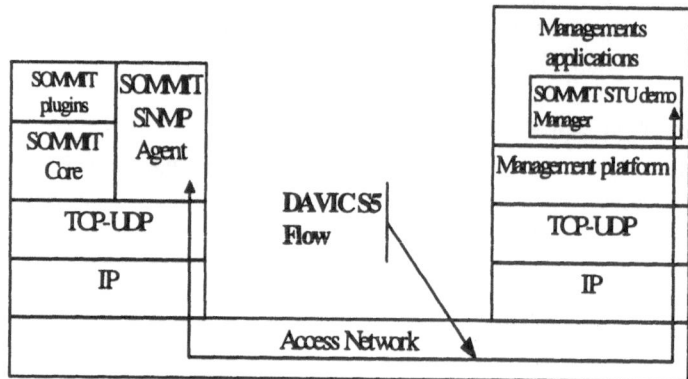

Figure 4: Simplified setup to make the management application running

4. Conclusions

Trials performed using the testing applications previously described, allowed the validation of the developed architecture. The performed validation tests revealed that the SOMMIT architecture provide a suitable environment to easily design scaleable and modular multimedia terminals that are able to support different kinds of multimedia applications. Furthermore, the partners learned a number of lessons during the project lifetime: first, CORBA is a powerful technology for defining and developing large object-oriented distributed architectures, but, under particular conditions, is not suitable for systems of objects that reside totally in a single machine. In fact the tests that we have performed prove that the CORBA environment may overload the terminal in terms of run-time memory and amount of processor time required. Thus, studies devoted to optimize the terminal performances are still running.

Another result of the project is the experience gained in designing an developing an architecture that is independent of the delivery media, allowing the terminal to adapt itself -transparently to the application- to different kinds of networks. This result allows future upgrade and development of the architecture and will likely raise the need of integrating the terminal with new technologies (not foreseen at the time of the beginning of the project) as MPEG-4. In fact, the adaptability of the architecture will allow future integration with the MPEG-4 environment (e.g.: communication capabilities with an MPEG-4 server).

The solutions proposed by SOMMIT will allow to provide contributions to the process of standardization inside international standardization bodies (i.e.: ISO/IEC). Also SOMMIT will provide a validation environment to new standard interfaces developed by such standardization consortia.

Acknowledgments

The Consortium duly acknowledges the European Commission DG XIII for having supported this research activities in the framework of ACTS Programme.

Acronyms

ATM: Asynchronous Transfer Mode
CORBA: Common Object Request Broker Architecture
DAVIC: Digital Audio-VIsual Council
EPG: Electronic Program Guide
IDL: Interface Description Language
MIB: Management Information Base
MHEG: Multimedia Hypermedia Expert Group
MPEG: Moving Pictures Expert Group
OMG: Object Management Group
SNMP: Simple Network Management Protocol
STB: Set Top Box
STU: Set Top Unit
VCR: Video Cassette Recorder

References

1. OMG Press, The Common Object Request Broker: Architecture and Specification, Rev 2 July 1995.
2. ISO/IEC 13818-1 (MPEG-2 Part 1) Systems, International Standard, 1994.
3. ISO/IEC 13818-6 (MPEG-2 Part 6) Digital Storage Media Command & Control, International Standard, 1996.
4. Software Open MultiMedia Interactive Terminal: Technical Annex, Project Confidential, 1995.
5. X/Open JMDI, Inter-domain Management: Specification translation - Open Group Preliminary Specification P509 ISBN 1-85912-150-0, 1997.
6. DAVIC, Specifications version 1.1, parts: 1, 2, 3, 5, 7, 9,11 1995.

Adaptive and Embedded Java-Codecs for Multimedia Terminals

Petri Mähönen
VTT Electronics, Technical Research Center of Finland,
Wireless Internet Laboratory, P.O. Box 1100, FIN-90571 Oulu, Finland

Abstract

The wireless multimedia terminals are integrating a large number of different advanced technologies. The terminal system must include technologies for e.g. access, applications, and data manipulation. Moreover there is need to support multiple technologies in the same terminal, or at very least guarantee that the R&D work done for a particular terminal platform is reusable for other products as well. In this paper we summarize some of our early results in the adaptive codec and operating environment research. Large part of the software is written in Java to test platform independence.

1. Introduction

The wireless networking, especially mobile telecommunications, has been one of the leading technology areas during last decade. The state-of-the art in the mobile telecommunication has reached a point, where new products are introduced within cycle of one year – being smaller and more energy conserving. These days you can buy a GSM phone, which weights less than 100 grams. In multimedia domain the standard PC technology is still dominant, with some exceptions in digital set-top-boxes for televisions. However, it is clear that the need for wireless multimedia delivery is compelling. We would like to point out that wireless multimedia is not just directed towards small hand-held terminals. The future Local Multipoint Delivery Systems (LMDS) among other access systems, support also stationary and more-powerful terminal equipment.

In any case, the need to reuse as much R&D work as possible and necessity to support multiple standards have lead to architectures that are based on software radios and adaptive codes. As an example we can think about video transmission. In Europe the Global System for Mobile Communications (GSM) is entering into the phase 2+ that will include packet switched data services [1]. The GSM2+ can support data-rates up to 384 kbps, which is enough for low quality video transmission. Around 2002-2005 we can expect to see the third generation mobile systems (UMTS) that will offer mobile data services with user bitrates up to 2 Mbps. Of course, full-scale broadband wireless networks, such as LMDS, can provide up to 8 Mbps for user. It would be very beneficial, if our "super terminal" could adapt for different networking environments, including not only point-to-point networks but also broadcast services like digital television based on DVB/DAVIC standardisation [2-5].

We have been looking at architectures to support maximal terminal adaptability in our software terminal research program. At the moment some of the results are already at the commercial level, but quite many aspects have been implemented using very powerful (and expensive) microprocessors and/or virtual reality design. However, as the silicon capacity is still increasing by Moore's law, most of the manufactures are telling us not to worry too much about today's economical limitations.

2. Configurable Terminal Environment

2.1 Terminals

The codec development we have been doing is based on component-design. All codecs and communication protocols are treated as modifiable and downloadable components. This approach does not deny the possibility to use joint-channel source coding approaches, but it is seen as one more module in this kind of reference model. We have used two different platform approaches for our case studies. First, multimedia set-top-box with PC environment was chosen as high performance hardware architecture. More demanding problem has been to study and develop wireless hand-held multimedia terminals. Some studies have been done with manufacturing partners with near commercial market products, but to have maximum freedom in public domain research, we have also used virtual reality based UMTS-platform called CyPhone as the prototype for the wireless terminal. The CyPhone exists in full capacity only in virtual reality engine in the laboratory. However, many separate *hardware* and *software* components exist also in reality.

We want to point out that the configurable terminal approach is not chosen just because we believe that the *software radio* architectures are important in the future. The wish to get our separate communications research groups working together and exchange ideas and results more smoothly, was one of the main reasons for the decision to begun the component based terminal architecture program. This has lead to quick testing implementation of our adaptive wireless video transmission system, where video source coding and channel coding can be changed dynamically.

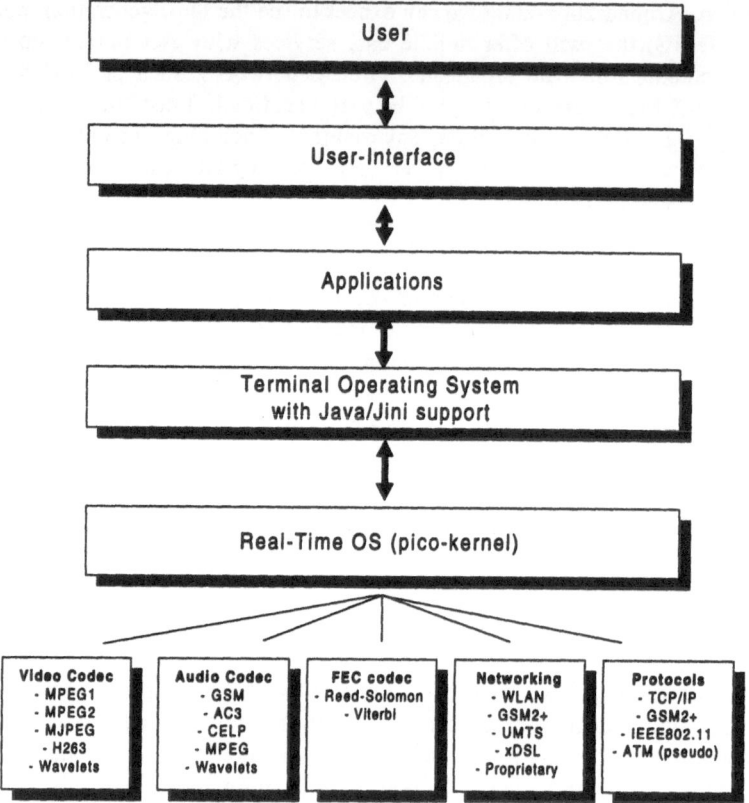

Fig. 1 The overall layered architecture of the software terminal architecture. Note that the basic real-time operating system is in different layer than the terminal operating system, which will do a real adaptivity and codec optimisation.

The overall architecture of the adaptive terminal system is shown in the Figure 1. A commercial real-time operating system, such as VxWorks, is used for task-management. The terminal operating system is an upper layer, which controls the software and hardware components. Each different component is build to be as hardware independent and downloadable as possible. During last year we have tried to use Java-technology to provide this independence. The approach is promising one, but it is far from "magic bullet" approach (which is claimed by some Java enthusiastic people). Nevertheless, we have found out by practical experience that Java can be effectively used also for real-time signal processing – with extensions such as PERC.

The adaptive terminal and dynamic video transmission research is done under this overall architecture. However, inevitable hardware specific optimisations have been done in several cases to implement workable system. The development of the architecture is still in infancy and will continue for some time.

Finally, we have added to our system possibility for encryption operations [7-9]. The privacy and information protection will be very important issues in case of wireless communications. The encryption system is studied in this context, since it needs dynamic adaptability because of (a) legal differences between countries and (b) not all encryption methods work well in wireless channel. The code downloading also implies the need of authentication and encryption to minimise security threats.

2.2 Codecs

The codecs for experimentation were partially chosen from our on-going and past research and industrial project. The main interest areas are shown in Figure 1. As we are working with multimedia terminals that are also capable to *transmit high-speed multimedia data*, we have had a lot of performance optimisation problems. The dynamic configuration aims to two goals: (a) we want to be able to support different networks and services in a single terminal and (b) dynamic configuration should be able to fight against changing channel conditions.

The embedded voice-codecs are based on existing well-known algorithms and implementation in Java is quite straightforward, as can be guessed. The memory requirements are small and swapping between modules can be done easily. A single graduate student developed the Java-GSM voice-codec in few months. The implementation is as stable as our previous C and assembler version. The Java-GSM packet is now in experimental use both in Internet/PC connectivity and wireless environment.

Two most interesting video source coding methods for us are ITU-T H.263 based low bitrate standard and MPEG-x family. The H.263 is suitable for limited bandwidth wireless channel, because its target bitrate is <64 kbps. The MPEG-2 is, of course, well adopted standard [6]. It is also interesting for our wireless-roaming terminal, as it will be widely used with future digital television systems. The problem with video is not only to receive video, but also transmit at least H.263 – level video from the terminal. As the bandwidth is very scarce and conditions are time-dependent, we have decided to use dynamic parameter configuration. Finally, we are developing more advanced wavelet-video coder that will be part of the test environment.

The dynamic video coding selection is planned as a two-stage operation in a final terminal product. First, the terminal decides what kind of access network is available; MPEG-2 is not an option in GSM2+, but could be used with UMTS or WLAN connection. This is fairly straightforward operation, which can be done in the network and operating system level. For each access method there are predefined channel coding parameters, e.g. Reed-Solomon codewords etc. The connection is opened with these initial parameters. However, the channel is monitored with test-traffic (and when a new transport stream is opened) for quality, congestion etc. conditions. If there are need for optimisation or quality control, this is done dynamically. As an example we can use hybrid ARQ-FEC method for channel coding with video streams. In this approach both transmitter and receiver have to co-operate to negotiate best possible *effort* for quality of service. We show in Table 1 some video coding bitrates as an example of huge dynamical range that the multimode multimedia terminal has to support [15].

Service	Downlink	Uplink
Broadcast video/TV (MPEG-2 coding)	1.5 to 8 Mbps	None
HDTV	~20 Mbps	2 – 9.6 kbps
Interactive HDTV		
Casual Internet access	14.4 kbps to 10 Mbps	14.4 kbps – 56 kbps
Multimedia, Heavy Internet use	160 kbps to 25 Mbps	64 kbps to over 2 Mbps
CD-quality audio broadcast	256 kbps	None
Video surveillance:		
H.263 coding	None to few kbps	28.8 kbps
H.261 coding; two-way	64 kbps – 2 Mbps	64 kbps – 2 Mbps
Video surveillance:		
MPEG-1	None to few kbps	1 – 2 Mbps
MPEG-2		1.5 – 8 Mbps

Table 1. Transmission data rate requirements per user for some digital services.

2.3 Reed-Solomon FEC as an example

The Reed-Solomon forward error correction is used with MPEG-2 in many standards and systems such as DVB and DAVIC [2-5]. The RS-code is part of our component library. It is a part of the backbone of our dynamic video link. RS-coder can be implemented cost-effectively by hardware, of course. But to learn more about the software and Java architecture performance, we chose do it by software.

The *modified extended Euclidean algorithm* was used for Java code [10]. This algorithm was derived from the standard Euclidean algorithm to be used in hypersystolic implementation [11]. It is also very well suited for SW-based implementation, partly because of the easy mapping from hypersystolic arrays to SW arrays and because the operations in single hypersystolic cells are preferred to be as simple as possible.

The Reed-Solomon codec was first implemented both in C and Java languages. The first commercial use for codec is in uplink of LMDS system at 40 GHz. The first independent C-version of the codec was compiled with Borland C++ version

4.52. The codec was thereafter translated into Fortran 90 language and compiled with Digital Visual Fortran v. 5.0. Translation was straightforward and easy because of the similarities of C and Fortran data structures.

The Java version is quite interesting. Because the language itself is highly similar to C, the translation from previous C-code itself was easy. Though it demanded some imagination in changing the functional data structures into objects. The development tool mainly used was Borland JBuilder version 1.0, although Visual J++ version 1.1 was also in use.

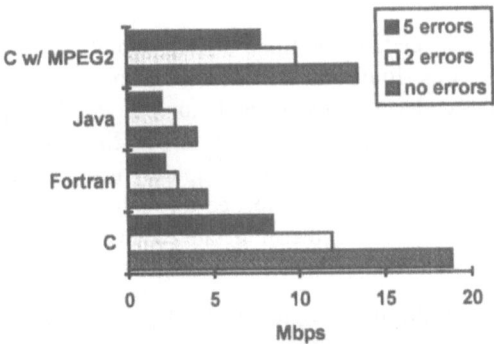

Fig. 2 Performance of downloadable Reed-Solomon codecs (without platform specific optimisation) available in the adaptive terminal environment.

The performance tests (Figure 2) show the decoding throughput of C, Java and Fortran codecs. The three lowest bars use LMDS upstream link parameters, that is (63,53) RS-code, but we also tested the decoding speed of C codec with (204,188)-RS-codewords, which meets the DAVIC standard for MPEG-2 transport stream. Tests were run with internal debuggers of each compiler and in a Pentium II under Windows CE operating system. For comparison of processor speeds, the codec was also tested with StrongArm/233 and Texas Instruments C4x processors. Results are shown in Figure 3. The Java does perform well enough even in the general purpose processor. The native java-processors will naturally increase the performance. The code is now use as a basis of dynamic video coding library, to provide robust MPEG-2 video link over wireless channel.

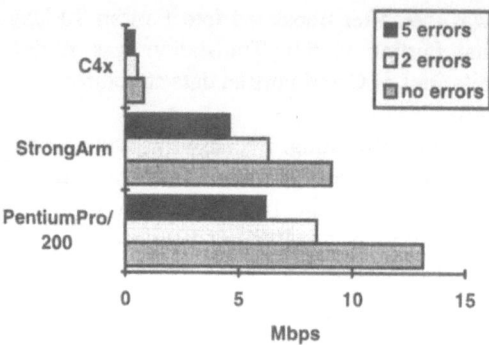

Fig. 3 The relative performance of adaptive RS-code in different processor architectures.

The architecture independent modified hypersystolic Reed-Solomon decoding algorithm [10,11] was found to be a solid choice for our purposes. The performance of the codec is good enough for our R&D purposes. In fact, the unmodified version is incorporated in to the up-link of commercial LMDS-system. The early tests with the cellular telephony using GSM-type connectivity show that the RS-codec can be downloaded through the network in less than 1s. The parameter exchange time in the moderate environment is less than 15 ms. A new optimized Java bytecode version 2.0 of the module is just to be finalized.

In figure 4, we show the futurological CyPhone terminal, which is designed in the consortium lead by VTT and the University of Oulu (Professor P. Pulli). The Cyphone is UMTS based communication device, which is providing augmented reality features to user and also includes a fast uplink for image transmission. The whole basic architecture and idea is done in virtual reality. There is no physical CyPhone available, but digital simulation environment with our component libraries can implement the functional model of the terminal.

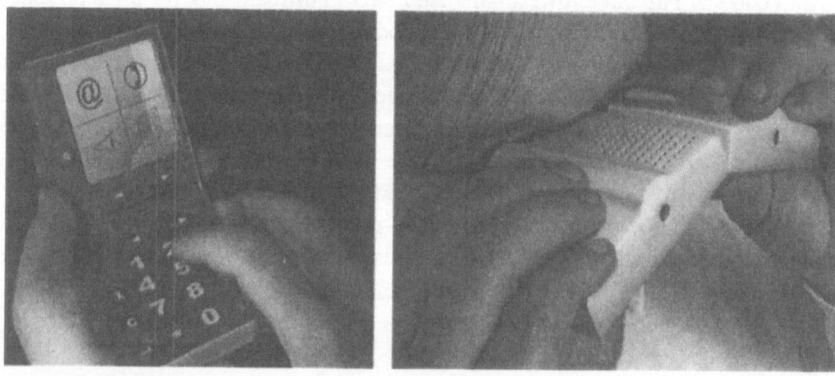

Fig. 4 The vision of the future CyPhone multimedia terminal, which is using fast UMTS-link to provide augmented reality information to user. It is also capable to send slow speed image information upstream (Photo by courtesy of prof. Petri Pulli).

4. Conclusions

The component and platform based design for terminal is a powerful paradigm. The preliminary results show that Java is a well-suited programming language for platform/processor independent codec design. However, as the C-language cross compilers are very common for the most embedded system environments, we would not like to urge too strongly for Java development, if there are no compelling need for e.g. downloadability of code. The Java has been a good choice also because it is *de facto* industry standard (unfortunately the ISO standardisation has been slowed down by "political reasons"). However, there is a lot of development to be done with embedded and personal Java, before it is truly suitable for real-time embedded use.

At the moment we have a good start with the Java and C based component libraries. In short we have

- Voice codecs
- Still image codecs
- Video codecs
- Encryption environment framework
- Error correction framework (RS, concatenated code, interleaver, Viterbi)

All the codec components can communicate through the common "software-bus" interface. The overall dynamic system is tested with different networks and protocols. The most notable wireless and wired links over which we are testing our approach are; GSM, IEEE802.11 and Hiperlan type WLANs, TCP/IP, DVB/Davic compatible LMDS-link and WCDMA.

In our laboratory we believe that the modular approach has to be implemented to fight against the increasing complexity of terminal and telecommunication business area. The advanced electronics will make software implementation of most functions feasible in near future. Thus the telecommunication and broadcast industry would move more towards computing. The computer is a general-purpose machine, you by different software to run applications in same platform. This should be aim in the (telecom) terminal design too. Not separate gadgets and standards for different services, but common platform for multiple services. In short, the modularity means reusability, configurability and adaptability. This will imply cost-savings (we have already seen *considerable savings*), move towards horizontal services/markets, better services and good possibility to work without global standard for each service. Especially the possibility to work without over political and complex telecommunication standards, just think about the needed standardisation time for GSM and UMTS, is a very tempting goal. We are hoping that our codec and software-bus research is a small step towards this future vision.

The outlined R&D test-bed approach is easy to implement. The digital simulation and virtual reality environment has been built for functional and workable prototype products. The dynamic code handling with adaptive channel coding has been already tested in set-top-box terminal, which is implemented for LMDS digital television and wireless broadband use. We have been able to do this because in set-top-box environment there is quite much memory, no serious power-consumption restrictions, and Intel Pentium II class CPU-processing power is available. Naturally, in case of portable wireless multimedia terminals the task is much more difficult. The work is still progressing towards the universal approach in this area. This R&D work will continue in VTT Wireless Internet Laboratory – research program with our academic and industrial partners.

Acknowledgements

The terminal and wireless communications research is funded in part by the technology Development Centre of Finland (TEKES), the Academy of Finland and VTT.

References

1. G. Brasche and B. Walke, 'Concepts, Services and Protocols of the New GSM 2+ General Packet Radio Service', IEEE Comm. Magazine, vol. 35, no. 8, Aug 1997, pp. 94

2. Digital Audio-Visual Council, DAVIC 1.2 and 1.3 Specifications, 1996

3. ETSI, 'Digital broadcasting systems for television, sound and data services; Framing structure, channel coding and modulation for 11/12 GHz satellite Services' ETS 300 421, 1994

4. ETSI, , "Digital Broadcasting Systems for Television, Sound and data services; Framing Structures, Channel Coding and Modulation for Digital Terrestrial Television", ETS 300 744, 1996; See also ETS 300 429 (definitions for Cable Systems)

5. Electronics & Communications Engineering Journal, vol. 9, No 1, February 1997, IEE Special Issue on Digital Video Broadcasting

6. L. Chariglione, 'MPEG: A Technological Basis for Multimedia Applications', IEEE Multimedia, vol. 2, Spring 1995, pp. 85

7. R. Kaksonen, P. Mähönen, 'Object Modeling of Cryptographic Algorithms with UML', Lecture Notes in Computer Science, 1998, 163, pp. 193-205

8. Schneier A., Applied Cryptography, 1996, John Wiley & Sons, N.Y.

9. P. Haskell and D.G. Messerschmitt, 'In favour of an enhanced network interface for multimedia services', submitted to IEEE Multimedia Magazine 1996

10. T. Niemelä and P. Mähönen, 'Configurable Reed-Solomon Codec for LMDS system', to be published in the proceedings of International Telecommunications Conference 1998

11. E. Berlekamp E., G. Seroussi G. and P. Tong, in Reed-Solomon Codes and Their Applications (S.B. Wicker and V.K. Bhargava), IEEE Press, 1994, pp. 205-241
12. G.M. Parulkar and J.S. Turner, 'Towards a framework for high-speed communication in a heterogeneous networking environment', IEEE Network, March 1990, pp. 19
13. N. Morinaga, M. Nakagawa, R. Kohno, 'New Concepts and Technologies for Achieving Highly Reliable and High-Capacity Multimedia Wireless Communication Systems', IEEE Comm. Magazine, vol. 35, No 1, pp. 34
14. For information in Java see www.javasoft.com
15. T. C. Kwok, 'Residential Broadband Internet Services and Application Requirements', *IEEE Comm. Mag.* Vol. 35, No 6, June 1997, p. 76

Part 5

Multimedia Services and Applications

Multimedia Applications and Services

N.S. Jayant[1] and S.K. Mitra[2]

[1] School of Electrical Engineering, Georgia Institute of Technology, Atlanta, Georgia
[2] Department of Electrical and Computer Engineering, University of California, Santa Barbara, California

Abstract

This paper describes the technical and business context for multimedia communications, summarizes the broader technical capabilities that support advanced applications and services, and outlines a brief overview of the multimedia applications and services that are either commercially available today or likely to be made available in the near future.

1 Introduction

The recent and on-going explosion of multimedia applications is one of the most visible aspects of an overall revolution in communications technology. In this introductory perspective, we describe the technical and business context for multimedia communications, and summarize the broader technical capabilities that support advanced applications and services.

A fundamental enabler of multimedia communications is the digital revolution that began years ago, with major leaps in technologies such as source coding, channel coding, and modulation. Because of the pervasive use of digital technology, media signals can be represented, processed, communicated and networked in the efficient common denominators of bits and packets. Digitization has also helped multimedia services in the indirect sense of catalyzing convergence in the information industry. Examples of converging elements include data communications and telephony, computing and video, and information and entertainment. The seamless availability and processing of heterogeneous media types is central to the convergent trends described above. These media types include audio signals such as speech and music, and visual signals that include still images, video, handwriting and hand gestures. In emerging multimedia applications, audiovisual communications need to co-exist with data communications, regardless of how information networks and devices are configured.

While trends in communications and user expectations are demanding sophisticated multimedia services, advances in several complimentary technologies are supporting the technical features needed for such services. Advances in networks, devices and media processing are together making possible multimedia applications that were unrealistic literally a couple of years ago.

From the point of view of total channel capacity, recent advances in optical, wireless and copper- and cable-based communications have all been exponential. Capacities available to an individual user have also seen unprecedented and continuing enhancements. As a result, the communication of media signals of high bandwidth is becoming increasingly viable, although significant challenges still exist in bandwidth-constrained situations such as Internet and wireless communications. Advances in networking have likewise enabled multimedia communications through improvements in capacity and throughput properties, as well as greater flexibilities in media integration.

The devices at the ends of the communication networks have increasingly become media-friendly, because of advances in general-purpose digital signal processing (DSP), native signal processing on a host computer (NSP), and application-specific integrated circuits (ASIC). These advances are generally measured after Moore's Law (a $2\times$ increase every 18 months, or a $10\times$ increase every 5 years). While the general exponential trend is quite real, Moore's Law is more appropriately regarded as an upper bound description in the context of multimedia time lines, for two reasons. First, the utilization of Moore's Law depends very much on optimal processor architectures which are generally unknown, or at least not generic or scaleable across applications. Second, our capabilities in media processing are not merely limited by MIPS or megabytes; they are also knowledge-limited, especially as we seek to achieve ever-greater levels of performance (sometimes without even an existence proof of these levels).

Advances in media processing algorithms depend both on efficient and affordable microelectronics, and on sophisticated algorithmic capabilities. The latter, in turn, depend on better understanding of media signals as well as on enhancements in algorithms that are sometimes media-specific, and often generic.

Fundamental scientific work has provided increased understanding of media signals from the dual viewpoints of signal production and signal perception by humans. Examples are models for speech production, object-based description of video, and psychoacoustic models for noise masking by audio.

In parallel with these inquiries, there has been steady, if not exponential, advancement of algorithms for media compression, synthesis and recognition. These advances have been due to increased interactions between scientists in areas as diverse as information theory, signal processing, linguistics and computational geometry. In general, however, the most compelling enhancements have been registered by media experts (for example, specialists in speech and image processing), working closely with DSP engineers and computer scientists.

In this paper we provide a brief review of the multimedia applications and services that are either commercially available today or likely to be made available in the near future. In Section 2 we outline applications and services which have naturally evolved with the development of multimedia technology. Many of these applications have been enabled by advances in media compression, together with advances in transmission technology (wired and wireless modems). With the rapid deployment of Internet communications, traditional versions of these multimedia applications have seen parallel and sometimes competing versions on the Internet. In fact, Internet and Intranet systems will very likely play the dominant role in the shaping of multimedia services and businesses. In Section 3, we outline and speculate on applications and services which are technologically feasible but not available presently. These are next-generation versions of the examples described in Section 2, making use of continuing advances in compression, transmission and networking. Also prominent in these new applications will be an increased use of technologies for media synthesis and media recognition, leading to increased use of human-machine communications in multimedia services and transactions Concluding remarks are included in Section 4.

2 Evolutionary Applications and Services

Several classes of multimedia applications and services have evolved naturally as multiple-media generalizations of existing media communication services such as analog telephony, analog television, and the Internet data network. As such, many of the currently available applications are based on first-generation products such as those that combine video with speech, and text with simple images and graphics. Also emerging rapidly are applications that use multiple media in more creative and synergistic ways.

2.1 Market Drivers of Multimedia Applications

The markets for multimedia technology are many, and some of these are listed below, with no claim of completeness.

Business Applications - By applying multimedia technology in the workplace and in employee's homes, businesses stand to enhance productivity, lower product development costs, eliminate middlemen and to increase sales. Generic applications in this category include multimedia telephony, multimedia videoconferencing/virtual meeting room, telecommuting, desktop collaborative computing, as well as electronic customer service, maintenance and repair.

Consumer Applications - Personal Communication Services (PCS) represent the biggest application to date, providing increasingly facile communications of both voice and data. Two-way pagers and wireless video are examples of major new opportunities for multimedia personal communications.

Residential Applications - Current services, based on low-level integration of multimedia signals, include electronic home shopping, electronic catalogs, and electronic banking. With new trends in home networks and user-friendly devices, the residence offers an increasingly fertile opportunity for the deployment of multimedia information appliances.

Entertainment Applications - The entertainment industry has made the maximum use of multimedia technology in developing new applications such as the generation of movies and video, multimedia magazines, and multiplayer interactive games. Another application which appears promising in this category is that of movies on demand. These, as well as digital broadcasting of audio and high-quality television (DAB and HDTV), depend extensively on technologies for media compression.

Medical Applications - Medical services can benefit tremendously from multimedia technology, although the profession has been slow to adopt it. Some of the steadily emerging applications include multimedia medical reporting systems, remote diagnosis, and telemedicine.

Education Applications - A potentially huge application of multimedia technology is in the inexpensive delivery of educational materials developed by integrating multimedia data to geographically distributed students. This approach is the key to distance learning and training of employees. Several of the technology solutions for audiovisual teleconferencing carry over to the application of distance learning. One of the specific challenges in the field is the digitization and media processing of legacy material in heterogeneous formats.

2.2 Descriptions of Some Specific Products and Services

Multimedia Telephony

A natural application of multimedia technology is in the development of telephone systems incorporating an integration of multiple media signals. An early example of a multimedia telephony product is the VideoPhone 2500 [3] introduced by AT&T in 1992. It employs a display screen, a video camera, and a standard speaker-phone. It multiplexes audio, image/video, and control signals over regular analog telephone lines. Long-term patients in hospitals constituted a prominent user-segment of this product. The Video-Phone, originally developed for person-to-person communication, is finding other variations, such as video answering machines, video messaging products, and systems that include person-to-machine communications. Many of these systems are likely to be PC-based.

A second example of a multimedia telephony product is the screen-based Picasso system [3] introduced by AT&T in 1993. It is a still-image telephone based on the VideoPhone 2500 architecture. It multiplexes over the transmission medium four different data types: audio, image, control signals, and annotation. The system provides simultaneous transmission of voice and high-resolution color images while permitting the users to point and annotate the image using a mouse. It can store up to 32 images which can later be displayed and manipulated using built-in image processing tools to improve the perceptual quality of the images and the transmission speed. Picasso includes an interface to a Windows-based PC, providing additional flexibility in the manipulation of images.

Desktop Video Conferencing

The second natural application of multimedia technology is in videoconferencing. One example of such a system is "DECspin", developed by the Digital Equipment Corporation [6]. This software-based system has been designed to allow users to see and hear one another from individual desktop computers The system can acquire audio, video and graphics data from two to eight desktops, and it can distribute these over local and wide area networks to implement real-time video conferencing. It employs an easy-to-use graphical user interface with push-buttons to activate pop-up menus.

Another example of a real-time desktop videoconferencing system is the PC-based Vistium^{TM} introduced by AT&T in 1992 [3]. The system includes a set of video conference cards, the software to run the videoconferencing, various communication interfaces, cameras, speakers and microphones. The system also provides an electronic whiteboard shared over a communication channel. It allows both users to view and control a single Windows application, which can be opened by any one of them, with a two-way transfer of files or still images. It permits manipulation of graphics, addition of banner heading, and circling or highlighting of numbers by means of an annotation tool.

Some typical applications of a videoconferencing system include remote consultation between users and experts, video help lines for providing access to data (files, photographs, and pages from manuals), personalized customer and vendor relationships, telecommuting and collaboration between telecommuters, remote training of employees, and development of project team organization between geographically distributed members.

Multimedia Medical Reporting System

The Siemens multimedia medical report (MMR) system is an asynchronous reporting tool for the creation of detailed descriptions of radiological examinations. It can be used for exchanging second opinions between physicians in situations where face-to-face discussions are not possible [8]. The system employs a technique called dynamic image annotation which permits the use of synchronized voice, drawings, and gestures (mouse movements), and image manipulation operations identified on one or more medical images.

Digital Libraries

There have been a large number of activities in several companies and universities to build prototype digital libraries for electronically accessing, browsing, indexing, sorting, and searching multimedia documents from very large databases. There also several image library systems providing content-based search based on low-level features. These include the QBIC [5], VisualSeek [9], and Photobook [7] . The CYBRARY system is part of a digital library project at AT&T [2]. It aims to provide the user effective Internet access over low-speed links to documents stored in the form of high-quality images and to reproduce these faithfully with full color on any screen.

Multimedia Magazines

CD-ROM magazines developed using multimedia technology are now commercially available. An example of a CD-ROM music magazine, available through subscription or at stands selling printed magazines, is the one being produced by Launch (www.launchcdrom.com). Each issue contains live concert performances, exclusive video interviews with artists, music reviews with song samples, movie, and interactive game. A recent review states that the Launch CD-ROM is basically "a music magazine placed in a virtual city."

2.3   Internet-Based Multimedia Services

As mentioned in Section 2.1, Digital Broadcasting provides a natural opportunity for multimedia services. Current systems, using terrestrial and satellite channels, are evolving towards the notion of Integrated Services Digital Broadcasting (ISDB). In parallel, there has been an even greater pace in the deployment of multimedia broadcasting over Internet and Intranet systems, along with related services for narrowcasting, multicasting and pointcasting.

Most of the multimedia services available today are Web-based and are of first-generation. Businesses are embracing the Internet faster than consumers. Around 90 million people worldwide are exchanging information or buying/selling products over the Net [1]. Today there are over 40,000 companies with their own Web sites. Online sales of goods and services to customers in the United States and Europe is expected to be over 5.1 billion U.S. dollars in 1998, according to Forrester Research [1]. Facilitating the trade over the Net are search engines provided by the Internet portals such as Yahoo!, Lycos and Excite.

Internet commerce, more popularly known as e-commerce, can be divided into two classes: business-to-business commerce and business-to-consumer commerce. According to Forrester Research, more than 43% of durable goods manufacturers will conduct e-commerce by the Year 2001, with sales close to 99 billion US dollars. Sales of office supplies, electronic goods, and scientific equipment is expected to be around 89 billion US dollars [1]. Other business-to-business commerce, involving services and transportation is expected to be much slower in embracing Internet, resulting in a much smaller sales figure by 2001. Likewise, business-to-consumer commerce in housing, foods and beverages, and health care will have slower acceptance of Internet based commerce.

Examples of currently available online services are travel reservations, automobile transactions (e.g. AutoByTel, Autoweb, Carpoint), computers (e.g. Dell Computer , Gateway), brokerage services (e.g. E*trade, Suretrade), books (e.g, Amazon, Barnes&Noble), toys (e.g. eToys), credit loans (e.g. GetSmart , E-Loan), insurance (e.g. InsWeb Corp, John Hancock), virtual flea market (e.g. eBay), and gardening services (e.g. Garden Escape).

3 Future Applications and Services

The next generations of multimedia applications will be characterized by increasing use of different media and modalities, as well as by the utilization of techniques for media processing beyond signal compression, media recognition and synthesis in particular.

Algorithms for media recognition and synthesis have steadily improved over the years and many of them currently claim levels of performance that are beginning to be acceptable in real and serious applications. Examples of algorithms for machine recognition of speech, printed text and handwriting, as well as algorithms for computer synthesis of speech and face, based only on text input, are provided in [4], together with some performance metrics.

User Interface technologies based on audiovisual recognition and synthesis can have a profound impact on several classes of transactions and information retrieval [1]. For example, it is expected that smart software will provide real-time parsing of customer's messages and provide automatic responses that are either preprogrammed, or in more intelligent paradigms that are user-driven. The possibilities in such natural language processing are fascinating, in both the speech and visual domains, as are the possibilities of codesigning the natural audiovisual interface with intelligent databases and network agents.

As in Section 2, new advances in multimedia communications will touch a variety of market domains, including business, consumer, entertainment, medical and educational segments of the information industry.

Linear advances in multimedia are resulting from steady improvements in the performances of media processing algorithms. These include algorithms for the compression, synthesis and recognition of media signals: speech and audio, as well as image and video, including handwriting and gesture. Of these, the most mature applications are those that depend on the interworking of compression with the disciplines of communications and networking. Internet multimedia and digital broadcasting of high-quality sound and television are prime examples.

Perhaps even more interesting are the nonlinear applications that would result from hitherto unpracticed combinations of the various subfields in media processing. Rather than depending on quantum improvements in the component technologies, these new applications may result from pragmatic combinations of admittedly suboptimal components in systems that the end user would find to be advantageous or even compelling. Initial examples of these may be email-to-speech conversion for portable messaging , web-friendly multimedia transcriptions of television broadcasts, and multimodal human-machine communications, as in internet browsing by voice and gesture.

4 Concluding Remarks

The papers in this special session provide in-depth descriptions of leading-edge systems for multimedia communications and services. These descriptions will serve to illustrate the multidimensional opportunities in this field. Implicitly, they also bring out the dual and complementary roles of basic research and concurrent system integration, for maximizing the timeliness and impact of multimedia technology.

References

1. "Doing business in the Internet age," *Business Week*, June 22, 1998, pp. 122-168.

2. R.V. Cox, B.G. Haskell, Y. Lecun, B. Shahraray, and L.R. Rabiner, "On the applications of multimedia processing to communications," *Proceedings of the IEEE*, vol. 86, No. 1, May 1998, pp. 755-824.

3. P.E. Crouch, J.S. Rodriguez, and W.C. Schwartz, "Screen-based multimedia telephony," *AT&T Technical Journal*, Special Issue on Multimedia, September/October 1995.

4. N.S. Jayant, B.D. Ackland, V.B. Lawrence and L.R. Rabiner, "Multimedia: Technology Dimensions and Challenges," *AT&T Technical Journal*, Special Issue on Multimedia, September/October 1995.

5. W. Niblack, R. Barber, W. Equitz, M. Flickner, E. Glasman, D. Petkovic and P. Yanker, "The QBIC project: Quering images by content, texture, and shape," Proc. SPIE Conference on Storage and Retrieval for Image an Video Data Bases, February 1993.

6. L.G. Palmer and R.S. Palmer, "DECspin: A networked desktop videoconferencing application," *Digital Technical Journal*, vol. 5, No. 2, Spring 1993, pp. 65-76.

7. A. Pentland, R.W. Picard, and S. Scalroff, "Photobook: Tools for content-based manipulation of image data bases," Proc. SPIE Conference on Storage and Retrieval for Image an Video Data Bases, II, Bellingham, WA, vol. 2185, pp. 34-47.

8. Siemens Corporate Research News Release, August 14, 1997.

9. J.R. Smith and S-F. Chang, "VisualSEEK: A fully automated content-based image query system," Proc. ACM Multimedia Conference, Boston, MA, November 1996.

Multimodal Human/Machine Communication

James Flanagan, Ivan Marsic, Attila Medl, Grigore Burdea, Joseph Wilder
Rutgers University
Center for Computer Aids for Industrial Productivity (CAIP)
New Brunswick, N.J.

Abstract

Natural communication with machines is a crucial factor in bringing the benefits of networked computers to mass markets. In particular, the sensory dimensions of sight, sound and touch are comfortable and convenient modalities for the human user. New technologies are now emerging in these domains that can support human/machine communication with features that emulate face-to-face interaction. A current challenge is how to integrate the, as yet, imperfect technologies to achieve synergies that transcend the benefit of a single modality. Because speech is a preferred means for human information exchange, conversational interaction with machines will play a central role in collaborative knowledge work mediated by networked computers. Utilizing speech in combination with simultaneous visual gesture and haptic signaling requires software agents able to fuse the error-susceptible sensory information into reliable interpretations that are responsive to (and anticipatory of) human user intentions. This report draws a perspective on research in human/machine communication technologies aimed to support computer conferencing and collaborative problem solving.

1. Introduction

Networked computers are becoming pervasive. With this progress comes opportunity for accomplishing collaborative knowledge work by participants who may be geographically separated (Fig. 1). The networked system takes on the role of both mediator and aide, as it supports activities of increasing complexity. Through embedded intelligence and software agents, the system dynamically allocates computing and storage resources, as well as monitors quality of service. For example, network-congesting calls might be avoided if client objects and server objects are migrated to a common host prior to the calls (e.g., downloading of JAVA applets). Under this architecture and control strategy effective communication between the human user and the mediating system becomes central to realizing the full capabilities for collaborative effort.

Humans favor natural modes of communication for information exchange. The sensory dimensions of *sight, sound* and *touch*, used in combination, offer capabilities that go well beyond mouse and keyboard for collaboratively manipulating objects in a shared workspace. (Fig. 2). An obstacle is that the individual technologies for human/machine communication are, as yet, imperfectly developed. But prudently applied they can be used to benefit. Next, we will demonstrate a prototype system for multimodal human/machine communication.

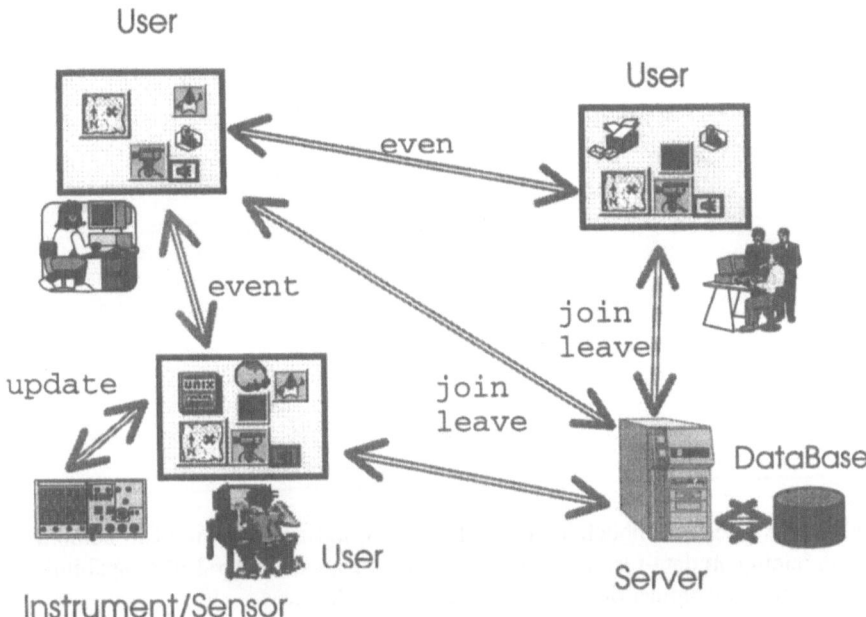

Figure 1 Networked computers with distributed processing, storage, archiving, and embedded intelligence support collaborative knowledge work. Probes installed in client and server Object Request Brokers monitor network traffic and automatically migrate objects whose frequent calls produce congestion.

Prior research at the CAIP Center has established several interface technologies that, in properly delimited application, can contribute naturalness for human/machine interaction. They include:

- *sight*
—region-of-interest segmentation
—gaze tracking and visual gesture
—face detection and recognition

- *sound*
—beam-steered microphone arrays
—speech and speaker recognition
—computer voice response
- *touch*
—tactile force feedback and grasp
—manual gesture

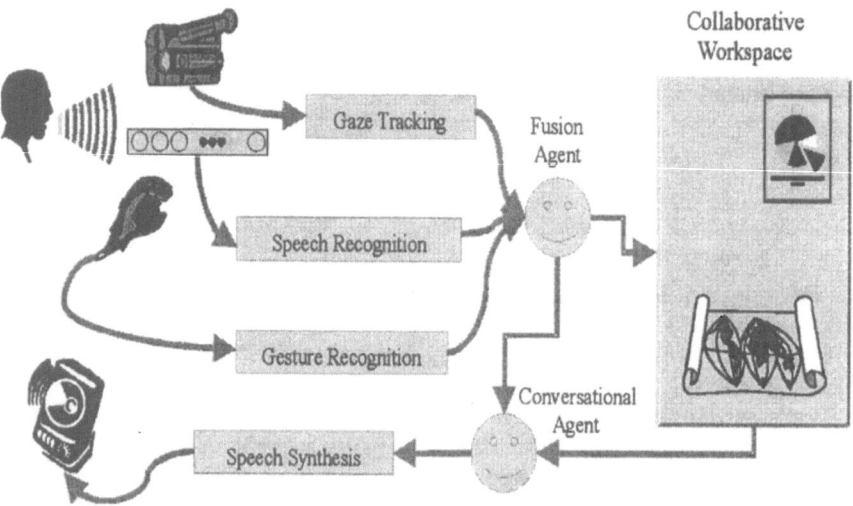

Figure 2 Interface modalities for sight, sound and touch — used in combination with intelligent data-fusion agents — provide enhanced, natural-like capabilities for cooperative manipulation of objects in shared workspaces.

Region-of-interest segmentation of images is included in an available suite of signal-processing algorithms which can be downloaded and locally executed. Automatic segmentation is based upon color, form and luminance, and is useful in identifying similar objects in complex images such as satellite pictures, blood-cell microscopy and MRI analysis. Gaze tracking provides an eye-controlled cursor derived from automatic face tracking and infra-red illumination (to determine the angle between the centroid of the pupil and the corneal reflection).

Speech recognition and text-to-speech synthesis permit conversational interaction with the system. Beam-steered microphones provide "hands-free" sound capture where the human is unencumbered by tethered or body-worn sound pickup equipment. Tactile force feedback is achieved from computer-controlled pneumatic thrusters applied to the finger tips of a close-fitting glove. Finger deflection is sensed by an LED and photo detector pair, located coaxially within each thruster. A Polhemus coil on the wrist provides absolute position.

2. System Design and Results

Building upon these techniques, a current activity is the development of a client station that supports more natural communication with the user (Fig. 3). The ingredients include automatic face detection and gaze control of displayed cursors and objects, "hands-free" conversational interaction using speech recognition, synthesis and beam-forming microphone array, and manual gesture and force-feedback grasp with a tactile glove. To employ these in combination (as the human does) the system must be able to fuse the often unreliable sensory inputs to reach a reliable decision and action that is responsive to the intent of the user (see Fig. 2).

In an initial implementation we have combined gaze, speech, and tactile inputs to serve a specific application (i.e., civil preparedness and mission planning) that requires collaborative manipulation of graphical objects on a topological map displayed in the shared workspace. (Fig. 4). In this contrained task the vocabulary and grammar for the conversational interaction are delimited for increased reliability. The microphone array is integrated into the workstation housing and fix-focused on the user position (seated at the keyboard).

For the constrained task, fusion of data from the gaze, speech, and tactile inputs is accomplished by a slot filler method in which a parse of the recognized text string is synchronized with the visual and tactile inputs. (Fig. 5). The figure illustrates the

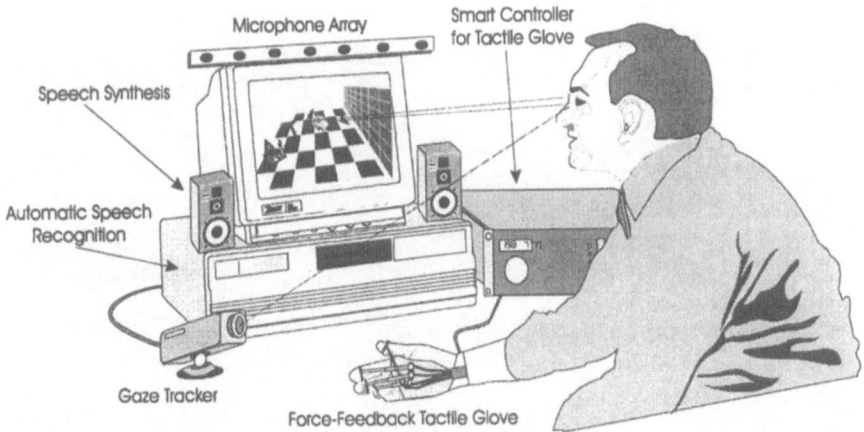

Figure 3 Experimental client workstation incorporating sight, sound and touch modalities for human/machine communication. The eye tracker provides a gaze-controlled cursor for indicating objects in the display. The tactile force-feedback glove allows displayed objects to be grasped, "felt", and moved. Hands-free speech recognition and synthesis provides natural conversational interaction.

command "move red circle to this location." The recognized text string indicates the operation to be performed ("move") and the object ("red circle"), but the destination ("this location") is determined from either the manual gesture or visual cursor coordinates that temporally coincide with the utterance "this". A preceding spoken command assigns the relevant cursor. Clearly, the sensory inputs can overlap in information content and exhibit redundancy. Clearly, too, in some instances a single modality is sufficient and natural, and can subsume the entire task. The fusion agent must cope with all combinations (a very natural utterance being "move this to there").

For conversational interaction, synthetic voice answerback is essential so that the system can advise the user of actions needed or taken. Text-to-speech synthesis is the appropriate technology to supply the user:

- answers to queries related to the dynamic state of the workspace
- requests for confirmation if necessary
- general error messages and warnings about unexecutable commands
- notifications about semantic or user-related errors

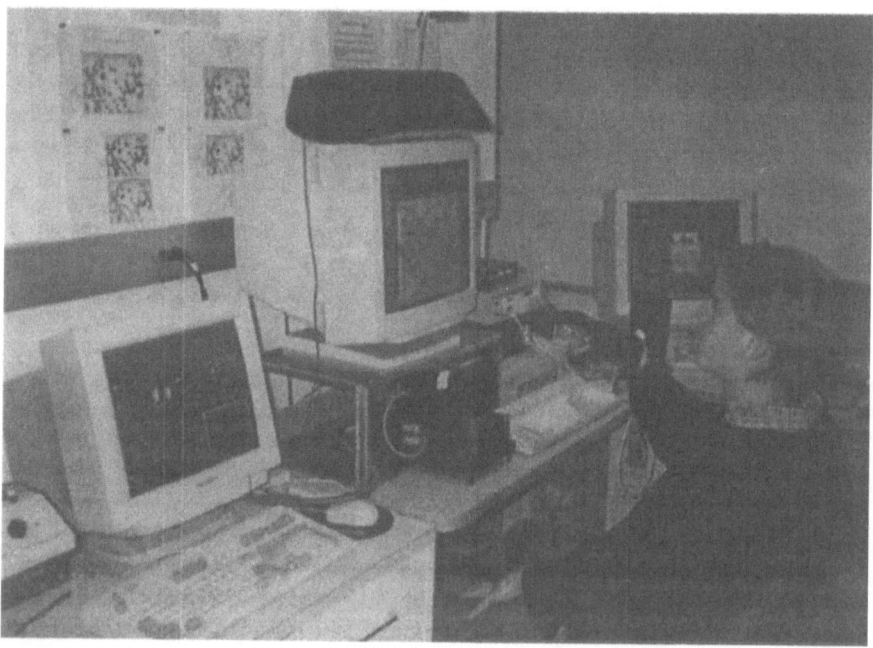

Figure 4 Implemented user station incorporating speech recognition and synthesis for conversational interaction, combined and synchronized with manual gesture sensing from a force-feedback glove and visual gesture sensing from a desk-mounted gaze tracker. A fixed-focus microphone array atop the workstation

captures speech from the user keyboard location while mitigating interfering acoustic noise and reverberation.

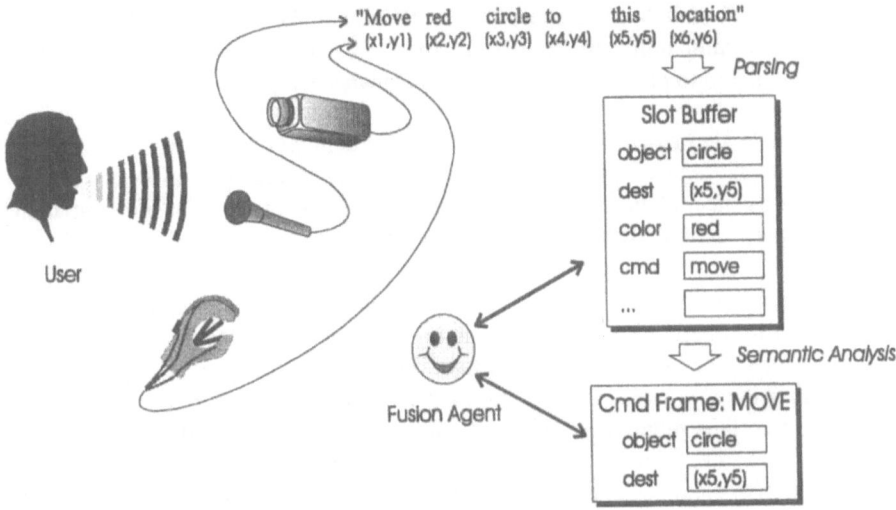

Figure 5 "Slot-filling" technique used to coordinate, fuse, and interpret simultaneous speech, visual, and tactile inputs.

The preceding has addressed single user communication at the client station. In some cases (such as situation rooms, command centers, or "smart-room" group teleconferences) multiple users must be accommodated at a client station. Speech sound pickup and face detection that can follow conferees about the meeting room are desirable under these conditions. Toward this capacity we have implemented an acoustic source locator which steers a line array microphone and points and focuses a video camera to the x,y,z coordinates of the source. (Fig. 6).

Source location is determined by two "quads" of omni-directional microphones placed preferably on adjacent orthogonal walls. The spacing is typically about 30 cm to insure that usefully correlated signals are obtained. For each pair in the quad an estimate of the time difference of signal arrivals is made. This estimate is obtained by computing an FFT for each microphone on a single DSP32C processor and forming the normalized cross power spectrum for each possible pair in the quad (i.e., 6). Inverse transformation gives the cross correlation function. A time difference of arrival estimate is made from the peak value. For the two quads, 12 such time difference of arrival estimates are made, each one defining the surface of a paraboloid on which a source could be positioned to produce the observed time difference. Ideally the intersection of the 12 surfaces would be a point. Practically, in most realistic environments it is a volume about 20 cm. in linear extent.

508

The 12 time difference estimates are passed from the DSP32C signal processor to a Pentium PC, which applies a gradient descent algorithm to find the x,y,z coordinates in the enclosure which provide the best least-squares fit to the set of time differences. (Computational speeds of DSP32C and Pentium presently limit the position determinations to 2 per second. Faster signal processors, under construction, aim to provide moving source estimates at 5 per second.)

Having made the x,y,z source location, the Pentium points and focuses a video camera to the source location. It also activates control hardware for an electronically-steered beam-forming microphone array of 21 first-order gradient microphones, pointing the beam to the source location.

Finally, because tethered equipment is confining, a wireless appliqué is being considered for tactile gloves worn by multiple participants. Also for multiple users, personal identification is usually important. The data-fusing agent serving the client station may therefore need identity information on enrolled conferences passed to it from the interface modalities. The technologies of speaker identification and face identification are advanced enough to serve this need and can be used in combination to achieve high accuracies of identification.

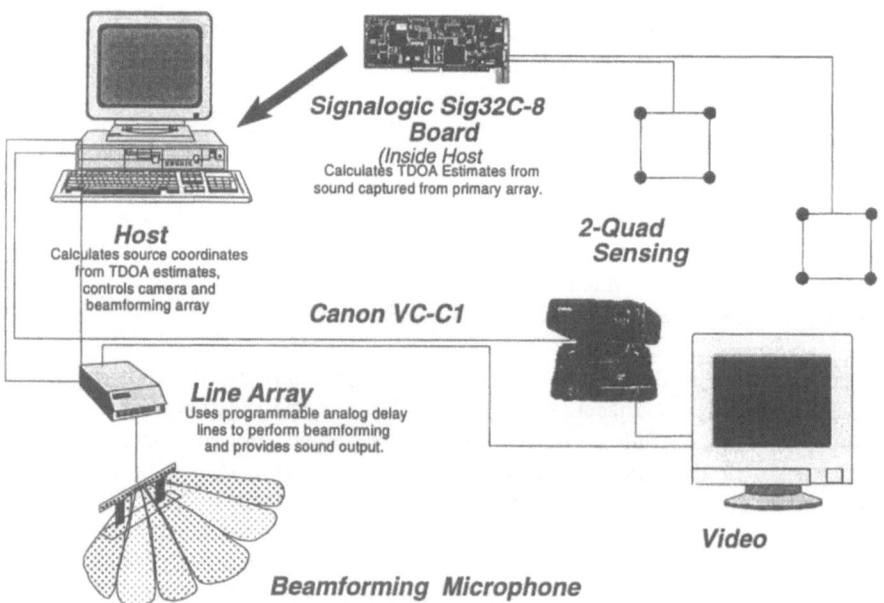

Figure 6 Auto directive system for image and sound capture in a "smartroom" client station. The system is able to locate and track the position of talkers moving in the conference room.

3. Acknowledgment

Components of this research are supported through NSF Contracts MIP 93-14625 and IRI-96-18854 and through DARPA Contracts DABT63-93-C0037 and N66001-96C-8510.

References

1. J. Flanagan and I. Marsic, "Issues in measuring the benefits of multimodal interfaces," Proc. IEEE Int'l. Conf. Acoustics, Speech and Signal Processing (ICASSP'97), Munich, Germany, April 1997.
2. A. Medl, I. Marsic, V. Popescu, A. Shaikh, M. Andre, C. Kulikowski and J. Flanagan, "Multimodal User Interface for Mission Planning," submitted to the ACM SIGCHI Conference on Human Factors in Computing Systems (CHI'98), Los Angeles, CA, April 18-23, 1998.
3. G. Burdea, Force and Touch Feedback for Virtual Reality, John Wiley & Sons, New York, 1996.
4. J. Flanagan, "Technologies for multimedia information systems," Proc. IEEE, v.84, no. 4, pp.590-603, April 1994.
5. J. Flanagan and E. Jan, "Sound capture with three-dimensional selectivity, Acustica, v. 83, no. 4, pp.644-652, July/August 1997.
6. D. Rabinkin, D. Macomber, R. Renomeron and J.L. Flanagan, "Optimal Truncation Time for Matched Filter Array Processing", In Proc. Of ICASSP'98, v. VI, pp. 3629-32, Seattle, WA, May 1998.
7. R. Renomeron, D. Rabinkin, J. French, and J.L. Flanagan, "Small-Scale Matched Filter Array Processing for Spatially Selective Sound Capture", J. Acoust. Soc. Am., v. 102, No. 5Pt. 2, p. 3208, Nov. 1997.
8. A. Waibel, M. Vo, P. Duchnowski, and S. Manke, "Multimodal interfaces," Art. Intell. Rev., v. 10, nos. 3-4, 1995.
9. J. Wilder, P.J. Phillips, C. Jiand, and S. Wiener, "Comparison of Visible and Infra-Red Imagery for Face Recognition", In Proc. Of the 2nd Int. Conf. On Automatic Face and Gesture Recognition, pp.182-187, Killington, VT, 1996.

Multimedia Processing for Advanced Communications Services

B. Shahraray, R. Cox, B. Haskell, Y. LeCun, and L. Rabiner
AT&T Labs – Research, New Jersey, USA

Abstract

The advent of digital multimedia communications has generated a growing need for powerful multimedia processing techniques to enable the generation of useful and intelligent communications services. Multimedia processing techniques play a significant role in creating communications services by; 1) enabling efficient transmission and storage of multimedia information through media compression techniques, 2) creating effective user interfaces through media conversion, understanding, and dialogue systems, and 3) providing intelligent information searching and browsing mechanisms based on media processing and understanding techniques. In this paper we present a brief overview of media compression techniques and standards, touch upon several media processing techniques. Then we give a brief overview of three prototype services based on these techniques.

1. Introduction

Communications systems depend on a wide range of hardware and software technologies to create useful communications services. The advent of digital communications, which was brought about by the digital representation of signals, has made it possible to represent different modes of communications involving textual, auditory, and pictorial information in a unified form. This has created a bridge between computing and communications that has enabled the application of digital media processing techniques to advanced communications services.

The Internet, the World Wide Web and the associated protocols and client/server software have revolutionized the way we communicate information. They provide the transfer mechanisms and the user interfaces that facilitate the creation of useful communications services. They have connected small and large heterogeneous sources of information to create a large distributed repository of multimedia information.

Remote communications between humans, or between humans and machines, involve the exchange of information at the endpoints, as well as the transmission of information between the endpoints. Media processing techniques are utilized at both of these stages. They facilitate the exchange of information between humans and machines through technologies such as speech recognition, text-to-speech conversion, natural language processing, and language translation. They make it possible to transmit more information using less bandwidth. They convert information to match the capabilities of the information appliance and the available bandwidth. They help users navigate through large volumes of information.

Coding and compression techniques are among the earliest applications of media processing to communications. Despite the sizeable increase in the network bandwidth and data storage capacities these techniques continue to play a major role in today's communications systems. Media conversion technologies have been used in communications services to convert the information to a form that is suitable for a particular communications terminal, or to meet the needs of users with special needs. Techniques for searching and browsing of textual information have been the basis of several information services for many years. More recently, the availability of large on-line repositories of multimedia information has resulted in a need to create technologies that would enable the searching and browsing of multimedia information. It should be noted that in most cases these media processing techniques are not independent of each other. For example, searching and browsing techniques have certain requirements that need to be taken into consideration when the media are compressed.

In this paper, we give a brief overview of media compression techniques. We touch upon several media processing techniques. Then we present an overview of three prototype services that have been developed in AT&T Labs, namely; the Cybrary Digital Library, the Digital Video Library (DVL), and the VideoTalks.

The Cybrary system builds upon novel document segmentation and coding techniques to create high-quality and highly compact representations of complex documents. Scanned images containing color pictures and text are processed and converted to a new format. This format enables efficient storage, fast retrieval, and flexible browsing of high-quality images. The system applies OCR techniques to these documents to create indexing information that allow content-based retrieval of documents on the Web.

The Digital Video Library (DVL) system applies robust multimedia indexing techniques to the pictorial information contained in video frames, as well as to the associated audio and text information, to create a searchable and browsable library of video programs. Content-based sampling of video coupled with text and audio processing creates a condensed representation of the information contained in video programs. Video information can be searched using the information extracted from the pictorial, auditory, and textual components by media processing. Selected video segments can be delivered in streaming mode over an IP network.

The VideoTalks system combines media compression technologies with video event detection, and network technologies to enable real-time delivery of technical presentations to remote locations. Presentations are decomposed into several audio and visual components. Each component is digitized and compressed with the appropriate resolution. Intelligent transition detection algorithms are used to generate indexing. The resulting data is transmitted over an IP network to recreate the presentation at a remote location. Multicasting techniques are utilized to deliver the presentation to multiple remote locations and to users desktops using minimal bandwidth.

2. Overview of Media Processing Techniques

In this section, we give a brief overview of several media processing technologies that are the basis of the prototype services discussed later in this paper.

2.1 Compression and Coding

By definition, multimedia services involve several media components that are presented in a way that certain spatial or temporal relationships between them are preserved. This imposes requirements on how to compress each media component and how to bundle the constituent media components of a multimedia source to achieve proper synchronization. In general, designing and using media compression algorithms involves making intelligent compromises between several attributes to meet the specific requirements or constraints of a particular application. These attributes include quality, bit rate, delay, and complexity. In general, better quality and smaller delay can be achieved at the expense of higher bit rate. Lower bit rate can be obtained by employing signal processing algorithms, which result in higher complexity (in terms of the computations required to do the coding or decoding), and/or higher delays.

Error resilience is another feature of coders that has gained more significance with the widespread transmission of multimedia data over the Internet. This has created a real need for coders that can maintain an acceptable quality in spite of the possible loss or delay of some of the packets.

For service generation, interoperability of hardware and software is a significant factor. Therefore, in the brief discussion of the media compression and coding techniques we focus on those that have been standardized.

2.1.1   Speech and Audio Coding

Speech and audio compression are an important part of many communications services. There are considerable variations in the requirements of different applications. For example, services involving live conversations between two or more people require low-delay coders, whereas voice and multimedia messaging applications do not have such requirements. Speech coders used in wireless telephony need to match the complexity of the coding algorithm to the power of the CPU chips that are feasible on mobile phones [Ref 1]. Services involving the distribution of high-quality music have much wider bandwidth requirements than those used for telephony.

There are two types of speech coders, waveform coders, and parametric coders. Waveform speech coders are used extensively in the telephone network. For example, the standardized G.711, pulse code modulation (PCM) coder is used in digital telephone networks and switches. The G.726, and G.727 adaptive differential PCM (ADPCM) coders are used in satellite links and underwater cables to increase their capacity. The G.722, wideband coder (7 kHz) is used over frame-relay and ISDN connections. Parametric speech coders use a model for the speech generation process with a simplified excitation signal. The parameters of the model are extracted from the input signal at some regular interval, sent to the receiver, and synthesized using the same model. Since only a few parameters are used to synthesize the speech at each given interval, these coders can achieve very low bit rates. An example of such coders is the Mixed Excitation Linear Predictive (MELP) coder which can operate at bit rates as low as 2400-b/s.

When the audio component contains non-speech signals, such as music, audio encoders are used to reduce the bit rate while preserving the quality of the signal. The Internet and the World Wide Web have created real possibilities for generation of audio-related services such as music distribution. The high bit rate of uncompressed audio (e.g., 1.4 Mb/s for CD-quality audio) precludes the cost-effective transmission of high–quality audio over IP networks and generates a real need for audio compression techniques. An example of standardized audio coders is the MPEG-1 with layers 1,2, and 3. Each successive layer generates a lower bit rate stream at the expense of added complexity. The respective bit rates for a single channel to make what is judged to be a transparent reproduction are 384, 256, and 96 kb/s. Another example is the MPEG-2 audio coder that was meant to provide multi-channel theater-style sound with one to five channels. A more recent example is the Dolby AC-3 coder that has been selected as the standard for the Digital Video Disk (DVD) and the U.S. High Definition Television (HDTV). The MPEG Advanced Audio Coder (MPEG-AAC), which is a perceptual coder is a highly flexible coder that can handle a wide range of applications. This coder can handle from one to 48 channels at bit rates starting from 8 kb/s per channel to 192 kb/s per channel.

2.1.2   Image and Video Coding

Various forms of pictorial information often constitute a large segment of multimedia information. They cover a wide range of data from bilevel images to full-motion color video and are an integral part of many communications services [Ref 2].

Bilevel image coding algorithms and the associated standards are the main reason for the widespread use of fax machines. These standards include the analog group 1, and group 2 (G1, G2) standards, and the digital group 3, and group 4 (G3, G4) standards. The G4 standard can achieve a 25 to 1 compression of simple text documents. For binary halftone images, the JBIG-1 standard was created. This standard supports progressive coding that allows for fast reconstruction of a low-resolution image that can be successively refined. Progressive encoding is particularly useful for image browsing applications. To provide optimal lossless and lossy compression of arbitrary scanned images, a new method known as JBIG-2 has been proposed. This method is based on "soft pattern matching" which eliminated symbol substitution errors by introducing a new symbol when a good match to the existing symbols is not found. The JBIG-2 forms the basis for a powerful and flexible document compression method that is used in the Cybrary digital library system described later in this paper.

When multimedia communications services involve the storage and transmission of continuous tone images, the JPEG (for Joint Photographic Experts Group) standard plays a major role. This still-image compression standard is heavily used in today's Web-based information systems. JPEG can provide both lossless and lossy compression of arbitrarily sized monochrome and color images. It supports progressive encoding modes. The computational complexity of this algorithm is reasonably low. JPEG-compressed images have excellent quality at compression factors around 10-to-1 and can be used at compression factors around 60-to-1 with fair quality when a really high compression is essential. The JPEG standards were established in the early 1990's. New research in still-image coding is being focused on the introduction of an improved standard known as JPEG-2000 which is expected to be completed by the year 2000. This new standard is expected to address the needs of applications such as the WWW, photo and art digital libraries, medical imagery, electronic photography, facsimile, and document imaging.

Video coding is the major enabling technology for many new Web-based services. It has applications in many services from teleconferencing to High Definition Television (HTDV). The video coding standards used for teleconferencing include H.261 for ISDN (Integrated Services Digital Network), H.263 for POTS (Plain Old Telephone Service), and H.262 for ATM/broad-band networks. The MPEG-1 video coding standard was mainly developed to support the storage of multimedia data on a standard CD-ROM with a bit rate of 1.4 Mb/s. The MPEG-2 standard, originally developed for high-quality encoding of television programs with bit rates from 4 to 9 Mb/s, is currently being used in digital satellite television service. The MPEG-4 multimedia standard, that is expected to be finalized by the end of 1998, has focused on object-based coding. It includes independent coding of audio-visual objects (AVO). Such AVOs are assembled at the receiver to create an audio-visual scene. MPEG-4 will integrate most of the features of multimedia into one standard, thereby facilitating the creation of a variety of advanced multimedia services. The significance of content analysis and indexing has resulted in activities to create the MPEG-7 standard. This effort is aimed at creating a standardized description of different multimedia content. Such standardization will be a major step in creating multimedia searching techniques.

2.2 Media Conversion

Media conversion is the process of converting information from one media form to another. It includes the conversion of text-to-speech (TTS), and inversely the conversion of speech-to-text. Many communications appliances, such as ordinary telephones, are incapable of presenting textual information. When a communication service involves the presentation of information stored in textual form, over these appliances, a conversion to speech is essential. TTS is also essential to enable the presentation of textual information to visually impaired audience. The current state of the art in TTS is capable of generating synthetic speech that is as intelligible as natural speech, with voice quality significantly below that of natural speech.

Accurate conversion of speech to text would go a long way towards creating speech-based human-machine interfaces. Although an error-free conversion of speech is not possible in many situations, the current state of the technology is sufficiently advanced to create many useful services. Services that require only the recognition of a small set of words (e.g., digits) have existed for many years. In the presence of error, spoken language understanding techniques can be employed to extract task-related information from whatever words are recognized correctly. When the goal of the system is to categorize the user speech input into one of several categories, correct categorization may be possible even with high word error rates.

2.3 Media Indexing

In this section we give a brief overview of media indexing techniques that can be used to index each media component of a multimedia information source. While the discussion considers each component independently of the other components, in practice a multimedia indexing scheme based on collaborative processing of several media is capable of generating better results that one based on the fusion of information extracted independently from each component.

2.3.1 Text-based Indexing

Text-based information retrieval systems have existed for many years. Extracting information from structured sources, such as databases with pre-defined data fields, has proven useful. However, a more challenging task is that of finding information in information repositories containing unstructured text, such as those in printed or electronic documents. Full-text searches based on word occurrences, morphological expansions, and Boolean operations are the basis of many search engines (such as AltaVista, Lycos etc.) used on the WWW today. More sophisticated text processing techniques can be used to extract fragments of information from many relevant documents, and piece them together into a single document that satisfies the specified information query.

2.3.2   Audio indexing

Auditory information can be indexed in several ways by event detection, audio classification, and speech recognition. Audio event detection refers to the process of locating the positions at which certain attributes of the audio signal change considerably. Examples of this include detecting sudden changes in volume, silence detection, and the detection of speaker changes. Audio classification refers to the process of classifying the audio signal into different categories such as speech, music, animal sound, noise, etc. When the audio stream includes spoken words, speech recognition can be applied to extract index information. An automatic and error-free transcription of all the spoken words will reduce the problem to that of text-based information retrieval. Such a transcription requires a large vocabulary automatic speech recognizer (LVASR) system with sophisticated acoustic and language models [Refs 3 and 4]. The existing recognizers (e.g., WATSON from AT&T, SPHINX from Carnegie-Mellon University, etc.) can generate transcriptions with word error rates ranging from 5% to more than 50% depending on the quality of the speech and the models used [Ref 5]. While such levels of accuracy may not be sufficient for generating readable transcripts, they can be used in conjunction with appropriate text-based information retrieval algorithms to produce acceptable results [Ref 6].

2.3.3   Image Indexing

Understanding and interpretation of pictorial content is a highly application dependent task. While our visual system is extremely good at performing this task, automated image understanding by machines does not remotely approach these levels of performance. Nevertheless, intelligent utilization of even low-level image features, when coupled with the capabilities of a user in a closed-loop system has proven to be effective. Features such as intensity, texture, color, and shape can be used to segment large collections of images into different categories. Image-based queries can take advantage of approximate matching techniques to extract similar images from large collections of images. Neural-net based techniques can be used to detect the presence of faces in images.

2.3.4   Video Indexing

The same techniques that are used to index still images are also applicable to image sequences that make up video programs. However, unlike still images, video sequences have a temporal dimension that gives rise to a much richer set of features. These features include motion information. Motion can be the basis for segmenting moving objects from the stationary background. Qualitative and quantitative information about the movement of objects in the scene mark significant visual events. Global motion information enables the detection of camera operations that were used during the production to convey important information. Professionally produced video programs are composed of many video shots that have been put together using various editing modes such as cuts, fades, dissolves, etc. A shot-based decomposition of the video program combined with the intra-shot analysis of motion information to detect significant events is the basis of a content-based video sampling method [Ref 7] that indexes the video and generates a compact pictorial representation of the video program. This method is used in the Digital Video Library system described in this paper. A good example of a video search engine based on color, shape, and motion is the VideoQ system from Columbia University [Ref 8].

2.4 Intelligent User Interfaces

The user interface is an important part of many communications services, such as messaging, customer service, and information services, that involve interaction between humans and computers. An effective user interface should be able to adjust to the capabilities and limitations of the communications appliance, the communications channel, and most significantly, the user. Many of the media processing techniques discussed so far in this paper (e.g., compression techniques, media conversion and understanding, content-based sampling etc.) are important elements of such an interface. When the capabilities of the information appliance are limited to audio, spoken language interfaces based on speech recognition, speech generation (TTS), and natural language processing are needed. Dealing with limited capacity of humans to absorb and remember auditory information requires the machine to present the information in manageable pieces and to conduct a dialogue with the users to help them achieve their goal. Such two-way interactions can be generalized to include exchange of multimedia information between the user and the machine.

3. Examples of Multimedia Services

In this section we discuss three prototype multimedia communications systems that exemplify how some of the multimedia processing technologies discussed earlier can be used to create useful communications services.

3.1 Cybrary Digital Library

An important purpose of communication systems is to give people easy access to information. A large percentage of the information that is currently available exists in printed form. The collection of artistic, cultural, and scientific information in printed form is growing continuously. These documents can be scanned and stored on computers and transmitted over communications networks. In doing do, image compression techniques such as JPEG are used to reduce the size of these files. Since document images usually contain both color images and text, achieving an acceptable quality that leaves the text readable requires low-loss compression and results in large file sizes. Moreover, retrieving and browsing these documents over the WWW requires making special considerations in the coding algorithm to allow for progressive and piecewise rendering of the

518

images of the user's display. Achieving these goals requires using specialized document image compression techniques. The DjVu compression technique, which is a component of the Cybrary digital library prototype service, is one such technique [Ref 9]. It can reduce the size of a typical color document page that has been scanned at 300 dpi from about 25 Mbytes down to 40 to 60 Kbytes.

Fig. 1 A gray-scale rendition of a DjVu-compressed color document image

The main idea of DjVu compression technique is to decompose the document image into three images, namely; the background image, the foreground image, and the mask image. The mask image is a high-resolution (300 dpi) bilevel image that contains text and drawings in the original image. The background and foreground images are lower resolution (100 dpi) color images. The foreground contains the color of text and line drawings. The mask image in encoded using JB2, which is a variation of AT&T's proposal to the JBIG-2 fax standard. The foreground and background images are encoded using an efficient progressive wavelet-based encoder called IW44. The efficiency of IW44 and JB2 is a result of a binary adaptive arithmetic coder called the Z-coder [Ref 10]. The compressed images are decoded and combined at the receiver to generate a high-quality rendition of the original image. Figure 1 shows an example of a gray-scale rendition of a color image compressed with DjVu. The compressed image file is only 67 Kbytes in size. Compressing the same image with JPEG with a quality factor of 30% generates a file size of 82 Kbytes, with noticeable degradation in the text regions. Work is underway to apply document analysis and optical character recognition (OCR) techniques to create indexing information for content-based retrieval of printed documents. More information about DjVu can be found at the following URL: http://djvu.research.att.com.

3.2 Digital Video Library (DVL)

Advances in video compression and network technologies have made it possible for video information to be stored in digital video libraries for on-demand delivery to users. As these digital video libraries grow in size, it becomes essential to create automated and intelligent ways of organizing and indexing their contents so that users can easily locate relevant pieces of information among many thousands of hours of video programs. The DVL system is an example of how the media processing techniques can be employed to achieve this goal.

The DVL system builds upon our earlier work on Pictorial Transcripts [Ref 11] and adds new text, audio and video classification algorithms to extract more information to enhance the searching and browsing capabilities. It uses a content-based sampling algorithm to segment the video program into small units with relevant content. First, it uses shot boundary detection to segment the video into individual shots. Next, it uses motion analysis algorithms to analyze each shot to detect significant events and further segment each shot into smaller units. A single representative image from each unit serves the purpose of representing the visual content of the unit. These images are used to generate a pictorial index into the video program. The textual component needed for generating tables of contents and indices for satisfying text-based queries is recovered from closed-caption text. Linguistic processing is used to refine the text for presentation with the images, create tables of contents, and generate hyperlinks to relevant material. Work is underway to employ AT&T's WATSON speech engine to perform automated transcription of the speech component of the audio track. Audio and video components are compressed and their temporal relationship with the textual and pictorial components is established. For each program, the images and the processed text create a textual and pictorial index that can be searched to find programs of interest. Image similarity measures are used to find occurrences of similar images in a program. Audio processing is applied to detect the presence of certain speakers (e.g., the anchorperson). Work is currently underway to detect story boundaries using a combination of text, video and audio analysis.

520

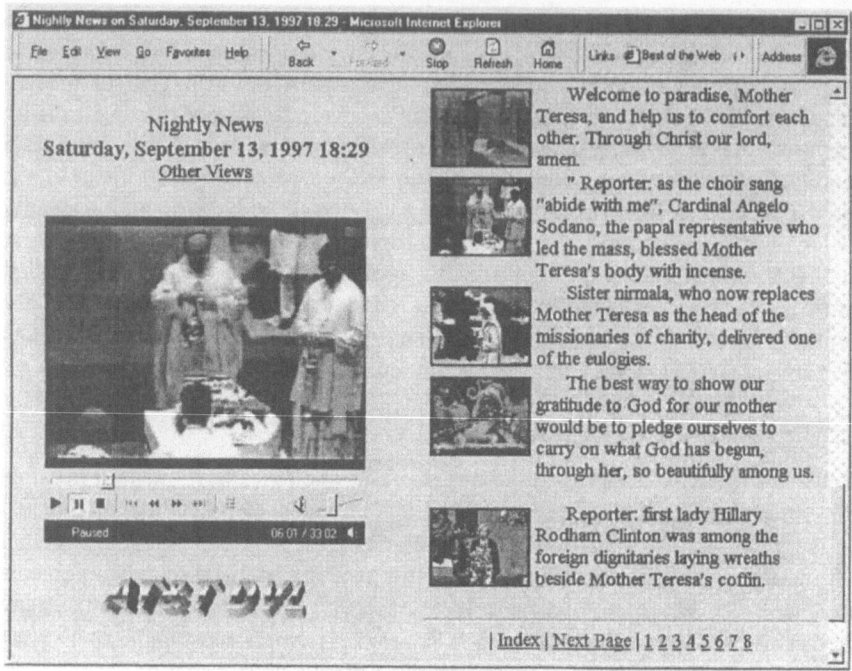

Fig. 2 A sample page of the DVL system showing the video player

A sample page of the Web-based user interface for the DVL system is shown in Fig. 2. The thumbnail images and the associated text provide a compact pictorial and textual representation of the video information that can be easily browsed to find relevant segments of the video. Selecting the image initiates streaming delivery of the video. Currently the system supports a full-text search of the text generated from the closed captions (or transcribed from the audio track by the WATSON search engine). Initial results of using natural language processing to enable natural language queries are encouraging. Work is also underway to use a dialogue system to help the user formulate and refine queries through a two-way dialogue with the system.

3.3 VideoTalks

Multimedia teleconferencing is an effective way to minimize the expenses and the loss of time associated with attending non-telecast meetings. The challenge of multimedia teleconferencing systems is to create a sense of presence for the remote participants. This can be substantially achieved by reproducing the relevant visual and auditory information from one location (or several locations) at remote locations. Moreover, when possible, such systems should reproduce information from remote locations to better integrate all the participants. Such teleconferencing systems can add more value by recording and archiving the information that is exchanged during the meeting and allow for time-shifted delivery. Meeting and presentation archives should also be indexed and be used for selective retrieval of information. An example of such a system is the VideoTalks research prototype developed at AT&T Labs – Research [Ref 12].

The VideoTalks system connects several main conference rooms. Each room is equipped with cameras that capture views of the speaker, the presentation material (viewgraphs), and the audience. Each view is compressed and coded with the appropriate spatial and temporal resolution, and is transmitted, together with the audio information, to the other locations over an IP network. The viewgraph images are compressed using JPEG at a resolution of 640x480. A redundant transmission of the same viewgraph is avoided by using a media processing algorithm to detect the viewgraph changes. Views of the speaker and the audience are compressed using motion JPEG at 320x240 at 12 and 8 frames per second respectively. At the receiving end, the video streams are decoded and presented on large video displays. Cameras and microphones at the remote location capture the view and audio of the remote audience for regeneration at the main location.

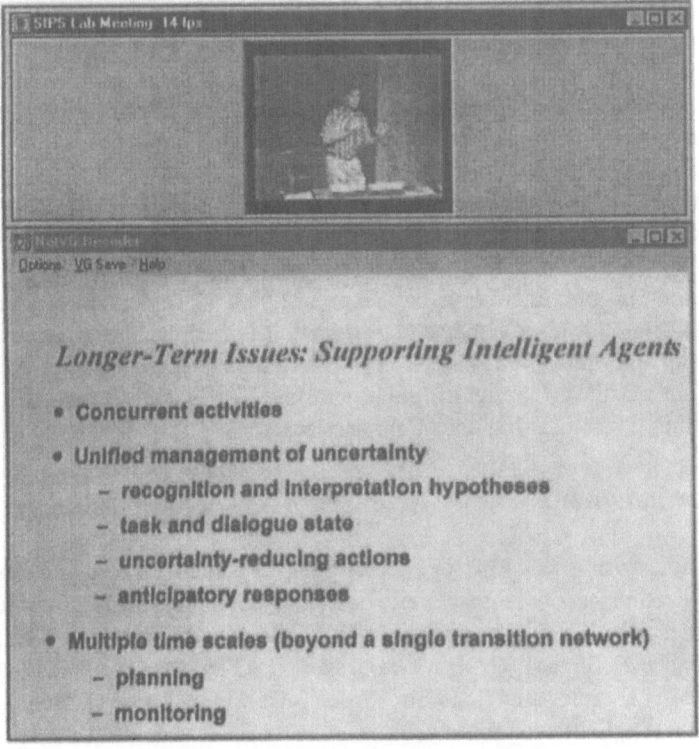

Fig. 3 A view of two windows showing the speaker and the viewgraph

To further facilitate remote participation, multicasting techniques are used to send audio-visual streams to office locations. An H.261 video encoder used for this purpose. Software modules render the view of the speaker, the presentation viewgraphs, and the associated audio streams on the user's computer. An example of the desktop view showing the speaker view (top), and the viewgraph (bottom) is shown in figure 3.

The system supports time-shifted and on-demand replay of the presentations. Time-shifted replay of the information uses multicasting to rebroadcast the recorded presentation to the desktops. The on-demand mode presents list of the available presentations, and allows for selective replay of segments of the presentation using the viewgraphs as the index.

4. Conclusions

Multimedia processing plays a central role in the creation of advanced communications services. In our view, three major applications of multimedia processing for communications services are in the areas of intelligent multi-modal human-machine interfaces, media compression and coding, and intelligent searching and browsing techniques. Among the three, coding algorithms have evolved through many years of research, development and standardization activities. Advanced multimedia communications services involving interactions between humans and machines are becoming more commonplace. This has increased the need for intelligent multi-modal human-machine interfaces to facilitate such interactions and enable the delivery of useful services using a wide range of communications appliances. Effective utilization of the vast repository of information made available by the Internet and the World Wide Web requires the application of multimedia processing techniques to organize and index information. While more remains to be done to advance the performance of media processing algorithms, the examples presented in this paper are indicative that the current state of the art in multimedia processing is sufficiently advanced to enable the creation of useful communications services.

Acknowledgment

The authors wish to acknowledge the original work on the prototype multimedia communications services that were described in this paper by the following researchers in AT&T Labs – Research: Y. Bengio, L. Bottou, P. Haffner, P. Howard, and P. Simard (Cybrary), A. Basso, G. Cash, M. Civanlar, D. Kapilow, R. Kollarits, B. Paul, and C. Swain (VideoTalks), D. Gibbon, Q. Huang, and H. Jafarkhani (Digital Video Library).

References

1. W. B. Kleijn, and K. K. Paliwal, Ed., Speech Coding and Synthesis. Amsterdam, The Netherlands: Elsevier, 1995.

2. A. N. Netravali, and B. G. Haskell, Digital Pictures – Representation, Compression, and Standards. 2nd ed. New York, Plenum, 1995.

3. L. R. Rabiner, and B. H. Juang, Fundamentals of Speech Recognition. Englewood Cliffs, NJ: Prentice-Hall, 1993.

4. L. R. Rabiner, B. H. Juang, and C. H. Lee, 'An overview of automatic speech recognition' in Automatic Speech and Speaker Recognition, Advanced Topics, C. H. Lee, F. K. Soong, and K. K. Paliwal Eds. Norwell, MA: Kluwer, 1996.

5. M. Mohri, and M. Riley, 'Weighted determinization and minimization for large vocabulary speech recognition' Proc. Eurospeech, pp. 131-134, 1997.

6. M. Christel, T. Kanade, M. Mouldin, R. Reddy, M. Sirbu, S. Stevens, and H. Wackler, 'Informedia digital video library' Commun. ACM, vol. 38, no. 4, pp. 57-58, April 1995.

7. B. Shahraray, 'Scene change detection and content-based sampling of video sequences' Proc. SPIE 2419, Digital Video Compression: Algorithms and Technologies, pp. 2-13, February 1995.

8. S. F. Chang, W. Chen, H. Meng, H. Sundaram, and D. Zhong, 'An automated content-based video search system using visual cues' ACM Multimedia Conference, November 1997, Seattle, WA. also Columbia University/CTR Technical Report, CTR-TR #478-97-12.

9. L. Bottou, P. Haffner, P. G. Howard, P. Simard, Y. Bengio, and Y. LeCun, 'High quality document image compression with DjVu', Journal of Electronic Imaging, 7(3), pp. 410-428, July 1998.

10. L. Bottou, P. G. Howard, and Y. Bengio, 'The Z-coder Adaptive Bilinear Coder' Proc. IEEE Data Compression Conference, pp. 13-22, Snowbird, UT, 1998.

11. B. Shahraray, and D. C. Gibbon, 'Automatic generation of pictorial transcripts of video programs' Proc. SPIE 2417, Multimedia Computing and Networking, pp. 512-518, February 1995.

12. M. R. Civanlar, G. L. Cash, R. Kollarits, B. Paul, C. Swain, B. G. Haskell, and D. Kapilow, 'VideoTalks – A Comprehensive Multimedia Conferencing System' to be published in IEEE Trans. on Multimedia.

Progressive Content-based Retrieval of Image Databases through the Internet

Lawrence D. Bergman, Vittorio Castelli, Chung-Shang Li, John R. Smith, Alexander Thomasian

IBM T.J. Watson Research Center, 30 Saw Mill River Rd., Hawthorne, NY 10532

Abstract. Content-based search has attracted the interest of numerous researchers as a promising paradigm for retrieving information from digital libraries. The *content* of an image or video segment can be specified at least at three different levels of abstraction, namely, *pixel* level, *feature* level, and *semantic* level.

In this paper, we describe an architecture and the initial implementation of a progressive framework to store and retrieve images (and, in particular, scientific image data) from a collection of archives, using the Internet as the communication backbone. The described framework allows search constraints to be specified at one or more abstraction levels. Content is described in terms of simple and composite objects: a simple object is a connected region which is homogeneous with respect to some specific characteristics (for instance, spectral reflectance, texture etc.), while a composite object is a collection of simple objects (or simpler composites) related by a set of spatial and/or temporal relations, which can be either sharp or fuzzy and deterministic or probabilistic. The framework is extensible in that the user interface allows the user to specify new types of simple objects and to construct composite objects using pre-existing or newly defined objects.

The proposed architecture is scalable thanks to a progressive framework, which relies on algorithms for efficiently managing collections of simple and composite objects, on high-dimensional indexing schemes, and on a paradigm for extracting information from the raw images, that combines the processing operators with the data representation. We are in the process of extending our framework to diverse application areas, including petroleum engineering, environmental epidemiology, forestry management and precision agriculture.

1 Introduction

As the cost of fast storage media and of powerful processors decreases and the infrastructure allowing users to connect to the internet becomes more pervasive, publicly accessible large digital libraries containing videos, images, audio and text data are being developed at an increasing pace. Such databases address the information needs of different communities, from the public at large, now able to access videos, photographic images, audio and text on line, to very specialized users, from both the scientific community and the humanities.

The key to making these archives useful lies in providing a content-based search capability. The exponential growth of digital imagery precludes manual labeling of all but a handful of the images that are produced. Search engines that can use information within the images themselves, either by searching based on similarity ("find images that look like this") or by automatically deducing content, is critical.

Most existing systems that support content-based search rely on low-level image features such as shape, color histogram, and texture to perform image or video indexing. Prominent examples for photographic images are IBM QBIC [12], VisualSeek from Columbia University [15], the MIT PhotoBook [13], and the Multimedia DataBlade from Informix/Virage [2]. Content-based search techniques have also been applied to medical images [8] and video clips [19, 17, 1].

Although some of these systems provide tools for specifying several of these features to be combined within a search, there has been little focus on a general, extensible framework for supporting search at a variety of levels of abstraction. Unified systems that allow the user to ask either, "show me all images with this color" (use of low-level features) or "show me all images containing forests" (use of high-level semantics) and to combine these different forms are yet to be developed. Current systems also tend to have limited extensibility of the query and search facilities provided.

An additional requirement that is central to image databases is the ability to handle large data volumes. The amount of data stored in digital libraries is growing at an exponential rate, and the currently existing storage and retrieval infrastructures are not adequately scalable. Thus, the need for new methods and infrastructures for storing, automatically indexing, and efficiently retrieving the information from the repositories arises naturally.

The SPIRE system, presented here, attempts to address these various needs. We have developed a progressive framework that allows indexing and retrieval of images and videos at multiple levels of abstraction, with a particular emphasis on efficiency. The framework provides for extensibility of the interface and the search engine, and enables search and retrieval over the internet.

2 A Scenario

We have recently constructed an application for content-based retrieval from archives of petroleum well-bore imagery using our framework. We will describe the application and the use of the retrieval tool as an introduction to the types of retrieval capabilities that our system supports.

Data are collected from an oil well by lowering a package of instruments ("logging tools") to the bottom of the well, and slowly pulling them back to

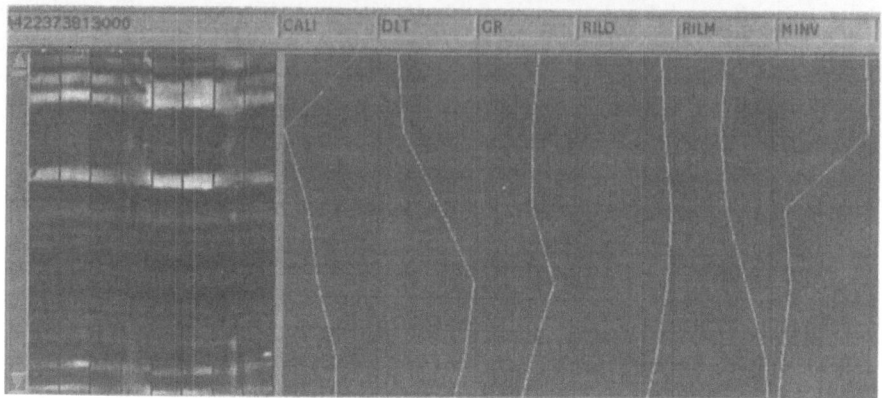

Fig. 1. Section of FMI and 6 single channel data from the Boonesville dataset.

the surface. While the package is being pulled to the surface, the instruments are measuring various physical properties of the rocks surrounding the borehole. The physical properties which are measured include electrical resistivity, sonic velocity, and natural and induced radioactivity. Different instruments can measure the physical properties to different depths into the rock formation. Most measurements are 'single channel': typically every 6 inches a single measurement is made of the surrounding rock by a given instrument. These measurements are transmitted to the surface via the telemetry cable holding the package of instruments.

Other instruments are more sophisticated. For example, the Formation Micro-scanner Imager (FMI) has four arms, each with two 'pads' which press against the surrounding borehole walls. The pads on the FMI have a very high density of electrodes which can detect subtle variations in electrical resistivity. With this tool, instead of one measurement being made every 6 inches, 192 measurements are made every 0.1 inches. The result of measuring 1000 feet of a borehole with this instrument is therefore an image 192 pixels wide, and 120,000 pixels high.

Figure 1 shows an interval in a borehole from the Boonesville, TX area which has been 'logged' with 6 single channel instruments sampling every 6 inches, and an FMI image sampled every 0.1 inches. This section of the borehole consists of alternating intervals of sandstone, siltstone, and shale.

Geologists studying well data are interested in identifying strata of particular types and/or particular characteristics. For example, it might be interesting to find all the coarsely bedded sandstone, or all the finely laminated shale intervals.

Bulk lithologies (rock types) can generally be distinguished using the single channel measurements. For example, sandstone is characterized by low gamma ray, while shale is characterized by high Gamma Ray. Electrical resistivity can be used to distinguish sandstone whose pores are filled with oil

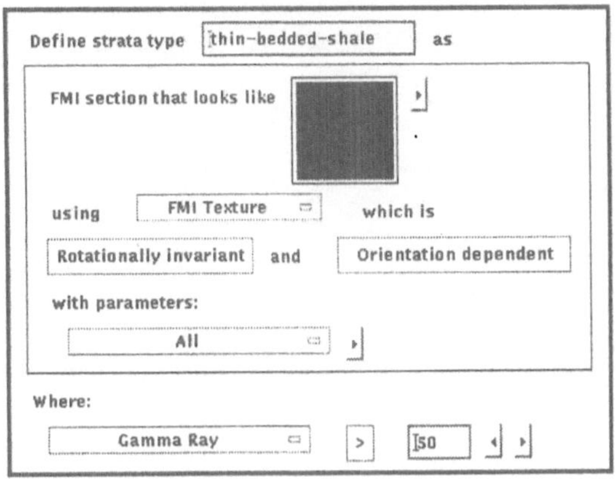

Fig. 2. Definition of a new search object type ("thin-bedded-shale") through a positive texture example and a constraint on a log measurement, using the query specification interface.

versus water. Once these gross lithologies have been identified, the higher resolution FMI images can be used to identify fine-scaled features within the rock. These fine-scale features can give the geologist clues as to the environment in which the strata were originally deposited: e.g., river, beach, desert.

Our application allows the geologist to identify stratigraphic intervals based on example strata extracted from FMI images, as well as constraining the search by specifying constraints on the accompanying single channel data. For example, shale might be defined as those regions with a gamma ray value greater than 50, while sandstone would have a value less than 50. Figure 2 shows an example of this strata definition which was constructed with our drag-and-drop query builder. A section of the FMI image has been selected that contains a sample of the desired strata. This sample image will be used to define a texture to be matched in any target images. The strata definition also contains constraints on the single channel measurements. In this example, we have specified that our searched lithology will be restricted to shale bearing intervals by requiring that the gamma ray value be greater than 50.

Figure 3 shows the results of using this strata definition to identify similar intervals in a single 1000 foot set of well data. It can be seen that a number of similar stratigraphic intervals have been identified that match the search criteria.

Often a geologist wishes to identify a set of lithologies which, when they occur together and in particular order, will characterize a geologic feature, e.g., a river delta is often characterized by a sand sequence coarsening in the upward direction, abruptly capped by shale. Using our system, the geologist can define a composite object by specifying a set of constraints between

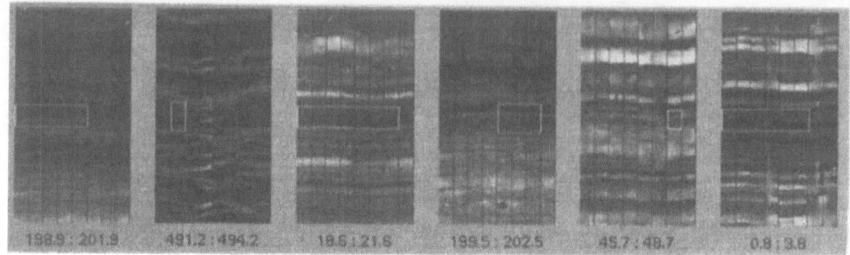

Fig. 3. Results of searching for "thin-bedded-shale" (as defined in Figure 2) over a 1000-foot section of the Boonesville dataset.

multiple component objects (potentially of different types). For example, the user might construct a query of the form "Define composite object called 'delta lobe', consisting of thinly laminated siltstone, overlain within 10 feet by coarsely grained sandstone, which in turn is abruptly overlain by shale". Such a query would be composed using the drag-and-drop query builder, with results returned as in the previous example.

3 System Architecture

The SPIRE system provides a client-server framework for specifying, processing, and fulfilling queries over the Internet. Figure 4 shows the high-level architecture of the system.

The client for a typical application constructed using SPIRE consists of a pair of browser applets. These applets, which are built in Java and run under a standard Internet browser, such as Netscape Communicator or Internet Explorer, consist of an image navigation component, and a query construction component. The navigation component is used for displaying imagery and performing simple image manipulations such as zoom and region selection and maintains an image history. For many applications, the navigation browser also displays sets of thumbnail images returned from a query, which may be individually selected for more detailed examination.

The query browser consists of a drag-and-drop interface, known as DanDE [3] (short for "Drag-and-drop English"). This interface provides for syntax-driven query construction and supports the inclusion of multimedia data types, such as images. Queries are constructed by dropping sub-phrases, selected from menus, into query phrase templates, with automatic enforcement of syntax by the interface. Query phrases may be wholly or partially reused, and can dynamically reconfigure based on database contents, or user actions.

The server for a SPIRE application consists of a search engine with an associated data repository. The search engine has access to a wide variety of search modules loosely packaged into a "data mining library". These modules include data access, filtering, classification, clustering, segmentation, and a

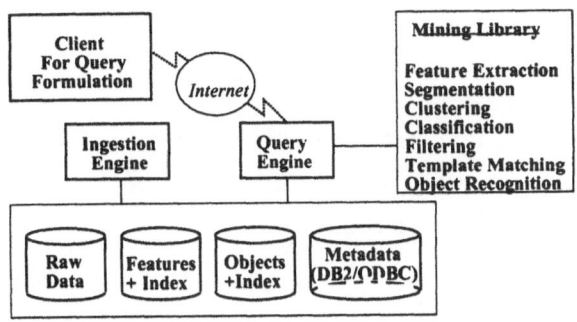

Fig. 4. High-Level architecture of the system.

variety of other image-processing operations. The data repository for SPIRE includes a relational database, used to store meta-data (we are using DB/2 for our current implementation), raw image data, and extracts from the data including features and object definitions (more on this in Section 3.2).

The SPIRE architecture relies on four important organizational principles. The first is an object-based model for query formulation, the second is the notion of a progressive framework, the third is iterative refinement, the fourth is extensibility. Each will be addressed in turn, and this section will be concluded by a discussion of scalability.

3.1 Object Model

Queries are formulated in terms of objects. Objects are of two forms: simple objects and composite (or compound) objects.

Simple objects are defined as connected regions within an image that are homogeneous with respect to some image features and that meet given constraints. In the first example in the scenario (Section 2) we defined a simple object that is homogenous with respect to image texture and where the texture matches that of the given sample (the texture extraction and matching algorithms will be described in Section 3.3). Furthermore this object is required to meet an additional constraint, in this case values in an associated set of data parameters are required to meet a threshold criteria.

Composite objects are aggregations of simple objects or other composite object types. Composite object relations are specified as sets of constraints, either on individual components or between groups of components.

Supported composite object constraints include spatial relationships (e.g., "distance from", "near", "between") and temporal relationships ("precedes", "follows"). Furthermore, the system can process both sharp constraints (such as "within 5 kilometers") and fuzzy constraints (for instance, "near").

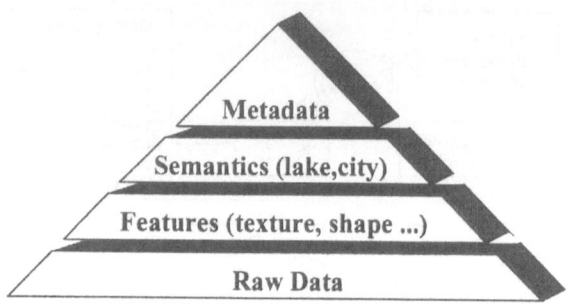

Fig. 5. The Information Pyramid.

3.2 Progressive Framework

An image database must provide query capability that can be used to retrieve results from large archives of information. Furthermore, queries should be expressible in terms of concepts that are relevant to the end-user, not in terms of low-level image processing operations. To fulfill both of these requirements, we introduce a concept that we call the "information pyramid" or InfoPyramid. The InfoPyramid is a two-layer structure designed to facilitate both query specification and processing.

The first layer of the InfoPyramid is an abstraction pyramid. At the bottom of the abstraction pyramid, diagrammed in Figure 5, is the raw data - pixels, in the case of images; data points for the wire-line logs described in our sample scenario; etc. Certain query operations, for example template matching [9], operate on the pixels directly.

The next level of the pyramid consists of features that are extracted from the raw data. Features, such as texture, local color histogram, or land-use category reduce the volume of information by compressing and abstracting the data. Features are typically computed from contiguous fixed-size regions (texture, or local color histogram), or by compressing multispectral information at each pixel (land-use category).

Texture, for example, consists of a number of measures (e.g., average gray-level, entropy, fractal dimension) extracted from fixed-sized rectangular regions computed using a moving-window scheme with a fixed number of pixels between window positions. When processing satellite imagery, we typically extract between 9 and 21 feature parameters from each 32 by 32 pixel region, with a 4 pixel interval between overlapping regions. Land-use categories are assigned by classifying the pixel's RGB or spectral values (using CART [4], neural nets [6], nearest neighbor classifiers [7] or any of a number of other classification techniques) into one of a small number of predefined categories.

Features can be further abstracted by segmenting the image and assigning a semantic label to each contiguous region. For example, we can aggregate land-use categories, and obtain regions that are labeled, "forest", "water", etc. Each semantic object so obtained is stored with boundary information and some associated attributes, such as size and centroid.

Meta-data, information about the raw data such as acquisition date, coverage region, etc., forms the most compact representation, and is represented as the tip of the information pyramid.

The abstraction pyramid provides a framework for query specification. The user may choose to request information by specifying an object type ("Find all forests"), by specifying a feature type ("Find all regions that have texture similar to this example"), or by specifying a pixel-level operation ("Find all images that contain sets of pixels that are a match to this example"). The net effect is a trade off between ease-of-use and specificity. Queries that are formulated using high-level abstraction are easy to express, but give the user little control. Specifications at lower levels of the pyramid enhance the ability of the user to formulate precise and specific queries for object types that were not anticipated by the system designer, but at the expense of a more complicated formulation that may require some expert knowledge of the types of matching operations being applied.

The second layer of the InfoPyramid is a multi-resolution pyramid, provided for any or all of the segments of abstraction layer. Within our system, we rely on the multi-resolution pyramid resulting from a wavelet decomposition of the data [11]. By using a wavelet-based image and signal compression scheme, and choosing appropriately the number of iterations of the wavelet transform, we obtain all the useful levels of the multi-resolution pyramid as a by-product of the data decompression operation. Both the abstraction pyramid and the resolution pyramids provide a framework for accelerating search processing. This will be described in the section on scalability below.

3.3 Iterative Query Refinement

Extremely important in any content-based search facility is the ability to refine a query based on a set of return results. In a system that supports example-based specification, inclusion of new results in the object definition is key to this. Our interface provides for adding and deleting both positive and negative examples.

Positive and negative examples are used to construct a similarity score function mapping from the feature space to the interval $[0, 1]$. From each example, we retrieve the set of texture vectors extracted from windows contained in the example spatial region, we construct their empirical distribution, and compute the empirical variances of the individual texture features. These variances are then used to construct a quadratic distance from the centroid (empirical mean) of the distribution. Distances from centroids are converted into degrees of membership to the example using a typical trapezoidal membership function. Positive and negative examples are used to train the parameters of the membership functions. To score a feature vector, the system first computes the membership degrees to the positive examples $(p_1, \ldots, p_k)$

and to the negative examples $(n_1, \ldots, n_h)$, then composes them through fuzzy logic connectors, as

$$score = (p_1 \text{ OR } \ldots \text{ OR } p_k) \text{ AND NOT } (n_1 \text{ OR } \ldots \text{ OR } n_h),$$

which, using the standard form of the fuzzy connectors, becomes

$$score = \min[\max(p_1, \ldots, p_k), \min(1 - n_1, \ldots, 1 - n_h)].$$

The DanDE interface provides the user with an iterative capability. Parts of query phrases can be reused via drag-and-drop operations, and query results can be imported into the interface, where they become a library a reusable components.

3.4 Extensibility

Extensibility takes a number of forms in our system. Elements that may be extended include the data model, the image processing library, the feature repository and the user-interface definition.

Data model. Although schemas are currently hard-coded on a per-scenario basis, we are currently working on providing schema definitions in a configuration file. The server contains a object type that we call a "data container" that provides support for general-purpose hierarchical data models. Data containers contain metadata, encoded as name-value pairs, a set of child data containers, and a data cache. Data containers implement the progressive framework described earlier, by responding to messages that initiate search on raw data, on features, or on semantic objects. Using existing methods, this class provides for a wide variety of search scenarios, and can be readily extended to add new search types.

Image processing library. The image processing library consists of a set of C++ modules with a predefined interface, which are dynamically invoked from a scripting language used to provide the high-level glue for server-side components. New image processing operators are readily incorporated into the existing framework by wrapping them in a thin interface layer, and adding the routine name to an internal lookup table.

Feature repository. The set of features that are defined to the system can be extended in several ways. We pre-extract features as part of the data ingestion process, and store them in our data repository. The set of features can be readily extended by adding code to the preprocessing module. Features are named (e.g., "texture") allowing a generic data access module to access them, and pass them to image processing modules. New image processing modules designed for particular types of features can be added as described above.

New feature types can also be defined at run time by the user. We provide a facility within our user interface for defining new types of features that are to be computed from the raw data. Feature definitions are scripted using a set of image processing operators including filtering, thresholding, classification, and arithmetic operations. These features are computed on a per-pixel basis and segmentation performed automatically to create simple objects comprised of homogenous regions (with respect to the values of these new features). Although we have not currently provided for applying run-time feature definitions during the preprocessing phase of data ingestion, such an enhancement would be a straight-forward addition to our existing framework. **User interface definition.** Our drag-and-drop interface, known as DanDE (short for "Drag-and-drop English") provides for syntax-driven query construction. The syntax and layout of DanDE interfaces are specified via a configuration file. This file contains a Backus-normal form (BNF) specification of the query language. The BNF contains embedded directives that specify the types of widgets to be used to represent the syntax, as well as information strings that are to be sent to the server when each phrase is submitted. Furthermore, certain forms of dynamic behavior can be specified. For example, typing into a type-in area can result in adding a menu item to a pull-down menu at another place in the query phrase. The interface definition can also specify Java methods to be invoked on objects contained within a global dictionary, either when a phrase is submitted or based on other actions such as selecting an in-phrase menu item. Finally, a Java API is provided which allows a developer to dynamically add information including image clips to a DanDE interface.

3.5 Scalability

There are a number of factors to be considered in creating a scalable architecture for image access. The primary ones that we consider here are: image storage/retrieval, image processing operators, indexing, and query planning.

Image search and retrieval is facilitated by storing information in a multiresolution format. In addition to a traditional wavelet representation, we also support a novel adaptive multi-resolution storage format known as the "SF-graph" [14, 16]. This format allows for rapid retrieval of image data at a variety of resolutions, with storage optimization based on usage patterns. The SG-graph further provides for progressive transmission of imagery with client-side caching to minimize the update cost when zooming or panning.

We capitalize on the multiresolution pyramids and abstraction pyramids used to store image information to provide speed-ups of image processing operations. Operations such as classification, texture matching, and template matching can be greatly enhanced by strategies that employ the progressive framework. An excellent example is our progressive classification scheme [5]. The progressive classifier operates on multiple levels of a multiresolution image pyramid. A progressive classifier is trained to distinguish between pixels that are homogenous at the next highest resolution (e.g., all 4 pixels at level N+1 that are combined to form the pixel at level N are of type "forest") as opposed to those that are inhomogeneous. Given the ability to so distinguish

534

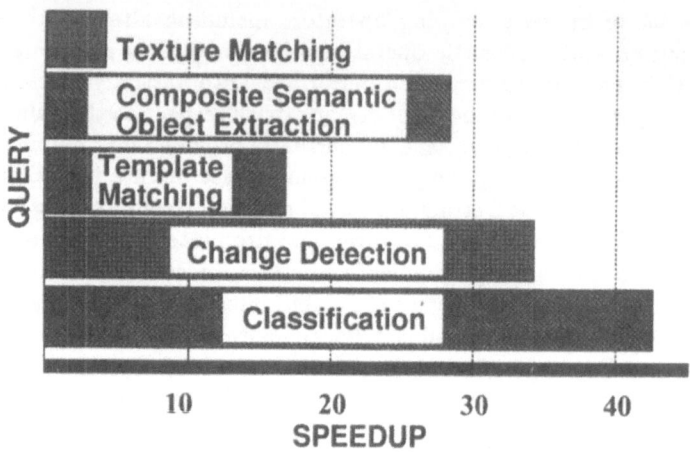

Fig. 6. Query speedup resulting from operating on compressed data.

at all levels of the pyramid, we can start at the top with a very high-level, low volume representation, and only where pixels are mixed, drop to a lower level for analysis. By performing such a search operation recursively, we can examine far fewer data items than would be required by a naive search at full resolution. In fact, the progressive classifier performs about 30 times faster than the full-resolution implementation on benchmark datasets.

Figure 6 shows speedups that we have obtained for several search operators using this type of scheme. Note that the numbers presented are multipliers, not percentages.

Similarity search requires extracting a representative set of features values from an example and comparing with features extracted from the images in the archive to be searched. The search process can be extremely time-consuming if a linear scan is performed. Furthermore, the feature vectors being searched are often of high dimensionality (we use up to a 21-dimensional vector for texture), causing the well-known "curse of dimensionality" to make traditional indexing schemes, such as R-trees almost useless. We have developed a recursive dimensionality reduction algorithm that we call "RCSVD" (Recursively Clustered Singular Value Decomposition) [18]. This algorithm alternates clustering of the features vectors, and singular value decomposition of each cluster, to reduce the dimensionality of the search space, until appropriate termination conditions are reached. The resulting index, which allows approximate searches, displays good precision-recall performances even for large compaction of the search space.

Although providing for both simple and composite objects allows construction of extremely powerful queries and gives SPIRE a great deal of flexibility, the processing of such queries poses problems for scalability. Conceptually, all candidate components of a composite must be joined, e.g., if the composite consists of type "A" and type "B", we must form the join of A and B and then evaluate constraints. Unfortunately, joins grow exponentially with the number of components, and polynomially with the number of candidate objects. Our solution is a sequential processing algorithm that we call "SPROC" [10]. SPROC consists of three steps. The first is to impose an ordering on the constraints to be evaluated based on whether the objects and relations to be evaluated have fuzzy or sharp, probabilistic or deterministic connotations. The second step divides the candidate set into fixed size blocks which are organized into a tree. The third step uses a linear programming technique to select candidate blocks for evaluation of the composite. The objects within the block are not instanced until the block is selected (since instancing the object may involve costly evaluation of similarity metrics), and the algorithm is designed to minimize the number of blocks that will be selected.

4 Future Work

A number of new directions are currently being pursued. The first is to have a deeper, more robust notion of time in our system. Although images may carry timestamps, and we support the simple notions of "before" and "after", we are not able to adequately handle queries that specify duration and evolution of phenomenon. We are currently exploring the notion of treating objects as 3-D entities, with time as one of the axes. Pre-extracted quantities might include time-boundaries for pre-defined phenomenon, and frame-frame attributes extracted from the objects. A variety of techniques for indexing such objects is currently under investigation.

Another topic that we are exploring is to how incorporate models into our framework. Such models might take a number of forms including physical models of phenomenon such as the rotation of the sun for processing solar imagery, and statistical models that allow us to relate phenomenon such as rainfall levels and vegetation greenness. We are extending the framework to allow easy specification and incorporation of models that use a variety of datatypes with seamless integration into our query facility. The goal is to be able to pose queries of the form, "Find all areas that are at high risk from Hantavirus", where an underlying specification states that risk is to be assessed at rural habitations, determined using areas of high vegetation surrounding dwellings which, in turn, are specified as bare areas larger than a specified size and within a specified distance of roadways.

An important related topic is how to expand our framework to accommodate a number of data sources and types, including time-series, 3-dimensional (and higher) data sets, vector data (such as definitions of roads or rivers) and point-source data. We are exploring a object-oriented framework that will integrate with our data containers, and allow for ready extension of the set of system-provided datatypes and accompanying search operators.

5 Conclusions

SPIRE provides a highly extensible, scalable architecture for querying image databases. Our query and retrieval strategy, based on a progressive framework, has allowed us to develop applications in a number of areas and using a variety of datatypes including color imagery, multi-spectral satellite imagery, and non-optical imagery.

The framework provides for ready extension of the query interface, of the search engine, and of the repository of data and information extracted from the data.

The key lessons for the image database system builder from SPIRE are: provide a variety of abstraction levels both for the system user and for the internal search engine; use these abstraction and resolution levels to optimize search engine performance; provide a rich set of operators including data access, data manipulation, image processing, and display that can be neatly packaged for the novice user or provided as tools to the expert; and provide a great deal of extensibility as part of the system framework.

ACKNOWLEDGMENTS

We wish to acknowledge our colleagues at Schumberger-Doll Research, Ian Bryant, and Peter Tilke. Peter, in particular, rewrote the scenario section to make it "more geological and oily" (his words). Also, we wish to acknowledge Anil Achuthan, at IBM T.J. Watson, for the help in implementing the system. This research was funded in part by NASA/CAN contract no. NCC5-101.

References

1. F. Arman, A. Hsu, and M.Y. Chiu. Image processing on compressed data for large video database. In *Proc. ACM Multimedia 93*, pages 267–272, 1993.
2. J. R. Bach, C. Fuller, A. Gupta, A. Hampapur, B. Horowitz, R. Humphrey, and R. Jain. The virage image search engine: An open framework for image image management. In *Proc. SPIE - Int. Soc. Opt. Eng.*, volume 2670, Storage and Retrieval for Still Image and Video Databases, pages 76–87, 1996.
3. L. Bergman, J. Schoudt, V. Castelli, L. Knapp, and C.-S. Li. Asimm: A framework for automated synthesis of query interfaces for multimedia databases. In *Proc. SPIE - Int. Soc. Opt. Eng.*, volume 3229, pages 264–275, 1997.
4. Leo Breiman, Jerome H. Friedman, Richard A. Olshen, and Charles J. Stone. *Classification and Regression Trees*. Wadsworth & Brooks/Cole, 1984.
5. Vittorio Castelli, Chung-Sheng Li, John J. Turek, and Ioannis Kontoyiannis. Progressive classlification in the compressed domain for large EOS satellite databases. In *Proc. of 1996 IEEE Intern. Conf. Acoust. Speech Signal Proc.*, volume 4, pages 2201–2204, May 1996.

6. Vladimir S. Cherkassky, J. H. Friedman, and Harry Wechsler. *From statistics to neural networks : theory and pattern recognition applications.* Springer-Verlag, 1993.

7. V. Dasarathy, Belur, editor. *Nearest Neighbor Pattern Classification Techniques.* IEEE Computer Society, 1991.

8. T. Y. Hou, P. Liu, A. Hsu, and M. Y. Chiu. Medical image retrieval by spatial feature. In *Proc. IEEE Intern. Conf. System, Man, and Cybernetics*, pages 1364–1369, 1992.

9. C.-S. Li, J. J. Turek, and E. Feig. Progressive template matching for content-based retrieval in earch observing satellite image databases. In *Proc. SPIE Photonic East - Int. Soc. Opt. Eng.*, volume 2606, pages 134–44, November 1995.

10. Chung-Sheng Li, Vittorio Castelli, Lawrence D. Bergman, and John R. Smith. Sproc: Fast algorithm for sequential processing of composite objects retrieval from large image/video archives. In *Proc. SPIE Photonic West - Int. Soc. Opt. Eng.*, San Jose, CA, Jan 24-30 1998.

11. Stephane G. Mallat. A theory for multiresolution signal decomposition: the wavelet representation. *IEEE Trans, Pattern Anal. Mach. Intell.*, 11(7):674–693, July 1989.

12. W. Niblack, R. Barber, W. Equitz, M. Flickner, E. Glasman, D. Petkovic, P. Yanker, C. Faloutsos, and G. Taubin. The QBIC project: Querying images by content using color texture, and shape. In *Proc. SPIE - Int. Soc. Opt. Eng.*, volume 1908, Storage Retrieval for Image and Video Databases, pages 173–187, 1993.

13. A. Pentland, R.W. Picard, and S. Sclaroff. Photobook: Tools for content-based manipulation of image databases. In *Proc. SPIE - Int. Soc. Opt. Eng.*, volume 2185, Storage and Retrieval for Image and Video Databases, pages 34–47, February 1994.

14. J.R. Smith and S.-F. Chang. Frequency and spatially adaptive wavelet packets. In *Proc. of 1995 IEEE Intern. Conf. Acoust. Speech Signal Proc.*, pages 2233–2236, Detroit, MI, May 1995.

15. J.R. Smith and S.-F. Chang. Visualseek: A fully automated content-based image query system. In *Proc. 4th ACM Multimedia Conf.*, pages 87–98, Boston, MA, USA, 18-22 Nov 1996.

16. J.R. Smith and S.-F. Chang. Joint adaptive space and frequency basis selection. In *Proc. IEEE Int. Conf. Image Processing*, Santa Barbara, CA, October 1997.

17. S. W. Smoliar and H. Zhang. Content based video indexing and retrieval. *IEEE Multimedia*, 1(2):62–72, Summer 1994.

18. A. Thomasian, V. Castelli, and C.-S. Li. Clustering and singular value decomposition for approximate indexing in high dimensional spaces. In *Proc. of Seventh International Conference on Information and Knowledge Management (CIKm'98)*, to appear, 1998.

19. H. Zhang and S. W. Smoliar. Developing power tools for video indexing and retrieval. In *Proc. SPIE - Int. Soc. Opt. Eng.*, volume 2185, Storage and Retrieval for Image and Video Databases, pages 140–149, Feb 1994.

Content-Based Searching and Filtering of Visual Information

Shih-Fu Chang
Columbia University, New York, U.S.A.

Abstract

Searching and filtering images and videos from large visual information sources calls for efficient algorithms and tools with new functionalities. Content-based tools have proven promising in recent research and development, as they provide powerful techniques that complement the traditional, text-based approach. This paper includes a brief survey of current work in this field, discusses promising directions of research based on experience in developing large-scale prototype systems, and finally describes our views towards the emerging multimedia content description standard, MPEG-7.

1. Introduction

Digital images and video are becoming an integral part of human communications. The ease of creating and capturing digital imagery has enabled its proliferation, making our interaction with today's on-line information sources largely a "visual" one. But, with the growing amount of digital visual media, how do we effectively search for visual information? With the growing myriad of image/video sources available from broadcast media and the Internet, how do we efficient filter through the information source and find content we need?

Digital imagery is a rich and subjective source of information. Different people extract different meanings from an image or video. Human response varies over time and in different viewing contexts. A picture has meaning at multiple levels, e.g., generic description, analysis, and subjective interpretation. Visual information is represented in multiple forms and often has accompanying data streams in different media like captions, audio, and graphics. These factors make searching and filtering of visual information an exciting and challenging topic.

In this paper, we will present a survey of existing work and open issues of content-based visual search and filtering. Several existing directions will be discussed, including those using textual annotations, speech transcription, key content extraction, storyboard summarization, human perception, user relevance feedback, subject categorization, and similarity matching.

We will also discuss novel functionalities of new tools and systems, based on our experience in developing several large-scale prototype systems, such as an automatic object-based video search engine, VideoQ, Web-based image collection and cataloguing systems, WebSEEk and MetaSEEk, and an interactive framework for learning high-level personalized query templates, Semantic Visual Template. We will discuss the lessons learned in developing systems using large collections (more than 650,000 images and 10 hours of video) and in working with actual users in multimedia applications.

2. Content-Based Visual Search

There has been a lot of research undertaken by academia and industry in developing search tools for images and videos. Representative image search systems include QBIC of IBM [1], Virage [2], Photobook of MIT [3], VisualSEEk of Columbia University [4], and the image search system of University of Washington [4]. Representative video search systems include Informedia of CMU [6], Pictorial Transcript of AT&T [7], the News Retrieval System of IBM [8], VideoQ of Columbia University [10], and Princeton's Digital Video Library [9].

Several commercial systems have been developed for image/video content management. In addition to those mentioned above, commercial systems also include Excalibur [11], Islip [12], and Magnifi [13]. It's expected that many more commercial systems will be launched in this area soon.

Most of the systems use traditional text-based search tools with some enhancement of unsophisticated content-based search tools. The application and user model for this type of content-based search is not fully understood yet. A few successful applications using content-based image search tools can be found for image collections in art [14] and other special domains, including remote sensing [15], stamps, clothing *etc.*

Challenges remain in applying the above content-based image search tools to meet real user needs. In the following sections, we discuss several promising research directions and some of our current research results in each area.

2.1 Automatic image /video analysis

Although today's computer vision systems cannot recognize high-level objects in unconstrained images, we are finding that low-level visual features can be used to partially characterize image content. These features also provide a potential basis for abstraction of the image semantic content. The extraction of local object features (such as color, texture, face, contour, motion) and spatial/temporal relationships among the objects is being achieved with success. For image indexing, the automated segmentation of image/video objects does not need to accurately identify real world objects contained in the imagery. Our goal is to extract the "salient" visual features and index them with efficient data structures for fast and powerful querying. A minimum set of initial user input and some domain knowledge may further improve the accuracy of the extraction process.

VisualSEEk/VideoQ - Combined Spatial-Temporal/Feature Search

One of our prototype systems for image search, VisualSEEk, uses a unique framework for searching for and comparing images by the spatial arrangement of regions or objects [4]. In a VisualSEEk query, objects or regions are assigned by the user. These are given properties of spatial location, size and visual features, such as color. The VisualSEEk system finds the images that best match the query. VisualSEEk uses fully automatic tools for region/feature extraction and indexing. VisualSEEk also resolves spatial relationships, which allows the user to position objects relative to each other in a query.

Examples of query include "find images including a blue region on top and a wide green open region in the bottom (looking for images with blue sky and open grass field)," and "use this spatial pattern of red, white, blue colors to find images containing American Flags."

Another prototype search engine, VideoQ, substantially expands the framework to video search by adding advanced techniques for automatic video segmentation and spatio-temporal matching [10]. With a large video collection (10 hours video, 2,000 video clips, 20,000 video objects), VideoQ is the first fully automatic video search engine that supports object-based spatial-temporal matching (query by objects that the video contains and motions they make). Figure 1 shows a snapshot of a video search specifying spatio-temporal features of a video object and contextual keywords

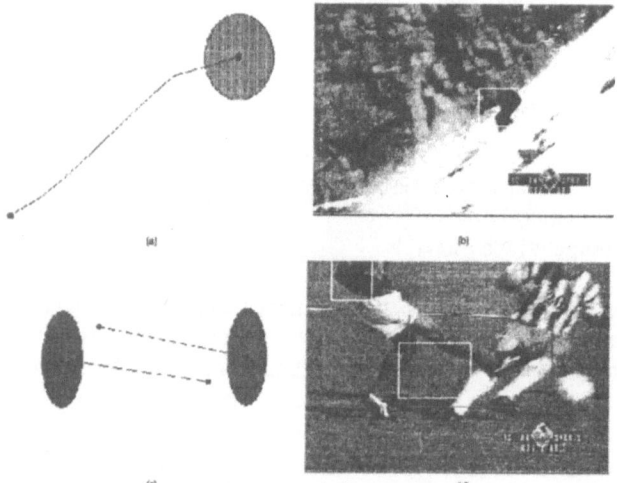

Figure 1: Video search examples using combination of video object features and keywords. By matching the color, texture, shape, and motion of the objects (e.g., query sketches shown in (a) and (c)), video clips containing most similar objects and scenes (e.g., (b) downhill skiing and (d) soccer players) are returned. (Video courtesy of Actions, Sports, and Adventure, Inc.)

In addition to query by sketch, the user can browse the video shots or search video by text. The video shots are cataloged into a subject taxonomy, which the user can easily navigate. When textual annotations are available, the user can also perform text search of keywords, or a more sophisticated search using textual and visual features combined.

2.2 Synergy between compression and functionalities

It's impossible to anticipate users' needs completely at the feature extraction and indexing stage. The ideal solution is that images and videos in their compressed formats are represented in a way that is amenable to dynamic feature extraction. Today's compression standards (such as JPEG, MPEG-1, MPEG-2), are not suited to this need. The objective in the design of these compression standards was to reduce bandwidth and increase subjective quality. Although many interesting analysis and manipulation tasks can still be achieved in today's compression formats (e.g., as in our MPEG video editing system, WebClip [20]), the potential functionalities of the images were not considered. However, recent trends in compression, such as MPEG-4 and object-based video, have shown interest and promise in this direction. The goal is to develop a system in which the video objects are extracted, then encoded, transmitted, manipulated, and indexed flexibly with efficient adaptation to users' preferences and system conditions.

2.3 Learning semantic-level search models through user interaction

The above systems have shown great promise for a new generation of powerful image/video search tools. However, a gap remains between the feature matching techniques provided by these systems and the ultimate search tools that general users desire — semantic-level interaction (e.g., find images of "sunset" or video scenes containing "crowds on the beach"). With traditional mindset, most people prefer direct interaction at the semantic level when searching information in a large archive. It's difficult for users to translate the desired semantic search targets to low-level graphic sketches like the types utilized in VideoQ.

Currently, we are developing a new framework, *Semantic Templates*, that learns representative visual/audio templates through interactive queries. The foundation is a two-way interactive learning system that allows users to define personal semantic concepts, and the system to generate an optimal set of visual templates for each concept. Figure 2 shows representative visual templates for the concepts "sunset" and "high jump". Once the system generates the optimal set of visual templates for each concept, users may search for possible images/videos of the concept by matching the templates to images/videos. The goal is to develop a large library of semantic audio/visual templates that can be used to describe various scenes and compose sophisticated multimedia queries.

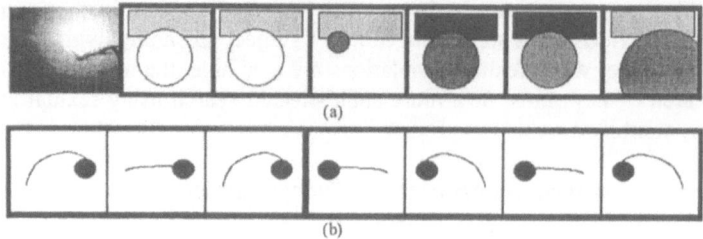

(a)

(b)

Figure 2: A new visual search paradigm using semantic visual templates. Image icons shown are subsets of optimal templates for each concept ((a) "sunset" (b) "high jumper") generated through the two-way interactive system.

2.4 Integration of multimedia features

Exploring the association of visual features with other multimedia features, such as text, speech, and audio, provides another potentially fruitful direction. Our experience indicates that, compared to video, it is more difficult to characterize the visual content of still images. Video often has text transcripts and audio that may also be analyzed, indexed, and searched. Also, images on the World Wide Web typically have text associated with them. In this domain, the use of all potential multimedia features enhances image retrieval performance.

WebSEEk – Content-based Search/Cataloguing of Web Images/Videos

WebSEEk [18] is a content-based image and video catalog and search tool for the World Wide Web. WebSEEK collects the images and videos using several autonomous Web agents that automatically analyze, index, and assign the images and videos to subject classes. The system utilizes combined text and visual information synergistically to provide for cataloging and searching for the images and videos. The complete system possesses several unique functionalities, including searching using image content-based techniques, query modification using content-based relevance feedback, automated collection of visual information, image and video subject classification and navigation, and flexible search results lists. During a three-month period in 1996, the system catalogued over 650,000 images and 10,000 videos from the Web.

One lesson we learned in developing and testing WebSEEk with public users is that mapping images/videos to a clearly organized taxonomy proves to be a very powerful way of managing large image/video archives. Users usually prefer to start image queries with simple ways of navigation among subject classes. However, automatic image classification using image features alone is known to be error prone. To solve this challenging problem, we are currently studying innovative methods of integrating multimedia features for image classification. Examples of such multimedia features include the image features (including color, texture, spatial layout, and motions), audio features (including physical features and psychoaccoustic features), and textual features (including associated documents, transcripts, and captions). Specifically, we are studying the effectiveness of various machine learning methods (such as decision trees, rule based systems, and Bayes networks) in image classification using a single medium (such as visual, audio, or textual alone) as well as combinations of multimedia.

2.5 Interoperable Content Description and MEG-7

Unlike text documents, visual materials do not share consistent formats, indexes, or meta-data. This issue was brought to the fore in our earlier experiments with meta search engines [19]. Dozens of formats are used for representing images and videos on the Web. Many different techniques are used for implementing content-based searching in the visual search systems. Furthermore, there is no standard for inter-operability between the search systems. For example, even at the semantic-level, the ontologies are custom-developed in the target search systems.

The issues of heterogeneity must be solved in order to improve inter-operability. Standardization of the representation of meta-data should provide a uniform method for labeling visual information at the semantic-level. Such standardization will enhance the inter-operability between different search systems and improve the effectiveness of the individual systems. Similar to the development of standards for meta-data for text documents, other efforts such as that of Dublin Core have been made to extend the meta-data schemes to images. Also, a standard taxonomy for graphic materials has been proposed by the Library of Congress. In addition, the audio-visual research community has started to investigate the development of a standard for the description of multimedia information in MPEG-7 [22].

MPEG-7 aims at standardization of content description of multimedia data, with a focus on audio-visual data. The standard may include specifications of features descriptors, description schemes, and description definition languages. The objectives are to facilitate content-focused applications like multimedia searching, filtering, media summarization and conversion. To achieve maximal interoperability and flexibility, we are using self-describing schemes (e.g., XML of W3C) to create interoperable content descriptions of images/videos. Such self-describing schemes will enable mediation among heterogeneous search engines, content searching using private examples, distributed intelligent content discovery, and dynamic content filtering at clients. One of our goals is to demonstrate the feasibility and efficiency of such self-describing schemes by incorporating such indexing schemes into our existing image/video search engine prototypes mentioned in earlier sections.

544

4. Conclusions

In this paper, we present a brief survey of the state of the art, and then discuss several promising directions of research, including automatic feature extraction, compressed-domain processing, learning high-level semantic models, multimedia integration, and using self-describing schemes for interoperable content description.

Development of efficient and effective multimedia search tools and systems remains an exciting and challenging task. Related efforts are emerging in various fields and application domains (e.g., Digital Library Initiatives and the MPEG-7 standardization effort). Collaboration among cross-disciplinary research groups, system developers, and content providers is one of key factors in achieving success in these efforts.

Acknowledgements

Many students have contributed to the work described here, in particular, Mandis Beigi, Ana B. Benitez, William Chen, Horace Meng, Seungyup Paek, John R. Smith, Hari Sundaram, Di Zhong. The work has been supported in part by the National Science Foundation, Office of Naval Research, industrial sponsors of Columbia's ADVENT project, and a UPP Faculty Development Award from IBM.

References

1   M. Flickner, H. Sawhney, W. Niblack, J. Ashley, Q. Huang, B. Dom, M. Gorkani, J. Hafner, D. Lee, D. Petkovic, D. Steele, and P. Yanker, "Query by Image and Video Content: The QBIC System," *IEEE Computer Magazine*, Sep. 1995, Vol.28, No.9, pp. 23-32.

2   J. R. Bach, C. Fuller, A. Gupta, A. Hampapur, B. Horowitz, R. Humphrey, R.C. Jain and C. Shu, "Virage image search engine: an open framework for image management", Symposium on Electronic Imaging: Science and Technology – Storage & Retrieval for Image and Video Databases IV, IS&T/SPIE, Feb. 1996.

3   A. Pentland, R.W. Picard, and S. Sclaroff, "Photobook: Tools for Content-Based Manipulation of Image Databases," Proc. Storage and Retrieval for Image and Video Databases II, Vol. 2185, SPIE, Bellingham, Wash., 1994, pp. 34-47.

4   J. R. Smith and S.-F. Chang, "VisualSEEk: A Fully Automated Content-Based Image Query System," ACM Multimedia Conference, Boston, MA, Nov. 1996. (ftp://ftp.ctr.columbia.edu/CTR-Research/advent/public/papers/96/smith96f.ps) (demo http://www.ctr.columbia.edu/VisualSEEk)

5   C. E. Jacobs, A. Finkelstein, and D. H. Salesin, "Fast multiresolution image querying," ACM SIGRAPH, pp. 277-286, August, 1995.

6   A. G. Hauptmann and M. Smith, "Text, Speech and Vision for Video Segmentation: The Informedia Project," AAAI Fall Symposium, Computational Models for Integrating Language and Vision, Boston, November 10-12, 1995.

7   B. Shahraray and D. C. Gibbon, "Automatic Generation of Pictorial Transcript of Video Programs," SPIE Vol. 2417, pp.512-518, 1995.

8   R. Mohan, "Text Based Search of TV News Stories," SPIE Photonics East Intern. Conf. on Digital Image Storage & Archiving System, Boston, MA, Nov. 1996.

9   W. Wolf, B. Liu, A. Wolfe, M. Yeung, B.-L. Yeo, and D. Markham, "Video as scholarly material in the digital library," Chapter 1 in Advances in Digital Libraries '95, Springer-Verlag, 1995.

10   S.-F. Chang, W. Chen, H.J. Meng, H. Sundaram, and D. Zhong, "VideoQ-An Automatic Content-Based Video Search System Using Visual Cues," ACM Multimedia 1997, Seattle, WA, November 1997. (ftp://ftp.ctr.columbia.edu/CTR-Research/advent/public/public/chang/videoq/acmpaper.ps.gz) (Demo http://www.ctr.columbia.edu/videoq).

11   Excaliber System: http://www.excalib.com/rev2/products/vrw/vrw.html.

12   Islip Media: http://www.islip.com

13   Magnifi: http://www.magnifi.com

14   Arthur: Art Media and Text Hub and Retrieval System,The Getty Information Institute, http://www.ahip.getty.edu/arthur/

15   C.-S. Li, L. Bergman, S. Carty, V. Castelli, S. Hutchins, L. Knapp, I. Kontoyiannis, J. Robinson, R. Ryniker, J. Shoudt, B. Skelly, J. Turek, "Scalable Content-Based Retrieval from Distributed Image/Video Databases," submitted to IEEE Trans. Circuits and Systems for Video Technology, 1997.

16   S.-F. Chang, J. R. Smith, M. Beigi, and A. Benitez, "Visual Information Retrieval from Large Distributed On-Line Repositories," Communications of ACM, Special Issue on Visual Information Management, Vol. 40 No. 12, pp. 63-71, Dec. 1997.

17   S.-F. Chang, J. R. Smith, H. J. Meng, H. Wang, and D. Zhong, "Finding Images/Video in Large Archives- Columbia's Content-Based Visual Query Projects," CNRI Digital Library Magazine (on-line), Feb. 1997. (http://www.dlib.org/dlib/february97/columbia/02chang.html)

18   J. R. Smith and S.-F. Chang, "Visually Searching the Web for Content," IEEE Multimedia Magazine, Summer, Vol. 4 No. 3, pp.12-20, 1997. (ftp://ftp.ctr.columbia.edu/CTR-Research/advent/public/papers/96/smith96e.ps) (demo http://www.ctr.columbia.edu/webseek)

19   M. Beigi, A. Benitez, and S.-F. Chang, "MetaSEEk: A Content-Based Meta Search Engine for Images," SPIE Conference on Storage and Retrieval for Image and Video Database, San Jose, Feb. 1998. (demo and document: http://www.ctr.columbia.edu/metaseek)

20   J. Meng and S.-F. Chang, "CVEPS: A Compressed Video Editing and Parsing System," ACM Multimedia Conference, Boston, MA, Nov. 1996. Demo: http://www.ctr.columbia.edu/WebClip ftp://ftp.ctr.columbia.edu/CTR-Research/advent/public/papers/96/meng96c.ps

21   S.-F. Chang, W. Chen, and H. Sundaram, "Semantic Visual Templates - Linking Visual Features to Semantics," IEEE Intern. Conference on Image Processing, October 1998.

22   ISO/IEC JTC1/SC29/WG11, MPEG-7: Context and Objectives (v. 5) Oct. 1997.

Internet Video-Streaming

Bernd Girod, Niko Färber, and Klaus Stuhlmüller

Telecommunications Laboratory
University of Erlangen-Nuremberg
Cauerst. 7, 91058 Erlangen, Germany
{girod,faerber,stuhl}@nt.e-technik.uni-erlangen.de

Abstract. The heterogeneous structure of the Internet is a great obstacle for establishing real-time video services. Scalable video codecs, generating bit-streams decodable at different rates, have been proposed to address the heterogeneity problem. In this paper, we will review standard-compliant and non-compliant codec architectures for Internet video-on-demand. For compression based on the H.263 standard, we have developed a compatible architecture that allows to switch between preencoded bit-streams of different bit-rates. This architecture provides excellent streaming performance for point-to-point communication scenarios that offer a low delay feedback channel. Then we present a non-compliant fully scalable video codec based on a spatio-temporal resolution pyramid that is available as a Netscape plugin for the World Wide Web. This approach can encode embedded lower bit-rate layers at the same overall bit-rate as needed by H.263 single-layer coding and can also support multicasting. Finally, we show how scalable video coding can be efficiently combined with a retransmission protocol achieving graceful degradation and investigate the streaming performance of the above architectures.

1 Introduction

Information systems and distributed applications for the Internet show a growing demand for real-time audiovisual services. Real-time audio services can already be found on numerous World Wide Web (WWW) servers, as well as within the Internet MBone tools. For video services the heterogeneous structure of the Internet is still a great obstacle. Reservation protocols alone, e.g. RSVP (resource reservation protocol [14]), cannot solve the heterogeneity problem. Consider an Internet video server with sequences encoded at a predetermined bit-rate. If the bit-rate is sufficiently low, nearly everybody connected to the Internet can view the videos in real-time at the cost of very poor quality. But those who can connect to the server at higher speeds do not benefit from higher bandwidth because the sequences have been encoded at a much lower bit-rate. On the other hand, if the sequences were encoded at higher bit-rates, those connecting to the server at lower speeds cannot view the videos in real-time any longer. Storing bit-streams for different bit-rates is possible but requires certain precautions to allow switching between different bit-rates during a transmission.

Scalable video coding has been proposed to address the above problem. A scalable video coder produces a bit-stream, decodable at different bit-rates. It allows computation time and memory-limited decoding on less powerful hardware platforms [5], and it can substantially improve the quality and acceptance of video services on the Internet. For video servers, sequences need only to be encoded once. Depending on the available bit-rate, the server selects more or fewer bits from the bit-stream to be sent to the receiver. For broadcast applications a layered multicast scheme can be used. The different bit-rates can be distributed into several layered multicast groups. Picture quality increases with the number of received multicast groups. For videoconferences with several participants, the RSVP protocol can be used for bandwidth allocation over each intermediate link. If necessary, the scalable bit-stream can be pruned easily at routers such that the allocated bandwidth can be utilized in the most efficient way.

In this paper we propose two different approaches for Internet Video-Streaming and compare their performance with H.263 simulcast. The comparison is based on a simulation scenario with two quality levels which is described in the following section. After comparing the resulting bit-rates for a given target PSNR we finally provide simulation results for the transmission over a packet network using a retransmission protocol.

2 Simulation scenario

The simulation results in this paper were obtained under the assumption of an Internet video-on-demand situation. We require that the server can provide two different picture qualities such that in case of network congestion the higher quality bit-stream can be discarded. The low quality layer is encoded in QCIF resolution at 15 fps, while the high quality layer is encoded in CIF resolution at 30 fps. For both layers the target PSNR is 34 dB which is achieved with only little variations for all investigated approaches. The sequences are encoded with fixed quantizers and no rate control. We require random access and resynchronization points at intervals of 32/30 seconds. A switch from high to low quality shall always be possible, while a switch from low to high quality shall be supported every 8/30 seconds. All results in this paper are based on the MPEG-4 test sequence SEAN, frames 1-96.

3 H.263 simulcast

We compare the proposed architectures with simulcasting of two H.263 encoded bit-streams. Since we desire a maximum delay of 8/30 seconds for switching from lower to higher quality, an I-frame has to be inserted in the CIF layer every 8 frames. To allow an immediate switch from the CIF to the QCIF layer, we assume that the QCIF H.263 decoder keeps decoding the low quality layer even if the higher quality layer is used for display. Note that this requires two running decoders at the receiver. H.263 coding simulations were carried out with mode decision according to TMN-5. No special coding options were used. Fig. 1 illustrates the simulation model used for simulcast and the dependencies of frames caused by prediction. The frame types are 'I' (intra) and 'Pt' (motion compensated, temporal prediction). Note that the illustrated sequence is continued periodically, such that the next frame in the QCIF layer is again of type 'I'.

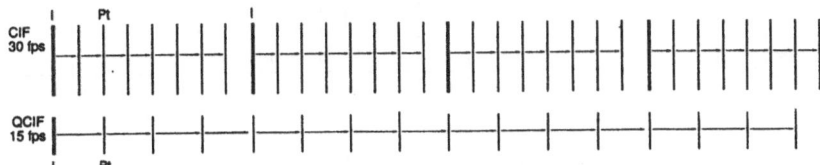

Fig. 1. Frame dependencies for simulcast.

Fig. 2 shows the transmitted bits during a switch from low to high quality. It can be seen that I-frames in the CIF layer require about 8 times as many bits as Pt-frames and therefore significantly contribute to the overall bit-rate. Because both bit-streams are encoded independently, no advantage is taken of the fact that the QCIF layer could be used to improve the coding efficiency of the CIF layer. This could be achieved by using interpolated frames from the QCIF layer to spatially predict frames from the CIF layer or by encoding side information more effectively. This approach results in a fully scalable codec as proposed in the next section.

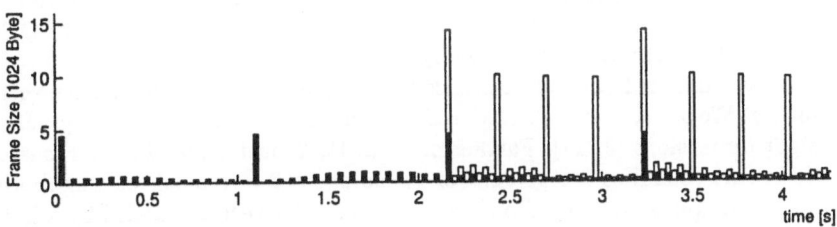

Fig. 2. Simulcast: Transmitted bits during a switch from low to high quality.

4 Scalable coding

Our approach to scalable video coding is based on a spatio-temporal resolution pyramid. This kind of a multiresolution decomposition was first proposed by Uz and Vetterli in 1991 [12, 13]. The frame dependencies are shown in Fig. 3 and are very similar to simulcast except for the two additional frame types 'Pst' and 'Ps' which are described below. As above, the illustrated sequence of frames is continued periodically for longer sequences.

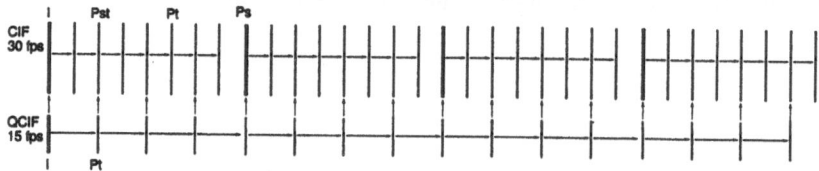

Fig. 3. Frame dependencies for the scalable codec.

A scalable codec superior to simulcasting has to exploit spatio-temporal redundancies of the pyramid decomposition by an efficient compression technique. In our approach we have combined low complexity downsampling and interpolation filters with highly efficient lattice vector quantization. Decoder complexity is sufficiently low to allow software-only implementations on today's PCs and workstations. Encoder complexity is mainly determined by motion estimation.

For I-frames, the original frame is successively filtered and downsampled by a simple averaging filter with coefficients (1 1), separately applied in horizontal and vertical direction. The lowest resolution layer is encoded by a DPCM technique [10]. For all other layers a spatial prediction is formed by interpolating the lower resolution layer by a filter with coefficients (1 3 3 1) again applied horizontally and vertically. Spatially predicted frames are denoted 'Ps' and may use any lower resoltion frame as a reference frame, e.g., a Pt-Frame instead of an I-frame. In the above simulation model Ps-frames are used to allow switching from the QCIF to the CIF layer and are inserted every 8 frames in the CIF layer.

The residual prediction error is encoded using an 8-dimensional lattice vector quantizer. We selected the E_8-lattice since it is the best known lattice quantizer in eight dimensions [3, 11]. Furthermore, in [1, 2] fast algorithms for nearest lattice point rounding are given. For encoding, a 2×4 block of neighboring samples is mapped into an 8-dimensional vector. This vector is scaled by a factor $1/s, s \gg 1$ where s corresponds to the quantizer step size in one-dimensional quantization. By varying s, the bit-rate of the quantizer can be controlled. The scaled vector, a point in R^8, is rounded to its nearest E_8-lattice point by using the algorithm given in [1]. From the obtained lattice point an index is computed which then is transmitted to the decoder. The decoder can reconstruct the lattice point from the received index either by computation or by a simple table-lookup. By rescaling the reconstructed lattice point with s, the finally reconstructed 2×4 input block is obtained.

Besides I-frames we are also using predicted frames in our coding scheme. Motion-compensated prediction is based on 16×16 blocks and works within each layer similar to motion-compensated hybrid coders [6, 7]. Motion vectors are estimated and coded in a hierarchical way. Motion vectors found for a lower resolution layer are used to predict motion vectors in the next higher layer. Among predicted frames, we distinguish between temporal prediction (frame type 'Pt') and spatial-temporal prediction (frame type 'Pst'). Spatial-temporal prediction is only possible when a lower resolution frame with the same temporal reference is available. Then each block can be predicted either spatially from the interpolated lower resolution frame of the same temporal reference or temporally from the previous frame of the same resolution.

Fig. 4 shows the transmitted bits during a switch from low to high quality. The performance of Pt-frames in the QCIF and CIF layer are very similar to H.263. Interestingly the sizes of Pst-frames and Pt-frames in the CIF layer are almost identical, indicating that only little advantage can be taken from the presence of the QCIF layer. However, a significant gain can be obtained by replacing the I-frames in the CIF layer by spatially predicted Ps-frames. The size of those frames is reduced from approximately 10 KByte to 7.5 KByte respectively.

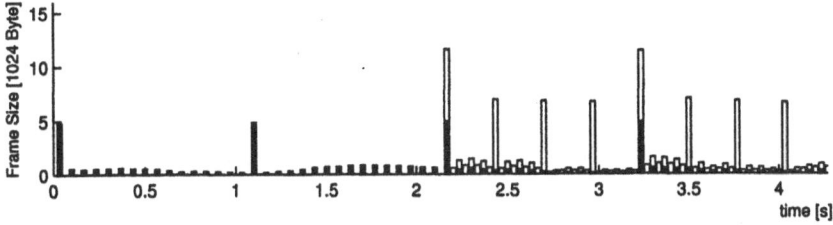

Fig. 4. Scalable codec: Transmitted bits during a switch from low to high quality.

Based on the scalable coder described above, we have implemented a software-only video server application called NetClip [8] and integrated it into the World Wide Web (WWW). The decoder is available as a Netscape plugin for various Unix hardware platforms [9]. The HTTP server machine has on its local disk several video clips which have been compressed by our scalable codec.

5 S-frame architecture

In both of the above approaches the periodic transmission of I-frames (Ps-frames) significantly effect the overall bit-rate for the high quality bit-stream. Also note that I-frames (Ps-frames) do not provide any additional functionality during normal transmission of the CIF layer. In [4] we have therefore proposed a new architecture for the standard compatible storage and transmission of video that avoids the overhead of periodic I-frames (Ps-frames). According to this architecture both layers are encoded independently using only the most effective temporally predicted Pt-frames. A switch to the CIF layer is possible by dynamically assembling the bit-stream at the video server whenever a switch is requested from the client. In between the last frame of the QCIF layer and the first frame of the CIF layer an especially encoded "Switch-frame" (S-frame) is inserted that is also preencoded and stored for each position a switch shall be supported, i.e., every 8 frames. Because this approach is feedback driven, it is only feasible in a point-to-point communication scenario such as Internet video-on-demand. In contrast to the above two approaches it would therefore not be suitable for multicasting or broadcasting. Furthermore it requires a reasonably low delay for the feedback channel to allow a quick response time and interactivity. Under those conditions, however, it provides excellent streaming performance as will be shown below. Fig. 5 shows the frame dependencies where 'S' indicates a switch-frame.

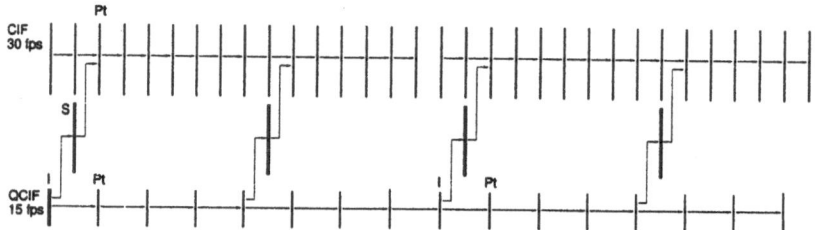

Fig. 5. Frame dependencies for S-frame architecture.

The purpose of S-frames is to resynchronize the frame buffers at the encoder and decoder before any following Pt-frames of the CIF layer are transmitted. Therefore, instead of encoding the original frames, the reconstructed frames from the CIF layer are used for encoding the S-frames. As a reference frame the last reconstructed frame of the QCIF layer is used. With this special encoding strategy, the content of the frame buffer in the CIF layer can be approximated with increasing accuracy by decreasing the quantization parameter Qp. For all simulations in this paper we use Qp=3, which results in a remaining loss of PSNR of less than 0.2 dB compared to the case that the CIF layer is decoded without any switch. After the switch, the loss of picture quality decreases over time.

Note that an S-frame does not require a new frame type but can be encoded as an H.263 compatible Pt-frame. The change in spatial resolution requires the use of the optional coding mode 'Reference Picture Resampling' that is part of a series of new options for H.263 that have recently be adopted by the ITU-T under the working term 'H.263+'.

Because an S-frame has to approximate the reconstructed frame from the CIF layer very closely, it requires a higher bit-rate than an I-frame. On the other hand S-frames are only transmitted during a switch from the QCIF to the CIF layer. Therefore their influence on the average transmitted bit-rate is very low. For storage, however, S-frames have to be considered every 8 frames. Therefore we need to distinguish between the effective bit-rate for storage and transmission. Fig. 6 illustrates that during a switch from low to high quality a high peak in bit-rate is encountered. After the switch, however, the S-frame architecture provides excellent streaming performance.

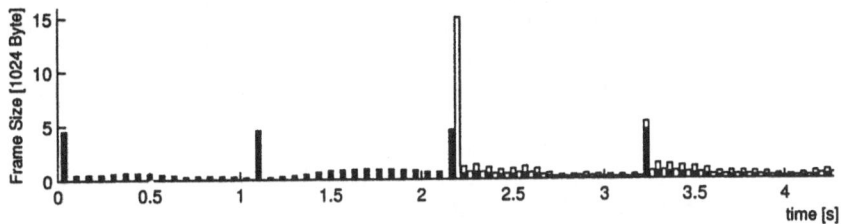

Fig. 6. S-Frame architcture: Transmitted bits during a switch from low to high quality.

6 Comparison of coding performance

Table 1 summarizes the coding performance of the three approaches. Note that the scalable codec can encode both layers at roughly the same rate than required by H.263 for the CIF layer alone. For the S-frame architecture the given rate does not include any S-frames since those are only transmitted during a switch. The rate that is needed to store the bit-streams including all S-frames is provided in brackets. The results indicate that the S-frame architecture is advantageous, if the network bit-rate rather than the storage capacity is the limiting factor. The peak in bit-rate may be a problem for packet networks and requires buffering as described in the following.

	Simulcast	Scalable	S-frame
rate QCIF [kbps]	90.94	88.56	90.94
rate QCIF+CIF [kbps]	527.96	350.72	218.39 (658.52)
PSNR QCIF [dB]	34.2	34.3	34.2
PSNR CIF [dB]	34.3	34.3	33.9

Table 1. Summary of coding performance.

In addition to the above results we have simulated the transmission over a packet network using a retransmission protocol. Though retransmission protocols like TCP are generally considered to be less suitable for real-time data, we will show how scalable video coding can be used to increase the robustness of video streaming significantly.

At the video client a receive buffer of size B is used that is divided into a CIF and QCIF partition of size B_{QCIF} and B_{CIF} respectively. Each partition stores only information of the corresponding layer. The proposed retransmission protocol assures that the QCIF partition is filled completely before any CIF layer information is transmitted. This is achieved by including buffer state information into the acknowledgments that are also used for error control. For simplicity we assume that the server gains instantaneous and complete knowledge of the receive buffer state and avoid detailed implementation issues. If the CIF partition suffers from underflow, the transmission of the CIF layer is interrupted and the decoder resorts to the QCIF layer. If sufficient data rate is available the CIF layer restarts at the next resynchronization point. Similarly, if the QCIF partition underflows, the transmission restarts with the next resynchronization point in the QCIF layer.

The channel model used is an "ON-OFF" channel, that is periodically switching from the ON state to the OFF state and remains in each state for a fixed duration of T_{on} and T_{off} respectively. During the ON state a data rate of R_{on} is available, such that the average data rate of the channel is $R_c = T_{on} R_{on}/(T_{on} + T_{off})$. The parameter T_{off} can be interpreted as the duration of network congestion, while R_{on} is the data rate of the weakest link in the connection.

By carefully selecting the size of the partitions, it is possible to achieve a graceful degradation of video quality. This is illustrated by the following example. Assume a total buffer size of $B = 800$ kbit and a partitioning of $B_{QCIF} = B_{CIF} = 400$ kbit. Furthermore assume that $R_{on}T_{on} > B$, such that the buffer is filled completely during an ON state. For a QCIF layer encoded at $R_{QCIF} = 100$ kbit/s and a CIF layer encoded at $R_{CIF} = 400$ kbit/s the CIF layer can be displayed without interruption for $T_{off} < 1$ second while the QCIF layer can be displayed without interruption for $T_{off} < 4$ seconds.

In Fig. 7 the streaming performance of the proposed scalable video architectures is compared when the above retransmission protocol is utilized. The ratio of (completely) received frames and encoded frames is used as a performance measure. The ratio is calculated separately for both layers. A value of less than one indicates, that a certain percentage of frames had to be skipped in the given layer because of buffer underflow. Subjectively, a value of less than 0.8 can already be very annoying. The described ON-OFF channel is used for the simulation. The average channel rate is set to the total video rate of H.263 simulcasting, i.e., $R_c = 528$ kbit/s. The rate during the ON state is set to $R_{on} = 1000$ kbit/s, while T_{off} is used as a parameter. The total buffer size is set to $B = 200$ KByte (1 KByte = 1024 Byte) and the partitioning for each architecture is designed, such that the CIF layer breaks down for $T_{off} > 2$ seconds. Therefore the performance in the CIF layer is very similar for all architectures. The availability of the QCIF layer, however, differs significantly for the three approaches. For H.263 simulcasting, the scalable codec, and the S-frame architecture the QCIF layer breaks down for $T_{off} = 4$, 5, and 12 seconds respectively.

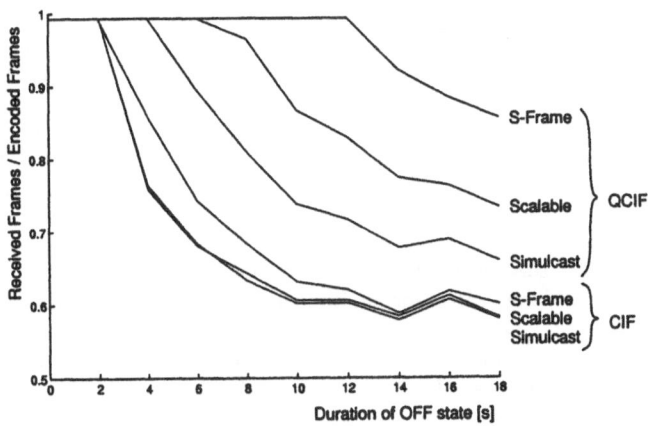

Fig. 7. Comparison of streaming performance for ON-OFF channel.

7 Conclusions

The heterogeneity of the Internet remains a great obstacle for providing reliable video services. Scalable video codecs can substantially improve the quality of video services for many applications. The three approaches to scalable video coding that have been compared in this paper show very similar performance in the QCIF layer where only temporal prediction is used. In the CIF layer the proposed scalable codec performs significantly better than H.263 simulcast. Both layers can be encoded at roughly the same rate as required by H.263 for the CIF layer alone. This gain is mainly achieved by using spatially predicted frames instead of I-frames to allow switching from low to high quality. The third approach is limited to point-to-point communication scenarios that offer a low delay feedback channel. A switch from low to high quality is accomplished by inserting especially encoded switch-frames that cause a high peak in bit-rate. After the switch, however, the approach offers excellent streaming performance. The S-frame architecture is fully standard compatible but requires increased storage capacity. Finally, it is shown how graceful degradation of video quality can be achieved when scalable video is combined with the proposed retransmission protocol. With this approach, the ability to maintain QCIF quality during long periods of network congestion can be improved significantly. For the given simulation scenario, the S-frame architecture can tolerate 3.0 times longer periods of network congestion than H.263 simulcast and 2.4 times longer periods of network congestion than the scalable codec.

References

1. J. H. Conway and N. J. A. Sloane. Fast quantizing and decoding algorithms for lattice quantizers and codes. *IEEE Trans. on Information Theory*, IT-28(2):227–232, Mar. 1982.

2. J. H. Conway and N. J. A. Sloane. A fast encoding method for lattice codes and quantizers. *IEEE Trans. on Information Theory*, IT-29(6):820–824, Nov. 1983.

3. J. H. Conway and N. J. A. Sloane. *Sphere Packings, Lattices and Groups*. Springer, 1988.

4. N. Färber and B. Girod. Robust H.263 compatible video transmission for mobile access to video servers. In *Proc. ICIP'97*, Santa Barbara, USA, Oct. 1997.

5. B. Girod. Scalable video for multimedia systems. *Computers & Graphics*, 17(3):269–276, 1993.

6. B. Girod, U. Horn, and B. Belzer. Scalable video coding with multiscale motion compensation and unequal error protection. In Y. Wang, S. Panwar, S.-P. Kim, and H. L. Bertoni, editors, *Multimedia Communications and Video Coding*, pages 475–482. Plenum Press, New York, Oct. 1996.

7. U. Horn and B. Girod. Performance analysis of multiscale motion compensation techniques in pyramid coders. In *Proc. ICIP'96*, volume III, pages 255–258, 1996.

8. U. Horn, B. Girod, and B. Belzer. Scalable video coding for multimedia applications and robust transmission over wireless channels. In *7th International Workshop on Packet Video*, pages 43–48, Brisbane, Mar. 1996.

9. U. Horn and J. Weigert. Netscape pyramid decoder plugin. www4.informatik.uni-erlangen.de/Projects/netclip/, Mar. 1997.

10. A. N. Netravali and B. G. Haskell. *Digital Pictures*. Plenum Press, New York, 1988.

11. T. Senoo and B. Girod. Vector quantization for entropy coding of image subbands. *IEEE Transactions on Image Processing*, 1(4), Oct. 1993.

12. M.K. Uz, M. Vetterli, and D.J. LeGall. Interpolative multiresolution coding of advanced television with compatible subchannels. *IEEE Trans. on Circuits and Systems for Video Technology*, 1(1):86–99, Mar. 1991.

13. M. Vetterli and K.M. Uz. Multiresolution coding techniques for digital television: A review. *Multidimensional Systems and Signal Processing*, 3:161–187, 1992.

14. L. Zhang, S. Deering, D. Estrin, S. Shenker, and D. Zappala. RSVP: A new resource reservation protocol. *IEEE Network*, 7(5):8–18, Sep 1993.

Software Design and Simulation of a DS/CDMA Multimedia Transmission System for Remote Video – Surveillance Applications

Paolo Piccardo, Carlo S. Regazzoni, Claudio Sacchi, Giorgio Sciani and Andrea Teschioni

University of Genoa
Department of Biophysical and Electronic Engineering (DIBE)
Via all'Opera Pia 11/A 16145 Genoa (Italy)
Phone: + 39 – 10 – 3532792
Fax: + 39 – 10 – 3532134
E – mail: carlo@dibe.unige.it

Abstract

The most recent developments in the research concerning the Video – Surveillance reveal an increasing interest related to the systems operating in remote modality. The applications of such systems generally concern the video – surveillance of unattended environments. A local elaboration system acquires images by video sensors, it processes the images and it sends the elaborated information to a remote control centre. A secure and noise – robust transmission link for the critical information is required. In this work, an innovative software design technique of a DS/CDMA multimedia transmission system is presented, together with the visual results related to the multimedia transmission simulation.

1. Introduction

The remote video – based surveillance of unattended environments (i.e. not surveyed by assigned personnel) is an emerging research and development field of activity in these last years. The tele – presence of the unattended site is provided by a data and image processing system, which acquires images from the surveyed areas by using video sensors (e.g. monochromatic or colour video – cameras). The multimedia data locally elaborated are then transmitted to a remote control centre, which is generally some miles away from the surveyed areas. The received information, incoming from different unattended sites, is processed by a central elaboration system and finally displayed on monitors to a human operator.

A critical aspect in the development of such a system is surely the radio-transmission link between the unattended local and the remote control centre.

In fact, two particular requirements have to be satisfied, i.e. the compression of the transmitted information in order to reduce as much as possible the bandwidth required for the image sequence transmission, jointly with the necessity to protect the information that is sent to a channel where other transmitters are present.

A possible technical solution in order to meet the security requirements concerning the multimedia information transmission is represented by *Spread Spectrum (SS) based communication systems*. It is well – known from the literature [2] that the Spread Spectrum communication techniques have been studied for military and high security applications to fight strong interference and to avoid listening of messages by non authorised people.

SS signals have some properties that make them useful for allowing multiple access to shared transmission means: the access for more users takes place in asynchronous way and sharing the same frequency band. The associated pattern with each signals permits to the receiver to detect the considered message and to reduce other signals to an interference noise. *Code Division Multiple Access* (CDMA) [2] is considered as a valid alternative with respect to the traditional narrowband techniques such as TDMA and FDMA in the future development of wireless [3] and wired (i.e.: cable network) communications [4].

In [1] a multimedia system for the surveillance of unattended railway stations is described. The local data computing systems of the unattended railway stations send multimedia information concerning potentially dangerous situations related to the presence of abandoned objects in the waiting rooms to a remote control centre, by a digital wireless communication system, based on the Direct Sequence Code Division Multiple Access (DS/CDMA) techniques.

In order to verify the feasibility of the digital transmission link adopted in [1], an innovative DS/CDMA multimedia transmission simulator, whose structure is presented in the following, has been developed using the MATLAB® SIMULINK® environment and some results concerning the transmission and the detection of some image data using such simulator are presented. The simulator has been implemented by using SIMULINK® blocks that emulate the functionalities of some hardware commercial components, such as shift – registers, sinusoidal oscillators, filters etc., so that the structure of the simulator is one – to – one linked with the hardware implementation of the real communication system.

The paper is structured as follows: in Section 2 a brief description of the remote video – surveillance system for unattended railway stations is presented, in Section 3 the software structure of the simulator is depicted. Section 4 is devoted to the exposition of some results from simulation. The conclusions of the paper are reported in Section 5.

2. Remote video – based surveillance system for unattended railway stations

The block scheme of the multimedia system for video – surveillance of unattended railway stations considered in this work, widely described in [1], is represented in Fig. 1.

The remote control centre receives data from K (generally $K \leq 10$) unattended railways stations. In every unattended station, a monochromatic TV camera acquires images from the surveyed environment/room. The images are then digitized and processed by a PC – based local image processing system, whose aim is to detect the presence of abandoned objects in the waiting-rooms or other unattended rooms of the station [1]. The results of the elaboration are sent to the remote control centre as soon as an alarm situation related to the presence in the surveyed areas of an abandoned object is revealed by the local image processing system [1]. The human operator in the remote control centre can see from some monitors the reference images of the surveyed locals in the unattended station (i.e.: in the considered application *background images* [1]). When an object, that is present in the scene, is classified as an *abandoned object* [1], an alarm issue is generated by the local image processing system. So the transmission system send immediately an *alert image* containing the detected abandoned objects to the remote centre, together with the numerical data related to their spatial 3D coordinates.

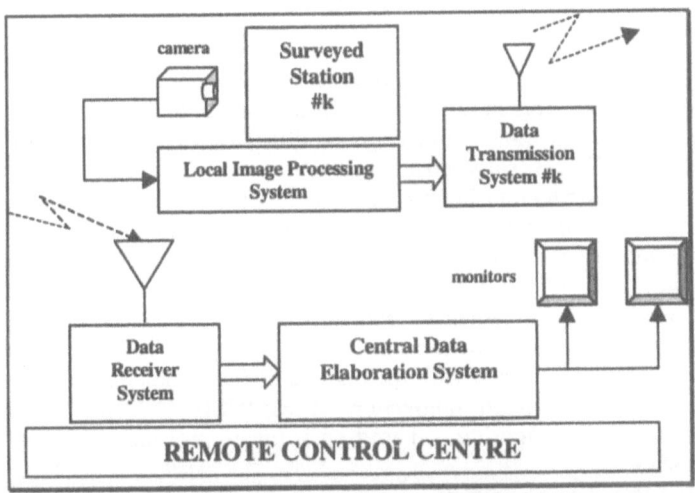

Fig. 1: Unattended railway station surveillance system

The monochromatic alert image is overlapped to the background image, whereas the spatial co – ordinates of the object are displayed on selected areas of the man – machine interface of the system, and a clearly visible graphic alert signal is shown on the monitor [1].
In [1] a *Direct Sequence Code Division multiple Access* (DS/CDMA) communication system has been considered in order to ensure a secure and noise – robust digital radio transmission link between the unattended stations and the remote control centre. The use of the DS/CDMA is almost profitable in the above exposed application, as the transmission from the local stations is completely *asynchronous*. In fact, when an alarm issue is generated, the transmission system is immediately operated and the alert information is sent. It is known that the CDMA techniques allows a good number of users to share the same bandwidth portion without fixed time or frequency slot assignments [2].
As the transmission of the background images and the one of the alert images from the unattended railway station are characterized by different bit – rates, and different quality requirements [1], two separated DS/CDMA transmission

channels have been provided: one for the background image and one for the alert image [1]. The two channels either can belong to the 2.4 GHz. ISM band (2.4 – 2.4835 GHz. [5]), and have the same bandwidth, but they are characterized by different processing gains and different source coding (JPEG compression and FEC coding), in order to provide a more flexible and efficient radio transmission link to the system.

In the next sections the details of the design and the simulation of the multimedia DS/CDMA transmission link between unattended stations and remote control centre, will be considered.

3. Software design of the DS/CDMA multimedia transmission system

In this section an innovative technique of software design and simulation of a DS/CDMA system for multimedia information is presented. The proposed approach exploits the potentialities of the SIMULINK® tool, belonging to the Mathwork's MATLAB® software environment.

SIMULINK is a simulation library whose components, called "blocks", emulate, among the others, the behaviour of commercial hardware components for signal processing and communications, such as: sinusoidal wave generators, analog and digital filters, Gaussian noise generators, etc. The SIMULINK® library blocks (conventionally called "elementary blocks") can be used in order to implement more and more complex blocks in articulated design and simulation structures, that are one – o – one linked with the hardware VLSI implementation of the real systems. This is the main innovation of SIMULINK®, with respect to the traditional simulation algorithms, based on software procedures typically written in C language.

The SIMULINK implementation of a multimedia DS/CDMA system for video – surveillance application, as the one described in [1], is reported in Fig. 2 (number of users $K = 4$):

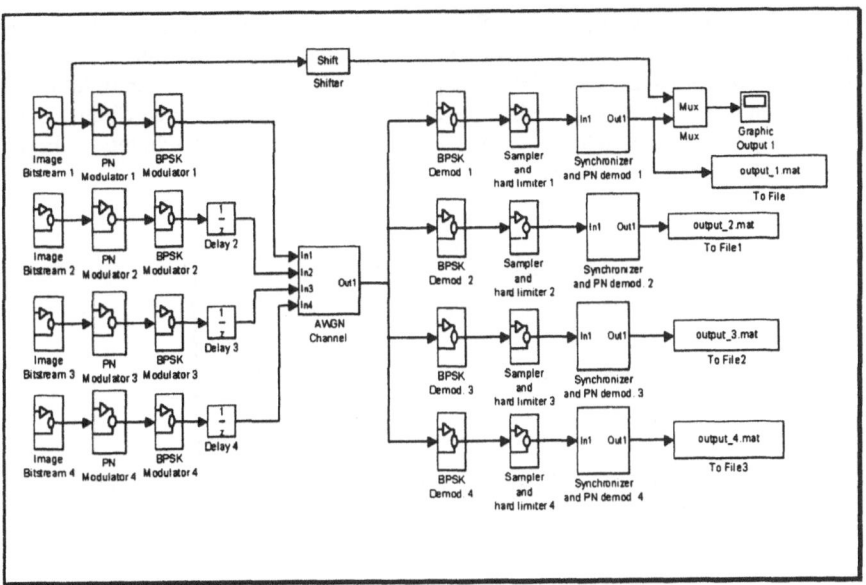

Fig. 2: SIMULINK® software implementation of a four - user DS/CDMA system for multimedia data transmission from unattended railway stations to the remote control centre

It is easy to see that the SIMULINK® implementation of the DS/CDMA system is complete. The block chain starts from the transmitting sources (i.e.: "Image Bitstream" blocks), that generates the JPEG coded stream containing the multimedia information to be transmitted, an ends to a graphic output monitor (i.e.: "Graphic output" block) related to the signal received by the reference user (i.e.: the user #1, for simplicity), that displays in real time the simulation results. The numerical values of the output signals can also be stored in binary files by using the "To file" SIMULINK® sink blocks (In Fig. 2, the output signals of the other users are sent to files, without displaying). It must be said that, at the present moment, the proposed simulation does not comprise the Forward Error Correction (FEC) coding of JPEG bit – stream. Hence, the results of the simulations are to be considered less good than the ones provided by the system completed by the source FEC coding.

It is known that FEC coding is not strictly compulsory in the DS/CDMA systems, but its employment allows to maintain very low the bit – error – rates, by transmitting signals at low power level, that is one of the most required features by the digital communication system engineering. For this reason, the use of the FEC coding in DS/CDMA system is fairly suggested by the recommendation rules concerning the Spread Spectrum communications [7].

The delay blocks placed at the end of the last three transmitters are the *asynchronous transmission delays* of the interfering users τ_k [6] with respect to the reference user (that is assumed not delayed), in order to simulate the behaviour of a real asynchronous Direct Sequence Spread Spectrum multiple access system [6]. The inner structure of each block of the DS/CDMA communication system

562

depicted in Fig. 2 is composed by SIMULINK® elementary blocks, corresponding to a commercial hardware components. In Fig. 3, 4, 5 and 6 the implementations, in terms of SIMULINK® elementary blocks, of the BPSK modulator and demodulator, and the ones of the spreading (PN modulator in the scheme of Fig. 2) de – spreading (Synchronizer and PN demodulator in the scheme of Fig. 2) blocks respectively are reported.

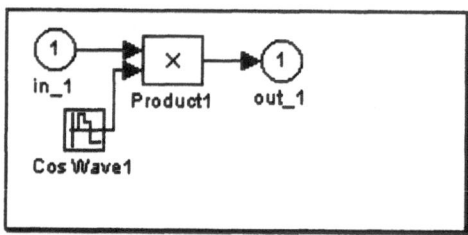

Fig. 3: SIMULINK® implementation of the BPSK modulator block

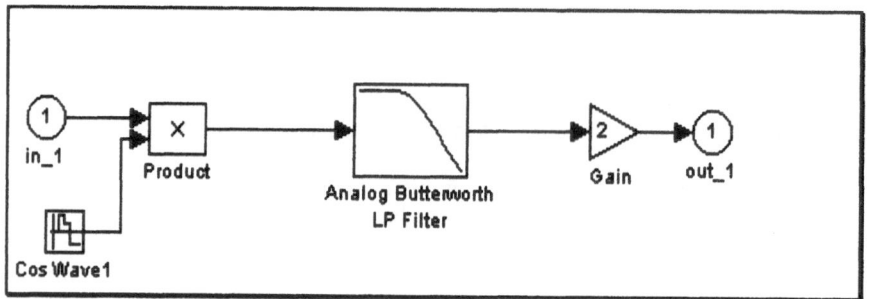

Fig. 4: SIMULINK® implementation of the BPSK demodulator block

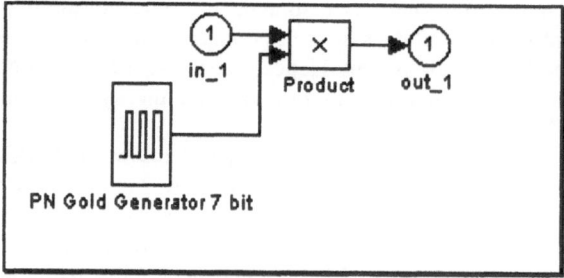

Fig. 5: SIMULINK® implementation of the PN modulator block

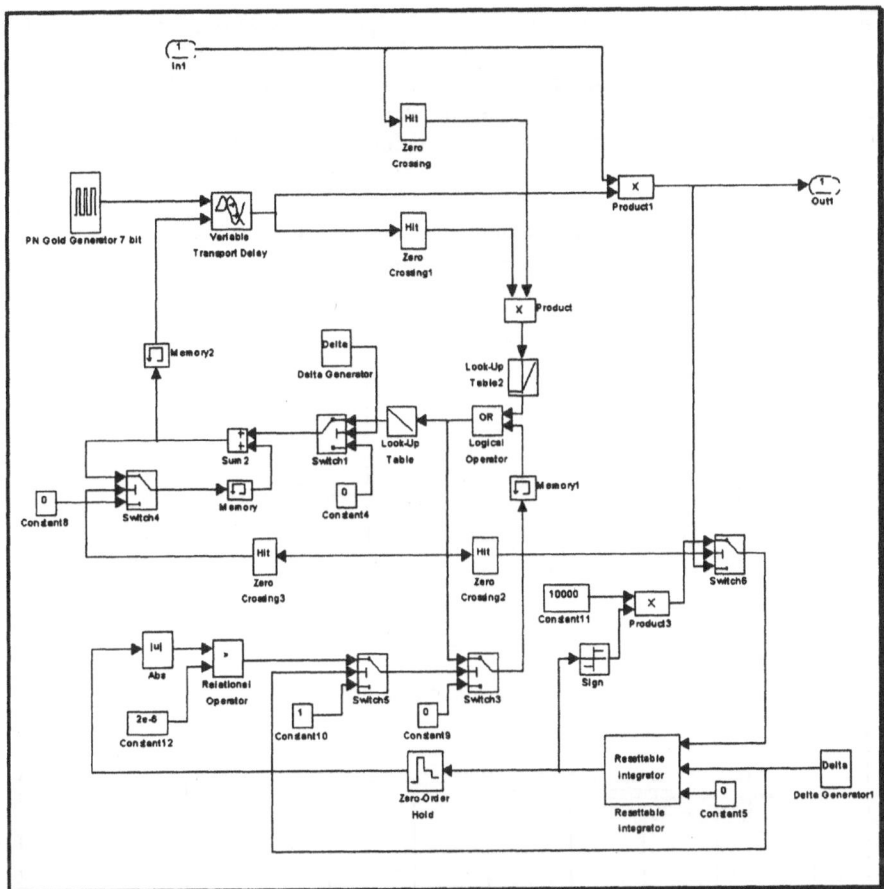

Fig. 6: SIMULINK® implementation of the PN demodulator block

The number of users of the system is configurable with the maximum flexibility, as the structures of the single users' transmitters and receivers are fully replicable. The AWGN channel block adds to the transmitted DS/CDMA signals a Gaussian noise sequence, whose energy level can be set by the SIMULINK® software as a function of the required transmission signal – to – noise ratio.

It is worth of noting that the PN demodulation block has been configured as a complete "de – spreader" device, comprising the Pseudo – Noise (PN) code acquisition and tracking operations. The PN acquisition strategy is based on the *serial search* concept [9]. The replica of the spreading sequence generated by the expected receiver is step-by-step delayed (see the Variable Transport Delay block in Fig. 6) and correlated with the incoming received signal, that is the noise-corrupted and BPSK demodulated digital Pseudo-Noise signal transmitted by the considered user. This signal comes from the sampler and hard limiter block of the SIMULINK® receiver scheme represented in Fig. 2 (the sampling time is equal to the chip-time). The operation ends when the point of correlation between the two

564

signals is found. Only when the received signal is synchronised with the PN sequence, the de-spreading operation starts. The PN demodulator operates a continuous monitoring of the cross-correlation between the received signal and the synchronised PN sequence generated by the receiver, so to reveal an eventual synchronisation loss and to start again a synchronization trial (tracking).

The employed spreading codes are *Gold Sequences* [8] [9], generated by two 7-bit shift register generators (see the "Gen PN" blocks in Fig. 7) (length of the codes N = 127). The SIMULINK® implementation of the binary sequence shift-register generator is shown in Fig. 8 (the polynomial vectors of the generator can be directly set by the SIMULINK® software).

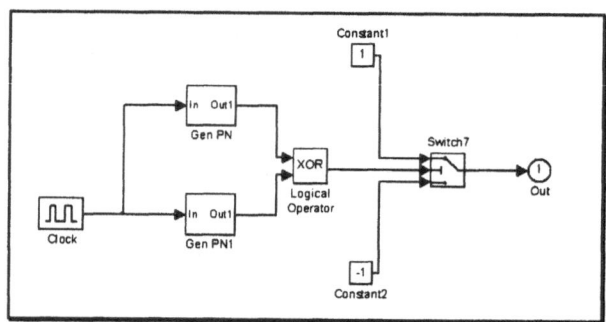

Fig. 7: SIMULINK® implementation of PN Gold code generator

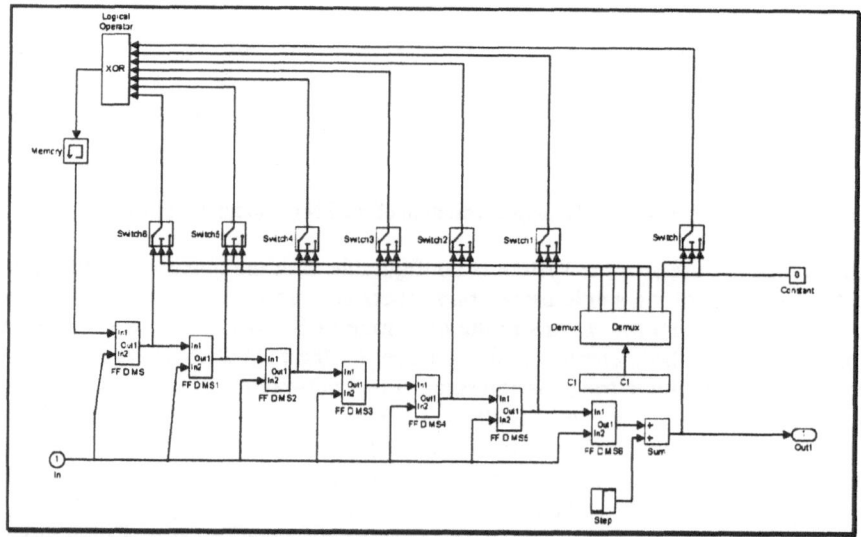

Fig. 8: SIMULINK® implementation of one of the 7-bit shift-register binary sequence generators belonging to the PN Gold code generator ("Gen PN" block)

4.   Simulation results

In order to achieve some numerical results from the simulator of the multimedia DS/CDMA communication system implemented by using SIMULINK® software tools, it has been stated that each users of the system transmits the JPEG coded background image of an unattended railway station premise. The rate of compression of the coded images is $C = 16$. This rate can provided an acceptable quality of the decoded images (about 75 %), considering that the background image transmission has not the same quality constraints of the alert image transmission.

The four transmitted images are shown in Fig. 9 (a, b, c, and d):

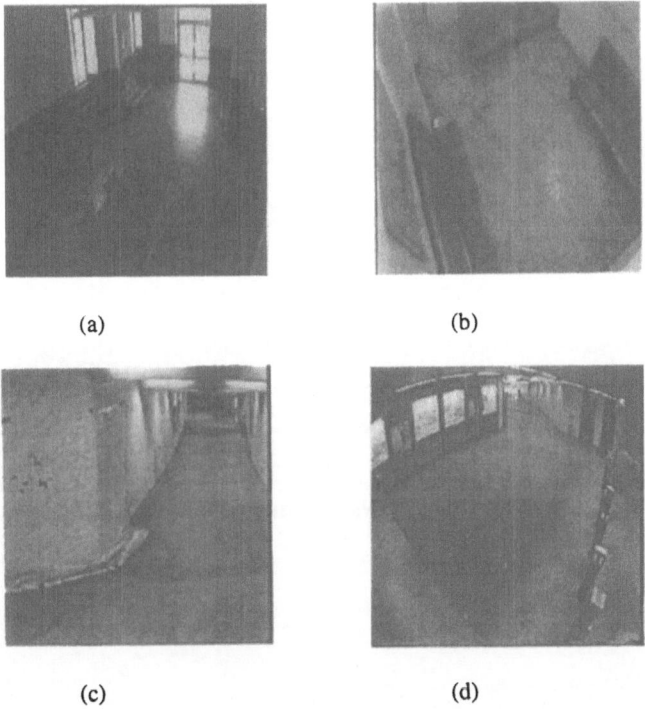

(a)   (b)

(c)   (d)

Fig. 9: background images of the unattended railway station surveyed rooms transmitted to the remote control centre by the user #1 (a), the user #2 (b), the user #3 (c) and the user #4 (d), respectively

The fixed value of processing gain of the DS/CDMA system is $N = 127$. The time of duration of a single bit is assumed as $T_b = 1.2192x10^{-5}$ sec., hence the chip – time is $T_c = T_b/N = 9.6x10^{-8}$ sec., corresponding to a bandwidth of the Direct Sequence Spread Spectrum BPSK modulated signal $B_{ss} \cong 10.5$ MHz. The transmission signal – to – noise ratio of each user has been fixed as $SNR = 5$dB and 10 dB.

Considering these values of the parameters characterizing the performances of a DS/CDMA communication system, the simulation has been started. The graphic output block shown in the SIMULINK scheme of the entire system (Fig. 2) displays on the PC monitor the temporary evolution of the transmitted bit stream and the one of the received bit stream of the reference user (i.e.: the user #1). So it is possible to see, instant by instant, the modulation/demodulation process. In Fig. 10 and 11 the results of the simulation displayed by the graphic output of the system, concerning the first two bits and the first two bytes transmitted by the reference user are shown (signal-to-noise ratio $SNR = 10$ dB)

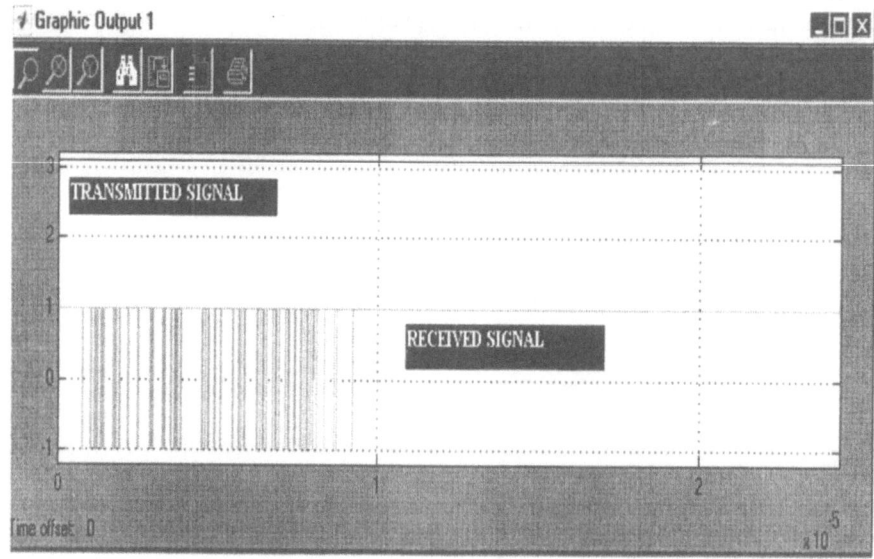

Fig. 10: simulation results related to the first two bits transmitted by the reference user ($SNR = 10$ dB)

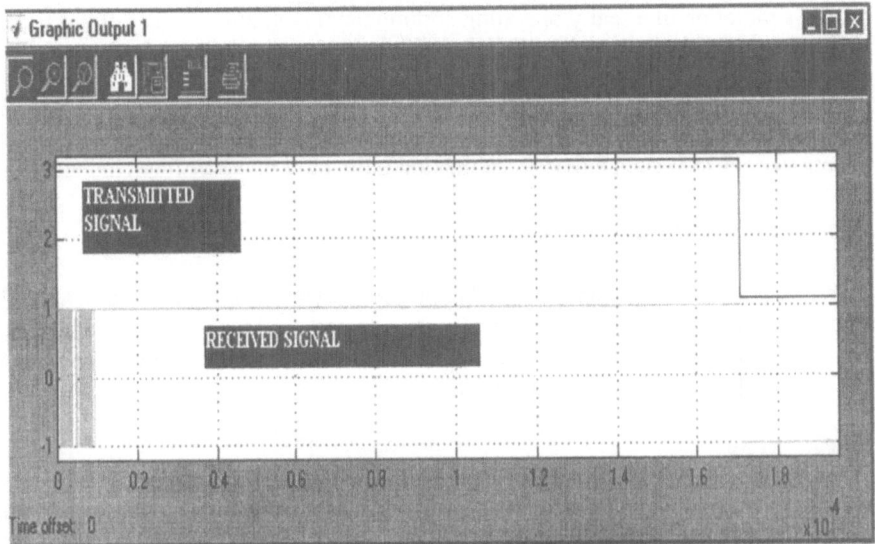

Fig. 11: simulation results related to the first two bytes transmitted by the reference user
($SNR = 10$ dB)

The values of the transmitted and received signal are all binary (i.e.: ± 1, using the notation assigning: "0" $\rightarrow$ "-1" and "1" $\rightarrow$ "+1"). The transmitted signal has been shifted by a constant value (i.e. +2.1) in order to visualise it together with the received signal without overlapping.

Fig. 10 and 11 show clearly the temporary evolution of the Pseudo – Noise initial acquisition process. The failed attempts of synchronization are marked by the vertical lines in the graph of the received signal. These lines represent the chips of the Spread Spectrum received signal, that is not yet synchronized with the PN sequence replica of the receiver. The time of PN acquisition achieved by the simulation is $T_s = 9.2x10^{-6}$ sec. (which is comprised within the bit duration).

The simulation results, concerning the transmission of 300 bits have proved that no bit error happens if the transmission signal – to – noise ratio is $SNR = 10$ dB, whereas a bit-error-rate equal to 0.005 has been observed if $SNR = 5$ dB. These results are quite in line with the ones shown by the literature concerning the BER evaluation of DS/CDMA communication systems, and are to be considered an upper bound of the BER performances that can be achieved by the complete system, with the Forward Error Correction (FEC) coding block included in the simulation chain.

5. Conclusions

In this work an innovative approach in the software design and simulation of a DS/CDMA system for multimedia information has been presented. The considered application is related to the remote video – based surveillance of unattended railway stations. The proposed approach allows to obtain an interesting tool of simulation of a really operating communication system. In fact, it has to be highlighted that the use of the SIMULINK® environment allows to have a perfect simulation of the behaviour of the proposed image transmission system by displaying to the user the temporary evolution of the modulation/demodulation process, and also allows to define precisely all the hardware components necessary to design such a device, so making quite easy its VLSI implementation.

Acknowledgement

This work is partially funded by the project CNR – PFT2 (Italian National Council, Transports Finalized Project), sub – project 3: "Technologic Supporting Systems and Infrastructures", theme 3.2.4 "Technologic Supporting System for Railway Transport – Low Traffic Lines".

References

[1]   E. Stringa, C. Sacchi, C. S. Regazzoni, "A multimedia system for surveillance of unattended railway stations", *European Signal Processing Conference (EUSIPCO) 1998*, Special Session on Multimedia Signal Processing, Rodhes, Greece, 8 – 10 September 1998, in press.

[2]   R. L. Pickholtz, D. L. Schilling, L. B. Milstein, "Theory of Spread – Spectrum Communication – A Tutorial", *IEEE Transactions on Communications*, Vol. Com –-30, No. 5, May 1982, pp. 856 – 884.

[3]   D. L. Schilling, "Wireless Communications Going Into the 21ᵗ Century", *IEEE Transactions on Vehicular Technology*, Vol. 43, No. 3, August 1994, pp. 645 – 651.

[4]   Y. L. C de Jong, R. P. C. Wolters, and H. P. A. van der Boom, "A CDMA Based Bidirectional Communication System for Hybrid Fiber-Coax CATV Networks", *IEEE Transactions on Broadcasting*, Vol. 43, No. 2, June 1997, pp. 127 – 135.

[5]   *"Radio Equipment and Systems (RES); Wideband transmission systems; Technical characteristics and test conditions for data transmission equipment operating in the 2.4 Ghz ISM band and using Spread Spectrum modulation techniques"*, European Telecommunication Standard Institute (ETSI), Draft pr. ETS 300 328, July 1996, Second Edition.

[6]   M.B. Pursley, "Performance Evaluation for Phase-Coded Spread Spectrum Multiple-Access Communication - Part I: System Analysis", *IEEE Transanctions. on Communications*, Vol. Com.- 25, No. 8, August 1977, pp. 795-799.

[7]   P. Karn, *"Spread Spectrum Rule Recommendations"*, ARRL Southwestern Division Convention, Long Beach (USA), September 1995.

[8]   D. Sarwate and M.B. Pursley, "Correlation Properties of Pseudorandom and Related Sequences", *Proceedings of IEEE*, Vol. 68, No. 5, May 1980, pp. 593-619.

[9]   R. C. Dixon, *"Spread Spectrum Systems With Commercial Applications"*, 3ʳᵈ Edition, J. Wiley & Sons, New York: 1984.

Rendering and animation of a proprietary face model into the MPEG-4 3D Player

Franco Casalino, Gianluca Francini, Claudio Lande and Aldo Poma

CSELT-Centro Studi e Laboratori Telecomunicazioni S.p.A
Via Reiss Romoli, 274 - 10148 Torino ITALY
Tel. +39 11 228 5100 Fax. +39 11 228 6190

Abstract

After a decade from its origin, MPEG, with its current MPEG-4 project, is now facing the challenge of providing a future-proof multimedia toolkit which aims at incorporating new and emerging technologies while ensuring backward compatibility with its previous and successful audio-visual encoding standards. This paper provides an overview of the standard focusing mainly on the system aspects that, by their nature, represent the most peculiar features of the future specifications which are scheduled to become an International Standard by the end of year 1999. The paper, after a first briefly introduction about the MPEG roots (MPEG-1 and MPEG-2), describes the detailed list of the group Synthetic Natural Hybrid Coding (SNHC) which has to do with the integration of natural and synthetic audiovisual contents into a unique scene. In particular the area of work of the Face and Body Animation (FBA) group is described; its goal is the definition of a set of parameters for the definition and animation of a synthetic human face and body. The CSELT proprietary facial model is described and how it has been integrated into the MPEG-4 3D Player is described. Finally the procedures of calibration and texture mapping to make the model similar to a real human face are described.

1. MPEG Overview

The Moving Picture Coding Experts Group (MPEG) was established in January 1988 with the mandate to develop standards for coded representation of moving pictures, audio and their combination. It operates in the framework of the Joint ISO/IEC Technical Committee (JTC 1) on Information Technology and is formally WG11 of SC29.

The first standard developed by the group, MPEG-1 (ISO/IEC 11172) [1] is titled "Coding of Moving Pictures and Associated Audio for Digital Storage Media at up to about 1.5 Mbit/s". This was motivated by the prospect, which was becoming apparent in 1988, of storing video signals on a compact disc with a quality comparable to VHS cassettes. The second standard, MPEG-2 (ISO/IEC 13818)

[2], has the title "Generic Coding of Moving Pictures and Associated Audio". Unlike MPEG-1, basically a standard to store moving pictures on a disk at intermediate bit-rates, the much larger number of applications for the MPEG-2 standard forced MPEG to develop and implement a "toolkit based approach". Different coding "tools" serving different purposes were developed and standardized. Different assemblies of tools, called "profiles", were also standardized and could be used to serve different needs. Each profile had in general different "levels" for some parameters (e.g. picture size).

MPEG-4 (ISO/IEC 14496), the current standardization project of MPEG, combines some of the typical features of previous MPEG standards, but extends the definition of systems for audio-visual coding in two dimensions:

- evolving from a "signal coding" approach to an "object coding" approach: defining new techniques for the coded representation of natural audio and video, and adding techniques for the coded representation of synthetic (i.e. computer generated) material;
- evolving from a *fixed* (though generic) standard (with a fixed specification of a single algorithm for audio decoding, video decoding and demultiplexing) to the definition of a *flexible* standard, where the behavior of particular components of the system can be reconfigured.

The driving motivations for this new standardization effort are derived from a requirement analysis embracing existing or anticipated manifestations of multimedia, such those listed below:

- Independence of applications from lower layer details, as in the Web paradigm;
- Technology awareness of lower layer characteristics (scalability, error robustness etc.);
- Reusability of encoding tools and data;
- Interactivity not just with an integral audio-visual bitstream, but with individual pieces   of information within it, called "Audio-Visual (AV) objects";
- The possibility to hyperlink and interact with multiple sources of information simultaneously as in the Web paradigm, but at the AV object level;
- The capability to handle natural/synthetic and real-time/non-real-time information in an integrated fashion;
- The capability to composite and present information according to user's needs, as in the VRML and computer graphics paradigm in general.

MPEG-4, started in July 1993, has reached Committee Draft level in November 1997 and reach International Standard level in November 1998.

2. MPEG-4 Terminal

The MPEG-4 Terminal is the client side of the MPEG-4 system. On the server side the information is composed, compressed and multiplexed to form an MPEG-4 stream. The stream is then conveyed in the client side; here it is subdivided into the Elementary Stream components by the demultiplexer. Among these, the first to be decoded is the *Scene Description* stream which contains the description of the scene which consists of a hierarchical tree whose nodes are named Media Object (the way of describing a scene will be explained in the following paragraph). A Media Object is any natural or synthetic audiovisual object (i.e.:a Video, an Audio, a synthetic facial model, ...) and in some cases it needs a stream of data to be updated. In these cases, the Media Object will instantiate its specific Decoder by which it receives its Elementary Stream. Finally, after the scene is instantiated, the Presenter, also named Compositor, will cyclically compose, render and visualize the information comtained in the Scene Description. The Compositor will also manage the user generated events and will update the scene accordingly.

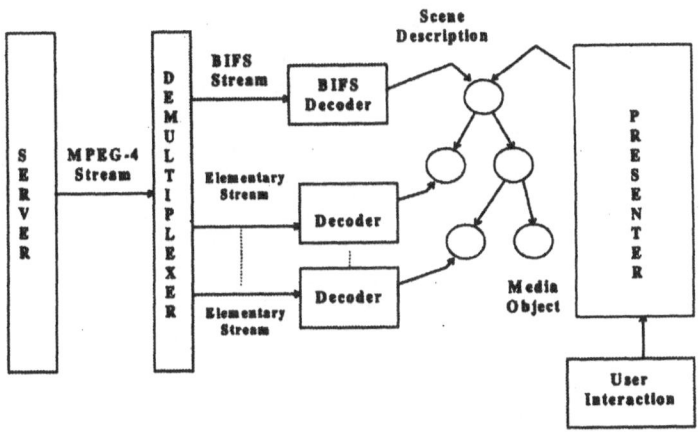

Fig. 1 MPEG-4 Terminal high-level system architecture

3. Scene Description and Composition

The description of the scene provides the information that the compositor needs to perform its task. The scene description provides information on what objects are to be displayed and where they are to be displayed (which includes the relative depth ordering between the objects).

It is envisaged within MPEG-4 that there will be two types of compositors. Firstly there will be simpler two dimensional compositors, combining all the objects in a two dimensional output frame. The second type of compositor assumes all the objects are three dimensional and are combined together in a three dimensional space.

The model adopted by MPEG-4 to describe the composition of a complex multimedia scene relies on the concepts defined by the Virtual Reality Modelling Language (VRML) [3]. VRML, currently being considered by JTC 1 for standardisation (ISO/IEC DIS 14772-1), provides the specification of a language to describe the composition of complex scenes containing 3D material, plus audio and video.

The main areas where MPEG-4 Systems has introduced new concepts according to specific application requirements are:

- dealing with 2D only content, for a simplified scenario where 3D graphics is not required
- definition and animation of (synthetic) human faces and bodies
- interfacing with streaming media (video, audio, streaming text, streaming parameters for synthetic objects)
- adding synchronisation capabilities.

The outcome is the specification of a composition format based on (a subset of) VRML tuned to match the MPEG-4 requirements. This description, known as BIFS "Binary Format for Scene description", will allow the proper description of complex scenes populated by synthetic and natural audio-visual object with their associated spatial-temporal transformations and inter-objects mutual synchronisation.

Multimedia scenes are conceived as hierarchical structures that can be represented as a tree. Each leaf of the tree represents a Media Object (Audio, Video, synthetic Audio like a MIDI stream, synthetic Video like a Face Model), as illustrated in Fig.2. In the tree, each Media Object is positioned relative to its parent object. The tree structure is not necessarily static, as the relationships can evolve in time, as nodes or sub-trees are added or deleted. All the parameters describing these relationships are part of the scene description sent to the decoder.

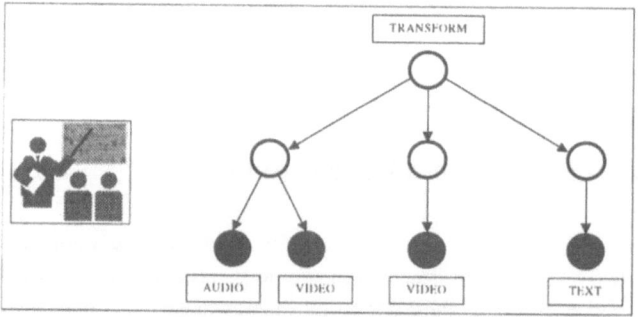

Fig. 2 Example of a Scene Graph.

The BIFS description concerning the initial snapshot of the scene is thought to be sent/retrieved on a dedicated stream during the initial phases of the session. It is then parsed and the whole scene structure is reconstructed (in an internal representation) at the terminal side. All the nodes and tree leaves that necessitate streaming support to retrieve media contents or ancillary data (e.g. video stream, audio stream, facial animation parameters) are logically connected to the decoding pipelines.

At any time, an update of the scene structure may be sent. These updates can access any field of any updateable node in the scene. An updateable node is a node that received a unique node identifier in the scene structure. The scenes can also be interacted locally by the user, and this may change the scene structure or any value of any field of any updateable node.

Composition information (i.e. information about the initial scene composition and the scene updates during the sequence evolution) is, like other streaming data, delivered in one Elementary Stream. The composition stream is treated differently from any other, because it provides the information required by the terminal to set up the scene structure and map all other Elementary Streams to the respective Media Objects.

As the regular media streams, the composition stream has an associated time base, which defines the clock to which Time Stamps in the composition stream refer.

3.1. Spatial relationships

The Media Objects may have 2D or 3D dimensionality. A typical Video Object (a moving picture with associated arbitrary shape) is 2D while a wire-frame model of the face of a person is 3D. Audio also may be spatialized in 3D, specifying the position and directional characteristics of the source.

Each elementary Media Object is represented by a leaf in the scene tree, and has its own local coordinate system. The mechanism to combine the nodes of the scene tree into a single global coordinate system is the usage of spatial transformations associated to the intermediate nodes, which group their children together (see Fig.2.). Following the tree branches from bottom to top, the spatial transformations are cascaded until the unique coordinate system associated to the root of the tree.

In case of a 2D scene the global coordinate system might be the same as the display coordinate system (except for scaling or clipping). In case of a 3D scene, the projection from the global coordinate system to the display must be performed by the last stage of the rendering chain.

3.2. Temporal relationships

The composition stream (BIFS) has its own time base associated. Even if the time bases for the composition and for the elementary data streams might be different, they must however be consistent except for translation and scaling of the time axis. Time Stamps attached to the elementary media streams specify at what time the Access Unit for a Media Object should be ready at the decoder input, and at what time (and for how long) the Composition Unit should be ready at the compositor input. Time Stamps associated to the Composition Stream specify at what time the Access Units for composition must be ready at the input of the composition information decoder.

4. FBA SNHC

The "object coding" approach enables to define synthetic objects which are described by a set of parameters and synthetised with computer graphics techniques. The parametric description of synthetic objects allows easier interaction with the contents than that possible with natural objects; moreover the required bandwidth is typically lower because synthetic object can be updated using small sets of parameters.

In the context of these last activities[4],[5][6], identified with the acronym "SNHC" -Synthetic and Natural Hybrid Coding -, it is of special interest the activity of the SNHC Face and Body Animation (FBA) working group that works to define a set of parameters for the definition and animation of a synthetic human face and body.

FBA group does not define a specific facial model, but the measurement units (ex.: eye separation, mouth width, ...) and a set of *feature points* which can be extracted from any facial model in neutral position. In addition, FBA defines the parameters for the definition and animation of the facial model by means of the two BIFS nodes *Facial Definition Parameter* (FDP) and *Facial Animation Parameter* (FAP). These nodes define:

- FDP: may contains either a new downloaded face model or a set of feature points for calibrating the proprietary face.
- FAP: each FAP describes the displacement of a particular feature point of the face model. FAPs are relative to eyes, eyelids, eyebrows, lips, jaw, nose, cheeks and head rotations and are defined in a neutral way with respect to face models. This feature allows having a single set of parameters that can be used regardless of the face model used by the MPEG-4 terminal. Most FAPs describe atomic movements of the facial features (e.g. open_jaw, depress_chin, etc.).

The face model is made as a mesh of polygons. The application of FAP *i* on the model involves the displacement of a given set of vertices of the mesh. The value of each FAP indicates the quantity to sum to the neutral position of a particular feature point; this quantity is calculated on the basis of the FAP intensity and of the measurement unit which has been used for the FAP. The modification of the vertices adjacent to a given feature point depends on the particular model; the general policy is that the farther a vertex is from the feature point, the less it is influenced by the feature point motion.

FAPs combinations give place to all the possible facial expressions. If a continuous animation must be performed, a sequence of FAPs must be applied to the face model: for each time instant (frame of the sequence) a set of FAPs must be used to update the model.

5. JOE: Join Our Experience™

The synthetic face model JOE is defined by a mesh of 780 triangles which are defined by a set of 548 3D vertices. The triangles are smaller and more numerous in the most expressive and curved area of the face, like eyes and mouth, and are bigger in the smoother areas as forehead and cheeks.

JOE has been created on the basis of the face model "Fascia" of Parke[7] and is described in VRML 2.0 format.

5.1. Calibration and Texture mapping

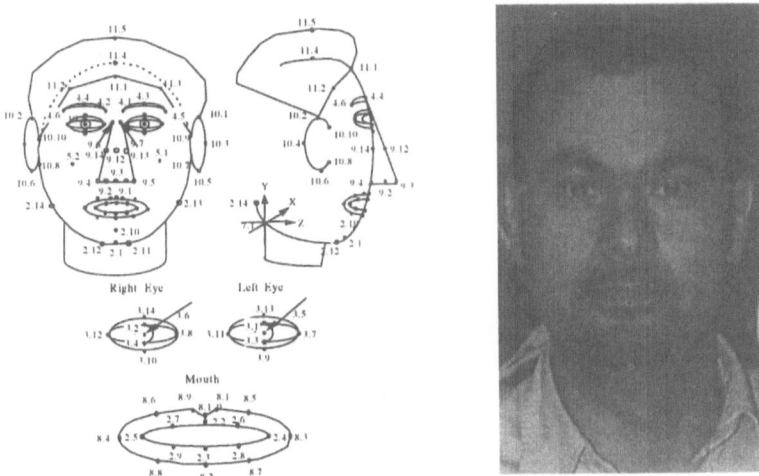

Fig. 3 (left) MPEG-4 feature points - (Right) The set of used points

The wire-frame of JOE may be calibrated on the basis of a set of feature points (see Fig. 3) which are extracted from a human face picture; in this way the face model physiognomy is made similar to the face of the person in the picture. The calibration is made with regard to the X and Y axis, whereas the Z axes is ignored. The target of the algorithm is to modify the wire-frame template so that the projection of the obtained model coincides with the position in the photo of the marked feature. In order to do this, a projection with the focus in the origin is performed (see Fig.4).

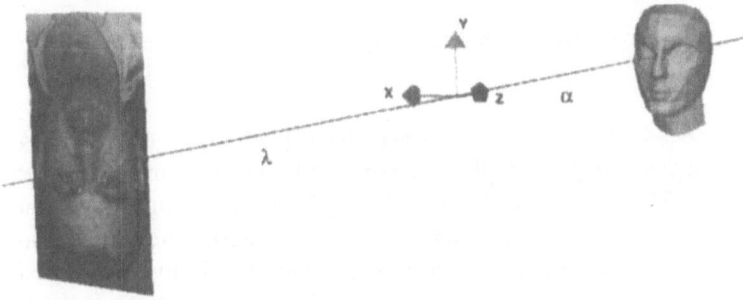

Fig. 4 Projection with focus in the origin

After the calibration, the texture may be applied to the wire-frame. In this way the synthetic facial model with human aspect is obtained.

6. JOE's animation into MPEG-4 3D Player

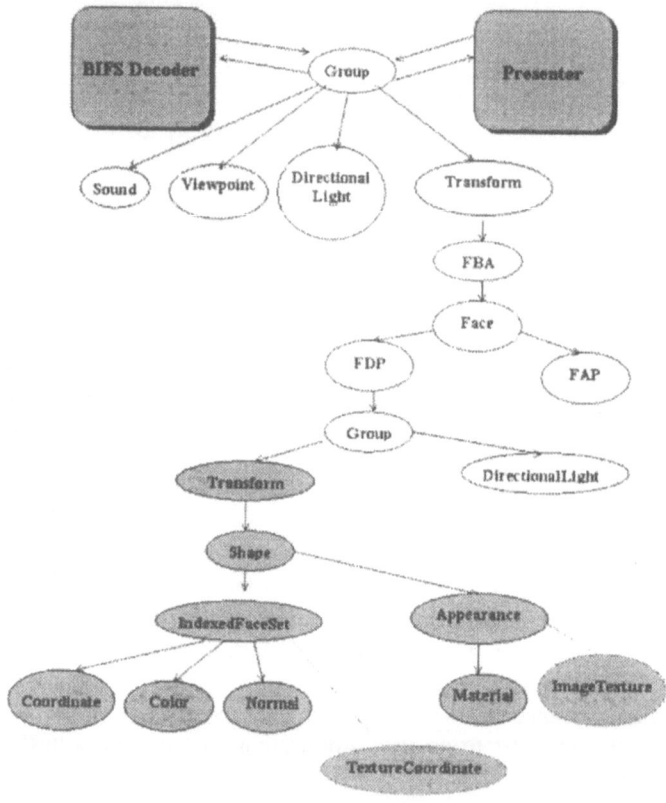

Fig. 5 The BIFS scene graph of JOE's Animation.

The MPEG-4 3D Player is the development platform of the MPEG-4 standard specifications. To perform the rendering and animation of the proprietary face model the PC platform has been used. It uses the C++ programming language of and the OpenGL library as graphic engine [8],[9],[10]. The software architecture is made of a set of threads managed by preemptive priority based scheduling and uses all the potentiality of the object-oriented paradigm (classes, templates, virtual functions,...)

In figure 5, it is shown the scene graph that has been created to instantiate and animate the proprietary face model. The BIFS white nodes are conveyed by an

MPEG-4 stream and the gray ones are instantiated inside the Player. In fact, when during the stream parsing the Face node is met, the Player instantiates the proprietary model that has been stored locally. The FAP node receives the FAP stream which has a given frame-rate and updates the Transform and Coordinate nodes. The first is a group node and it is used to apply global transformations to the model (i.e.: head rotation), whereas the second is used to store the coordinates of the model.

To compose the scene description the FAP, FDP and IndexedFaceSet have been implemented. The IndexedFaceSet node defines a mesh of polygons with its colors, normal vectors and texture coordinates. All the nodes are derived by the MediaObject classes and must implement the Render() virtual method. This method performs the rendering of the node and then calls the Render() method of the children nodes. Into the Render() methods, all the calls to the OpenGL library occur.

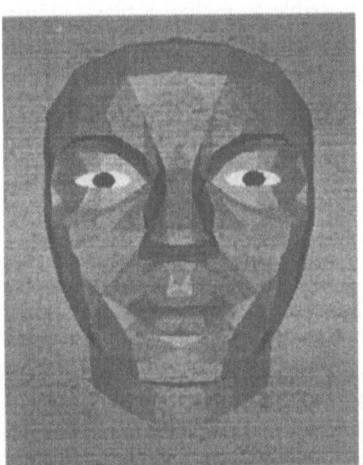 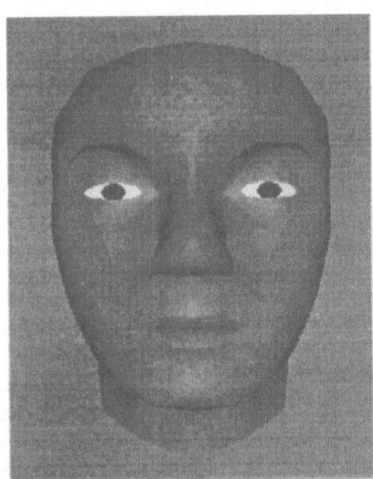

Fig. 6 (Left) Joe with Flat shading and normals per vertex– (Right) Joe with Gouraud shading and normals per vertex

Finally a module has been developed that, starting from the VRML description of the facial model, performs the automatic calculation of the normal vectors per face and per polygon. The normals for vertex are calculated like the average of the normals (normalized to 1) of the polygonal face to which the vertex belongs.

The visualization of JOE in the Player appears as in the above pictures; in both of them an infinite viewpoint, a DirectionalLight node (parallel beams of light) and one color for each polygonal face has been used.

7. Conclusions and future work

This work has led to the visualization and animation of the proprietary facial model JOE into the MPEG-4 3D Player. Now it is possible to instantiate a proprietary model and then to animate it by the MPEG-4 stream which has been previously described and which contains a FAP stream. According to the kind of stream it will be possible to visualize either the synthetic facial model or the texture mapped facial model.

The animation of the facial model into the MPEG-4 3D Player can run in real time on medium level PCs (i.e.: Pentium Pro 180 Mhz) with an OpenGL graphics board; hardware acceleration increases the speed of the rendering of the face models so that the correct frame rate can be ensured.

The future work involves the interfacing of the animation system of the Player with a Text to Speech module (TTS) which can generate the vocal synthesis starting from a written text so that it will be possible to realize a facial animation driven by synthetic speech.

The applications of the face model will be all the ones which must use a low bit-rate of transmission (some Kbps), as for example the virtual meeting or the videotelephone. Other applications will be the automatic e-mail reader, the friendly interface, the story teller and the personification of intelligent agents.

References

1. MPEG-1 (ISO/IEC 11172), "Coding of Moving Pictures and Associated Audio for Digital Storage Media at up to about 1.5 Mbit/s", 1993.
2. MPEG-2 (ISO/IEC 13818), "Generic Coding of Moving Pictures and Associated Audio",
3. VRML (ISO/IEC DIS 14772-1), "Virtual Reality Modeling Language", April 1997.
4. MPEG Systems Group "Text for CD 14496-1 Systems", ISO/IEC /WG11 N1901,
5. MPEG Video Group "Text for CD 14496-2 Video", ISO/IEC JTC1/SC29/WG11 N1902,
6. MPEG Audio Group "Text for CD 14496-3 Audio", ISO/IEC JTC1/SC29/WG11 N1903,
7. Parametrized models for facial animation, IEE Computer Graphics and Applications, November 1982.
8. OpenGL Programming for Windows 95 and Windows NT, aut.:Ron Foster, ed.: A.W.
9. OpenGL Programming Guide: The Official Guide to Learning OpenGL, Release 1. Aut.:Neider, Jackie, Tom Davis, and Mason Woo. Ed.: Addison-Wesley 1993.
10. Computer Graphics principles and practice, second edition, aut. Foley, van Dam, Feiner, Hughes, ed.:Addison Wesley

Author's Index